Redesigning the Planet
Foundations

Version 5.0

Date: 14 March 2014

Redesigning the Planet
Foundations

Reshaping the Constructs of Civilizations
Through the Use of Ecological Design & Other
Conceptual & Practical Tools, such as Common Sense,
Deep Ecology, Totemism, Systems Theory, Metaphor, Holistic
Science, Thought Experiments, & Eutopian Strategies

Version 5.0

Alan Wittbecker

Urania Science Press
Mozart & Reason Wolf, Ltd.
Sarasota
2014

In memoriam: Arne Naess, Buckminster Fuller, Garrett Hardin, Paolo Soleri
　　　　Merv Wilkinson, Paul Shepard, Thomas Berry, & Mark Wittbecker

Cover Design: 2011, Rian Garcia Calusa
Front cover photo: Terra cotta brick works near Agra, Uttar Pradesh, India
Back cover: Hittite city of Hattushash from Ringling course ES100 (original unidentified)
Graphics and photographs (unless otherwise noted): Alan Wittbecker

Published by Urania Science Press
for Mozart & Reason Wolfe, Ltd at SynGeo ArchiGraph
Mail: Post Office Box 370, Tallevast, Florida 34270
Email: Design@SynGeo.org or Mozart@ReasonWolf.com

Publisher's Cataloging in Publication Data
Alan Wittbecker 1946—
 Redesigning the Planet: Foundations / Alan Wittbecker
Includes bibliographical references and index.

ISBN-13: 978-1499336078
ISBN-10: 1499336071
  1. Human Ecology. 2. Ecological Design. 3. Deep Ecology. 4. Architecture. 5. Geography.
I. Title.
GF55.W51 2014

Book Design by Rian Garcia Calusa

Printed in the United States of America

FOUNDATIONS CONTENTS

0.0. Introduction

This work is about sharing the local, regional and global resources and services of the planet to meet the needs of all living beings and their community patterns. It uses ecological design to create a simple method to implement and manage the sharing. We start by assessing what the planet needs to develop in a stable flow, then we set aside a satisfactory area of the planet to ensure the continuing operation of evolution in wild systems. Next, we measure the ranges of productivities of wild ecosystems as well as agricultural and urban systems, then use those results to determine optimum human populations for local places, regions and the planet. Finally, within human systems, every culture would claim a share of local resources and global services not set aside for wild regeneration.

The equal apportionment of 'resources' to all cooperating participants in the global commons is supported by the practice or doctrine of recognizing and honoring the legacy of the entire planet that hosts its legatees as tenants and is supported by the political 'rule' of all beings, although in the human legal system, humans represent the interests of all other beings, much as they are starting to do now. This reapportionment is enhanced by the wisdom of harmony and the drawing and making of ecological zones, which emphasizes the relative isolation of wild and artificial areas. This reapportionment of 'resources' that human communities have already claimed, as well as of resources that have been badly distributed as a result of theft or violence, may cause some degree of discomfort or suffering for wealthier people, but that is minimal compared to the suffering and death under the current industrial system, which encourages overconsumption and large, immoral differences in the distribution of wealth.

Ecological design would work on global and regional scales, as well as the local scale. For example, the Colorado river would be allocated a percentage of water to keep the river and its downstream ecosystems (including shallow ocean canyons) healthy—this may require 50% or more of all the water flow. The remaining water would be divided between resident cultures sharing the river environments upstream. This approach promises a fair way to deal with carbon emissions, toxic wastes, and energy use, also. Like metaphysics, ecological design has a vision that exceeds its bounds and a reach that exceeds its grasp. And, we have to use it to explore possibilities of local and global harmony, without having complete knowledge or experience. Ecological design requires participation and cooperation to accomplish its ambitious goals. It has to be flexible and adapt to changing environments.

This means understanding challenges and problems, as well as natural and artificial ecosystems, histories and cycles, before using a variety of physical and conceptual tools to create ecological designs on local scales, but considering the regional and global implications. This means trying to design places, ecosystems and landscapes, as well as cycles and processes. It means redesigning flows of minerals and gases, wetlands and streams, domestic and wild forests, and animal paths and reserves. It means redesigning human patterns, from transportation corridors to traditional and modern cultures. It means redesigning agriculture, cities—traditional cities and proposed arcologies—buildings, neighborhoods, vehicles, industries, and medicine. It means trying to redesign social traps, cultural adaptations, corporate goals and responsibilities, formal commons, styles of conflict, economic frameworks, political forms and sizes, religious applications, and even advertising. And, the purpose of all this is to restore harmony to systems that support health and development.

0.1. *Preliminary Thoughts on a Framework for Redesign*

Redesigning places, regions and the planet is a conscious effort to correct the massive 'unconscious redesign'—perhaps 'unbalanced reordering' is a more appropriate phrase—of local and global systems. This 'unconscious redesign' is an unconsidered effect of the activities of changing economic and political structures devoted to self-growth and private profit. This unbalanced reordering is rapidly changing the social and environmental orders that represent the natural capital of social and environmental evolution. The reordering is also constructing an accidental trap that leads to massive catastrophes, which will destroy that capital it needs as the basis for its renewal. Only an equally massive response of a conscious ecological design framework can lead to some form of balance.

Buckminster Fuller suggested that the appropriate tools, such as eutopian design frameworks or advanced technologies, might help human worlds to work, given the right emergency. We have been encountering the right emergencies for well over a hundred years, and we need to refine and try these tools, now. The emergencies that he considered are the human responses to a variety of catastrophes, such as fires or earthquakes. But, we are finding that not all catastrophes are fast, human-scale or visible. The effects of those large, slow, invisible, long-term changes make us uneasy, but not really adrenaline-ready. The changes are reflected in starving children, hotter summers and stronger storms, failing food supplies, and collapsing infrastructures. We must learn to recognize and respond to these other catastrophes. And, we have to do it now, before they crest and become overwhelming. We can do it. We have the evidence that things are taking a downward turn (the original meaning of the word catastrophe). And, we have the tools for the appropriate responses.

0.1.1. *What this Work is Not*
This work is not a list of emergencies. It is not catalog of losses. Many comprehensive lists and catalogs already exist. It is not a linear analysis of sudden problems and surprises. It is not a larger, more data-rich, compartmentalized model of change. It is not an abstract discussion of possible shallow changes. It is not a plan to avoid or to control catastrophes. This work does not propose solutions to stop changes or to prepare for emergencies. It does not offer a way to save the planet or our global, industrial, capitalistic, semi-democratic civilization. It is not about saving the world, or nature, or humanity. This work is not about preserving the status quos of industrial societies or of formal democracies.

This work is not about greening our civilization with more efficient cars. It is not about the continuity of market forces into private lives, or about breaking with tradition to counter market intrusion. It is not about profiting with green businesses, after certifying mild improvements. It is not about sustaining our current styles of living or levels of luxury by tweaking light bulbs, battery recycling, engine efficiency, or the color or depth of rooftops; it does not address sustainability (a word now meaningless in its generality). This work does not focus on the symptoms of a single small environmental crisis, or even several crises. This work is not about adapting to drier climates or finding immense forms of alternate energy.

This work is not about big institutions and big successes. It is not about creating spectacles or spectacular successes. It is not about creating virtual worlds in a utopian fantasy of voluntary good will to minimize exploitation and waste. It is not about the management of the planet or a detailed design for the planet. It is not about the stewardship of remaining resources or of other species, domestic or wild. It is not about a partnership or cooperation with nature—the planet is much too large, complex and dangerous for such a pretense. It is not about the invention of global cooperation or a map to achieve it.

Many recent works herald the smallness of nature, the end of nature or the death of wildness. Other books recount the themes of recent times, from the horrors of war to the triumphs of science and technology. Many authors have emphasized the urgency of responding to climate change, to extinctions, to trends of violence and victimization, and to growing inequity and financial dishonesty. Others have identified further emergencies, gender differences, border insecurities, travel dangers, and natural disasters such as hurricanes, tsunamis and earthquakes. Some have emphasized the importance of miscommunications, misinformation, violent attacks, terrorism, misuse of resources, epidemics, potential asteroid collisions, energy shortages, and desertification and ecosystem collapse. Others have recognized the momentum of short-term self-interest, self-deception, perceptual limits, overconsumption, polarization, and destabilization. And, they are all certainly right to do so—but these urgent problems are *interconnected*!

0.1.2. *What this Work Is*

The problems are interconnected, so this work has to address everything *simultaneously*, because solving one problem will only offer a temporary, limited solution as other problems affect the one in the center of the focus. This work is a recognition of the problems, as well as of relentless change and radical uncertainty, that now confront many species as well as human civilizations. It is recognition that we do not have adequate understanding, knowledge or control to solve any problem once and for all.

This work is a way to adapt to ignorance, limits and chaos. It is a call for immediate action to solve significant problems to balance the whole planet, rich nations and poor nations, methane and carbon dioxide levels, and bad designs and good designs. It is a call for large kinds of public investments through the UN to avert global risks and to avoid some kinds of catastrophes. It proposes a global framework for simultaneous changes in consciousness and action, and a platform for people to self-organize and to create responses to challenges. It is a framework for personal, immediate action.

This work is a process of assembling ideas into a nonlinear narrative for the development of interacting patterns of design. It proposes creative alternatives to the business of business and the technological imperatives of progress. It is about reforming cities, agriculture, and transportation for survival. It is about reordering economics, putting constraints on businesses, but removing other restraints, such as taxes, from the goals and ends we desire. It suggests radical strategies to confront disruptive environmental and climatic changes. It urges preparations for the immediate future, using ways to survive and prosper that are exciting, adventurous, imaginative, and joyful, but that also may be hard, painful and demanding.

This work is about creating meaningful images and goals. It presents strategies to implement the images and goals. It attempts to create a framework for details, because we cannot know those details in advance or plan for them. It is about replacing the idea of sustainability with that of flexible creative fitness (in context) or cultural stability (in scale or tempo). It is about understanding the underlying causes and reactions to ecosystem stresses. It offers sketches of possible directions for ecological and cultural development.

This work is a thought experiment for a design framework to contain the creative coevolution of wild nature and human culture. It is an experiment on how to invent a new approach, given the immense body of knowledge that humanity already has, from the present and the long-term past. It is about recreating civilization from the bottom, by reducing our impacts and reducing our vulnerability to catastrophes. It is about managing and constraining humanity at appropriate levels (local, regional, global).

This work asks neglected questions about big problems, about the long, deep, wet history of life, about how many wild ecosystems we need, as well as what humans really need—about our history with its images and goals. It asks questions about our cultures and our interference with natural systems and cycles. It urges people to act heroically and exuberantly to live joyful, meaningful lives under stressful conditions. It is about respect and engagement with the planet, so that we can accommodate changes. This work is a call to imagine, participate and then to act through a common global design project.

0.1.3. *What we Need & Why it would Work*

There may only be a small window of opportunity, between 1 and 20 years—unless it passed from 1 to 20 years ago—for changes to be effective. Otherwise, losses due to external changes will be unacceptable economically; the losses are already unacceptable ecologically, especially with extinctions and habitat collapses. Fast changes are needed, requiring fast investments, mobile capital and social capability.

Our assumptions have to be made explicit. This is an emergency, requiring large-scale, multiple approaches, with new technologies, massive conservation efforts, and micro-energy solutions (which require participation), but not using old, unconscious assumptions and design traditions. There is no time to wish politicians into patterns of good behavior. There is no time to beg corporations to stop cutting forests or poisoning lakes and ecosystems; as Garrett Hardin pointed out, conservation goes against their self-interests, which are focused on profits. Tree hugging is good; orgasms for peace are also good, but we need a large-scale, coordinated approach to be effective. The economy is based on rapid destruction for commodification. We might eventually convince everyone that ecosystems are part of larger human selves and that our long-term self-interest is served by conservation and preservation, but it is necessary to demand and force people to do 'the right thing' before everything collapses. Although we can learn from the many past collapses, if some cultures or civilization collapses now, it will be too late to learn. And, recovering from any collapse is going to be more difficult with globalized cultures in a completely occupied planet.

According to Garrett Hardin, many of the ideas necessary to fitting humanity into the pattern of nature are known but not yet popular. For instance, exponential population growth (or economic growth) cannot be maintained very long. Human communities cannot grow at four percent per year without disastrous consequences to the infrastructure and to the quality of life. Growth cannot be continued because the landscape is limited, in terms of productivity, energy, and resilience. Thus, we need to fit our population into the limits of the landscape (although some limits can be expanded by technology or by lowered expectations). The carrying capacity of the area is not only a function of the limits of the community, it is equal to the number of people multiplied by the level of comfort (or quality of life style). Having more energy and space means having fewer people. What is required for dangerous times on the planet is a sense of common purpose. We need authoritative leaders, who can command. Courage and imagination are needed obviously. Intelligence and passion, rightly so. But, also we need faith in our abilities and in nature's process of regeneration. We need to accept failure and uncertainty. And, we have to have the willingness to proceed with realistic optimism, despite limits to knowledge or cooperation.

Although nature and our human nature, are not enemies to be vanquished, the current situation has similarities to war. Massive changes threaten our lifestyles. Resources are removed from our reach by thoughtless or inefficient use. Growing insect and animal populations seem to be attacking our food supplies. Species are being forced into extinction. Habitats are collapsing. Dangerous chemical wastes are accumulating. Ozone holes are growing, extreme climates pressing, and the entire planet seems to be wobbling. Changes in climate and ocean balance, as well as renewed diseases and infiltrated toxic chemicals, threaten our lives. And, it is happening everywhere, at once.

Although we want to respond with a warlike approach, we have been fooled by the fact that we cannot see an enemy—and fooled by thinking there is an enemy. We have been misled by the slowness and subtlety of the penetration of our defenses. We have been betrayed by our own desire to continue our industrial dreaming at any cost. Some people have noticed changes and have been crying alarms, but they have not been loud enough or persuasive enough. Everybody needs to be awakened; everybody needs to participate, everybody needs to sacrifice and work towards peaceful solutions. The big problems seem insurmountable, and simple actions will not save our civilization from catastrophes. Part-time participation will not be enough to reverse the degradation of ecosystems, and partial business greening will not stop the unraveling of global cycles.

What is needed is an immediate, peaceful, comprehensive approach to this situation—wise actions in a framework of intentional, reflective design. We have some experience; thousands of people have collaborated on computer operating systems or online encyclopedia. So society might benefit from a push to redesign society itself, as well as to redesign the environment and planet. We have the time: One billion people in affluent countries have 2-6 billion hours to spare every day. We have the money to do it, although the costs of change are going be relatively high, from 1 to 24 percent of the productivity of every culture. The cost of collapse, however, would be much higher and more painful.

We need to act on a global scale. We have acted on a large-scale before, in times of world wars. We were able to treat war as an emergency and to encourage or enforce remarkable changes, such as rationing or redefining jobs. We were able to take these actions without destroying our citizens or our cultures (or most of the planet).

Why would ecological design work? Because life has over three billion years experience with changing and adapting, because human life and cultures have over 50,0000 years of practical experience adapting and making changes, and because humans are immensely adaptable—if they can adjust to poverty and suffering, they can adjust to a few good changes. Perhaps it is already too late—limits may have been passed and the catastrophes cannot all be reversed. We do not know, and may never know, but we can still act as if we were wise, as if doing the right thing makes a difference. And, we can work together to help others, to improve things and to make good places. If we act *now*, this month, this week, this day, then the changes might be more effective.

0.1.4. *What can you Do?*
Read this work. Correct it and supplement it. Extend it. Apply it. Grow up, mature. Lose your fear of speaking and acting, or of losing time or everything. Prepare to fight for ideas and actions that you know will benefit the complex forms of life and cultures. We are trapped in an industrial megaculture that is embedded with us in an ecological network of trillions of interdependent living beings that are being destroyed by industrial processes and financial practices. You—we—have the power to walk out of the trap and to commit our considerable energies to redesigning and rebuilding our places in the planet.

Is this work complete or original? No. Certainly hundreds of people have had ideas that are more original. Sometimes however, originality is aided by having many ready-made pieces to put together—for instance, Charles Darwin needed the works of Thomas Malthus and Adam Smith for his ideas. And, these were put into a new pattern. Earlier design work is almost always useful in later work and has value for later design. All our revolutions have to do with recombining ideas in new patterns rather than discovering a single, new, independent idea.

Is this work an adequate response to the pressing emergencies? Can it be effective? Maybe it could be, as a frame of ideas or as a matrix that could guide action. What's wrong with this work? It was written in a hurry. It is sloppy and unfinished. Why should you read it? Because it has many good ideas and it attempts to put them into a very large synthesis. People sometimes object to one person delivering finished material and ideas to a larger number of readers. That cannot be the case here. That is why this work is a crude outline. It is constructed as a frame that others, like you, can add to. In fact, think of this as a workbook to which anyone can contribute, as a sketchbook where you can fill the sketches, or as a dynamic corrective exercise that can be corrected, improved, and expanded.

The attempt at local ecological design means accepting the opinions of everyone, regardless of their motives or level of understanding. The attempt therefore becomes the beginning of reeducation, which may be a long process, longer than single human lifetimes. This work should aim to be an extensive community of cocreation, like the Grameen bank or Wikipedia. This work is an open framework that addresses and limits local designs within regional and global designs. The work is a participatory process. That may be the only way it is useful. So, please think about it, but then become part of it and contribute to it. Otherwise, it will just be another impoverished, unread book, with a snazzy title.

0.2. *Redesigning Ecosystems & the Planet?*

How can we redesign ecosystems with ecological design? What is ecological design? How can it defined, much less applied to ecosystems? The title and subject of this book refer to the intentional design of the emergent aspects of the planetary system, especially living forms and more especially human forms, who are becoming a growing regional and global influence. Of course, a few people have been doing it for decades and more people are practicing and promoting it regularly.

What does the word local mean? The word local refers to the unique features of whole systems. The word design refers to the entire range of human intentional behavior, as it is applied to local systems, from benign neglect of some systems to complete control of other systems. What does the word 'global' mean? In the Old English Dictionary, it meant 'spherical' referring to the earth. By 1945, the word was used by the United Nations to describe political action, referring to actions to commonize costs. The United Nations used it thus. Marshall McLuhan and others used it in reference to a global village. Then, it began to mean ubiquitous, or everywhere. Garrett Hardin contrasts the ubiquity of potholes around the world with the specific meaning of a property of the planet earth, such as the atmosphere. What are global things? Certainly the whole of the planet, as well as the atmosphere and ocean, are global; perhaps life is a global phenomenon. Global things are quite different than things that appear globally. The global system, for instance, has unique characteristics that emerge from local interactions. What does it mean to be global or local? These terms are terms of scale, used to describe a field.

The word global refers to the unique features of the whole system; it is not used to mean things that are ubiquitous, such as potholes or soft drinks, or large, such as human trade and business exchanges between nations on separate continents. The word design refers to the entire range of human intentional behavior, as it is applied to the entire planet, from benign neglect of some systems to complete control of other systems.

We have not designed patterns or structures for ecosystems so far, much less the planet, possibly because of the challenges of scale and complexity, or possibly because we have trouble thinking about large, slow, long-term processes. Local systems are connected with and have interactive histories with other systems. Every system has unique properties and limits that have to be addressed in any design. Every system exhibits very-long term trends (called gigatrends) compared to a human lifetime.

0.2.1. *Starting with Design*

The vernacular designs of many cultures are superb adaptations to their local environments. Many early cities were planned and built as a response to difficult conditions. Many of our modern buildings are models of beauty and efficiency. However, our modern cities are generally examples of wild, stochastic growth. Very few people, other than Paolo Soleri, are designing cities. Our habitations, fields and transportation seem completely haphazard. Instead of placing cities on mountains or barren ground, we are filling fertile grasslands and swamps. Instead of placing roads around vital breeding grounds, we are bisecting and fragmenting ecosystems. Instead of integrating our food areas with wilderness, we are simply converting productive ecosystems to temporarily subsidized monocrops, which we then convert to roads and city squares, without regard to place.

One way to play with design is through ecological thought experiments. The design itself must be considered with regional and global factors, including the biogeochemical spheres of the planet, cycles, forests, and animals. Other factors, such as wilderness and the extent of human cultures, must be an integral part of any design. Wilderness areas, of many kinds from autopoetic to neopoetic, have to survive in appropriate sizes and shapes to thrive and provide services to all living beings.

Human cultures have adapted to different environmental conditions over tens of thousands of years. These cultures need to operate within local and global processes, as well as within human structures. Cultures become critical design factors since their adaptive patterns, such as agriculture and technology, create opportunities and problems on local, regional and global levels. Human populations have a significant impact on the planet, especially through conversion of ecosystems and the addition of exotic elements.

In fact, culture and nature are converging in a new pattern that can be referred to as 'domiture.' Domiture is concerned with local, regional and global designs that have to begin with a foundation of wild species in wild systems and continue to the common and artificial places that human cultures create and maintain. This new pattern is supported by ecological design and ecosystem medicine. Design operates on many levels and works upwards and downwards, as well as inwards and outwards. It has to incorporate numerous other factors relating to religion and urban shapes, as well as to economic and political forms. Corporations and the growth of inequality must be addressed in any designs, due to their negative characteristics, which can distort actions.

Creating designs has to be done in a large political framework, complete with new political and corporate structures. Cultural solutions to local and regional problems have to be applied, but global problems resulting from conflict and inequity have to be addressed in every culture and especially through a larger framework. Local designs have to fit global patterns and limits. Global designs have to accept local processes. Ecological designing has to be a coordinated, constrained series of actions that build good local places and adjusts them to global patterns.

Ecological design rests on a foundation of ecological knowledge, as well as on an appreciation of historical problems and philosophical ideals. This work outlines how to create designs consciously, using our knowledge of how things work, but being aware of our ignorance, and by being careful and respectful. The design process has to address many problems, from problems of scale to political inappropriateness.

0.2.2. *A Snapshot of Ecological Design*

Ecological design can create appropriate images for ecosystems to replace the harmful image of the machine. We can build on archaic cultures, with the image of a Turtle, or expand a mythic image scientifically, as with the Greek Goddess of the earth, Gaia (James Lovelock's theory). We must create an image of a whole living planet as a context for the local designs we want. Lewis Thomas suggested that the planet was like a living cell, and the atmosphere was its outer membrane that can edit the sun to promote life.

Design can formulate specific goals to be associated with a common image. These might begin with working towards a civilized human existence on a dynamic planet, characterized by interhuman and interspecies equities. They might include the reservation of the wild, to keep the minimum conditions and processes of the planet in motion. They might include the definition of optimum human populations based on normal ecosystem productivities and the limits of geochemical cycles. They might include a plan for the independence and relative isolation of independent cultures. They might include new kinds of cities and immense areas of common lands for domestication.

Design could also develop strategies to discuss, refine and work towards these and other popular goals. Some of these goals would increase our ability to anticipate and respond to new catastrophes. Others would work towards equity of human societies, within important ecological limits. And, still others would reduce our impacts and disruptions on wild ecosystems and cycles.

One strategy would be to create a eutopian structure for human societies that emphasizes self-reliance and international cooperation on global resources and cycles. It would allow for coordination and planning, yet it would encourage compartmentalization of cultures as regards food and shelter. It would redesign economic and political systems so that they would become smaller and more flexible, in order to respond better to environmental uncertainties and challenges. Other strategies could be coordinated to accomplish further common goals. A few of these would include:

- Empower the UN (or global government) to support itself by coordinating and charging for use of the commons. Encourage the membership of independent cultures and other groups.
- Create a steady state (in Herman Daly's term) framework of local economies, that is, a homeorhetic state constantly changing with ups and downs, but without large massive disruptions or differences, a state limited by optimum numbers and sizes and processes.
- Limit the use of fossil fuels to strategic uses, not for cars, burning or fertilizers. Limit human use of everything to within ecosystem limits. Conserve 80 percent of remaining energy and resources by limiting heating/cooling, driving, and other uses.
- Manage the human end and impacts to avoid large-scale technological fixes that require constant monitoring and control.
- Preserve small scales of everything, from cities and farms, to roadways and air transport. Deindustrialize agriculture. Make it more labor intensive.
- Monitor possible catastrophes, from asteroids to earthquakes and tsunamis. Insure readiness for dealing with them.
- Keep options open. Keep flexibility high. Flexibility is crucial for dealing with changes.

Equally important is creative co-constraint.

- Compartmentalize systems more by following an ecological model, reducing connections and overintegration. Increase diversity and redundancy at multiple levels.
- Require a diversity of sources of materials and services for all kinds of manufacturing, always favoring local ones.
- Restructure financial institutions from interest and profits to fees and community donations. The times and reasons for profits are gone; that age and those needs no longer exist. At this point the concern is not exploration or new trade, it is enoughness, support and equity.
- Create massive stores and vaults for food, seeds and genetic information. Create long-term seed vaults, like the Norwegian doomsday vault. Create seven-year food banks of cereals. In fact, create long-term banks of needed technologies, tools, and physical materials. Keep reserves of everything needed for catastrophes. Make sure supermarkets have more canned goods. Pharmacies, medical supply companies and hospitals need more warehouse supplies for catastrophes. Keep oil reserves. This in fact is a pressing need for any Homeland Security office.
- Accept all of the effects of living on a dynamic planet: extinctions, death, suffering, destruction, life, accomplishment, beauty, joy. It has always been a wild ride and it will continue to be. The music is wild, the dance is wild.

These and other strategies will be expanded throughout the work. Some of the strategies will be constant and permanent, while others will be temporary suggestions for transiting to desired states.

0.2.3. How should we Regard Design?

We can do this now! What would it be like, though? It might feel similar to some times during World War II for some people in some nations. People could react in days. The economy could shift in months. Some things would be limited or rationed. Most spirits would be high because people would sharing the losses and risks, the challenges and stresses—and they would be doing it for the common good, to avoid catastrophes that could affect everyone and to save their part of civilization. The world would not suddenly become peaceful and utopian; there would still be conflict and crime, cheating and violence. But, these problems would occur within a smaller scale, within the context of redesigning human structures to be in tune with the natural processes and scales of the planet.

The thing is, we do not want a uniform—and likely boring—planetary industrial culture. What we want is a framework to insure the development of local cultures. An analogy might be the atmosphere of the earth, which is essentially the same, but allows many kinds of life to develop. Designing systems is a process of assembling ideas, a narrative for the development of interacting patterns. It is not a comprehensive model, but a sketch of possible directions and developments. It is, in John Thackera's words, an incomplete score, because the composer, player, audience, site, all vary with every performance. What is important is the framework for action, with its understood constraints. The work is coadaptive with the participants.

0.3. *Design Science as Poetic Production (Conversations with R.B. Fuller)*

Buckminster Fuller dedicated his life to solving problems. The problems were big then, as they are now. And, what daunting problems they are now: Humans deaths on an unimaginable scale, perhaps over 200 million people in the last 100 years, from shortages and distribution failures to wars and diseases, as well as the conversion of vast areas of vital ecosystems to agricultural fields and deserts, resulting in the deaths of animals, species, habitats and ecosystems. All of these problems interact, so that it becomes difficult to isolate separate problems. These long-standing problems result from long-standing challenges to human health and happiness: Disruptions in distribution, violence, stupidity, greed, forgetfulness— not just personal forgetfulness but larger forms of cultural amnesia—old diseases and new, lust for power, and the lust to consume. Some of these problems are the result of detachment. Corporations, a trademark form of our industrial civilization, contribute to this detachment. Fuller named an imaginary corporation, 'Obnoxico,' as a foil for his ideas on addressing these problems. Fuller dismissed the flatscapes that resulted from uninformed design and recommended a form of eutopian design—Eutopias is from the Greek word meaning 'good places.' This word, with Outopias meaning 'no places,' was the source of Thomas More's pun 'Utopias,' which was also the title of his fictional account of human development to the beginning of the industrial era. This work is an extension of eutopian design to deal with long-standing problems.

We tend to think of problems as unwanted 'side-effects' of the wanted main-effects, but all effects are equal, as Fuller noted, and must be addressed as equal. A problem (from the Greek words 'to throw forward,' which is what we tend to do with them) can be considered as a question proposed for solution. Most things identified as problems are embedded in a network. Nothing is simple; there is not one problem, there is not one solution. Problems could be considered also as challenges that we must respond to continuously, in the process of living, not as puzzles that have to be solved once for all time. A challenge is a calling into question or a demanding task (a challenge is defined as 'a call to take part in'). It is about consciously choosing to see what can be done, rather than dismissing a conflict as terrible and unsolvable. When challenged by some situation, we react by habit, although this may be disconnected from other habits. Habits protect us from many problems. Addressing a problem often has to do with a power struggle, which becomes part of the problem. If problems are regarded as challenges that require a social response, then some conflict can be avoided.

The problems of cultures, of natural ecosystems, and of modern, industrial, corporate, urban civilization, have been documented quite thoroughly. We have identified most of the problems in the problematique, from erosion, pests, and fertility loss, to population migration and diseases, and we have addressed them separately, using technological innovations or political adjustments. But, we have not dealt with them in a whole pattern. We have not understood them as equal parts of complex large dynamic systems.

Sometimes we forget, moreover, that the decisions of our ancestors saddled us with losses, just as our losses will encumber our heirs with deforested landscapes on depleted soils, despoiled by exotic chemicals and hazardous wastes, in a network of impoverished habitats with an unstable climate, and of course, compounded by large intergenerational financial debts. This network of problems can be grouped under large categories, each of which contains a multitude of related problems. Each category also extends numerous threads to the other categories. The loss of nature and the wild, that is, the loss of the whole and the pieces, ecosystems, habitat, species, and individuals, may be devastating to all self-renewing systems.

The remaining losses, such as the loss of place and the loss of design, are fundamentally human losses. These losses, especially of uniqueness and diversity, tend to flatten our image of the planet even more than economic connections.

0.3.1. *Allowing Flatscapes & Killing Living Systems*

Our habitations, fields and transportation seem completely haphazard. Instead of placing cities on mountains or barren ground, we are filling fertile grasslands and swamps. Instead of placing roads around vital breeding grounds, we are bisecting and fragmenting ecosystems. Instead of integrating our food areas with wilderness, we are simply converting productive ecosystems to temporarily subsidized monocrops that we then convert later to grazing areas, roads and city squares, without regard to place or appropriate use—and then let them degrade to deserts, ruined landscapes or placeless flatscapes.

Placelessness begins with the adoption of a modern attitude, an abstract and a geometric view. With this inauthentic technique, places can be treated as interchangeable and unremarkable, where nothing is significant. Cutting historical roots and eroding symbols contribute to an awful placelessness, an alienation to place, and an inability, finally, to have a home and to live there. This becomes the fate of millions, and it increases.

After shaping ourselves to technology and necessity, we have lost the knowledge of how to care. We have learned to be dispassionate (uncaring), objective (uninvolved), and unattached (placeless). Other animals have used languages and tools, so it is not those things alone that account for the lost knowledge. We 'not-care' because we are confused. Our confusion results from being out of place and not having an identity.

People used to identify with place. The gain of monoplaces or flat spaces has led to a loss of identity for many. The commonness of industrial culture, at a low common denominator, has led to fewer identities to choose from. The mass production of homes, as well as of music and art, has led to fewer creations.

A number of scholars have noticed that we are creating flatscapes, devoid of depth and providing only mediocre possibilities. Flatscapes, Norberg-Schulz's appropriate term, have allowed landscapes and places to become monotonous and dull, displaying fewer unique qualities and fostering weaker attachments. Edward Relph outlined the disappearance of variety that results in placelessness. Places become homogenous and interchangeable, as people everywhere share limited ideas and limited ways of relating to others.

David Brower urges that the human sense of place needs to be revived. Places cannot be preserved, however, without cutting their vital connections and making them lifeless. Places cannot be restored to some Arcadian fantasy, without severely limiting their movement and development. Places cannot be created using a machine metaphor that boasts the substitution of anything for anything.

There have been attempts to define and design places. Christopher Alexander decomposed environmental objects and activities into their constitutive elements, to be reconstructed into designs that could fit local places. These formal solutions can improve strategies for design and can provide a matrix for the making of places. But, they should not assume that human variables can be manipulated to achieve a predicted response. And, they cannot ignore specific psychological or cultural patterns.

There must be a way to define places, using an ecological approach, and considering cultural modifications and adaptations to places. The approach has to be tradition-based and partially self-conscious. It has to be responsive to the genius of place, as well as to human meaning derived from the existential and phenomenological significance of the place. The concept of place incorporates physical, biological, and cultural dimensions. The solutions cannot, and do not, need to be precise; they can be unfinished and ambiguous. They can be fuzzy. The design does not need to guarantee rootedness or workability, but it can identify limits. It can provide possibilities through a matrix that allows tradition and richness. It can provide direction from the understanding of the parts. It can understand how to make a fertile kind of soil, as a metaphor, where things can live in and develop. Living beings synthesize the parts and find meaning in living there, and in doing so revitalize a place.

0.3.2. *Designing Real Living Places: Eutopian Experiment*

Perhaps the problem lies in our images; we use old images of 'frontiers' or simple, inadequate images, like the 'machine' metaphor for nature. Perhaps the problem is in a failure of imagination. Ideal images of the world have been offered as schemes for social and political development, but are dismissed as being unrealistic and flighty. On the other hand, adopting these images would require too much change—rejecting the past or refusing the cultural present, before creating new institutions from nothing. Some of them are thoughtful and imaginative, but most of them are rejected as irrelevant dreams and self-indulgent imaginings. Yet, as Pierre Dansereau has said, the failures of pollution, poverty, and urban decay are failures of imagination. Rejecting the solutions of imagination, therefore, can only make the suite of crises worse. Dreams and imagination are needed to describe desirable futures, to support plans, and to outline goals.

These rejected utopian designs can be replaced by realistic and achievable ones, based on a new image of the earth and humanity, that use an ecological perspective, to ease human societies into partnership models, to restore wilderness and common places we have destroyed, and to change international relationships into a poetic framework capable of limiting war and permitting the unique human expansion of cultural expressions.

This work outlines how to create places consciously, by using our knowledge of how things work, by aware of our ignorance, and by being careful and respectful. Certainly this seems less radical than continuing to surround ourselves with nuclear weapons and habitat destruction in the name of an isolated political reality. A eutopian process, quite different from the utopian, would solve many problems, from problems of scale to political inappropriateness.

Buckminster Fuller was one of the first to consider this other meaning of utopias, as good places. In the 1920s, Fuller began work on a Dymaxion Air-Ocean World Map, as an alternative eutopia that sought to define civic and ecological order through maps of known areas (Dymaxion is a word formed from the words dynamic, maximum and ion—an ion is an atom or molecule that has lost or gained one or more electrons). On this map, the North Pole is the neutral center around which land mass and ocean unfold; this deemphasizes Europe, China or America as centers. In the dymaxion series of maps, Fuller superimposed a spherical icosahedron grid onto the earth's surface to limit the distortion of the relative size and shape of its components, so his projection represents all areas with equal weight.

Later, he sketched the World Town Plan, a map that preceded and directly influenced his subsequent maps. The map, in its projection method and form, highlights the connectivity of the "one world island." Fuller saw his map as an operational tool to be used by the members of the global citizenry. He intended to establish a map that does not prioritize cardinal direction, political entities, or hemispheric organization. Instead, the Dymaxion map provides a base for presenting larger global themes such as human migration, natural resources, and population distribution. Temperature replaces politics as an organizing feature. World climate is shown in a range of coloration from warm reds to cool greens and blues. Even with these specific theme, Fuller's map emphasizes wholeness across the global surface.

The proposed framework of *ecological designs* is based on eutopian design, an ecological and cultural framework for making good places on local, regional and global levels. Eutopian design is a way to preserve what is good and useful in human cultures and sciences, and to reserve what is necessary for nature to keep regenerating itself, while addressing the cascading problems of the modern expansion and development with an emergency approach. Eutopian design is a practical framework for allowing the creative anarchy of traditional-size cultures to be able to implement appropriate technology to deal with their resources and with other cultures through a revitalized and empowered international body that has the power of taxing global resources and properties for its own support, as well as the power to disarm and neutralize the unhealthy influences of large nations and corporations. This framework limits human expansion to domestic and artificial areas, by specifying responsibilities and duties, while permitting the free operation of nature on the majority of the planet. It saves neopoetic areas and reserves wilderness, and encourages respect for natural and cultural capital. It recommends recognizing limits and planning for them using an ecological perspective and a metaphorical approach—it is the framework for local, regional and global ecological design.

Eutopias can integrate tools and designs. Tools and designs are important extensions of the human mind. Their purpose is to foster and assist survival, not to make it more difficult. Tools and designs can be made appropriate to environmental limits and cultural preferences, both of which are often ignored by industrial approaches. Eutopias can suspend the designing of noplaces or flatscapes by promoting the understanding of the inadequacies of bad characteristics and bad designs; it can stop the plague of uniformity and paucity. Through an understanding of the consequences of human ambitions and actions, eutopias can avoid many of the evils that result from a civilization set on technical autopilot. Design can be used to reduce impacts from catastrophic events; for instance, by denying building permits on floodplains. The losses from some events, such as droughts resulting from El Nino, can be ameliorated by design and planning, e.g., by having surplus food and supplies stockpiled.

A single utopia would not work for the whole globe. That is why many local patterns, eutopias, are necessary. Most utopias expect a perfectly rational humanity in a stable, ordered nature. Eutopias accept the imperfect nature of humanity and the changing ambiguity of nature. Most utopias have a finished, closed, completely planned society. Eutopias would encourage building an open, progressive, partially planned society. Utopias are the dreams of reason, while eutopias are dreams of small traditions and cultures, reasonable or not. Where an imagined utopia offers revelations promising a desired future, eutopias offer references from selves and cultures for producing good places on earth now.

The first strategy would be to create a eutopian structure for human societies that emphasizes self-reliance and international cooperation on local resources and cycles. Eutopias would be less vulnerable to downward drift or collapses than modern nations because of its new approach to values and limits. Through the United Nations (UN), the framework would allow for global coordination and planning of local associations of communities.

The eutopian structure would create new goals and images that would be applied to priority approaches, starting with survivability and protection of the wild. Eutopias would offer redesigns of our cultural and economic systems so that they would become smaller and more independent and flexible, in order to respond better and faster to new environmental uncertainties and changes. Eutopias would provide a repository for the diversity of human knowledge (cultural knowledge) as well as the diversity of crops and animals—and of course the genetics of wild beings. It could catalog cultural legacies from archaic cultures, similar to what is being done for some indigenous peoples and cultural heritage, for example the Foxfire books on Appalachian folk knowledge, to preserve it from being lost from the transition to cities. Archaic societies have experience adapting to local conditions or catastrophes. A data base from a variety of cultures, each adapted to different environments, would provide resiliency and functional redundancy to human survival.

There is no mechanical prescription for designing and making good places, nor is there a blueprint or timetable. The current institutions cannot create good places; the market has not been able to create health and equity; even scientists and humanitarians have not been able to create a way—Eutopias is a fourth way. It is not an institution that benefits only the rich; nor is it a schedule of temporary handouts. It is a plan for a framework for local self-reliance and exchange, which is respectful of traditional cultures and ecological networks.

Eutopias has a low—but not too low—political feasibility. The benefits must be worthwhile to justify the costs. Benefits cannot be vague and unsatisfying when the costs are immediate and painful. Poetry and education must show the existence of benefits, so that the eutopian alternative can begin. This code emphasizes its flexibility. This strategy would avoid eventual hardening of the choices, but it must be instituted at once. The crisis caused by exponential growth and destruction cannot be solved just after some final limit is passed and a great catastrophe has begun. The crisis of ignorance cannot be solved by hurrying and creating more problems.

Eutopias can reduce the losses of nature and culture by creating a framework to protect them. Eutopias can reduce the losses of health, fitness and accord, by emphasizing them and creating circumstances for their continuity. Eutopias can reduce the losses of equity, renewal and design by offering new designs that allow for a normalization of equity and for the normal processes of renewal. Losses from accidents and diseases can be reduced by preparedness. Losses from earth and climate changes can be reduced, also, with preparedness for 'normal' regular events, such as hurricanes, earthquakes, and droughts.

Eutopias can show how to preserve and restore, design and plan. Eutopias can provide an ecological planning process that offers a structure of limits and divisions for the planetary systems that would permit the preservation and restoration of natural cycles and places. An ecological design process would be applied to ecosystems as well as to cities and fields.

0.3.3. *How can we Redesign the Experiment?*

For his commitment to a design science revolution, Fuller made "Ten Proposals" (from Earth Inc.) that addressed the problems of our civilization. One of the most important is to learn the mathematical coordinate system of the universe. In fact, an educational revolution, based on synergy, should be the highest priority. It would start with an inventory of all known principles, using Fuller's world game for theoretical exploration. To inventory the resources of the planet, we would convert general accounting systems to a planetary ecological accounting system, intergenerational and cosmic rather than annual and agricultural. Wealth would be refined from a scarcity model to an energy model. By making ownership onerous, excess property could be eliminated, and we would be liberated from our slavery to 'thingness.' World sovereignties, with their suite of barriers, would be modified. Humans would apply their unique skills as problem solvers by realizing our competence at design science.

Fuller made many brilliant suggestions, and applied many ideas in brilliant inventions. His technical focus, however, allowed him to overlook many of the needs of people and cultures. One sovereignty, replacing many hundreds, would make things worse, for example. As Leopold Kohr noted, bigness is the source of most problems, such as misery and conflict, and smallness is the simplest and most elegant solution.

For Fuller, the discipline of design science proceeds from a subjective search through experimentation and feedback to generalization, then development, including practice and regeneration, to evaluation and back to the subjective search in a loop. Fuller notes than humans have evolved from the local 'rejuvenation' of agrarian farming to a collective nonlocal rejuvenation of industrialization. However, industrialization has proved to be nonrejuvenating, as it has been stuck too long in a flow-through pattern of production, where everything is treated as commodity. Something that is capable of self-renewal is needed.

Fuller also argued that we have displaced ourselves as specialists and must again become generalists and 'comprehensivists.' We need greater degrees of freedom to increase the probability of cross-fertilization to solve design problems. We need new skills in design that can come from considering the local systems within a regional context. We need to try new approaches and correct the inevitable mistakes. And, we need to ask questions about the global systems, before we can try to answer them.

Fuller emphasized the importance of global communications, through radio signals, and global transport, complete with a computerized global transport system and traveling cartridges that would be assembled at local stations or loaded onto planes or ships. Fuller also proposed a global energy grid (the Global Energy Network International), which would connect all areas, although we would have to be careful not to create connections that are too rigid. The dymaxion house was to be a global dwelling service, which would optimize performance and efficiency.

A design science revolution, using a comprehensive anticipatory design science, would reform the environment. But, Fuller thought we could reform the environment without reforming people. This focus on technical solutions has resulted in more damage to the environment, instead of less. Therefore, design cannot be limited to simple mechanics, but has to extend to human cultures and behaviors, to ecology, economics and politics.

Fuller does suggest that a design scientist has to take the initiative rather than being retained by a client to carry out a limited design. The design scientist has to perform the fundamental invention, underwriting, development, and experimental proof of a project. Design science has to provide effective anticipatory strategies for formulating and managing the regeneration of industrial organisms in the same way that the medical profession deals with human metabolic regeneration, according to Fuller (WDSD No. 5, 1962). Design has to be holistic. Fuller states "the design scientist would not be concerned exclusively with the seat of a tractor but with the whole concept and distribution of food." For the field of local ecological design, this means fitting the total human industry within the context of the local, regional and global system.

0.3.4. *Poetic Production*

According to Aristotle, productive science, or poetic knowledge, has a product as its end, not knowledge or action. Productive, or poetic, science derives many propositions from theoretic and practical sciences, and because it involves more elements, poetic science is the least exact of all. Yet, in a sense, poetic science is more basic than the others, since all sciences produce an end.

Poetic knowledge is inseparable from the power to make. Poetic knowledge is a kind of knowledgeable activity with a product for an end. Aristotle defined poetic art at various times as having to do with creation. Poetic science is a making, a fashioning of random data into a whole. The whole is structurally unified into a complex thing. This unity is based on organic themes, for Aristotle. Art is a mimesis of nature, but it does not copy nature's products. It presents unified wholes, in the manner of nature. Works of art are structured like unified living things; when things exhibit unity, then they have order, and when things have a definite size and order, then they are beautiful, in art and in nature. Fuller once said: "When

I am working on a problem I never think about beauty. I only think about how to solve the problem. But, when I have finished, if the solution is not beautiful, I know it is wrong." In fact, only by being organically structured can art be mimetic; and only by being mimetic to life can art works be organisms. Art, design and poetry imitate, not fragmentary reality, but the essential whole.

Our designs of ecosystems, of wildernesses, of cultural forms, and of the planet need to be poetic productions. By following the principles of ecology and applying them to the characteristics of good places, we can make local designs. According to David Orr, certain design principles work with ecosystems and nations: Small units dispersed in space, redundancy, short linkages between modules, simplicity, diversity of components, self-reliance, decentralized control, large margins, and immediate feedback. A megaframe like Eutopias would allow this scaling of design from the local to the global.

Ecological design, for instance, is the design of whole communities. We can design places as organic wholes to promote the well being of individuals and the common good. But, we can only really do so as participants of ecosystems (and this takes us back to the fundamental lessons of physics: that we cannot not be part of an experiment, that disorder creates order which creates disorder, and so on). Humans need to recognize that they automatically participate in everything. Furthermore, due to the uncertainty in dealing with large, long-lived systems, we have to learn to accept that the system needs most of its own productivity and to limit our use of the systems to well below critical limits, that is, within the flexibility of the system.

Ecological design is the design of human impacts on unique places in the planet. It may require patterns of constraint more than specific technological applications for that level. It requires the participation of everyone, dealing with small actions and very large cycles. But, the designs and actions, and certainly, our awareness, have to extend to the region and perhaps beyond the planet.

1.0. Preparing for Challenges to Design

Human consciousness of our effects on the planet and on each other has gradually increased, as shown by recycling programs and by celebrations such as Earth Day, which has been getting larger every year—as a celebration. But, celebrations do not seem to have lasting effects. Environmental deterioration has worsened in most places. Levels of consumption have increased; populations have increased. Wastes and pollutions cannot be controlled. Death and destruction cannot be counted.

We need to propose and execute local policies to steer change. We do not have time to look at all the information that we have collected, or to convert it to knowledge. We never have had time. We cannot connect with all the information flows. So, we will have to act as if the information we have is enough for wise decisions.

The earth does not need to be saved or healed, as if we could do either. The ways of life that we remember and prefer, the places that depend on other species and natural processes—these can be saved. Our own divided minds, that let the poor be enslaved by the wealthy, that let 'good' animals be domesticated and 'bad' animals be eradicated, can be healed. The sacrifices will have to be great; the changes will have to be radical. But, only then will the congratulations and celebrations be meaningful.

1.1. *The Rape of Gaia: Facing a Damaged Planet & Knots of Problems*

Every year numbers are collected in every area of human interest, across the entire planet. Those numbers indicate the deterioration of water and air quality, the erosion of soil and land, the destruction of forests, the decline in health and longevity, the deaths of people, the fracture of cultures, and the wobble of planetary cycles. In the past few years, the numbers have worsened dramatically.

1.1.1. *Human & Ambihuman Deaths*
The numbers of human deaths are very disturbing. There have been massive human die-offs in the past 106 years since the start of the Twentieth Century. These numbers tell a story of big death. From democides—the intentional killing of races, nations, tribes or communities—total deaths may range from a minimum of 150 million to a possible 350 million people, who were shot, knifed, burned, suffocated, poisoned, starved, crushed, drowned, hanged, bombed, or buried alive, in a plague of violence.

Famine, once the greatest producer of deaths in agricultural nations for thousands of years, may have killed 32 million in this century. Disease, once the greatest producer of deaths during early phases of globalization, has killed many millions. In the Twentieth Century, the influenza pandemic in 1918 killed over 20 million people, and possibly as many as 50 million. The recent AIDS epidemic, from 1978 to 2001, counts for over 23 million.

Disasters, from drought, flooding, earthquakes and other regular planetary events, killed 21 million people. Accidents, from transportation from horses to space shuttles, killed half a million people.

Many of these deaths overlap by category. For instance, ethnic cleansing can start from changes in the distribution of food that lead to famine and disease. Poverty is never listed as a cause of death, but over 15 million, perhaps as many as 30 million people, die from a lack of clean water, food, medical services, or shelter, every year, including 2006.

The numbers on the living environment are more disturbing and critical. Although species are still being identified at the rate of 8,500 new insects species and 100 new fish species per year, probably 400 species are driven to a premature extinction and 1,000,000 species are threatened every year. Statistics for habitats are almost incomprehensible. Twelve million hectares of forest are cleared annually and 10 million hectares are degraded. Marshlands are filled in; coral reefs are mined; and grasslands are paved over.

Possibly 59 percent of arable land has been degraded; 50 percent of fresh water co-opted for human use; 50 percent of the planet's wetlands modified, drained or destroyed; 50 percent of the coral reefs damaged and perhaps 20 percent destroyed; individual deaths and species extinctions are uncounted, perhaps uncountable.

1.1.2. *Catastrophe & Progress*

Many deaths, collapses and extinctions result from catastrophes. The word catastrophe means 'down turning,' from the Greek words. A catastrophe is a down turning, literally. Most catastrophes are assumed to be fast, sudden, and manageable, like floods or fires. But, catastrophes come in all ranges of speed, size, temporality, visibility, and combinations. Our language is poor in its terms for catastrophe. A slow catastrophe, such as climate change, could be called a bradycatastrophe. A long-term catastrophe, such as chronic drought, could be called a chronocatastrophe. A large catastrophe, like a large meteor impact, could be called a megacatastrophe, a global catastrophe perhaps. An unseen catastrophe, such as an extinction, could be called a cryptocatastrophe. A multi-pronged catastrophe, such as the collapse of an entire human culture, could be called a polycatastrophe. Unfortunately, catastrophes can and do occur in many combinations. Thus the loss of many species and ecosystems by the planet might be called a polycryptobradochronomegacatastrophe.

We know of many catastrophes that have happened in the past. Such catastrophes may have included an asteroid strike that ruined the dinosaur dominance or the recent ecological collapse of freshwater systems, but until they happen to us, knowledge of them will not be adequate to inspire change and preparation.

People have been predicting catastrophes and global shifts for the past 50 years. They have identified discontinuities ranging from weather patterns to disease patterns to political upheavals and collapses of alliances. The trends they identify are unsustainable. The responses are unknown or uncertain. The predictions are uncertain. So, we need some kind of direction, as well as an understanding of catastrophes.

Many kinds of catastrophes are possible and that raises questions. What are the contributing causes? How can they be changed? How can catastrophes be diverted or lessened or stopped? Disasters regularly arise from natural and economic systems. They are part of the process of biological or cultural development. Recently, we have a new catastrophe, called the 'U.S. problem,' the 'U.S. emergency,' or the 'U.S. crisis,' even though it is a human crisis or a possibly global capital crisis. Other catastrophes that befell other cultures, from the Mesopotamians to the Greeks to Rapa Nuins, were generally local and did not affect the remainder of humanity. Now, we are more connected. Regional and local problems can become global.

With our sophisticated instruments, we possesses incredible scientific evidence of environmental wobble, biological imbalances, and the unfitness of entire domestic species, but knowledge moves few to action. With our wonderful communications, from radios and televisions to computers and cellular telephones, the poor are able to see everything that the rich have and do, and they want equal benefits from the progress of our industrial financial culture, but this knowledge allows everyone to want to stay in the trap of affluence. Probably nothing will be done about traps and collapses until catastrophes become common experiences.

Catastrophes concentrate attention on a landscape and its people and that is their benefit on human affairs. Ideally, it should not take catastrophes to precipitate corrective measures. Instead, we might resent the necessity to change. William Catton recommends that we do not indulge in resentment. Our bad present circumstances result from the innocence and hope of our ancestors; they are the result of decisions to have babies, fires, televisions, tractors, and status. The understanding of catastrophe may let us avoid it or at least ameliorate it. Catton makes the biological analogy that die-off is a signal to overshoot, and overshoot leads to habitat damage. The agent of a post-irruption crash may be starvation, war or just behavioral stress.

If civilization collapses, the struggle back to a technological society will have greater limitations. Accessible minerals will have been scattered; the gene pool will have been greatly reduced. Then it may be too late. Our species may die. It is hard to imagine all life on earth dying. Even the worst of catastrophes would leave a simplified ecology of mosses and slugs. Weeds and pioneer species would prosper. Habitats would be ruled by the natural laws of ecology again. Perhaps herbivorous animals would build up large populations again. James Lovelock worries that an exclusively bacterial and fungal planet, however, would not be able to support the biogeochemical cycles and atmospheric composition that larger animals and plants require.

Fortunately, most catastrophes do not occur in completely destructive patterns. Limited starvation occurs before total starvation. There will be uncomfortable smogs before acid rains destroy most crops. But these things could signal the necessity of immediate corrective actions. Possibly, our attention needs to be global, since humanity has provoked a global crisis of local crises, typified by tremendous losses.

1.1.3. *Eight Terrible Losses*

The problems of cultures, of natural ecosystems, and of modern, industrial, corporate, urban civilization, have been documented quite thoroughly. We have identified most of the problems in the problematique, from erosion, pests, and fertility loss, to population migration and diseases, and we have addressed them separately, using technological innovations or political adjustments. But, we have not dealt with them in a whole pattern. We have not understood them as complex large dynamic systems.

Sometimes we forget, also, that the decisions of our ancestors saddled us with losses, just as our losses will encumber our heirs with deforested landscapes on depleted soils, despoiled by exotic chemicals and hazardous wastes, in a network of impoverished habitats with an unstable climate, and of course, compounded by large intergenerational financial debts. The losses indicated by these numbers can be categorized simply, under eight large categories, each of which contains a multitude of related problems. Each category extends numerous threads to the other categories.

1.1.3.1. Loss of Nature (Habitat Species Cycles & Nets)

The overuse of ecosystems results in deforestation, devegetation, and desertification, then in the depletion of raw materials and the depletion of agricultural productivity. This overuse results in the attendant losses of many habitats, species, and individuals. Many species are not hunted to extinction; they are being lost as a result of the simplification of their habitats for human uses, from the expansion of agriculture to the pavement of cities and road systems. The rates of extinction are now far higher than the natural, unassisted rates. The loss of habitat also decreases any new speciation processes. Demands from a growing human population, as well as their violent conflicts, accelerate the losses. As the human world becomes more abstract and separated from the wild worlds, it becomes more isolated and human specific. As our habitats become more removed from natural processes, we tend to think they are all self-created and controlled, at least until a dramatic hurricane or earthquake overwhelms our abilities at control.

1.1.3.2. Loss of Culture

With the globalization of trade and the domination of national languages, smaller cultures are being absorbed into larger ones. Although people retain some of their traditions, the loss of their language means that the whole perspective of their culture, related to the uniqueness of place and human adaptation to place, is being lost. Knowledge of plants, animals, and processes is being lost at a time when modern science is half-heartedly trying to catalog knowledge of every place. A lack of knowledge is often related to illiteracy—or the lack of numeracy or ecolacy, to use Garrett Hardin's words, but this kind of loss is personal and cultural—it had been learned once and is lost as a result of a set of circumstances where it is devalued and ignored. Smaller, less-competitive cultures are being abandoned by new generations of people or subverted by global comparisons of luxuries and expenditures.

Population pressures, resource shortages, and manufacturing 'side-effects' cause instability in many societies. Militarism, intolerance, crimes, and health problems are symptoms of the instability of cultures. Confusion and misinformation contribute further to the destruction of cultures. If cultures are lost and new forms cannot fit themselves to the patterns and uncertainty of natural systems, then people may not be able to adapt to the continued development of the earth and its ecosystems.

1.1.3.3. Loss of Health

The instability of cultures, as well as stress, insecurity, and insufficient diets, results in physical and psychological problems for people. Individual powerlessness and disillusion provoke further disintegration. With disruption of agricultural and manufacturing productivity, as a result of conflicts, people have less food and fewer necessities. Hunger affects over half the people in the world, in every culture, including many of the dominant cultures in China, Russia, and the United States. Extended hunger not only alters behavior and intelligence—that is the ability to evade hunger—but it allows diseases to flourish. Many diseases have made dramatic comebacks in unsanitary conditions with inadequate food. Health can be reduced by long-term trends in lifestyles, from sedentary work to family size. The size of families is crucial for mental health, for instance. Basically, extended families allow more outlets for communication and spread the stress among members. Small families have to bear much more stress and members often become unhealthy. Loss of health leads to other losses.

1.1.3.4. Loss of Fitness

Fitness is the ability to function under normal environmental conditions. Stress, obesity, illness, weak sperm, social conflict, and other kinds of dysfunctions, can reduce fitness. Some of this reduction, e.g., evidence of poor personal or reproductive health, is caused by chemicals, especially persistent chlorinated organic chemicals. Some of these chemicals also reduce our ability to reproduce, raise young, make good decisions about consumption and social conflict, and to persevere in general. The cost of fitness ranges from lower fertility and sperm viability to early death. However, a lifestyle dependent on physical and energy slaves, not to mention a diet riddled with addictions to cheap fats and sugars, decreases our fitness. All of the elements of a true addiction are present: We get a short term high, and we suffer long term health problems. The air pollution, the sedentary life styles, and the global climate chaos are all injurious to health. Exposure to toxic chemicals and stressful city environments makes things worse.

Perhaps this fitness problem is a result of a long-term trend. As scavengers we had to gather seeds roots and fruits, as well as dead animals. As hunters we moved quite a bit; as farmers, we had to work harder and longer, planting, harvesting and storing food. Then we started using animals and better tools and fossil fuels; we invented labor-saving, and activity-saving, devices. So, now exercise is simply an activity outside the education or work day. As human populations expand and shift to cities to live, they are losing the ability to fit with natural ecosystems.

1.1.3.5. Loss of Equity

Although many resources are distributed unequally over the globe, as a result of different kinds of historical geological processes, trade can allow access to those resources. However, as a result of long-term processes of inequity, from keeping people enslaved to cultural hoarding, many people have far less than others. This has resulted in permanent overclasses and underclasses, which are maintained by physical force, as well as by the force of economic and religious myths. These myths tell all people that they participate in the "best possible economic system" regardless if it justifies the differences of inequity based on history or on perceived racial abilities. Most people are hungry; few people are fulfilled. Even low average levels of food and fulfillment can be maintained only through theft from other species and from future human generations, and through the degradation of billions of humans as well as the ecosystems on which they depend.

Other myths, such as the "market is free" or "growth is beneficial to all," suggest to people not to try other economic forms. But, the market is not free; it is controlled to benefit the rich. Continued growth is amoral and pathological, benefiting the elite of authoritarian regimes as much as the oligarchs of democratic ones. It refuses to recognize, much less to pay, all of its costs, such as depletion, loss of security—which may be most important—or extinction. The entire system perpetuates mass poverty and justifies it by blaming individuals, but the system itself fails to reduce inequity or poverty. This loss reduces effort; it is responsible for tiredness and low kinds of health, productivity, and esteem, things that are necessary for personal and systemic renewal.

1.1.3.6. Loss of Renewal

Self-renewing systems renew themselves through a process of renewal—yes, the definition is self-defining and circular. But, renewal is limited when too much of the system is lost through waste. Manufacturing for a large population in "free" economies results in the mass production of waste, and in the storage of solid wastes in landfills. The waste is unavoidable in the current system, which is unable to renew itself and has to force a linear flow, trusting that natural recycling systems will be able to adjust to immensely higher volumes. Much loss in the current economic system is wrongly identified as waste. It is simply not economical to recover the waste from this system, due to the volume of "free" resources and goods from nature. People themselves are treated as a waste product of capital-producing system, although the system could be viewed as a waste-producing system that also produces capital. Separated, waste cannot contribute to the renewal of system.

1.1.3.7. Loss of Accord

As a result of the unequal distribution of natural resources, including unincorporated waste and pollution, and the unequal distribution of materials and wealth between people, economic conflicts arise, often becoming violent political conflicts. Accord, by definition, means agreement or the concurrence of will or action; it can mean harmony of mind, as well as harmony of sounds. As an agreement between the parties in a controversy, by which satisfaction for an injury is stipulated, accord allows people to reconcile their injuries and interests. Because many national boundaries were drawn as a result of colonial expansion and contraction, many cultures are artificially combined in large territories or stretched across several traditional territories. This has created the conditions for continuing cultural and political conflict. In addition to the normal conflict between different cultures with different ways and values, usually resolved by trade or distance, this new conflict resembles small permanent wars over large numbers of territories. This kind of hot conflict not only destroys habitats and resources, but it causes immense human suffering. As the rules change, and conflict includes noncombatants, as well as plants and animals, there is less accord between conflicting groups. Accord requires the ability to trust, which requires self-reliance and confidence. The loss of accord leads to other related losses.

1.1.3.8. Loss of Design

During the evolutionary development of human responses to environmental challenges, people tested many kinds of designs for their tools, houses, and living habits. These initial designs often worked well in situations of limited resources or of limited power. With the application of more power and the acceleration of the interchange of dominant designs, we have been able to force standard designs to work under almost any conditions, often replacing locally adapted designs.

Many of our current problems are design problems, Victor Papanek judges—that is to say, bad design causes all varieties of suffering and waste; it causes death, from accidents or collapses, and it causes immense waste by relying on power to overcome its flaws. Many of these problems could be solved by better design, not just by more comprehensive industrial design or political design, but also by the ecological design of entire processes and places. Of course, some problems are caused by gravity, wildfires, or earthquakes, which are a natural part of the planetary system. While they may seem unsolvable, we can adapt to them with better designs, for instance, by reserving flood plains or landslide areas as parks without houses or buildings, and by adjusting the houses that are built to local conditions. Other problems are caused by human demands and the design response to them.

1.1.4. *A Bad Combination of Losses*

All of these problems interact, so that it becomes difficult to isolate separate problems. These longstanding problems result from longstanding challenges to human health and happiness: Food shortages, disruptions in distribution, violence, stupidity, greed, forgetfulness—not just personal forgetfulness but larger forms of cultural amnesia, old diseases and new, lust for power, and the lust to consume. These problems result in the death of many individual plants and animals, as well as entire species and whole habitats and ecosystems.

People die for many reasons: Environmental catastrophes, such as earthquakes, accidents, old age, lack of food, diseases, or personal violence. We know that people also die "indirectly" as a result of the destruction of natural ecosystems, specifically, as an example, the lack of clean, drinkable water. People die for economic reasons, such as not having a job to afford food, or not having a place to live, as a result of war or dislocation. It is almost impossible to calculate how many die in each category. It is difficult to say how many die directly from hunger, since hunger leads to disease, such as pneumonia, and disease gets the credit; or, how many die from unclean water, since the contaminant or the accident gets the credit. It is also difficult to calculate deaths from theft, economic dislocation, lack of planning, bad designs, or depression. Only specific diseases, accidents or bullets seem to be unambiguously fatal, and even these are often causally linked with many other contributing factors. We need to use better tools to understand and solve these problems, as well as to understand how a cosmology or economic system can promote the problems and then fail to solve them. One of the biggest challenges to ecological design, beyond these losses, are the gifts that came with industrialization, more than just premature celebrations. We thought we wanted these gifts, but they seem to have many kinds of costs associated with them.

1.1.5. *The Gift of Bigness* (The Curse of Corporate Bigness)

Early in our history, when the very success of the species was in question, we humans learned to reproduce more rapidly than our rates of mortality. To extend our families, we have increased our numbers and our rate of increase exponentially. To ensure the success of our species, we have appropriated the places of other species. Our overpopulation has led to aggression against other cultures and species, then to indifference at their suffering. Even low levels of food and fulfillment at our current size can be maintained only through theft from other species and from future human generations, and through the degradation of billions of humans as well as the ecosystems on which they depend.

To provide for the needs of many and for the extravagant luxuries of some, we have produced waste and pollution on a geological scale, from islands of garbage to skies of acid rain. Manufacturing processes result in the production of new dangers, such as recombined genes and new substances, which are not easily incorporated into natural cycles. The overuse of ecosystems results in deforestation, devegetation, and desertification, then in depletion of raw materials and depletion of agricultural land. Economic and political pressures, derived ultimately from population pressures, force farmers to intensify their efforts to increase crop production, instigating a dismal cycle of population expansion, environmental deterioration, and poverty.

To provide for our needs efficiently, we have increased the scale of our activities. But we have decreased the diversity of habitats by filling in wetlands, felling forests, plowing grasslands, and irrigating deserts. Agribusiness has caused widespread landlessness; people who try to grow their own food are forced onto marginal lands or off the land. Acquiring fossil fuels also creates landlessness; coal mining in the Black Mesa mountains of the United States, for instance, may force the resettlement of twenty thousand Hopi and Navajo people. Without land, and the independence it allows, cultures are more likely to disintegrate.

Our local communities are proud to attract more people and larger industries, but do so thoughtlessly, without regard for the limits of population size or the rate of energy use, without sufficient consideration of the effects on the quality of our lives or on the quality of the environment. Although we make plans for people and their activities, the plans are usually reactions to growth and change. The formal development from planning results in a complex of problems, from pollution to ugliness.

Our societies are big. Our corporations are big. Our impacts are big. This creates discontent because people feel powerless. They feel powerless because they feel that democracy is not working. The country is run by the rich, rich corporations, rich people, and rich politicians, and they secure their needs for more money first. In fact the decline in earnings, relative to costs of goods, forces people to struggle with jobs and increased conflict everywhere. Now, at the perceived "end of history" and the "victory of democracy," it is difficult to imagine a better future, at least without questioning the rich, the ideals of democracy, and the corporate good will to society.

We have economic growth; we can see the numbers. But, the growth is premised on saving costs by forcing down wages, or by reducing the number of workers in the name of efficiency, and forcing overqualified workers into service jobs. The growth promotes inequality, improvement for a few and impoverishment for most. It is like the growth of a tumor, issuing a healthy glow from a fever and the false image of health. Profits go up, but public services decline for lack of funds. There is no money for schools, none for libraries or parks, little for private institutions, and little for national, state or local governments. Where did it go? Profits? Profits for individual corporations, profits for individuals? Could we find them, can we track the money? We should be able to, since the management revolution has made paper trails everywhere. Perhaps the trails are too complex.

Bigness overwhelms the ideals of cultures, which is why large nations such as China or the United states cut their forests, regardless of respectful religions or special cultural values regarding nature, regardless of the desires of local communities. The nations are willing to sell the wealth of the provinces, even though the provinces want to conserve them.

Something is wrong. But who is to blame? Where is the target? People rail against liberals or conservatives, against corporations or protesters, against big government, or big corporations, big permissiveness, big violence, the lack of faith or lack of prayer, the failure of responsibility or the failure of nerve, against guns, leniency, illegitimacy, bad rock, bad lyrics, bad welfare, bad politicians, bad people, and bad police. But, the real enemy is unseen. Who can argue against democracy or corporate wealth or the general vague feeling that things have improved? Against a reasonable system? Against the bigness of the system?

The course is downward. How can it be turned? How can we diagnose the problems? How can we suggest a path to health? Charles Reich says that we need a science of social change. Ecology is science-based. Conservation is science-based. Even management is science-based. Now economics and politics needs to be science-based, instead of dominated by an old ideology and weak mythology. Reich suggests that between citizens and government is a third entity, economic government, to which has been ceded power to determine the direction and type of economy. The only knowledge we get is knowledge from the economic government, and the numbers look good. We have no other knowledge, except for a weak self-knowledge and vague social knowledge, that tells us something is wrong.

We have always tried to exceed the physical and biological limits of places rather than recognize them and be guided by them. Every advanced country is now over-technologized. Past a certain point the quality of life diminishes, not improves, with each advance. Big science serves big technology, which supports and is supported by big government. And there is no science like big science, and no administration like big administration. But this enthusiasm is misdirected. Scientific advances and technological changes result in unforeseen consequences, good or bad. They cannot be controlled or legislated against before the fact. But the investment seems too big to abandon.

The planet, including human society, is threatened by the bigness of things. Although nature is big, it has evolved slowly; human size is new and sudden. A snail's pace is good enough for nature most of the time, but with our brief life span, we argue that we need quick changes. The technological advances have not been paid for yet, although the cost in pain and death is incalculable. We will be paying for hundreds of years for those past advances. The economic style is too great and reckless for ecological systems to absorb its impacts. The scale of things is an independent problem, which can ruin the best intentions of policy. A bigger system to control systems that are now too big might be a mistake.

But some solutions are even bigger. Buckminster Fuller, Alvin Toffler, and the Club of Rome favor a supertechnocracy. Science fiction visions predominate: Gerard O'Neill's orbiting cylinders or Simon Calder's floating domes. R.A. Smith, in "Unibutz," claims that we could achieve a pantheistic-humanistic-cosmic awareness, in achieving technology without materialism, plenty without selfishness, and community without tribalism. His Unibutz is a global goal for leaving the earth and reaching the stars. The voyage would help shape new world structures and give a purpose to humanity. But to what end? Bigness and wealth elsewhere?

Perhaps big science and big technology have too much momentum. Theodore Roszak acknowledges its schizoid attraction and repulsion, with the twin promises of glorious accomplishment and hideous death. Who could escape being torn between yes and no, if even our end would shine with radioactive, Promethean grandeur? Our image of big science—the scientist as tragic hero, isolated in chaotic nature, but strong in his proud individuality, perhaps driven to research by hubris and madness—is a barrier to any new vision, especially a small vision.

1.1.6. *The Gift of Speed* (The Curse of Momentum). To achieve even greater efficiency, we have increased the speed of our activities, converting materials and cultures into new designs without consideration of the meaning of, or need for, efficiency. The speed of our economy is too great for many cultures to adjust to; and the thoughtless transformation of cultures may result in great, irreversible mistakes. The speed of our conversion of wild habitats to domesticated lands is too great for many species to adapt.

Alvin Toffler foretells a dramatic redistribution of power, from slow countries to fast. Speed is the critical factor for Toffler, who states that, historically, power has shifted "from the slow to the fast," whether speaking of "species or nations." Certainly, being faster to the industrial market has advantages for many international corporations. But, this kind of speed is not applicable to species. Slow species have survived as well as fast, either adaptively or neurally; for example, many fast dinosaurs perished before their slower mammalian contemporaries.

Toffler notes that the industrial revolution stepped up the metabolism of economies, but does not seem to make any distinction between good or bad metabolism—fever as well as excitement speeds up a metabolism. Truly, we are speeding up our use of resources without knowing where they are coming from or going to. Modern economies, embracing the idea that "nature is capital," draw on the accumulated "capital" of ecosystems for production. By ignoring the real cost of the capital, as well as the costs of natural services, such as nutrient recycling, soil building, and atmospheric renewal, these economics create a temporary wealth—similar to the healthy flush of a fever, perhaps—and a long-term imbalance. When an economy falls out of balance with its local environment, massive disruption often results; industrial economies have only avoided disruption by trading advantageously with other economies, by using fossil fuels, and by promoting general institutional inequality.

Continuing his paean to speed for its own sake, Toffler states that fast economies generate wealth and power faster than slow ones. But, what kind of wealth? Financial or cultural, agricultural or symbolic? And, what kind of power? Mechanical or organic, political or personal? Industrial economic wealth is merely a small part of the wealth of the earth and humanity, most of which has little value to that economy.

Toffler describes an acceleration effect that makes each unit of time saved more valuable than the last, creating a positive feedback loop— inadvertently identifying the archetypal problem of modern economics: Runaway positive feedback loops leading to catastrophe. The fast economy he describes seems to depend on fleets of hypersonic jets racing around the world with the elite and their tonnage of possessions. Telecommunications, transportation, and tourism will accelerate, blithely unaware of their impacts on family structures, biogeochemical cycles, including the ozone layer, and wilderness. Have we learned anything?

Toffler sees revolutionary consequences in new management methods, but he does not see the negative effects. Managerial decisions regarding resources are made often on short-term economic grounds and lead to material shortages and human and environmental degradation. Newer methods seem only to offer a higher degree of impersonality.

The new wealth creation system, Toffler claims, holds the possibility of a better future for the vast populations of poor, if their leaders anticipate changes. The new system for making wealth consists of an expanding global network of markets, monetary, and production centers in instant communication with increasing flows of data and information, but not necessarily wisdom or understanding. He argues that the availability of this information flow gives more power to consumers, voters, workers, and small businessmen, taking it away from a centralized few. The potential is there, but Toffler does not go far enough to envision alternate economies and communities. The power still operates under the old assumptions and divisions in his speedy synthesis.

Of course, Toffler is right to recognize the problems of nonindustrial countries, many of whom depend on cheap labor or strategic military location for foreign investment. But, where does that "investment" go? To the local poor or to remote, rich politicians? Wealth could be distributed fairly, depending on many factors, such as synergy, generosity, reciprocity, and cooperation, but it is not. The gaps are growing. Toffler acknowledges that they will keep growing. But, we can redistribute wealth without industrializing every culture. We should try to achieve economic justice before accelerating to new glories.

Toffler concludes that a great technological and cultural wall will separate the slow from the fast, making problems for joint ventures. But, what are the products of these ventures? The debris of advertising fads, such as mink toilet seats, or the tools of real needs, such as evaporative water purifiers? Toffler foresees the emergence of an electronic neural system for a global economy, without which any nation will be doomed to backwardness. What kind of backwardness? Lack of fast things? Lack of professional enslavement? Lack of art, play, or culture? Lack of food, tradition, freedom, or happiness? Perhaps these have already been stolen.

There are other gifts of course: The gift of abstraction, of rationalization, of centering, whether of the self, family, city, or culture, the gift of delusion, of progress, perhaps of blinders, and many more. These gifts also have costs that we did not notice or calculate when we accepted them. And, these gifts have real consequences in terms of death and destruction, but now we are afraid to refuse them, in case we might be unhappy or suffer failure.

1.1.7. *The Scope of Failures*

Acceptance of limits is not a kind of failure. One speaks as best one can. Awareness of inadequacy, as of ignorance, is a positive accomplishment. True failure is indifference to inadequacy. The failures in our character or group or national character, can be seen to be responsible for the problems identified by Konrad Lorenz as the seven deadly sins of civilized humanity, from destruction of nature to the loss of civility. These failures can be described as a series of failures, from perception to intelligence, imagination, integrity, will, and charity.

1.1.7.1. Failure Of Perception

Perception is defined generally as the mental ability to grasp objects or qualities through the senses, resulting in understanding or knowledge. The number of things that we grasp, however, is limited by our senses. We see very little of the entire spectrum; we are limited by our size, our positions, and by our life span. These limits make it difficult for us to perceive very slow or large changes or to anticipate them.

In a market economic system, we have the perception of progress without the recognition of the costs of bigger cities and bigger farms. We have a perception that efficiency is related to maximum productivity, without regard to human health and happiness. Wealth and power are treated as primary needs, more than health or self-actualization. Failure of perception results in the inability to see the long-term results of economic actions that maximize profits through direct competitiveness. The same inability shows itself in the political frame where political cycles are limited to two, four, or six years.

Much of the environmental crisis is caused by the failure to understand patterns of cycling. This is especially true with industrial agriculture, which tends to break up cycles. Like most human endeavors, agriculture ignores cycles, as well as physiology, metabolism, and diversity. It fails to accommodate the reciprocity of the living environment—life does not adapt to a passive prior environment, it produces and modifies its surroundings.

We fail to see the incredible interdependence of humanity and nature, of diversity and success. We do not seem to be able to see others as feeling human beings, or animals and plants as feeling beings, or rocks as experiencing beings. Domination and conflict reflect our inability to perceive other human beings as equals. This failure reflects our inability to perceive the complex operation of nature, from the links of fungi to large-scale developments. The failure of perception leads to the disappearance of strong passions and emotions, as we avoid unpleasant realities and pamper ourselves with visions of separateness and superiority.

1.1.7.2. Failure Of Intelligence

Intelligence is the ability to learn from experience, retain that knowledge, and use it effectively in new situations, with common sense. Although we learn limited lessons from the short-term experiences of some people, we sometimes apply them unthinkingly to other situations. For example, British and U.S. farmers tended to take their success with monocropping to every other ecosystem, from tropical forests to deserts, with disturbingly bad effects. Explorers and scientists failed to learn from earlier archaic cultures that had adapted to places over thousands of years, again with disastrous results. Intelligence by itself is obviously not enough to guarantee our success. We even have special behavioral sinks for intelligence when it is too active but unattached—habits, computer games, and television, for instance. Beliefs, also, such as the belief in progress and technical improvement, can lead to the failure of intelligence. Therefore, intelligence has to analyze beliefs and myths, to make sure that they reflect ecological realities.

Common sense is the combination of intelligence with feeling and everyday experience. Common sense flows from the way people live in place and expresses what they want. Intelligence also grows out of living, but it is abstracted. Common sense is part of a conversation that results in cooperative behavior in the face of environmental and social challenges. Of course it is not perfectly transparent; people sometimes do not communicate exactly what they mean, which is why the context and body language of conversations are so important.

It is common sense that allows us to realize that our bodies and minds are real, that the world is a strange, wonderful, and dangerous place, that the earth has existed for a long time, with many radical changes—the shift from a methane atmosphere to an oxygen atmosphere was certainly radical—and, that people in other cultures have different, equally real and important experiences. It is common sense to realize that gardening is more productive than war. It is common sense to realize that enlarging a place is better than destroying a place where others live. It is a failure not to.

1.1.7.3. Failure Of Imagination

Imagination is the act of forming mental images of what is not present or has not been experienced to deal with new experiences. The world of many people is simple because their image is simple, so they think there are simple solutions to simple problems. Many people believe that energy and food increase automatically as people multiply, and that simplifying ecosystems can increase their productivity. This exemplifies the failure of imagination. We should not confuse the limits of our mind with the limits of the world, as the philosopher Schopenhauer warned. We seem not to have the ability to see what we have lost, in our rush to be civilized and big.

This failure of imagination limits our understanding and visions of a future. We have the ability to explore planets and modify genes, but cannot seem to offer functional education or meaningful jobs, dignity in retirement, or goals for living. Oddly, we seem to have adequate imagination to describe space colonies and interstellar migration. Will we develop them just to take the same inequities and problems with us? Humans can even create virtual worlds by limiting what could be received; for instance, if a being could see in the x-ray part of the spectrum. Yet, human imagination is as limited as human knowledge.

The failure of imagination leads to the inability to recognize or use the good ideas of others. This is obstructive to our adapting to a changing ecological contexts—thus the idea "protect the ecological basis of life" is never considered. We think that we have to address things one at a time, that we cannot see ourselves or our actions in the whole ecological system, partly because we are interested in continuing our immediate pleasure, even if over a short-term human lifetime, without regard to the indirect and shared costs to the system and to others within it, and partly due to a failure of imagination.

We do not try to imagine the connections between things that we do not know about. Not knowing how trees provide wood or how people cut the trees and process them, we feel no responsibility, we feel no connection. When we do not feel the connection to land, or understand what it does, in terms of cleaning water or providing food, we do not create groups to protect the land, or a constituency or a leadership. Many organisms exist of which we know nothing. Their worlds have little meaning in a human world. We know what it is to be human, but spend little time imagining other forms of existence.

We never imagine long stretches of time, thus we fail to anticipate the changes that occur in long time periods. Can reason comprehend deep history fully? Tending to think things too complex, or too expensive to change, we are not able to make good decisions to adjust to large or slow changes.

Perhaps this is a failure of ability in general. This inability to imagine the differences in situations and the consequences of our actions leads eventually to tragedy, which is the failure of a guiding image of the world, often referred to as a cosmology. In a theatrical play, the tragic hero triumphs at first, and incorporates the successful behavior that led to the triumph. This behavior, employed in new circumstances, however, leads to disaster. The hero refuses to give up a particular role or strategy or to imagine how to change in new circumstances. Hence, the failure to give up a chosen pattern of behavior leads to great loss later. This reflects our actions in the environmental play. Real-world tragedies result from the failure of our working images, the products of our imagination: Humans are responsible for the consequences of actions based on certain images, not on chance or fate. We can choose between the tragedy of the commons or the tragedy of total control, or we can expand our cosmologies. We are tragic because we have to accept responsibility for our actions.

Ultimately, many failures, as Dansereau has said, the failures of pollution, poverty, and urban decay, are failures of imagination. Rejecting the solutions of imagination, therefore, can only make the suite of crises worse. Will the failure of imagination condemn people to partial solutions or ignorance? We seem condemned to the weak trials of the past. Utopias are dismissed automatically as imaginary places. We can only imagine a society without any history or without a real place as being desirable.

1.1.7.4. Failure Of Integrity

Integrity is the state of being complete or whole. The term is applied to art, music, ecosystems, wilderness, computer databases, or people; it implies that these things have not been corrupted by natural processes or by human actions, a form of natural processes. Integrity can be related to the general character of a human being, having to do with the integration of the self into an identity that represents things beyond it, but also referring to a way of acting morally. Acting with integrity on a particularly important occasion could best be explained by the general presentation of that character and life.

The failure of integrity leads to both the breakdown of tradition, as we pretend that ubiquitous behavior forms a global culture, and to the destruction of natural habitats, as we take key elements from the ecosystems. We must acknowledge the failure of our remedial efforts, our failure even to address the flaws of our ideologies. The failure to value those things necessary for life, of a person or a habitat, is a failure of integrity. Value, as an expression of worth or exchange, cannot be limited or ignored. This keeps values and morality as local effects. We do not extend respect or love to distant others. Our personal values and beliefs do not let us.

1.1.7.5. Failure Of Will (& Courage)

Will is the power to make a reasoned decision, with a strength of purpose, with a firm attitude to control one's actions, with courage or nerve. Courage means an attitude of dealing with anything recognized as dangerous, difficult or painful. Nerve means strength, emotional control, or endurance. Will fails daily. We refuse to share or to help others even when it is easy.

If we cannot imagine extremes, it is hard for us to have the will to sacrifice things to avert it or ameliorate it. For all our cleverness, we still emulate flies and grasshoppers, when it comes to acting always in short-term self-interest. During the good weather and the good crops, we expand to or past the limits of water and food. We lack the political will for sacrifice or even planning. The failure of will leads to susceptibility to indoctrination by governments and even by advertising schemes. Policies are not implemented, due to social differences, corruption or war. Too much fear can lead to the failure of will. Afraid of failure or of being unpopular, many politicians, perhaps all politicians, exercise too much caution. They refuse to stray from their opinion polls and say what needs to be said.

Few people are without some goodness in their hearts, so that may not be the problem. But, few people have long-term, large-scale, ecological intelligence. Almost nobody wants to try to plan for some kind of mutual constraint to allow the rest of the planet to breathe. After all, it's just one more pair a shoes or a filet for dinner. What harm? And the answer is very little harm at all—it's just that on a massive scale, little sins can kill numerous species and destroy large habitats. And, few of us see the results, other than fewer bees and songbirds. The Gulf of Mexico west of here is dead for hundreds of square miles. Why? Garbage, phosphate dumping, or pesticide runoff? Who knows what combination exactly, but it killed the fish, crabs, and almost everything except red tide bacteria.

We know what we have to do, really. But, it seems that only in times of war or great catastrophe do we have the nerve to do something—although we do not seem to have the nerve to avoid those catastrophes or wars that could be avoided by planning and conversation. Perhaps it is fear of pain or change. Perhaps it is a character flaw in the species, the failure of nerve at critical junctures. Perhaps it is just the lazy habit of mob thinking. We may not have the courage for some to give up their lavish lifestyles and extreme comforts, their extreme profits or gargantuan excesses.

1.1.7.6. Failure Of Charity

Charity is love of others or the act of goodwill towards others; it refers also to the feeling of benevolence or kindness in judging others. In traditional hunting societies, charity was expressed through reciprocity, or mutual sharing. In urban societies, sharing was more than just an exchange; it was often characterized by generosity. In early industrial societies, the views and practices of others who shared resources or places were met with tolerance. In archaic societies, charity may just be sharing food. In more stratified societies, charity tends to be part of an institution, that may be modified by greed, laziness, or the arrogance of charitable workers. The failure of charity leads to the rat race of competition between human groups, at the expense of many human and ultrahuman groups.

As money and power lead to detachment from agricultural and natural cycles, from reciprocity and concern, charity disintegrates. Humans have the power to alter vast processes in nature, but do not care enough to refrain from trying. Power without charity is a satanic theme. Sigmund Freud wrote that Satan desired to be father for himself, an agency without community. Many governments have the same desire. Governments also have knowledge without charity, which was a demonic theme for the Christian church (the word 'demon' is from Greek word for knowledge). Power without knowledge, and knowledge without charity, are frightening possibilities.

1.1.7.7. Endless Combinations

Of course these failures occur in endless combinations. Maybe the failure of our modern civilization is a failure of imagination compounded with a failure of nerve. We cannot imagine an alternative to war, and we cannot act beyond emotion. We cannot imagine beauty in the old and messy nature, and we are afraid to try to do without luxuries or to try to sacrifice anything to try to change the momentum of industrial civilization.

Combined with our own failures of charity, imagination, will, and courage, using piecemeal knowledge, false models, and unrelated ideals without regard to traditional knowledge and practical experience, we have created common faceless cultures to sit on our flat, placeless utopias. To create our noplaces, we have accepted the gifts of bigness, speed, uniformity and allowed the thefts of life, intelligence identity, and choice to negate any gain from them.

1.1.8. *Three Grand Thefts*

Despite our bigness and speed, most of the world's human population goes hungry; even fewer are fulfilled as actualized human beings. The utilitarian aim of greatest good for greatest number has been vulgarized to mean the greatest number of goods for those who can afford them. In our attempts to manufacture the good life for those who can afford to buy it, we have deprived everybody of clean air and water, quiet nights, darkness, open spaces, and other indefinable qualities. Soils are destroyed, wildlife is killed. We devour nature to assuage our disease; we try to fill our emptiness with goods. We can only gain past a certain point before our gain causes the loss of something else that we need to be healthy.

Modern technological society ravishes nature and mutilates humanity with the products of its materialism. Industrialization has distorted people's lives and cheated them of bread and justice—during the time it takes to read this manuscript, over 10,000 people will die as a result of nutritional deficiencies. The cost of the paper and processing for this manuscript could feed twenty people for one day in many poor countries. Reluctantly, we rationalize that an idea is a better and more effective solution to global problems of production and self-reliance than a one-time aid or just ignorance, but the rationality does not sit well in the dreaming mind. Oppression darkens the mind and narrows the spirit. In a mass consumption society, people impoverish themselves spiritually while impoverishing others materially. This is theft. As with the Christian commandments, most loss can be reduced to theft, whether of a life, mate or name. Most of our modern problems can be considered the consequences of forms of theft, such as of life, common sense, and finally choice.

1.1.8.1. Theft Of Life

The relationships of humans and animals have changed drastically. The increase of humans and the destruction of animals have unbalanced the relationships. We kill millions of domestic species every year and many millions of wild species by accident or by deign. The destruction of habitats is accelerating. Up to a point, niches can be enlarged or increased, but with so many humans, it becomes a case of supplanting other species entirely.

The loss of cropland and soils, and the disappearance of genetic stocks essential for crop breeding, may eventually cause the collapse of the biological basis of our food supply. Huge quantities of fertile soil are washed away each year as a result of deforestation and poor land management. It has been estimated that Colombia losses 400 million tons of soil per year and Ethiopia 1 billion tons per year. In the U.S., in spite of its soil conservation service, almost 5 billion tons are lost annually; these losses over the past decade may have cut the potential to grow food by ten to twenty-five percent. This increases our dependence on oil imports, since $1.2 billion of fertilizer was needed to replace the nutrients lost through soil erosion in the U.S. in 1978. Increasing amounts of fuel equivalents are used every year to offset erosion. Key species in the soil are killed by erosion and by excess fertilizers and biocides.

In developing countries, hundreds of millions of rural people strip the trees and shrubs around their homes for fuel. In Gambia, it takes 360 woman-days per year per household to gather wood. Even when firewood is available for sale, it may be beyond the budgets of the poor. In South Korea, it can cost up to fifteen percent of a household budget; in the Sahel, it can consume twenty five percent of a budget. When wood is scarce, the poor are forced to use millions of tons of crop waste and dung, which should be used to regenerate soils. The poor are destroying their means of survival, in order to survive now. Then, the soils on which they live become even more vulnerable to erosion. Deforestation causes siltation, which cuts the useful lifetimes of reservoirs in half, decreasing hydroelectric and water potential. Deforestation causes floods, which devastate settlements and crops, incurring even greater replacement costs. Lack of soil and forest conservation contributes to rising energy costs and the financial costs of providing essential goods and services.

Tropical rain forests, genetically the richest land environments on earth, are being felled and burned at the rate of 27 million acres a year—50 acres a minute (in 1972). At that rate there will be no more tropical forests left by the year 2050. The lowland forests of

Malaysia, Indonesia and the Philippines are being ravaged the fastest. The rape of the tropics is endangering uncounted thousands of species. Over 25,000 known plant species and over 1,000 species of mammals, birds, reptiles, amphibians, and fish are threatened with extinction as their habitats are eliminated. Many of these species may be economically and culturally important; others are different and unique. Unknown species, perhaps half of all species on the planet, are not valued at all because of our ignorance.

Deserts are expanding at the rate of 23,000 square miles per year (for contrast, the nation of Belgium is 12,000 square miles in area). Eight million square miles more are seriously affected. Desert conditions jeopardize the survival of almost eighty million people now; 550 million more could be threatened. Desertification by overuse of land afflicts almost seven percent of the earth's surface.

Even the sea is not invulnerable to human impacts. The most productive areas, which are close to shore, are being polluted, overfished and destroyed. Estuaries and coastal wetlands are being destroyed. Overfishing has deprived people of millions of tons of seafood. Overfishing is destroying the fisheries' support systems. In the U.S., losses to fisheries from shore "improvement" and degradation cost $86 million a year.

The relationship of humans to humans has also changed in the past thousands of years. Human lives are stolen, not only through war, but large-scale murder, as well as through diminution of human value and denial of resources.

1.1.8.2. Theft Of Common Sense (& Humility)

The industrial machine is out of balance and defective, but we have been making it run faster to process more material. Western civilization still selects cultural changes in terms of wealth and power. This is destructive to most individuals, as well as maladaptive for societies under ecological restraints. Over the past two centuries, industrialized countries used great quantities of raw materials to create luxuries. Then they disseminated the ideas of wealth, equality, opportunity, and indulgence to many countries without industrial opportunities. There can be no peaceful future for civilization when such disparities, and popular knowledge of them, exist. The cult of competitive consumption seems to be the universal solvent of the modern world. Everyone wants what some have. The industrialization of Asian nations is seen as a solution to shortages of manufactured consumer goods, although imbalance and pollution are down-played. Even worse is the unavoidable waste. Probably over 50% of the productive effort in United States goes into making things that contribute nothing to the material standard of living.

Consumer desires must be satisfied promptly or despondence results. Although life, art and science depend on organization, the civilized consumer feels that organization implies stasis and death. Believing the line of succession ends with his claiming of his inheritance, the industrial consumer feels no obligation to provide for the next generation. The consumer justifies narcissism as a preliminary condition in the search for consciousness, truth and morality. Consumers become embroiled in their own causes, locked in the idealism of adolescence and repudiating the lessons of history. Hegel noted that the greatest lesson of history is that nobody ever learns the lessons of history. And, Cicero wrote that those who do not know the past are like children. Americans and the people of some other nations seem intent on validating these insights.

This repudiation of history creates a historical amnesia. Consumers retain little more than a dim notion of the past. Universities report a lack of interest in events that occurred before the current year's athletic season. The sense of time falls in upon itself, collapsing like an accordion into the present. Knowing nothing of history and expecting nothing of the future, people cannot escape the fearful isolation of the present. They join together in a melancholy herd, clutching at everything, but holding nothing fast.

Without the depth of history, experience is shallow and short, and intelligence is thin. Technology has reduced the globe to a single, closed system, which humans can share according to their financial powers. Our direct experience of the world has become shallow, in spite of faster travel. Travel used to broaden the mind, but now it narrows it. We travel in sealed corridors like boxed goods, comforted by homogenized foods and the English language. Our cultural adaptations to the pressure of homogenization throttles individuals and groups.

Unbridled consumption in a laissez-faire economy for fifty more years will probably carry humanity beyond a point of no return, leaving industrial society with insufficient resources to maintain itself and insufficient flexibility to retract. If lack of planning permits rapid increase in population and consumption, then our options may be diminished to two: A nasty, overplanned existence, or a squalid collapse. But the future seems as unreal as the past. The historical origin of the ecological crisis is in the failure of people to use their intelligence and common sense to anticipate the long-range consequences of their activities; this is a perennial human problem.

1.1.8.3. Theft Of Choice
Our civilization is dominated by all of the ideas of the industrial revolution, which form the outlines of a tragic world view: The primacy of humanity, anthropocentrism, the supercesion of the individual, the achievement of happiness through the accumulation of things, the perception of the incompetence of nature, the requirement of humans to control the environment, the expansion of the frontiers of technology and opportunity, and the identification of solutions for every problem. But, these ideas are false; programs that assume them for conditions are doomed to fail eventually, and perhaps destroy the kinds of environment that humanity needs. Cities and factories reproduce themselves exponentially through an unexamined industrial genetic code. This is an addicts' dream of affluence using science to create a dream pill from the dust of a bare earth. But, simply removing the causes of unhappiness may not produce happiness.

Industrial culture has been distorted by the modern emphasis on the scientific method, with its devaluation of philosophical concepts and emotional values. Rational values are exaggerated and spiritual events are ignored or suppressed. Most scientific studies seem specialized or irrelevant. Some parts of problems are identified and analyzed, but the conclusions are trivial and weak. The ethical impulse to solve the problems is even weaker. We choose to let them slide, or to label them as nonproblems.

The unified direction and responsibility of science is nonexistent; there is not even a concept of what is good. Control has been the goal of science and technology. Scientists are obsessed with the treatment of weeds and vermin. The industrial cosmology has put humanity at war with the planet, which always misbehaves; weeds and vermin threaten us. Our path has been worn so deep that it would be difficult to leave it. Stability has been raised to sacred scientific state. The interlocking of technologies and institutions makes it impossible to reform policy in any one part, separate from the other parts. This interlocking also makes people powerless to choose any alternative. The systems managers are preparing to operate the planet, according to systems principles applied in routine methods. Yet, the safety of the environment is too important to be left to scientists, even to ecologists. The crude history of science shows that scientists fall willingly under the dominion of money and power, like Christianity, communism, and most movements. The movements of narrow science and industrial culture reduce the possibilities of choices.

1.2. *Recognizing Perceptual Challenges to Design*

Many challenges to, or problems with, civilization are caused by our classical logic, which, in the western civilization is deductive and bivalued—that is, we arrive at decisions from an overall image, and develop categories that are mutually exclusive. This binary approach results in an 'either/or' situation, rather than a 'both/and' situation. For instance, the argument is tossed back and forth about whether our approach to resources should be anthropocentric or ecocentric, when in fact it should be neither. This argument occurs due to a misunderstanding of the concept of frame and focus. This section examines the kinds of challenges caused by our perceptual limitations.

Most problems can be divided into categories of too much or too little. There may be too few trees, no parks, too little clean air or water, no birds or animals, not enough resources or jobs, and not enough wealth or diversity. There may be too much dust, erosion, pollution, garbage, pests, change, surprise, or people. For instance, we think of problems as having a cause, but often there are many contributing factors. For instance, no trees is a problem in many kinds of landscapes. The reasons may include agricultural conversion, firewood use, bad landscape design, or psychological demand for treeless areas. A shortage of energy may also have many reasons: Overuse of energy for 24-hour lighting, inefficiency of lighting or engines, suburban spread, and demand for constant air-conditioning. A similar pattern emerges with pollution, which can be caused by dirty processes and technologies, demand for inexpensive energy, atmospheric inversion, and tolerance for the effects.

In this sense, the problems have to do with balance or limits. Problems can also result from randomness or bad luck. Causes can be immediate, proximate, contributing, or ultimate. Design is bad when it neglects to consider things such as the frame of interactions, or scale or limits.

1.2.1. *Focus/Frame*

Everything that we focus our attention on can be seen to be of a larger framework. Focus and frame can be understood metaphorically—and, in fact, metaphor itself can be understood as consisting of two parts, according to Max Black: Focus and frame (see Section 1.3.3.). The focus designates the figurative term signified through the process, and the frame refers to the subject or context. Using this distinction, it can be seen that most of the fuss in ecosystem-based activities, such as forestry, has occurred at the focus level. Modern foresters have so long focused on trees that they forget that the forest is a frame that holds many foci, or points of view. Alan Drengson is fond of saying that we do not see the important operations of nature because we are looking through the wrong paradigm—a paradigm acts like a pair of glasses, focusing on what we want to see.

The elements of a forest are related psychologically, by foresters, as focus or frame, as contrast or uniformity, as dominant or recessive, or in a number of other pairs. For instance, forests can be considered by scientists as either matter systems or energy systems, but the focus on either frame permits subtle differences and limitations in interpretation. Some scientists describe organisms as being configured by energy through time, but organisms are material patterns in space as well. Furthermore, it is not quite right to say that ecological forestry is ecocentric, because it is actually concerned with the frame and not the focus, the periphery and not the center; perhaps it should be called 'ecoperipheral' instead.

The anthropocentric needs of people can be contrasted with the needs of the forest itself. This dilemma can be answered by considering the focus/frame character of the situation. Both of the two basic views are merely aspects of one view of good forestry. Anthropocentric values that focus on commodities can only be considered in the context of the values that are contained in the whole, that is, they are derivative from the frame. Furthermore, it is not necessary to ascribe conscious purpose to the latter view. Massive disruption often results when a community falls out of balance with its local forest environment, and in fact industrial forestry only avoids the penalties for such disruption by trading advantageously with other communities in less powerful areas. Emphasizing balanced use, including natural forest requirements, rather than perpetual growth, as is done now, would permit the frame to stay pluralistic and multidimensional, and still provide resources to human populations.

1.2.2. *Small/Large Scales*
Switching our attention from a focus to the larger frame usually involves a change in scale. Scale has several meanings. Basically, as it is used here, it has to do with the level of measurement in a space/time energy/mass (STEM)context. For instance, in forestry the measurements of leaf litter by foresters can be made at several scales: Single tree, stand, annual measurements, or stand life measurements. Processes that are unimportant at a small scale might be vital at a large scale. Too much litter in one stand in one year, for instance, might suppress a soil cycle; too little litter over a century might interfere with several local, regional or global atmospheric cycles.

Some patterns in forests are scale-dependent; for instance, hemlock trees may dominate small clusters, but be scattered all across the entire forested landscape. That is to say, the pattern changes with the scale. This is true of processes in forests as well. Canopies shade the understory annually, but fires in a lodgepole pine forest may increase dramatically the amount of light to forest floors once every 200-300 years.

The scale of the system defined, for example, wooded patch, ecosystem or temperate broadleaf forest, depends on the scale of the phenomenon being addressed. Mangrove forests are in phase with the frequency of hurricanes, although hurricanes may not influence the life histories of short-lived organisms as much as daily or seasonal cycles. Microbes, for instance, are affected more directly by short-term cycles of precipitation and temperature.

An inappropriate time scale, its shortness and urgency, is a cause of many other problems in industrial land-use. Although ecosystems are considered renewable resources, they are slowly renewable, requiring hundreds or thousands of years to renew from catastrophic disturbance; this time is far longer than any economic plans, and really is nonrenewable on a human life scale. This has important implications for exploitation and continuity. Really large scales, bioregional or global, are not considered often at all. At large scales, other factors have to be considered.

1.2.3. *Global/Local Fields*
Differences in scale are often called local or global differences. Concepts of scale apply to physical and biological fields. Local fields or events are those separated from other fields or events and that exert minimal influence on the others. That is, the internal connections inside a local field are stronger than the connections between local fields. Both quantum theory and relativity consider that local fields are generally noncausally and nonlocally connected—that is, no communication is instantaneous—through higher dimensional realities.

Local ecosystems are unique and original. In an ecosystem, each locality supports a segment of the total species population in a unique context, with a particular set of predators, competition, food, or physical habitat. The environment requires an enormous amount of minuscule local adaptations between the system and its users. Production of pinecones, for example, is adapted to the local microsites; Loblolly pine planted fifty miles north or south of the seed source are less vigorous. The chemical DNA seems to reflect local conditions. Local ecosystems are separated from one another, not only in space, but also by differences. For example, Dutch elm disease is a local problem, even if it seems ubiquitous in all forests with elms, produced by local actions and requiring local solutions.

Local systems, however, can affect global systems. Cutting down local forests may contribute to the discharge of greenhouse gases, such as carbon dioxide, into the local atmosphere. This may have the effect of increasing quantities in the global atmosphere, creating a runaway increase in atmospheric temperature.

Local systems emerge from a global system, which has characteristics that emerge from the interplay of local systems. Global systems have properties that no local systems have, such as an overall atmospheric temperature or global biogeochemical cycles. James Lovelock suggests that the interaction between hardwood forests and softwood forests may act as a global regulator of oxygen for the planet. Many things such as human poverty and species extinctions only seem global because they are happening locally at the same time, and may affect global cycles.

1.2.4. *Minimum/Maximum Limits*

Our modern civilization has become enamored with the idea of the maximum; it is the goal for many kinds of planning and operations. Agriculture, for instance, is concerned with maximizing its products and profits. When that happens, the balance or harmony in a system is altered, and although it may take decades or centuries for the consequences to be known, the system is affected. We also strive to maximize the diversity or yield from nature, without a good grasp of what a maximum or minimum is.

In some senses, nature does try to maximize or minimize certain conditions. For instance, a bubble minimizes surface tension by assuming a spherical shape that contains a maximum volume. Of course, we could also say that the bubble is an optimum form for the least possible surface area for a given interior volume. In principle, mathematicians reduce the question of maxima and minima to a geometric construct. In trying to apply the idea to a tree, however, the calculations become incredibly complex. A tree appears to create a maximum leaf area to collect radiation and a maximum number of seeds for reproduction, but it also tries to minimize evaporation and energy for its metabolism, both of which would set a smaller leaf area. The tree puts out an optimum or at least a satisfactory number of leaves. A forest is even more complex. In reality, the processes of trees and forests have many local maxima and minima; these can be represented graphically as surface potentials on a catastrophe field, as Rene Thom has done.

Many scientists and ecologists argue that we should maximize diversity. Diversity is an emergent property of ecosystems, arising from the activities of multitudinous beings learning to use the productivity of the system to augment their own flesh. The ecosystem is self-organizing, but diversity is not a goal of the system. Diversity is just a characteristic of mature ecosystems. It is never an independent thing by itself. While it is meaningful to speak of an optimum diversity, as the result of limits and the interaction of many factors, a maximum diversity may never be reached in any system. As Paul Weiss noted, the patterns of organic nature are a combination of order and diversity; order involves constraint while diversity requires freedom for difference. Maximum order would result in a static universe, whereas maximum diversity would create chaos.

Few mammals—and humans are mammals—try for a maximum condition. Few mammals even strive for an optimum; most accept a 'satisficing,' that is a satisfactory amount, according to Francisco Varela. Some animals, individual rats for instance, seek substandard, or more challenging, conditions. Varela analyzes the evolutionary process as satisficing rather than optimizing; that is, a suboptimal solution is adequate to continue living; striving for an optimum or maximum does not pay off for investment of effort.

1.2.5. *Generating Problems*

Unfortunately, trying to solve problems often leads to new sets of problems. Also, the solution is usually limited to one category, often technological or political. For examples, the problem of no shade from trees to protect seedlings is solved by forestry technology with the use of shade cards or genetic changes. It could also be solved on a cultural or behavioral level by planting trees or cutting fewer trees in place. It could be addressed on the economic and political levels with landscape design. But, sometimes the solutions, if not approached ecologically, can result in new problems, such as dying forests from improper planting.

Any of the problems need to be addressed at all the levels. In another example, when bird populations are noticed to be getting low, we automatically think of putting numbers of the species in zoos. On the behavioral and political levels, we could be outlawing shooting or poisoning, or accommodating flyways in our landscape designs. But, we have to be aware of new problems, such as loss of wild context, or shifting migrations, from these actions. There are many other instances, such as lack of houses or settlements, which we can solve with new housing; sometimes, however, the solution is special housing projects, which result in dissatisfaction. The technical solution to industrial wastes can be as simple as new filters or traps. But, a complete response may include new regulations and taxes, which may result in higher costs and prices. Too many people in one place may lead to crowding, personal disorders, and more rigid planning. The lack of security in society may lead to extreme beliefs. All of these surprises and synergies might be ameliorated by adaptive design and new ways of thinking, such as deep ecology.

1.3. *Fashioning Tools for Planetary Challenges*

A tool is a material object, or idea, that extends the human body or human thought. Technological things, such as tools or artifacts, are extensions of our physical being. The spoken language, using sound, engages and extends all the senses. The written language, using visual symbols or glyphs, can also engage and extend all senses, if the body resonates with them and recreates the full environment. This section examines the consequences of tool use and suggests some general perceptual tools for designers.

Tools may have first been used to acquire or store food, then to extend the efficiency of those processes. Tools have affects on human beings, from psychological distancing from things, as the tool is intermediary to a thing, to intimacy with the tool. There are physical effects as well: Loss of use of teeth, muscles, memory, loss of hand-eye coordination and perception of wild, but increase of the hand-eye coordination of the tools. Each tool has an ideological bias, as skills or behaviors are amplified or ignored. To a man with a hammer, everything looks like a nail. To a woman with a grade sheet, everyone is a number. To a man with a computer everything is data, and to a woman with a camera everything is an image.

For tool-using cultures, tools had to do two things: First, they had to solve specific physical problems, and second, they had to serve a symbolic dimension in the construction of things or buildings. The complexity of tools leads to rules for their use, and for those that use them, such as trade unions and clubs, like masonry. Numbers become a tool to assess someone's behavior or worth. And, there are effects of tools on ecosystems, from the disturbance of soils to scale effects. Tools play a central role in the thought world of a culture, according to Neil Postman. Tools are not simply integrated into culture; they can actively reorder the culture. They may undermine the ideas in a culture or dominate them completely. At an extreme, tools may eliminate traditional world views and reduce the meaning of life to machinery. New technologies compete with old ones for dominance in a world view. The medium of technology contains an ideological bias. Tools can 'attack' tools, according to Postman—printing attacks manuscripts, television attacks printing, painting attacks rock art, and photography attacks painting. Tool use has effects, so care must be taken to use them consciously aware of their impacts and limits.

1.3.1. *General Tools: Language, Myth & Logic*
Some tools, such as words or logic, we rarely think of as tools. Others, such as habits, stereotyping, myths, and even mathematics, we often use to hide or dismiss troublesome problems. Myths, for instance, are spread and remembered through stories told in every culture. Remembered stories become myths, accepted and not examined. Myths become part of the cultural heritage. The recital of myths is of great importance, though; myths provide a cognitive order for people. Myths are a major way of explaining the universe as it exists and communicating the cultural view of nature. Myths explain that which is otherwise incomprehensible, like death and disorder. To be understandable to members of a culture, the elements of a myth must be taken from the features and values of that culture. Neighborhoods share a common store of general myths of a larger field of culture or a nation, in which their local areas are embedded.

1.3.1.1. Myth
Archaic peoples translate events in the natural world into the language of myth. Being a narrative, a myth is aesthetic as well as intellectual. Myths develop in terms of their own

internal logic, drawing together observations of the world. Claude Levi-Strauss described the process as bricolage, fitting the bits together, identifying impressions of life as sets and forming them into mythical systems; the world picture is a metaphorical puzzle. Bricolage is the mentality of synthesis, a technique for learning, creating, and expressing understanding, using whatever is available from the past and in the present to achieve an integrating form. This is what mythological thinking does, and what scientific thought might do. Mythologies are of major importance in the life of civilization.

Myths change and evolve. Our old myths, that reality is hierarchical, that the earth is passively female, and that 'man' is lord over the earth, have proven to be dangerous. They should be retired. The eighteenth century held that a golden age would appear if the priests and kings were deposed. In the nineteenth century, a central myth, founded at the 1851 exposition, was that industrialization would bring universal peace. A mythology can affect every element of our social, individual, spiritual, ecological and political life, all at once. Myths can be traps, if they are never renewed by every generation. Continued belief in myths about nature in the face of contradictory facts is a trap. Facts are always changing, as are myths, as is nature; and, myths and facts are based on nature.

Contemporary thinkers, such as McKinley, argue that we need a new mythology of humanity in nature. New myths can be developed, such as the myth of participating in organic beauty, where development, not growth, is without limit. The universe is a frontier; the mind is a frontier, but these frontiers are based on a whole and healthy planet and a nature of which humanity is a special part among many special parts. To achieve transformations of human culture, we must go beyond the authoritarian conspiracies and technocratic elitism, to create new myths. We will only learn to treat land as part of our community, and not as a commodity, when we have new myths.

Mythology is a tool that can shape ideas and behaviors, good or bad. Many cultural or ecological problems can be examined using simple tools such as questioning or metaphors. Good words or good metaphors can illuminate a problem better than long explanations and tables of numbers.

1.3.1.2. Logic

As our words need to reflect a larger sense, so our logic needs to be aware of its scope and limits. Most Western science assumes a predicate logic and is constrained by that logic. This Aristotelian logic is deductive and binary, where categories are mutually exclusive, and substance and identity are permanent. In this logic, contradictions are false by definition. The difficulties with this logic are evident in many instances in physics and ecology, e.g., light cannot be a wave and a not-wave, or wolves cannot be a keystone species and an unnecessary species in an ecosystem. Neither substance nor identity seem to be permanent in any human sense of that word.

Our science and theories are beset by logical problems. Would a more flexible logic solve the problems? A nonpredicative logic? There is another form of this logic that is multivalued, but it is also based on the same epistemology of a hierarchical universe in which values can be rank ordered. Magorah Maruyama calls these logics, and the epistemologies that contain them, homogenistic.

Other logic systems exist, although they are not as predominant. Maruyama distinguishes two other epistemologies that have a typical logic based on them: Homeostatic, with a complementary logic, as exemplified by Chinese thought, and Morphogenetic, with a logic incorporating change, harmony, uniqueness, and irreversible processes, as demonstrated in Mandenka, and Inuit thought.

Morphogenetic logic is quite different from homogenistic logic. The morphogenetic logic can be characterized as relational, qualitative, symbiotic, heterogenistic, reciprocally causal, and interactionist. Reciprocal causal processes can increase structure, differentiation and complexity in natural systems, according to Maruyama. Such a logic is more useful for addressing complex operations in ecosystems. As a many-valued logic, it can address values other than truth and falsehood. It can avoid certain fallacies embedded in two-valued logic.

No superlogic can combine categories, however, since they are based on different epistemologies and cosmologies. Any researcher, using any logic, will filter and distort data to some extent, as a result of epistemological 'selection.' A broader morphogenetic logic could work at a metalevel and avoid many of the problems of polar thinking.

1.3.2. *Words as Tools*

Words are tools for communicating ideas and feelings, and for building thought structures that can impact the physical structures of our worlds. The etymological meanings of words sometimes describe the intentions or limits of their original uses. We should perhaps heed these limits when we use words in this kind of work. The history of a word sometimes yields clues about how it has been used, as well as nuances of thought. The word local, for instance, means belonging to a place. And, a place, as we know, is an ecosystem claimed by human emotion. The local is likely to be the center of identity and loyalty for human beings.

Beyond being fun, making new words can force us to reconsider certain things. Established words often are bound with connotations and cultural baggage. A new word can call attention to gaps in our knowledge. A new word can suggest new directions for conversation, yet also remind us—if the word is well-made—of its origin. In this sense the construction of words shares the operation of metaphors, by bringing together things from a different context.

Words, such as 'world' and 'nation,' describe whole things. A world is that part of the planet designated by one culture, and the perspective of the inhabitants (from the German word for 'man-image'). It can also mean the entire planet. A nation is a people or tribe or a people living in a territory united by a single government, a stable, historically-developed community of people with a distinct culture occupying a common territory, although it has also come to mean an artificial designation as a result of political violence.

The neologism koinomics means 'managing sharing'—in this case the apportionment of resources to cooperating participants in the global commons. It is supported by 'Legatism' in a Panocracy' as described by Harmosophy, which is the wisdom of harmony. Related to this is ecological 'Zonagraphy,' the drawing of zones of relative isolation around the planet. Panocracy means the rule of all, although in a human legal system, humans have to represent other species—perhaps like a scientific totemism. Legatism refers to the legacy of the planet, which is respected by the legatees and tenants. It is sort of the holdings of all the occupants of the planet. Of course, this approach is not a panacea—or maybe it is (See Section 8.8.4).

Several metaphors have been used to describe the human place on earth: The earth is a storehouse, property, or a spaceship. But, the earth is not a spaceship or storehouse; it is home. Victor Ferkiss proclaimed that: "The world and humanity are one entity, one system in equilibrium. Earth is humanity's only home; humanity is one people ..."

1.3.2.1. Words like Place & Home

Human situations unfold in place. The word 'place' comes from the Latin meaning 'open space' and originally from the Greek word for 'broad' as it was applied to a way or street. The use now should reflect the kind of space modified by human interactions, whether cities, homes or pathways. The word 'environment' is from the old French word for 'surroundings.' In normal use, it implies the immediate nonhuman surroundings. We need to make sure that it includes larger scales that incorporate the large cycles in nature, to understand that our local neighborhoods are linked in many ways by atmospheric and hydrological cycles, as well as by elemental cycles. These words are not used in some cryptic way, but just to illustrate what they originally described, in order to understand how people thought, and then to extend those thoughts.

The place is an ecosystem, the subject of ecology, often regarded as a home. The study of life in place is 'ecology' (from the Greek word *oikos*, a dwelling place or house. The word economy derives from *oikonomia*, household management. Ecology is knowledge of the house, as economy is its management.

Insects and animals displayed a powerful attachment to places, as Adolf Portmann observed; this attachment was best understood as home. The fundamental ambiguity of existence is that humans have different capacities for feelings and awareness. Some feel strongly about a place or home; others never do. Several metaphors have been used to describe the human place on earth. The earth is a storehouse, property, or a spaceship. But the earth is not a spaceship or storehouse; it is home. Victor Ferkiss proclaimed that: "The world and humanity are one entity, one system in equilibrium. Earth is humanity's only home; humanity is one people in relationship to the earth."

There are a wide variety of meanings of home. It is a place of family residence, the family social unit, habitat, and place of origin. The word 'home' comes from the Middle English word *hom* and Old English *ham*, Old Norwegian *heimr*, Greek *kome*, and Sanskrit *kayati*, meaning village or home. The Old Norwegian word for home meant village or world. The word can be traced through the Greek to the Sanskrit, which meant "he is lying down." Its spectrum of reference is enormous. The word is used to describe house, village, city, bioregion, cultural world, and the earth. Its content is also ambiguous. Home can hold a single person, family, relatives, pets, domestic food animals, neighbors, or others.

Home living is simultaneous on different levels; the importance may shift from city to nation, or nation to state, or house to bioregion, or state to habitat. Each level can be a metaphor for the next. There are parallels between nature and a house, as the basis for home. Solar space is like the landscaping; wilderness is the foundation; conservation areas form the shell and provide services; and each bioregion is a unique room. The analogy cannot be carried too far, but it shows that a house is not, as Le Corbusier said, "a machine to live in." It is a matrix for home. Home is not just a house, either; it is a complex of significant events centered in place. It is the foundation of our individual identity on one level, and our role in the community, on another. What makes home different from house? Participation in its making, or commitment. People invest parts of themselves in a place, to make a home.

The concept of home has a mixed reputation. Paul Shepard finds humanitarians obsessed with the 'homelessness' of stray pets and wild animals. He points out how the fixation on shelter is taken over by the advertising of wood industries, who describe their meager reseeding efforts as 'creating a home for wildlife.' He sees protective organizations swaying to the tunes of propagandist lullabies. Perhaps, he is right. But, there is a misunderstanding of home. A home is not the house, not an undifferentiated place. Animals accustomed to rich woodlands are not at home in a replanted clearcut. The use of the word is a cheap advertising device, yet, it shows the importance of the meaning. A home is living thing, a house is not.

A home is a part of the environment claimed by feeling. Emotion creates an 'in-place.' A place must be found and made. Humans, like plants and animals, identify greatly with local environments. Maybe this is a function of the limbic system of the brain, a function we share with many territorial mammals. So far, no psychologists have studied what happens when a person sees her place, her very context, destroyed. These kind of catastrophes may be the basis for diseases, depression, or cancers, which also may be caused by the invasion of the home place by exotic substances.

The human ordering of the world makes places from wilderness. A place changes qualitatively; it becomes structured. Natural complexity decreases as the human complication increases, although the two are not mutually exclusive. Fitness is achieved after slow, progressive, reciprocal adaptations; it requires a stability of relationships between societies and place. Human places are complex integrations of nature and culture that develop in particular locations. The place precedes knowledge of it. The knowledge of place is one of the first links in a chain of knowledge. This knowing is essential to our existence. Being human is having and knowing a place. Only learning flowing from hospitable presence can promote life and enhance human existence.

Paul Shepard suggests that for each individual the organization of thinking and meaning is intimately related to specific places. Experience focuses on a place, which acts as the background for specific events. The features of the world are experienced meaningfully. The place is a matrix for ordering experience. The specificity of place is important. The earth extrudes itself into particular plants and animals; flexes mountains; and sweats weather. Places animate people (from the Latin word 'anima,' meaning 'inspire').

The inspiration of the sentiment of dependence is called impregnation. Animals and humans are imprinted early in life to particular places. Each difference in the landscape has meaning, which can be perceived, for instance, as when the aborigines of Northwest Australia perceive physical differences and even a symbolic landscape. In fact, they structure space according to myth, where Europeans use buildings and roads for structure. Every place has a unique identity as a result of combinations of factors.

Being apart from home can result in a disease, nostalgia. The word nostalgia was coined by Johannes Hofer a Swiss medical student in 1678 to describe an illness characterized by insomnia, palpitations, stupor, fever, and the persistent thought of home. The disease could result in death. For the Northern Aranda in Australia, as well as for émigré Russians and other groups, it is not possible to stay away from home indefinitely and still live. Nostalgia can be a fatal disease. Thus far, the sense of place cannot be gleaned from an analysis of the human nervous system. Yet a place shapes the nervous system, somehow.

Permanence is important element in idea of home. In English the term for dwelling is 'to stay.' This is the symbolic opposite of moving or changing. It means to 'withstand time.' Dwelling resists and persists. The Royal Commission on Local Government in England and Wales found that people's attachment to 'home area' increased over time. Gaston Bachelard has written much about the significance of home: "For our home is our corner of the world. As has often been said, it is our first universe." The home is a springboard to understanding the universe. The concept of home is proposed as a metaphor for the development of appropriate attitudes and appropriate ways of living. The Pueblo Indians, for instance, see their American desert as a providential home, because of their attachment and knowledge.

1.3.2.2. Words like Reality Wilderness Ethics & Design

The word reality comes from the Latin *res*, from which we get the word 'reality,' and it means 'thing.' Being real was being a thing. Reality was thinghood in general. The word thing goes back to Old English words that mean 'object,' action, event, condition, or meeting. It may have originally meant 'something occurring at a certain time.' The word might have been used originally as a metaphor for some event under certain conditions. It might have been used for describing any form of existence, permanent or transitory, limited or determined by conditions. It is now a very general word indicating a form of existence limited by conditions. According to David Bohm, the Latin word res comes from the verb to think, such that things are what are thought about. Perhaps thought is a dance of things in the mind.

The Romans did not have one word for wilderness. Words or phrases they applied to it included *vastitas*, meaning emptiness or waste, *solitudo*, referring to loneliness, and *locus desertus*, meaning deserted place (the etymologies of words are useful for tracing traditional thought). The definition of nonhuman nature changed slightly over centuries. The modern word wilderness comes from the Old English, *wilddeoren*, meaning "of wild beasts." Wilderness was understood as an uncultivated or empty region, or as a confusing multitude.

The word ethics is derived from the Greek word meaning 'custom,' which itself came from the Sanskrit word for one's 'own doing' (Greek *ethos*, Sanskrit *svadha*). Since it was used in the plural, it meant 'doing together.' By comparison, the word moral was derived from the verb 'to measure,' as to measure one's way. Morals means the 'way of going together.' The word 'morality' comes from the Latin word for will of the people.

The word 'design' means 'marking off a pattern.' Our effort should be to make sure that we are using the whole patterns. Design is a human project in which, as Oliver Lucas says, "visual and physical parts are assembled in order to achieve a specific end result." Ecological design is the creative modification of ecosystems to repair or enhance their ability at self-organization and the maintenance of their complexity and diversity. The word design comes from Latin word meaning 'to mark.' Nature is self-making and self-designing, but we humans now influence every natural system, taking what we need from some ecosystems, enhancing a few, misusing others, and interfering with the rest. We need designs to restore the balance between human needs and natural processes.

Ecological designs focus on whole communities that work in the same self-sustaining and self-limiting ways as nature. By consciously creating meaningful ordered patterns, we can develop ways of producing widespread community wealth while positioning the community for a long, satisfactory future in a healthy environment.

Words themselves can be considered design units. Meme is a term originally defined by Richard Dawkins as a noun that conveyed the idea of a "unit of cultural transmission," or a unit of imitation. He meant the term to be a unit of imitation for cultural transmission. Memes are cultural replicators. But the medium is ideas not genes or bodies. Ideas, of course are emergent phenomena from the activities of brains; they are physically patterns in the brain, although they can be expressed in other media, like stone or sound. Dawkins gives examples: Tunes, clothing fashions, agriculture, cities, domestication, arches, pots, faith, and free speech.

This idea can be useful in discussing units of design that are transmitted. It is important to note that the idea has been changed from Dawkins original idea. For instance Dawkins suggests that memes are inherited like genes; it is more likely they are communicated like the rest of culture. Dawkins suggests that they spread like a virus, but perhaps only in the sense that there is a new pattern of synapse connection. Unlike a virus, a meme can die easily if no one uses it. It may be replicated, but it is done so by brains, not by the meme itself.

But, there are problems with the concept, in terms of reduction and reification. Memes are not living beings; they are emergent patterns. Memes are not selfish. Memes cannot self-replicate themselves any more than a stone ax can copy itself. Nor do they respond to selective pressures like genes, although they may be filtered by use (and they can filter other ideas or such). Of course memes can generate behavior, as do symbols, metaphors, and signifiers. Although the word is a discrete unit, the meme itself may not be so discrete. Fuzziness and ambiguity is a problem with all models and labels, and the meme is used as a model. And, a meme is useful for thinking about design. Much like genes, however, memes can interact with each other and with other models. They can coevolve, like genes, with bodies, environments and cultures.

Of course, like any models, memes can occur in complexes, referred to by others as memeplexes. These can diverge from their original context and recombine in other contexts. Memes could also be used as units of cultural transmission of design. If we think of cultural ideas as memes that can be designed, then we can approach them as designers not simply as inheritors. For instance, the memes we would consider range from culture to nature: Culture, technology, agriculture, domestication, cities, economics, politics, wilderness, and nature. All of these are human constructs that have boundaries and centers, but all of them are also separate from the human sphere, although they interpenetrate. And, these can be treated as separate memes.

A unit of dwelling could be called a 'deme.' William Irwin Thompson used this word when he advocated a meta-industrial village he called a 'deme,' but the word could be applied to any species, from termites to wolves, and to any size unit, from villages to cities. The deme is a useful unit of design. A unit of wilderness could be called a 'wene.' This would be a regional ecological complex, similar to a 'rene,' which could be a unit of ecosystem renewal. The pathways through and between ecosystems could be described in terms of 'venes.'

A human ethnic group could be represented as a 'pene,' which would not have the cultural baggage of nation or republic. The specific goals of a group could be a series of 'zenes.' Which would be accomplished through specific actions, 'phenes.' We might even designate units to represent limits ('lene") and principles ('tene'), in order to discuss the importance and commonalities of these things.

Design could combine these units into new patterns that might be more flexible or resilient. Thus, renes could be combined to make up a single restored wene, within a matrix of demes and penes.

1.3.3. *Naming & Modeling with Metaphors*
Many words began as metaphors, useful for extending the understanding of something by calling it something else. The concept of metaphor has been defined and used for over twenty-five centuries. Before that, metaphor seemed to characterize language itself. Metaphor is used in all advanced languages.

Metaphors emphasize likenesses between things, living things, languages, or human constructs. A visual representation of a metaphor can be a model. Modeling can be done with metaphors as well as with shapes and numbers.

1.3.3.1. Metaphorical Models: Boxes, Traps & Filters
Mario Bunge used black boxes as metaphors for systems. The idea of the black box was originally conceptualized by electrical engineers to describe certain unknown systems devoid of structure. Black box theories include kinematics, thermodynamics, information theory, scattering-matrix theory, and circuit theory. The black box approach is useful for all theories whose variables are external and global, and are also simple and have a high degree of generality. As theories are supported by observation, black boxes become translucent. Some empirical theories have root metaphors that range from black to almost transparent: 'man is an animal' according to Karl Pribram, or 'the brain is a hologram' for Michael Arbib. When

we fill in the box, and make it transparent, we see the interactions of the components. In an ecosystem, if we make the box itself transparent, then we are left with a food web.

The box is a productive metaphor. The box can also be equated to a cage or trap. Karl Marx contended that we live in cages, part natural and part made, although human actions could modify them. The word cage is a metaphor; it implies being trapped. It is, however, a metaphor that can be expanded with a description in space as a four-dimensional box. Perhaps there is a better metaphor, since we depend on nature and society as a foundation for life, that of a trap. The word 'trap' is from the Old English 'to step.' A trap is a device for catching and holding animals or a stratagem for catching people. The idea of a trap can be related to the idea of closeness to limits or to the overconnection of links.

Taken in four dimensions, a trap can be a serial trap: The use of resources by a people, where the replenishment rate is constant and the rate of use exceeds it. This trap results in ecosystem degradation that is less reversible. The industrial age mistakes the rate of discovery for the rate of recovery. Agriculture is an energy trap, because it allows the concentration of energy, that is, higher yields, but then it requires more energy be put into the system to maintain it. The system has to produce more energy than the environment can tolerate. And it has to produce a surplus for trade. Instead of being free from economic want to develop their potential as creative human beings, people are trapped in a consumer cycle. Any flow towards a center, such as urbanization, can form an implosion. The implosion of any pattern, such as a cultural collapse, can form a trap.

A filter is an interesting metaphor. For organisms, the environment is a filter that allows some characteristics to continue, by providing opportunities and challenges. Selection operates as a survival filter that passes any structure than has sufficient integrity to persist. Organisms put together structures based on historical patterns, and move through a filter of limits like minnows through a fish net. Perhaps, the idea of a filter is too passive. Perhaps a better model might be a kind of dynamic mutual sorting filter, a mutual filtering of organisms and environments. Organisms and an environment co-order each other, sorting things out in a mutual process of activities and adjustments. In the next largest system, ecosystems are also filtered or interactively sorted.

That may be a very basic filter. These mechanisms are only parts of a culture. There are some parallels with biological evolution, especially if we use memes as the unit of transmission. The structure of cultural filters is a meme construction that is robust. Memes are still filtered by the mental environment for fitness to that environment, although in this case the filters are partly due to memes. The scale of replication is related to positive feedback, as when a work of art fetches great sums of money or is popularized from controversy. The information that is transmitted is in form.

Culture is more than a simple filtering or sorting process. It is more than a formation process. Because it occurs with physical, living, conscious, and social systems, culture is more like a novation process, that is, it recombines things into new orders. The models of box, trap, filter, and sorter are used in subsequent discussions.

1.3.3.2. Conceptual or Mathematical Models

Models can be constructed with a series of verbal statements or with a combination of statements and mathematical equations, using data. There have been many such models of the world system recently. Each model gathered data, analyzed it, and made a set of predictions based on the data and analysis. Each of the models also recognized having very general problems. We have separate kinds of models related to plant productivity, human population, and world games. We can try to generate models dealing, not only with the more constant of natural and human characteristics, but also with the limits of certainty and the vagaries of change.

For instance, *The Limits to Growth* model had vastly more information than it could use in an orderly way, due to deficiencies in theories of structure. J. W. Forrester emphasized that modeling projects should be global or national, but not regional, and should draw heavily on mental databases. The Forrester model distributed values homogeneously over the globe, disregarding specific sociocultural and ecological systems. It also seemed to tolerate an expansionist view of industrial human civilization. Dennis Meadows and others recognized the constraints on the planet with limited resources, and concluded that partial solutions could make things worse.

Mihajlo Mesarovic and Eduard Pestel established the necessity of using models to represent the objective aspects of world development, and defined a model as "a coherent and systematic set of descriptions of relevant relationships." The report used a strict scientific methodology to treat actual data and to avoid the feeling of an abstract academic exercise. But, it reduced everything to measurements, perhaps invalid measurements. Although they carefully contrasted their model with the Forrester-Meadows thesis, there does not seem to be any essential difference. Their multilevel model is still a reductionistic, as is their attempt at regionalizing cultural areas of the world. They claim a complete set of descriptions, but the subsystems sets are grossly oversimplified, and the results are starkly primitive graphs, reduced to meaninglessness by uniform, noncontextual computer decision-making processes. In summary, their model assumes basic industrial values and loses most human values in the shuffle of facts.

The book, *Reshaping the International Order*, coordinated by Jan Tinbergen, recognized a need for a new international order. The book identified problem areas and examined the progress toward their solution. Tinbergen noted the difference between a forecast, as in Limits to Growth, and a plan, which is a human effort to control destiny. In fact, the plans described here are based on many forecasts. The second part of the book described the architecture of the order and discussed strategies for steering change. One aim of development was towards an equitable social order, by reducing the differential between rich and poor. He explored the implications of his own plan, which accommodated the tension between developed and underdeveloped nations. Tinbergen recommended that poor nations negotiate with rich nations through collective bargaining to obtain a greater opportunity and right of equals. This tactic could bring some success.

Where the Blueprint group suggested more drastic steps, disregarding the adaptive dynamics of supraeconomic and ecological factors, Ernst Laszlo's alternative was to offer policies that mobilized adaptive capacities in these dimensions. Focus should be directed toward dynamic equilibrium models achieved without undue trauma, and sustained without repression. The ethical dimension of order relates to attitudes about violence, satisfaction of needs, and sociopolitical conditions compatible with equality and worth of human beings. Laszlo proposed a model that accounted for biospheric and sociocultural inputs from the total environment, recognizes sociocultural systems as converter and subsystematic, and, relates outputs—material and social technics—to positive and negative feedback. This system is more comprehensive and consciously planned than most.

We have separate kinds of models related to plant productivity, human population, and world games. We can try to generate models dealing, not only with the more constant of natural and human characteristics, but also with the limits of certainty and the vagaries of change.

1.3.3.3. Making Scenarios

Instead of describing models and thought experiments, perhaps, we should be creating 'scenarios,' a word associated with plots and summaries. In trying to redesign the planet, we need to incorporate stories and social understanding with the sciences and designs. A scenario could let us explore plausible approaches to complex patterns. It would let us incorporate the perspectives of other cultures as well as quantify common requirements. A shared scenario would emphasize the transformation and construction of healthy future states and then allow everyone to participate in the creation and make sure that their perspectives and values are represented.

An important first part of scenario construction is the process of imagining alternatives and then filling in the principles related to transformation and management. Another part is to create strategies to describe the course of action through specific practices. In Our Common Future, from the United Nations world commission on environment and development (the Brundtland report), the authors concluded that the positive effects of technological adjustments, such as recycling and energy efficiency, were insufficient to realize any kind of 'sustainable' basis (the word 'sustainable' is avoided due to its plethora of meanings).

Scenarios have been used before, traditionally in threes, beginning with the "business as usual" scenario, probably ending in overshoot and catastrophic collapse, and the "streamlining the existing system" one, also ending in some form of collapse. Usually the third scenario presented is "transformational utopian change," which involves completely regimenting society and then preserving the environment. Usually this is recognized as idealistic and unrealistic. The alternatives range from capitalistic individuals or a new tribalism to technological utopias or sustainable empires. However, by basing redesign on current cultural standards within a global cultural framework, we can avoid having to engineer human existence. We can modify world-images and behaviors, within a frame of standards and practices.

Scenarios have to consider all the factors determining the future, including change (the speed and rate of change), the momentum of the present (masses of people, habits and investments), and conscious adjustments to environment and society. Scenarios have to recognize the complexity of civilization and technology, as well as the increasing costs of the maintenance of the complexity. Scenarios have to address risks from external environments well as from technological change. They have to anticipate surprises from trends and catastrophes. These factors can be ameliorated by design. Some can be avoided by design. Maybe the following short, general scenarios are nonsense, but it would be interesting to compare them to the developed final future.

1.3.3.3.1. 50 Years from Now

Let us make a picture of the planet in 50 years. Physically, the ice sheets in Greenland and Antarctica are still mostly all there, except around the margins. The Arctic Ocean is generally open in the summer and closed in the winter. Some low-lying places have been flooded, although a few heroic constructions have helped a few. Many other low-lying areas are having troubles with periodic tides and floods. There are more storms, some with greater intensity. Australia, northern Africa and other delicate systems are drier and droughtier. El Nino and other patterns still exist and affect climate. More species have gone extinct. Many more are struggling with ecosystem transformations and destructions.

Because of the costs of fossil fuels, many kinds of crops are too expensive to grow at large scales. The costs of foods are much higher. Fewer foods are shipped around the world, although at small scales the groceries of the rich still offer a complete diversity of expensive foods. Businesses that rely on trucks and airplanes for deliveries have shrunk in number.

Oil is almost gone; coal is dominant. Pollution increases again, especially from coal, which is not used to power airplanes or most cars. People travel less to other countries. Although international services for youth allows many young people to work in other countries.

The human population of the planet has decreased to 4 billion, due to several changes, including the linking of population to resources and plans that guaranteed adequate cultural representation. A billion people live in traditional farming villages. Another billion live in industrial cities. A growing number, 350 million at least, live in new arcologies. A growing number, perhaps 10 million live in archaic cultures in wild zones, such as the Amazon.

Satellite views of the planet look different now. Much less gas is burned near wells; cities turn off most of their lights at night; most road lighting is connected to motion detectors. Alternate sources of energy, such as solar, are used for home heating and electricity, as well as several kinds of large buildings. Hydrogen and nuclear have increased, but most of the changes have come about through reduction and conservation. Artificial photosynthesis has been experimented with. Fewer clothes are available. Cotton and wool have become more expensive. Polyester clothing is restricted due to limits on the production of artificial fibers and plastics, due to the need to cut back on plastic wastes and the difficulty collecting older plastic wastes threatening the productivity of the oceans.

People have continued to move into cities for the advantages of urban living. Many cities have become more ecologically oriented. In industrial centers, most jobs have continued to grow in the service areas. Intelligent computers and hybrid robots have pioneered some kinds of research and are becoming prevalent in richer societies.

Some smaller areas have concentrated on labor-intensive production of permanent kinds of housing, furniture and electronics. People still fear change, despite living in an ever-changing world that surprises everyone with its losses and rebirths.

1.3.3.3.2. 550 Years from Now

The human population has stabilized at 630 million. This reduction has accompanied similar reductions in the numbers of cities and the extents of agricultural and industrial areas. Although the number of areas set aside as formal wilderness have increased, much of the planet has been converted back to common areas, dominated by forests and grasslands.

Archaic zones, for traditional hunting/gathering, fishing, foraging, herding, and basic agriculture have larger populations than in the 20th Century. The Masai population, for instance, stabilized at 440,000, and has autonomous control of parts of the nations of Kenya and Tanzania. Altogether, about 10 million people have remained or returned to foraging, although in some places where traditional knowledge and language had been lost, it had to be recovered consciously. Ultra-efficient stoves are used for cooking and heating.

Many traditional cities have been retrofitted to be ecologically balanced with their locations. As cities on faults and floodplains were deconstructed, the number of arcologies on stable, nonarable sites increased so that almost 40 percent of all cities were arcologies. Many highways, dams, and power lines have been deconstructed, also, allowing freer range to animal routes. Cities are powered by nuclear fission and fusion plants

Medical advances, combined with a revolution in common sense, have helped people to be much healthier; although they are not living much longer, there is less disease. Hybrid human computers and living robots are able to apply data more effectively to systems.

1.3.3.3.3. 55,000 Years from Now

A new glacial period started 500 years before. Possibly, it had been postponed by 35,000 years due to anthropogenic carbon dioxide and other human impacts on land, seas and species. In order to support increased mobility, efficiency and luxury, the human population reduced itself to 270 million. The population is concentrated along low areas along the Equator, since high elevations in Africa, South America, Tasmania, New Zealand, Antarctica and elsewhere have developed new or deeper glaciers.

New land area has been exposed in places like Florida, which is expected to triple in the next 1000 years. Siberia and Alaska have reconnected two continents with long exposed shorelines. The Sunda Shelf has connected Southern Asia to Indonesia, New Guinea and Australia, allowing endemic species to compete with new invaders (biodiversity initially increased but then decreased with extinctions). The planet as a whole has become more arid, with expanding savannahs and deserts, and shrinking closed forests and wetlands.

Parts of the northern hemisphere have been abandoned, although several deep cities have been mined in basalt (although they might be vulnerable to compression from growing ice sheets). The neopoetic land areas are being designed to have new wild areas as well as new agriculture. In North Africa and central Australia, salt-water rivers have been punched through deserts to create new wetlands and agricultural ecosystems. Most agriculture is based on a mosaic of perennial species, requiring less disturbance but more labor. Solar-powered networks create large oases (and cities) under adjustable microconditions. Many artificial lakes facilitate hydropumping for energy storage.

New cities have been created in areas of low biological productivity, freeing fertile areas for wilderness or agricultural systems. A new kind of arcology, modeled after a radiolarian, has been developed for use in rapidly changing ocean levels in the southern hemisphere. With its intricate mineral skeleton and embedded differential membranes, the arcology can be flooded in various degrees or tolerate desert conditions.

1.3.4. *Questioning Things*

Questioning has a long history in human experience. The Greek philosopher Socrates was one of the more renowned questioners. He approached teaching through a disciplined, rigorous dialogue with people he met on the street, sometimes by accident or often by design. Socrates tried to get others to recognize the contradictions in their ideas; he assumed that incomplete or inaccurate ideas would be corrected during the process of questioning, and hence would lead to progressively greater truths. He never seemed to reach an end to questioning, however, perhaps because there was no end, that is, the process of questioning could refine any kind of knowledge or ignorance, indefinitely, or perhaps because by itself questioning could only do so much with definitions and concepts. His method was a common search, through conversation, for the goal of truth.

1.3.4.1. The Socratic Method

Socrates asked questions as part of a conversation with others. He seemed concerned with discovering what the opinions of others were based on, an invisible truth that could be made visible with questioning. The questioning forced the other participants in the conversation to try to agree on the truths beneath the opinions. Socrates professed ignorance of the truth himself, in fact or in pretense, ignorance being the first step in the pursuit of knowledge. He expressed skepticism that the other conversationalists actually had real knowledge. The process of questioning subjected opinions to real examples from real experiences—an empirical method—leading to a more general concept—this is the process of induction used in a scientific, homogenistic logic—and then the consequences of the definition were drawn out, through deduction. These definitions were refined by further questioning until all members of the conversation had a better grasp of the concepts. Through thorough questioning Socrates demonstrated that knowledge was quite often uncertain. There was no absolutely certain knowledge. Questions were also meant to examine life as well as belief and truth, and to show that often people were ignorant of their ignorance. Socrates held that disciplined questioning enabled the other to examine ideas logically and to be able to determine the validity of those ideas.

1.3.4.2. The Hardin Extension

For Socrates the goal of knowledge was the acquisition of concepts, such as justice, courage or wisdom. He thought that the truth could be contained in a correct definition. And, he was groping for more abstract definitions. This became a problem, as abstraction became removed from the specifics of living. Socrates was most concerned with examining concepts, but concepts are a small part of reality. Of course, constant questioning of concepts can expose the psychological basis of concepts, and perhaps that is what Socrates meant to do. But, questioning concepts often reaches limits fairly quickly. Socrates never offered any answers, although he assumed that an absolute certain knowledge was possible to become established eventually.

The ecologist Garrett Hardin used questioning to illuminate partial knowledge and to track connections between things. Questions establish the limits of assumptions and perspectives. They can clarify the focus of a problem and test evidence related to any problem. Questioning can be used as a device to focus on a specific problem, not only the extent of the problem, but its aspects. Questioning can also be used to explore specific aspects of the dimensions of thought. Of course, questions can refine the process of critical thinking and can allow refocusing in a wider or narrower context. This type of questioning arrives at answers as workable hypotheses or guidelines for making decisions about operating in the world. Without certain knowledge, however, we can make adaptive decisions, based on partial knowledge. His questioning took on the form of asking what happened then, after an answer was arrived at. "And then what?" Hardin asks. Questioning works in a conversational way by weaving ideas. Konrad Lorenz decided that humanity would indeed have destroyed itself by its first inventions, were it not for the very wonderful fact that inventions and responsibility are both the achievements of the same specifically human faculty of asking questions.

More than an annoying part of a conversation, questions are legitimate ways to approach a known or unknown situation. More than just a way to turn around a conversation, questions are tools that allow you to surround a topic and define it more completely. More than simply an admission of ignorance, questions can form a phenomenological spiral that allows you to return to a subject from different perspectives with different levels of understanding.

1.3.4.3. The Maslow Questioning of the Norms of a Society

Before designing a good society, certain normative questions must be answered. These tentative answers are based on those questions first suggested by the psychologist Abraham Maslow. The questions have been slightly modified before being answered briefly.

Is the norm to be universal, national, subcultural, familial, or individual? The norm could have universal elements, not only for humans but also for all species impacted by humans. Maslow assumes that different norms must be on different levels, depending on the context. For example, there would be some universal human behavioral standards, but special local expectations, to conform to various cultures.

Should society be selective or unselective? The society should be unselected. It must account for all human variability; and accept it when possible and treat it when necessary. It would have to account for prisoners and misfits. The society should be pluralistic, and accept and use individual differences in constitution and character. Humans are not interchangeable; the insane and aged must be considered. It must integrate all people into a society or work in that direction.

Should society be pro-something or anti-something else? Society could be pro-industrial and pro-scientific, within the set limits of the society and the planet. But, industry must be properly scaled; and science must be cautious. The size of community cultures could be limited by function.

Should it be centralized or decentralized? The global unit could be centralized, electronically, at least, and socially planned; but individual cultures could be autarchistic, based on self-reliance and interdependence. Both should be flexible. Regions could be centralized for some functions.

Should society be tolerant? Society must tolerate all cultures in the nations. Each culture would determine styles and complexity for its individuals. It could aim for taoistic noninterference, but be available for help.

What should be done about injustices? Biological injustices exist and can be ameliorated; social injustices can be rectified, but the society would have to have the apparatus in place.

Should society determine family attitudes? Family or sexual attitudes can be institutionalized by the culture. All group adaptations would be determined by culture.

Should society be open to more than one religion? Society can tolerate any institutionalized religion, so long as it does not impinge on other groups. The spiritual life is a necessary part of a society.

How should leaders be chosen? Leaders within a culture could be chosen by traditional means, within the limits of human laws. Leaders of the international framework could be chosen by global referendum from the ranks of cultural leaders. The leaders would determine the relation of truth to people—who shall know how much about what and when.

For what is an individual responsible? The individual is responsible for the style and simplicity of their life and for its effects on nature and society. The individual is responsible for being tolerant of others, and is free to make many kinds of choices.

How these questions, and many others, are answered partly determines the shape of this project.

1.3.4.4. Questioning Ourselves

Things seem so confusing and contradictory, we have to keep asking questions about how things are done, for instance about old standards for the sizes of pipes or the lifetime of a tool. We have to question lifeboat politics, with its implications of gender, class and race values and inequities. What are the full effects of plastic wastes in the ocean or on ocean species? Behavior is determined by immediate personal consequences (short-term egoism), regardless of long-term consequences in modern and ecological systems. How stable is poverty in the midst of wealth? Why are we not trying radical designs like arcologies? Some of these questions can be answered by further questions. Will the technology harm the environment? Will the process waste energy? Does it use exotic, composite, dangerous materials? Does it really contribute to our welfare?

How can land and people be used well, or what is a good use? What is good knowledge of good work? How can one be honorably native to one place? When will we learn that neither art nor science that can be neutral? What are the risks of consuming now? Losing self-respect, seeing the collapse happen? There is an implicit larger question, once asked by Arne Naess: What is the role of humanity in the destiny of the planet? Then again, what are the risks of acting in unpopular ways? Being ignored? Losing status or being assassinated?

One serious question is what to do with those nonindustrial cultures that choose to continue to be nonindustrial. Should we make a park for them? Should we isolate them in some way or have some kind of boundaries that sort the technology that they can use? Maybe the word Park is not the best way the best word since they are not zoological specimens and maybe we cannot save them that way, but what we could do is allow them to create the boundaries they want.

Urban intensification leads to the question: Is there a limit to human numbers? Perhaps space, but is there a psychological limit? People in cities seem to do well with high-contact, high-proximity living. What happens when people are crowded or feel crowded? Physical complaints, emotional complaints, sexual dysfunction, or feelings of fear, seem to be expressed often. There may be limits of crowding. Are there social limits, in terms of the number of people one can tolerate? We may have a requirements for personal space, home space, and wild space.

So one question is why do we consume so much? Why do we consume some things and not others? Overconsumption is a very destructive pattern and the pattern is his defined by deception and lies. Our entire society now focuses on spending. For psychological reasons as well as economic ones. The banks do not help either with their campaigns to push second mortgages that higher interest rates. Tim O'Reilly suggests that more than real estate bubble we need to be concerned with the reality bubble.

As Speth notes, the fundamental question is how can the operating instructions for the modern world economy be changed so that economic activity protects and restores the natural world. He uses modern capitalism in the broad sense as the actual existing system of political economy as it stands, not as an idealized model.

Maybe, we need a big, formal program for questioning. Clarke suggests that we parallel David Hilbert's program for advancement and mathematics—the list of 23 problems to be solved by the mathematics community. We should have our own Hilbertian program for the Earth system. This includes analytical questions, normative questions, operational questions, and strategic questions. For example, a first analytical question is: what are the vital organs of the ecosphere? Another analytical question is: what are the characteristic regimes and timescales of dietary variability?

A normative question would be: what are the general criteria for distinguishing non-sustainable paths? Or, what kind of nature to modern societies want? An operational question is: What is the optimal mix of adaptation and mitigation measures to respond to global change? A strategic question is: What are the optimal goals for dividing the planetary surface into natural reserves and managed areas? What might be the most effective global strategy for generating and integrating and data sets? A final example of a strategic question is: What is the structure of an effective and efficient system of global development institutions?

We want the conventional system, with its 'humans first!" motto to lose its hold on our minds. To get to the heart of it, to weaken its grip, we have to attack the weakest point. And that is the questions of meaning and value. How meaningful is it to shop? Where is the value in eating Kobe beef? Is this radical enough? Ecological enough?

Perhaps we have become trapped by too much cheap energy to question things. We have benefited too much from cheap materials and cheap labor. We have benefited too much from too cheap food. This lets us break our faith with places. So how do we establish standards for places? Instead of profitability we should have health; instead of professional excellence we might have the durability of the community. Along with Wes Jackson, the idea of homecoming must include homemaking, especially now, dealing with remnants in ruins.

Can we design our way out? How can we act if we are ignorant? So far, we have not reduced problems and dangers by gathering more data, creating more theories, or by being cautious about industrial production. Ignorance is a problem. Another is that we cannot live without acting. We have to act on the basis of what we know, which is incomplete. So the question of how to act in ignorance is very important. We can act on the basis of incomplete knowledge when our culture has an effective way of telling us that the knowledge is incomplete and how we should act in the state of ignorance. Unfortunately it is also possible for incomplete knowledge to become the basis of arrogant behavior and dangerous actions. The standards of our behavior should be derived from understanding of our place and communities before incomplete technological knowledge. One conclusion we could make especially regarding uncertainty is that we could reduce the number of problems if we simply live that levels slightly below the lowest level of uncertainty, or the lowest absolute level productivity, of ecosystems.

Design cannot be separated from social and political questions. Designs, as Winston Churchill recognized, especially buildings, steer our experiences and actions. According to Langdon Winner, design includes the deliberately chosen, enduring forms of both material and intangible entities that affect human relations. Ecological design is a field that aims to recalibrate what humans do in the world according to how the world works as a biophysical system and a cultural entity. Design in this sense is a large concept having to do as much with politics and ethics as with buildings and technology.

Design has to be about more than making things, it has to deal with systems in context. Or poses questions for design: does society have to be organized differently to be capable of doing ecological design? What would this society look like? Do we need to did redesign institutions to be capable of doing ecological design?

Perhaps there is a model of questioning that could serve a large movement. In terms of procedures for management, the U.S. Navy has a workable program for questioning. Because the officers and enlisted men are in for a very limited time, the Navy system seems to be chaotic, but it's been tested and it has evolved over generations the system manufactures safety and efficiency even though there are numerous mistakes under new circumstances and there is a wide range of skills among the crew. Having a crew participate in question the actions of officers serve several purposes. One of them is to keep everyone in touch with the activity, providing redundant checks on actions. This increases safety, so that errors can get detected before they become much larger. But newer crew have a lot to learn in these discussions and criticisms of service training exercises. New crew also do not have the experience of older ones. They are not a sufficient and they're not always knowledgeable about what to do they need guidance. This process is quite sensible. It minimizes accidents. Considering that the Navy operations are dangerous he often at a fast pace and under high stress there are relatively few fatal errors. If the Navy were to follow formal procedure in a strict hierarchy that accident rate would probably rise. One of the problems is that were not taught this in normal situations and so in normal situations we seem to feel that we can make errors in the end they will not do have a high price.

Some people will want to decide boundaries through culture, watershed, or political power. These questions can be answered in meetings. This outline seeks to improve people's circumstances by enlisting them to save their own environment and their own way of life. Democracy itself is kind of like a thought experiment, where questioning and conversing about disagreements and disasters allows us to experiment with them, and our responses to them, before there is real conflict and real suffering. Even if the real disasters are not prevented at least they have been addressed. Dissent occurs within the context of loyalty. Freedom is expressed within a context of law that limits it and protects it. The democratic system avoids runaway feedback. Anthony Barnett suggests that there is a fourth kind of democracy now, reflexive, direct and large scale, made so by rapid news and easy communication.

Questions widen a narrow field. Hardin points out that concerns about narrow issues, such as pollution, can cause a deep examination of the process, such as distribution theory, that cause or contribute to the issue. Human activity simply produces things that we want and things that we do not want, such as pollution. As we ask questions about who pays and who benefits, we are able to think or rethink about these things.

Questioning can get to the basis of any conversation. But, questioning can also frame and direct the conversation. Questioning also provides feedback for any answers. Therefore, questions will be a critical part of the approach to design. In terms of stimulating learning and creativity, questions are sometimes more powerful than answers.

1.4. *Noticing Gigatrends Over Time*

Things change, sometimes rapidly, sometimes slowly. We often refer to rapid change as a revolution or as a catastrophe; we sometimes refer to slower changes as trends or fate. Trends are directional changes. Some trends are readily evident in the human present; other processes or cycles take many human generations to complete. Many trends in economics and politics can be fit into the megatrends identified by John Naisbitt over a decade ago. Naisbitt identified ten larger patterns in society, including the move from industrial society to information society, the economic interdependence of the human world, and the restructuring of society from short-term considerations to long-term time frames, that is, he says, from two-year horizons to a "very long-term" time frame of "six to ten years"—long by economic standards, but not long enough for trees and ecosystems. Some other counter-megatrends (we should resist calling them negamegatrends), such as the denigration of reason and science in the popular press and media, or the incredible explosion in information— without a corresponding increase or spread of wisdom—are ignored by Naisbitt. Negative trends in general are left out of the picture. The things that are most popular to society, such as money, real estate, insurance, and politics, are the things that are treated as the most important. Italian Foreign Minister Gianni De Michelis identifies the most important megatrend of the century as the availability of free time; he claims the US economy will remain the most important economy in the world because its GNP is increasingly geared to entertainment, communications, education, and health care, all of which are about individuals 'feeling well.' The things that will ultimately be most important, such as directing the course of civilization, limiting human activities, or preserving wild ecosystems, are ignored or relegated to a sideshow. This section presents some of those gigatrends for considerations for design.

The real long trends in ecosystems, forests, human history, and earth history occupy the entire human calendar and long before. They are sometimes invisible in the present. Often the changes are very slow, taking more than one lifetime to notice; slow change is rarely noticed until too late. We might call them gigatrends, since they are larger and more involving than megatrends. Gigatrends ("giga" from the Greek for very large or giant) are long or very-long-term trends, usually ignored by science and economics, such as atmospheric temperature increases or global deforestation (and perhaps cosmic trends, lasting millions or billions of years, should be called teratrends, although for the sake of comparison, they will be included as gigatrends). Some gigatrends may only cover a span of fifteen years, but some of them have continued for thousands of years. In general, the larger the unit of study, the longer the trend; for example, changes in the physiology of a tree occur in days and months, while changes in the forest ecosystem can take decades or centuries. Some gigatrends are beneficial, although most of them seem to be detrimental to human well-being as well as to the health and stability of ecosystems.

Because of their length and scale, most of these trends can be represented graphically with simple lines, which show gradual or rapid (exponential) changes. Several trends seem contradictory or inconclusive in the short-term, but are evident with long-term study. For instance, atmospheric carbon dioxide increases and decreases with seasonal change in vegetation, but has been rising slowly and steadily for at least 35 years, according to Keeling and Whorg. Ecosystem succession is also misleading in the early stages, as pioneer species take advantage of a disturbance—here the short-term abundance of intolerant nitrogen-fixers prepares the site for a mature forest. Other trends are actually complex. For example, wood use as fuel decreased from the 1800s until 1970, when it started to increase again, due not only to the fashion for stoves in industrial countries but to the rising prices of petroleum and coal.

Many of the negative trends have been noticeable for thousands of years, but nothing has been done to halt them. Environmental factors have shaped the course of human history to a greater extent than has been realized. The decline of Rome demonstrates that ignorance of forest ecology can have important consequences. There have been other environmental catastrophes in the Tigris and Euphrates valley, Greece, Khmer, Maya, Cahokia, and other places. These civilizations were very successful before they failed. Failure from success is tragic. For the Greeks, the operation of tragedy resulted from success taken to great lengths, that is, where successful behavior in one context is applied to all contexts, with the result that the opposite result occurs from the one desired. For example, humans in moderate numbers were able to take what they needed, such as wood, from natural ecosystems without interfering with the processes. Our dominance, once so successful because of our big brains and tool-using hands, has now become self-destructive. When human cultures adapted to ecosystems over long periods of time, the ecosystems also adapted to human cultures; when the human impact has been rapid and intense, as it has been in North America recently, the ecosystems collapsed or stabilized at a simpler state.

It has been argued that humanity is not adapted to live everywhere. Since the human species emerged in a subtropical climate, where it acquired certain biological characteristics, it may lack some degree of fitness to survive in the tropics, the arctic regions, or even temperate forests. The species may remain genetically best adapted to a certain type of subtropical savanna. Rene Dubos presents this development as explaining some of our present behavior patterns: Subconscious fear of forested wilderness (where good vision was little use for avoiding danger); the commonness of design features in landscape architecture; the preference for a narrow range of temperature; and the biochemical similarity of nutritional requirements. Perhaps this may explain why people modify their surroundings the way that they do, as well as why forests are less valued than lawns and gardens.

Although many gigatrends are interrelated, they can be discussed in categories, such as general human populations or ecosystems and forest ecosystems. Positive trends, often smaller and more recent, are discussed after negative ones. This list is not meant to be complete or detailed. These trends can be represented graphically with lines. Horizontal lines indicate constancy or stagnation. Straight lines indicate linear growth or decrease. Curved lines show accelerated growth or decline, as well as stabilization around equilibrium, rhythmic regulation, growth to a limit (asymptote), and oscillation and fluctuation.

1.4.1. *Ecosystem Gigatrends*

Ecosystems build up information. There are at least three different channels of information in an ecosystem: The genetic (in replicable individuals); an ecological based on interaction between cohabiting species (expressed in changes in their numbers); and the cultural, transmitted through individual learning based on experience. Feedback within the interaction of species is expensive memory with little storage capacity. Whenever succession starts again, after a volcanic eruption for instance, old information in the form of interactions has not been saved. Genetic memory has much larger capacity and is long-term (but its use might be limited by environmental conditions). In higher vertebrates, such as wolves or humans, cultural memory is enlarged. The unconsidered use of information results in still more long-term trends.

Forest destruction was first associated with hunting, when fire was used as a technique to drive animals into the open. Grasslands maintained by burning allowed safer hunting. The technique is still used today by a few small groups of people. Recently, centuries of burning in Africa reduced forest cover to less than 40 percent of its original cover by 1948, according to Eckholm. With a shift to agriculture, forest destruction increased dramatically.

Agriculture and forestry have been related in many civilizations. The expansion of agriculture is directly related to the shrinkage of forests; other trends in agriculture, such as opening southern lands, have resulted in forests reclaiming some of their northern territory. Ecosystems are simplified and degraded; deforestation, desertification, and exotic take-overs occur on a large scale. Humans simplify ecosystems and keep them at early seral stages to harvest the increased productivity. Much vegetation becomes a social artifact. In Scotland, for instance, forest cover was reduced from 55% of the total area to 5% by simple stock-keeping and agriculture; the moors decreased by half, but meads increased eight-fold.

Table 141-1. Sample Gigatrends (read down each column)

System Conversion	*Energy use*	**Animal use**	*Disease progression*
Burning	*Solar*	Domestic animals	*Infection*
Deforestation	*Human Labor*	Companion animals	*Respiratory*
Simplification	*Natural fire*	Food animals	*Childhood*
Specialization: Food	*Animal Energy*	Labor animals	*Malnutrition*
Fiber/wood land	*Water energy*	Pets	*Stress / diet diseases*
Desertification	*Wood energy*	Research animals	*Contagion (travel)*
Vista creation	*Fossil fuel*	Preserves	*Ecosystem diseases*
Preserves (refuges)	*Nuclear production*	Biotech creation	*System collapse*

1.4.2. *General Gigatrends for Human Populations & Needs*

With the success of the human species has come human domination of ecosystems. Human beings have modified animal and plant associations, simplifying patterns of energy and chemical exchange, and solidifying ourselves at the end of many food chains as a dominant species. Our domination is related to our large biomass, our large annual increase (over 2 percent annually), our high energy use, and our high structural organization (information and matter). These very large-scale effects relate to basic gigatrends having to do with population size and dominance:

Human populations increase exponentially—at 2 percent per year the doubling time of the entire population is 35 years. The growth of human populations for 500,000 years was minuscule; the agricultural revolution (10,000 years ago), which increased food supplies, and the industrial revolution (in the 1800s), which decreased the death rate, led to dramatic increases in population numbers and the rates of growth.

Are there big patterns of human cultures? Perhaps this sequence holds for all culture: Successful exploitation of a place > reproductive success > overshoot > ecosystem conversion and degradation (drought) > stupidity and violence in response > stagnation and collapse. When humans reach a new place, they take over as tertiary predators. Over time they take over all predatory roles. They exploit the resources to the best of their ability, using old and new techniques, even when they know they are overhunting. Then they degrade the habitats. Would this explain agriculture? It was the only option after degrading a habitat?

Human population pressure pushes some trends negative. The pressure from exponential population growth means that remaining forests will be depleted to meet basic human needs. The existing financial and political resources may not be enough to stop it. Eric Eckholm concludes that the United Nations must identify, analyze and marshal world resources against negative trends. A scientific method would take a long time, however, and poor countries cannot wait. They must attempt a rural regeneration of some kind, to stop urban drift and ecosystem destruction. Eckholm interprets negative trends as indicating the sinking of marginal peoples on marginal lands into a quiet helpless poverty, later leading to urban deterioration—which may perhaps be less quiet.

Table 142-1. Sample Gigatrends (read down column)

Cities	Specialization	Wealth	Ownership
Intensification	Differentiation	Wealth Creation	Owned by land
Urban living	Specialists	Skewed distribution	Territory tribal
Creativity	Classes	Accumulation	Private Property
Illness	Permanent classes	Invented wealth	National territory
Physical Decrease	Individualism	Useless wealth	Corporate ownership
Population Increase	Decl. participation	Dangerous wealth	Global ownership
Physical Increase	Refuges	Anti-wealth	Mixtures

1.4.3. Gigatrends in Economics & Technology

Eric Eckholm describes also how economic and political pressures, which are derived ultimately from population pressures, force farmers to intensify their efforts to increase crop production (and foresters presumably to increase their production). This instigates an "utterly dismal cycle" of population expansion, environmental deterioration, and poverty: as the population expands, forests are cleared for land, arable lands are used to capacity—and sometimes beyond; as the soil deteriorates, it requires more fertilizers that cause more hazardous conditions that decrease agricultural capacity; people starve, but the population increases, and marginal lands are used to meet increased demands—or food is imported from other lands—which are also experiencing stress.

Eckholm describes how the usage of such marginal lands could result in dust-bowls, when climactic conditions change. After the US dust-bowl in the 1930s, national conservation programs were able to restore some of the mythical fecundity, through pasteurizing, strip cropping, terracing, and contour plowing. However, current production efforts are causing greater losses of topsoil, and farmers are abandoning some of the conservation methods for economic reasons. Eckholm concludes that free market conditions encourage dangerous trends. The lesson of the latest dust bowl may be forgotten until the next one. As forestry copies the agricultural model, so it contributes to soil losses and destabilizes ecosystems.

There are economic trends related to forestry. First, there are shortages of timber. Most Old-World civilizations faced shortages of high-quality wood, from the Mesopotamians, Egyptians, Greeks, and Romans to the modern countries of Europe. Pressures on British woodlands in the 1300s forced people to turn to coal as a fuel source (a source regarded as inferior to wood). The timber famine reached Europe in the 1700s. It had existed in China and India over a thousand years before. Countries that exhausted their wood supplies had to invade other countries or find substitute such as coal or water power. Each substitute required more energy to produce. E. Eckholm warned that there is a serious firewood shortage over most of the earth due to population pressures on the remaining woodlands. There is an overall growth in wood consumption, and an increasing per capita demand for wood. An emphasis is on price for wood.

Energy use has been increasing, as more technical sources are being exploited. The average energy use per person per day has grown from 2000 Kilocalories (likely around 11,000 years before the present) to over 230,000 in the USA.

Some trends have stopped because planetary limits have been reached. Like ownership, territorial expansion is at an end. There are no 'new worlds' to discover and no virgin forests to be logged. Some of these negative gigatrends are hard to see, much less stop or reverse, because they are based on misunderstandings, fallacies, myths, and psychological blinders. For example, forest planners often treat exponential growth rates the same as linear rates. Thus, if our forests were to last for 400 years at our current rate of use (before extinction), they would only last for 75 years with a demand growing at 3 percent per year, or 50 years at 6 percent (demand is growing now at over 3 percent per year). Also, Dennis Meadows points out the speed with which surpluses disappear with increasing population and increasing per capita demand.

Many of these gigatrends are based on myths, such as the economic myth of forestry, which as it is related by Chris Maser, is based on the rationale of "soil rent" theory, a classic economic theory that assumes, fallaciously, that all ecological variables are constant so that capital investment is easily calculated from the rate of growth of the crop species. Due to the uncertainty of natural processes, we should limit our take to far less than a maximum rate.

1.4.4. *Social & Political Gigatrends*

Although culture has allowed humans to adapt to almost every ecosystem on the planet, it is a human construct, and as such subject to misdirections, fads, misperceptions, and superstitious behaviors. We can recognize large-scale trends in culture. Civilization in general has become more intense and complex, although on some local level there have been booms and busts of individual civilizations.

There has been a continuous disarmament of civilian populations. In England in 1995, murder rates are only a tenth of what they were 800 years ago and 1/2 what they were 300 years ago. The disarmament took place in many small steps from seizures to licensing, production controls, and fewer public exhibitions. The reach of the state into individual lives. From tribute to lack of arms, ownership, and licensing, to information-gathering

There have been slow changes in leadership from leaders of bands to chiefs, kings, emperors, presidents (including actors and puppets), and corporate boards. Slavery has become specialized also, moving from opportunistic to institutional slavery, and then from humans to animals, machines, and energy.

These trends are partly the result of our unconsciousness of large-scale, long-term events, partly the result of out cultural amnesia about things that make us unhappy, and partly the result of our cultivated indifference— doubtless due to our remoteness from wild nature, remoteness from the forest as a result of our tools, and the general romantic abstraction of civilization.

Table 144-1. Sample Gigatrends (read down each column)

Leadership	Ethics	Art	Religion
Leaders	Ethical Inclusion	Magic	Natural Gods/forces
Chiefs	Kin	Seeing invisible	War Gods
Kings	Tribe (friends?)	Control others	Regional Gods
Emperors	Nation	Decoration Pers.	Global Gods
Presidents	Animals	Abstraction	Universal Gods
Corporations	Humanity	Communication	Machines
Actors	Ecosystems	Express itself	Mixtures

1.4.5. Positive Trends

None of these gigatrends can be reversed until the remoteness is re-educated into participation and attentiveness. There are already a few trends flowing against the tide. Some positive gigatrends include:

- Adaptation of human cultures and ecosystems over time in Asia, Europe, and parts of Africa, resulting in diverse, domesticated landscapes.
- Setting areas, such as preservation of ecosystem processes, reservation of archaic cultures, and conservation lands, aside from industrial interference. Restoration of forests to abandoned fields, anthropogenic deserts, and ruined ecosystems.
- An increase in the scope of ethics, from family, tribe, nation, humanity, to include reverence for all living beings, identified by Schweitzer. An increase in the scope of ethics to include land and forests, identified by Aldo Leopold.
- An increase in the scope of law to include legal rights for forests and ecosystems, identified by Christopher Stone.
- Increasing globalization leads to more efficient production and to the recognition of the uniqueness of cultural creations.
- Increasing localization counters the threats of globalization to destroy local identities and homogenize cultural lifestyles. It can lead to knowledgeable governance and can respond faster to environmental destruction.

Considering other new trends in housing, such as arcologies (Paolo Soleri), ocean arks and bioshelters (the Todds), and in agriculture, including agroforestry, permaculture (Bill Mollison), and tree crops (Russell Smith), positive trends seem to be expanding. Large-scale ecological designs should be able to increase the number of positive trends.

1.4.5. *Summary: How Will Trends End?*

None of the negative gigatrends can be modified until our remoteness from nature, from places and regions, is re-educated into participation and attentiveness. By making long-term trends visible and immediate, we can understand how they shape our use of nature. The intent of describing large-scale patterns is to have human patterns fit with observed patterns in nature. Patterns have a form, sometimes repetition, and sometimes regularity, but each of these is caused by some limiting factor. Fitting the pattern can lead to both continuity and predictability, and both of these are needed to adapt human activities to natural limits.

With these gigatrends possibly ending in tragedy for humanity, we must ask many questions. What kind of ecosystems and forests do we want? Wild or domestic or both? Small or large? Managed or unmanaged or preserved? How shall we use those forests? For wood products? To protect watersheds and maintain global biogeochemical processes? As a home for other beings? Recreation? As some kind of balance for domestic landscapes? How many ecosystems do we (or all cultures or the earth) need? What kinds, in what forms? How many should be wild? These questions lead to new strategies for living on the planet, strategies that we can test with thought experiments.

Linear thinking leads to great short-term successes, but also to long-term problems, since nature produces many beautiful nonlinear curves and explosions. What these trends show together is that humanity is using more ecosystems, and more of each ecosystem, to produce more goods that cost more for more of us. These trends are not fated to continue. Some of them can be slowed, redirected, changed, or reversed. For instance, rather than be converted to plantation forests, remaining old-growth forests could be preserved or harvested at very low sustainable rates.

We have spent most of our infancy fighting nature. Up until the seventeenth century Europeans regarded untamed nature as a vast, hostile desert. Wild nature still remains unwelcome in our cities and gardens. We might understand the historical failure of cities and walls to lock out the forest or nature. After the Sumerian king Gilgamesh killed the great spirit of the forest, Humbaba, he became possessed with the fear of death and tried to lock out nature with the great wall of Uruk. It did not work. Like any ecosystem, the forest must be wooed. The forest will haunt us until we give it a new life in the heart of modern culture.

Thinking we have conquered nature and are omnipotent, we have quit thinking. Satisfied with our comforts, we do not ask enough of ourselves. We seem to be confused between luxuries and necessities. There may be enough forests for necessities, but not for luxuries. We also act confused by the distinction between temporary good and durable goods (nothing is permanent); temporary good are things like cars, entertainment, guns, and drugs (any kind), while durable goods are things like reforested areas, organic farms, well-designed roads, and healthy buildings.

Garrett Hardin used to say that the essence of ecology was found in the question, "and then what?" meaning that everything you did had a primary effect (there were no side-effects) on the system, every action a reaction, or as it has been rephrased by Barry Commoner and others, "you cannot do just one thing." We have to consider the consequences of our actions as much as we can. Even good actions, taken in isolation, can have tragic consequences. For instance, what if, in setting aside forests in North America and reducing the load on them, we put more pressure on forests in Malaysia, causing them to be cut faster and more disastrously? What is the solution then? Social equity with other cultures or peoples? Voluntary simplicity? Global laws? These questions lead to one reason why design has to be global—to coordinate the consequences of good actions used to direct trends.

1.5. *Performing Thought Experiments*

Humanity is engaged in a great experiment with the planet. We are replacing large, old, complex ecosystems with young, simple fragments, in which fires are suppressed, large predators are removed, large herbivore populations are encouraged, exotic species are introduced, soil is compacted, and excessive biomass is removed—all for the purposes of increasing the amount that can be harvested for human use. Our radical cultural experiment to dominate the planet comes from strengths that allowed us to survive rapid climactic changes during and after the ice ages. The ideas that are guiding us, the strategies we are using now, however, are obsolete; they worked effectively under less complex conditions, at lower population levels, but not at high populations on a global scale, where domination leads to ecosystem interference. This is the making of tragedy—applying strength in an inappropriate context.

We are also burning massive quantities of fossil fuels; this is a one-time, large-scale geophysical experiment that could not have happened in the past and can never be reproduced in the future (according to R. Revelle and H.E. Suess, 1957). Our actions are experiments, whether we want them to be or not. Unfortunately the experiment is not only bad science—there is no control planet—it is ill-considered. This experimental course, which may be global and irreversible, cannot be unmade, not by planning or science, much less by our standard methods of ignorance, cupidity, or denial. This section suggests the kinds of thought experiments that might guide large-scale, long-term designs, which could alter the direction of the great experiment.

There must be a way to refine the experiments, to minimize our impacts, to be less reckless, and to anticipate the outcome of our experiments before we finish performing them. Not all experiments must be physically implemented. Albert Einstein and Leopold Infield suggest that knowledge of laws can be gained through the contemplation of idealized experiments created by thought, *Gedanke-Experiment*. For example, to address the equality of inertial and gravitational masses, that is, how the problem of general relativity is connected with gravitation, Einstein imagined an elevator at the top of an incredibly high building, and then imagined what research could be done in this local environment. Such experiments might seem "fantastic" in his words, but they might help us understand things.

Although ecosystems and political orders are orders of magnitude more complex than physical systems, perhaps we could imagine and use such experiments to help us understand what is happening with our complex planet that is composed of many interlocking ecological systems. Thought experiments can give us clues about what can happen and what is the likelihood of that happening. "And then what?" asks Garrett Hardin again. Unlike medical doctors or scientists, we cannot either wait or directly experiment within a realistic time frame or scale. We cannot experiment at all in a traditional sense, where we hold most variables fixed, while changing one or two variables in experimental runs. Ecosystems operate over very long time spans; furthermore, their historical nature means that they cannot be restarted for tests. These limits are obvious on the scale of the planet.

1.5.1. *Sample Small Thought Experiments*
We can outline a few kinds of thought experiments. The first two, on wealth and roads, are for the nation of Bulgaria, but could be extended to the Balkans. A final experiment calculates the services and the number of arcologies needed to house the entire human population. The thought experiments presented below are incomplete, but suggestive of the kinds that we could be creating and manipulating to guide our plans and models.

1.5.1.1. Could We Replace All Forest Functions with Machines?

Could we replace ecosystem functions with mechanical devices? The machine metaphor dominates modern forestry. This metaphor, and the agricultural model, result in tree plantations, in which many of the functions of the wild forest have to be taken over by human ingenuity. On some tree plantations, many of the functions of a wild forest have to be duplicated. For instance, shade cards are used to protect young shade-intolerant species; plastic sheaths are used to protect bark from predators; fertilizer is used on young trees; and some trees are doped with mycorrhizal fungi. Extending this trend, modern forestry eventually may try to create an artificial forest. By taking this to a ridiculous extreme (the argument known as *reductio ad absurdum*), we might try to create an artificial forest with just one living organism, Douglas-fir trees. For example, we could replace the functions of nurse logs with gigantic nylon sponges. Imagine the kinds of artifacts or tools we would have to invent to replace the functions of woodpeckers, bats, insects, fungi, shrubs, or snags in a mature forest. Wild forests produce many useful things through their growth and change, from pure water and air to flood control and weather moderation. Suppose that we had to replace every function an ecosystem provides with a human service. Could we afford it? The following discussion uses very approximate numbers. Pure water can be bought in supermarkets for $0.50-$2.00 per liter; canisters of clean air can be bought (in Tokyo for instance) for about $10 a liter, although air can be cleaned for much less; flood control can cost $11,000-$90,000 per linear meter; wind protection increases the costs of buildings; solar protection can cost $300 per square meter; fertilizer and pesticides can cost $40 a gallon; waste recycling costs thousands a day; climate moderation, i.e., cloud seeding, can cost $44,000 per day; recreation can cost $25,000-$25,000,000 per park construction; genetic modifications can run into millions. As you can guess, for an area of 50,000 people, the costs of replacing some of the basic ecosystem services could be billions of dollars per year (and maybe $3 trillion for the whole planet). Many of these functions, like clean water, seem affordable to replace, but many such as climate moderation or soil creation are not affordable in any practical way. Taken to an extreme, the entire output and wealth of the market system might be directed towards propping up decaying forests and fisheries, instead of making televisions and shoes.

Perhaps one can get a rough estimate of these services if we extrapolate the costs from the shuttle Endeavor or from Biosphere II, both of which were decent attempts to create the conditions for living independently from the planet, for a short period of time at least, a week or two years. For example, the cost of the Biosphere II, with its 1.25 ha area and 190,000 square meter volume, was $200 million. The cost of operation was about $4 million per year. If we had to repair the entire planet, with 5.1 x 1014 square meters including oceans, at the same rate as Biosphere II, would cost $816 quadrillion. To operate it would cost less, at $16.3 quadrillion. Kevin Kelly suggested that the global infrastructure of the econosphere was worth $4 quadrillion (in 1998 dollars); this amount is quite a bit more than Gretchen Daily and associates estimated for the value of annual planetary ecosystem services ($32 trillion), but small compared to the cost of operating the planet (less than 25 percent; even taking all money from billionaires and millionaires would add only $29 trillion or so). These large amounts are for operating the planet, not just replacing a few services, such as clean water. Is that what we want to do, assuming we knew how to do it? After all, Biosphere II was unable to function long without importing clean air. And, we are experimenting with the only natural source of clean air.

1.5.1.2. What if we Calculated that Bulgaria is Wealthy?

Recently I was asked what I thought about poverty in Bulgaria, the lack of cash that makes almost every Bulgarian feel poor. I did not know what to say. From what I see, Bulgaria is rich. Perhaps we misunderstood each other's ideas of wealth.

Economics has always been concerned with measuring wealth. Wealth once meant tangible things, such as land, ships, and houses. Later, it was measured by labor and production. Now, it has come to mean negotiable symbols such as cash and stocks. Lately, information itself is considered a form of wealth.

The first economies depended on their own food and minerals. The mass economy rose after the industrial revolution, when networks of governments and institutions were created to hunt and acquire vast quantities of material in order to manufacture products that could be sold at profit. This economy is being altered by increasing populations and by increasing difficulties in finding cheap resources. Some of our perceived wealth and assets are disappearing in the process. But, symbols such as cash are considered new forms of wealth. And with computers, information itself is valued as the ultimate resource and source of wealth.

Information is apparently boundless. Yet it can be manipulated. It is information that defines the use of resources by people. For example, hydrogen is worthless unless technologies exist to transmute it to helium and manage the released energy.

Land and resources are considered less important. But information without "form" is nothing. Information lets us use resources and land more efficiently. Land and resources still are part of the basis of wealth (a material dimension)—as many native peoples have found out when coal or pharmaceutical plants were discovered on their lands.

The narrow definition of wealth (as just one thing, resources or information) means that it can be increased only by producing a bigger supply of goods or reducing the demand for goods. Wealth is defined as supply divided by demand. If supply is limited then wealth can only be increased two ways: reduce the expectations of individuals (smaller pieces of the pie) or reduce the number of individuals (fewer larger pieces of the pie). Supply may be mostly material things—but not status, for instance—while demand has the more psychological dimension. Wealth has a psychological dimension. This dimension is not limited by strict logic. Wealth can therefore be expanded without being limited to material supply.

The assessment of personal or cultural wealth, for instance, is mostly psychological; wealth may be measured by how many valuables one has, which may be physical, like feathers or salmon or gold, or by how by much status one has, which may be behavioral, as when enjoying deference or a good reputation.

Rich sensory experiences can be derived from direct contact with nature. But economists rarely mention these values. Light, wind, dirt, plants, birds, all act on us—but not with the meaning of crops or vehicles, which is for their utility—they just are. People do not live without these things. They are valuable to us.

Until now, economics has required growth to increase wealth. Growth has been a substitute for equality; it seems to be necessary to avoid revolt—even after 400 years, growth has not brought wealth or equality to most people. Ecologically, the goal of economics should be mature development, not growth. Development means the introduction of an innovation. Economic development will still require technology and new forms.

A mature economy would be like an animal, or plant, or like a mature ecosystem, like an old beech forest. In its early stages, a beech tree can still be stable. Growth in trees can delay the onset of senility by ridding it of waste products in more diluted form. However, too much growth produces a strain on tissues and early decay. Later stability must result from limits and metabolism. When it reaches a size that fits its genetic and environmental limits, it is mature. It continues to change, but that is development of new relationships and forms.

It seems that the wealth of a country is a function of its physical attributes (resources including information) and its culture (application of information by people). In fact, the attributes are only possibilities until appropriate cultural perceptions and technologies exist. The inclusion of nature as a source requires an ecological dimension to wealth.

Bulgaria has wild lands with all kinds of unique wildlife and plants. Bulgaria has fertile fields that could grow all the food that her people need. These things are an important basis of true wealth. Bulgaria also has an educated and exceptional people, who have the knowledge and information to make sure that some of the natural wealth is transformed to human wealth. By this most advanced definition of wealth, Bulgarians are indeed wealthy.

1.5.1.3. What If We Limited the Number of Roads?
Bulgaria has many roads, although politicians, as well as business people and drivers, are asking for more, wider, smoother, and direct roads. Cities want better roads leading to their centers. Forestry managers want more roads into the rich beech and pine forests. Resort areas want more roads leading to the sea or mountains. Economically, roads can stimulate income, at first. But, there are other ways of looking at roads.

We can look at roads with an ecological perspective. As a science, ecology describes the interrelationships of organisms and environments, that is, the experience of living together in the biosphere. But, ecology is also a way of "seeing" that human beings are participants in nature, as part of the food chain, for example. People, like most mammals, use roads (or paths) to get from one place to another, to get supplies, to visit others, or just to look around. This is a fine use of our technology. It allows us to increase our horizons and better our lives. Better roads make traveling more efficient. There is less waste of oil and gas, and less wear on vehicles and their occupants. But, every technological innovation in vehicles either requires or makes roads. Roads have effects that go far beyond the movement of people along them.

New roads lead more people to new places, thus changing the characteristics that often make those places attractive (e.g., being off the road). Roads increase the flow of things between points. But, too much flow (of matter, energy or form) can destroy biological relationships and diversity. Roads are a major force in fragmenting the habitats of plants and animals. Many animals cannot live near roads or noise or human activity. Many animals and plants need large areas to roam and roads cut into their areas (although, highway routes and underpasses can be modified; for instance Britain builds underpasses for frogs). Roads directly affect natural and human communities in many ways, causing:

• Changes in populations of animals or plants that cannot cross them (isolation)
• The spread of organisms that use roads to colonize new areas with plant or insect or animal pests (that is, things that are out of place)
• Problems with erosion
• Problems with spreading trash (other things out of place)
• Changes in hydrology and wetlands
• Changes in social circumstances. For instance, private cars changed public morality in America. Many kinds of crime are increased, for instance, bank robberies, if there are fast roads nearby with easy access.
• Changes in economies, as new roads bypass old routes.

Many countries answered the demands of their citizens, business people, and politicians (and ignored the environment) by building bigger, faster roads. Then as people crowded on the roads, they get more crowded and slower, and the demand for bigger roads rose again. Many countries have found that building more and larger roads does not solve the problems of congestion, accidents, and danger. These problems have a lot to do with the kind of transportation on the roads, that is, cars, trucks or buses.

Maybe Bulgaria should have a new autobahn and maybe all the roads should be upgraded. But, it would be better if it were part of a plan for the entire country that considered population movement, the needs of all the people, and the best forms of transportation. Buses and trains are far more efficient than private cars, and many countries are rebuilding their train and bus routes, from Brazil to France and Japan. By concentrating on a good public transportation system, and limiting the influence of private cars, Bulgaria could become a good model for them to follow.

1.5.2. *A Larger Experiment: What If All Urban Concentrations were Arcologies?*

At present over three billion people live in cities, about half of the total world population of 6.3014 billion (revised and estimated for 1 September 2003). What if essentially all people lived in arcologies, except for small traditional communities living in wild areas?

Historically, human hunting altered ecosystems, then human gathering and planting converted forest, grassland, and wetland ecosystems to agroecosystems that had to be managed. Now, the expansion of urban areas with roads, power grids, and other infrastructure, is interfering with the basic functioning of many ecosystems. Modification, conversion and destruction of ecosystems disrupts the complex interactions within and between ecosystems, the hydrology, soil structure, topography, and the predominant vegetation; it changes the complement of species, and it causes a loss of diversity. The new replacement systems are simpler, less mature, and less diverse.

We have achieved great horizontal growth, much like a gigantic fungus. However, if we want to be like a smarter fungus, slime molds for instance, we need to learn to cooperate to grow up and be more dense and vertical. The larger metropolitan regions are covering wild ecosystems and agricultural fields with single-family houses, malls, building, recreational areas, and roads—all of which are car-centric or auto-morphic. This means that energy and goods are also spread thin. Such systems of things are hard to control, hard to keep safe, and hard to remain interesting. In fact, it might be worthwhile to compare human systems to mature ecosystems; we are creating pioneer individuals that do not live well in concentrations. We are creating edge individuals and not those who can live in interiors and share resources, or can develop new resources with cleverness and intelligence. City designs do exist, however, which incorporate the properties of mature systems and ecological thinking.

An arcology, as defined by Paolo Soleri, is a city which embodies the fusion of architecture with ecology. The arcology concept proposes a highly integrated and compact three-dimensional urban form that enables radical conservation of land, energy and resources. Arcology eliminates the automobile from within the city, and with it, the fifty percent of land devoted to automotive needs. The multi-use nature of arcology design would put living, working and public spaces within easy reach of each other and walking, supplemented by elevators and airport things, would become the main form of transportation within the city. An arcology would use passive solar architectural techniques such as the apse effect, greenhouse architecture and garment architecture to reduce the energy usage of the city, especially in terms of heating, lighting and cooling.

The small footprint of an arcology, combined with many built-in gardens, would allow rural space and agricultural fields to be closer to the city, and a part of the immediate urban environment. Wilderness, also, would be much closer to population centers in arcologies. Psychologically, the intelligent design would be more conducive to inspired living, the kind found in traditional culturally-significant cities at certain times. The proximity of agriculture and wilderness would allow people to participate more in them, with the full range of benefits that comes from growing and cultivating plants, as well as being able to immerse in the otherness of wild ecosystems.

The sizes of arcologies range from 250,000 to almost a million people, although smaller or larger ones are possible. For the sake of argument, assume that the average arcology is the size of Soleri's proposed Novanoah, at 400,000 people. At that size, it would take 15,734 arcologies to house the planetary population; this number is less than the number of cities in the United States in 2004, at 19,354. Assuming that the area under the arcology is about 5 square kilometers (almost two square miles), the surface area taken up by arcologies would only be 78,768 square kilometers, which is only 0.00054 percent of the land area of the planet—that is half of one thousandth of one percent—(149.45 billion square kilometers, or roughly 0.0167 percent of the land area currently under concrete and asphalt now, which is 4.71 million square kilometers or 3.15 percent of the land area of the planet).

The number of roads would be significantly reduced, from 676,750 square kilometers (roughly half of one percent of the total land surface in every country, according to the International Road Transport Union) to 230,000 km² (a generous number to be sure, but it could be as little as 50,000 km²). Over 2 percent of the land is under road surfaces in the United States, with a smaller percentage in Europe, by comparison. Land under agricultural production would also be reduced, from 33 billion square kilometers (22 percent of the land area of the planet), partly due to improved practices, partly due to the integration of many kinds of agriculture into the city, and partly due to the carefully limited use of wild populations, without domestication or containment.

The shapes of arcologies would be as diverse as any. Many of Soleri's shapes are geometric. They could be large pyramids, filled with living and working spaces, connected by transits and illuminated by light wells. The traditional ziggurat modernized would offer a good ratio of sunlight and truck gardens to size. Arcologies could be empty tubular pyramids with modular dymaxion attachments that could be moved between arcologies or new sites. They could fit the shape of the landscape, as does the Palouse Arcology (Wittbecker 1992). Arcologies could be built around small mountains, in bridges crossing canyons, or threading through coastal seas. What would such a change mean to most people? Probably, few people would have the need for private cars. A typical day, for most workers, whether administrating, grading papers, policing, or making steel, would start with a walk to work, past local stores and businesses, playgrounds, and microfactories. Work would involve fewer layers of hierarchy; the pay range would only be from 1 to 7 times the minimum salary.

Since Soleri's heroic designs, arcologies have been confined to computer games and have become elements in science fiction and cyberpunk films. With more prototypes being designed, one may be built in the next twenty to thirty years. A proposed project for Tokyo Bay, the Shimizu TRY 2004 Mega-City Pyramid, if constructed, would become the largest artificial structure on the planet; it would be 2004 meters tall and house 750,000 people. The external structure of the pyramid will be an open network of megatrusses, supporting struts made from carbon nanotubes to allow the pyramid to stand against high winds, earthquakes, and tsunamis. The trusses will be coated with photovoltaic film to convert sunlight into electricity and help power the city. The building will be zoned into residential, commercial and leisure areas. Separate buildings for housing and offices would be suspended from the supporting structure with nanotube cables. Transportation would be provided by accelerating walkways, inclined elevators, and a Personal Rapid Transit system inside the trusses with individual driverless pods.

It seems that arcologies would be part of a whole package of changes, brought about by ecological planning on a global scale. The experiment would not require that arcologies replace archaic populations living in human-modified ecosystems, or even all low-density habitations or traditional cities. But, they could be new cities situated in infertile areas.

Many cultures could live in optimum configurations in their territories, as part of wild and domestic landscapes. But, we also need heroic architecture. Heroic design and extravagance in life is needed in general. It is not contradictory or antithetical to frugal lifestyles or to restoring a healthy environment. Life is exuberant; energy is used, lives are lived and used, not wasted or saved. Life is the accumulation of individual experiences that cannot be saved, stored, or owned. The heroic things in life are often those most admired or remembered by subsequent generations.

1.5.3. *Conclusions: And Then What?*

We could think of more large-scale experiments. What might the planet look like by 2060? What would it look like at a new hot stable state 5-10 degrees C warmer? Stable does not mean being without violent storms, just regular ones. There would be no ice, so the remaining few cold species would be kept in zoos. What might civilization look like in 2060? Would it be concentrated in fewer cities, with urban agriculture? Would it be recognizable?

Thought experiments can give us clues about what can happen ("And then what?" as Garrett Hardin always asks) and what is the likelihood of it happening. Unlike medical doctors or scientists, we cannot either wait or directly experiment within a realistic time frame or scale. We cannot experiment at all in a traditional sense, where we hold most variables fixed, while changing one or two in experimental runs. Ecosystems operate over very long time spans; furthermore, their historical nature means that they cannot be restarted.

Large-scale, long-term experiments are expensive and relatively few. Most experiments are short-range, small-scale, isolated, and detail dense. They do not present the hypotheses required for the management of ecosystems. Ecosystem management, because of uncertainties, lack of controls, age, and uniqueness, is an uncontrolled, large-scale experiment. Thought experiments can refine the design of our larger experiments by suggesting better hypotheses.

Thought experiments can help us avoid being overwhelmed by details. Thought experiments can help formulate goals and interpret information appropriate to scale. The idea of science is to manage our experiences with generalities. Once the thought experiments are started they can be refined with conceptual or mathematical models, which can simulate the changes and evolution of changes. Computer-based models can permit complex explorations, as well as suggest new patterns and further hypotheses. Through thought experiments and models, many of the dangers and expenses of our activities can be avoided.

Thought experiments are vital to understanding the complexity of ecosystems and global cycles. In practice, erring on the side of preservation—the prudent and conservative course—means minimizing the influence of human activities on the land. It means experimenting cautiously with new approaches to forestry and being properly skeptical about claims for sustainability. It means drastically reducing our demand for natural products, through conservation, reuse, recycling, and human population control, so that the greatest number of ecosystems can be left wild and degraded lands have time to be restored to health.

Thought experiments can also be used to examine possible scenarios of the future based on our actions. For example, if we continue the current trend of inequity, how might things play out? For example if the rich keep getting richer, how will they have to protect their wealth from the poor? Will laws be enough? Will they need ever-larger armies of security personnel? With already four times the number of civic police, will they need even more?

Will corporate police protect the wealth of their stockholders after the civic police give up? Will the poor collapse leaving rich enclaves that have to grow their own food? At what point will the gap be wide enough that the poor have to harvest the wealth of the rich by force? At what number of poor? Will the poor prey on each other first? At what point will the environment be used entirely for a few more years of life for rich or poor humans? At what point might the environment collapse?

Will Local groups, such as the northern hemisphere, form alliances to keep going, after writing off other regions or the southern hemisphere? Will this block be able to defend its resources? Will that extend the time of any collapse? Or accelerate it? Will the United Nations be able to coordinate some kind of peaceful reorganization? Can a revitalized UN guarantee a rational economic and political strategy for all nations? Should a China or a United States dominate this UN, so that it may operate without as much discussion? Is it utopian to think of such reorganization or redistribution for equity? Is this less naïve than allowing the market to sort out entire cultures and regions and consign them to poverty and violence?

Thought experiments will be suggested throughout this work. Some of the chapters on wilderness, population and design are, in fact, thought experiments. Most of the suggestions for designs are thought experiments. The best response to a question about what would happen as a result of some actions under some circumstances may be a thought experiment. Through that, you can create explanations and discover answers in a dialogue with others.

1.6. *Using Analysis Synthesis Systems & Other Tools*

This section suggests how analysis can be used appropriately to inform early stages of design. Basic analysis, in a scientific context, means taking something apart to understand how it works. The word is based on the Greek words meaning to 'loosen up.' Once a subject is identified to analyze, the subject is deconstructed, literally. To analyze a rock, it is necessary to break it apart and identify specific elements. To physically analyze a human being, you must cut into the flesh, weigh the blood and organs, and determine the different kinds of organs, fluids, and moving parts.

However, that is rarely enough to tell how a human operates. For that, you must undertake a functional analysis. You would have to observe a working specimen first: Make a list of characteristics, hair, eye, and skin color and shape, weight, shape, completeness; list inputs, such as air, water, and food; list outputs, such as carbon dioxide or manure; make another list of needs, from shelter to place, security, and social demands; describe the kind of places humans are found, from gardens to factories, and their relationships with other beings.

A lot of time would have to be spent analyzing human communities, at a family and neighborhood level, as well as working relationships and the larger relationships with the environment. Working relationships could easily be expanded into economic analysis. Conversations could be analyzed with linguistic analysis that would have to delve into the history of language and cultural shifts. A historical analysis of a person could trace the movements of ancestors and their interactions with others and their environments.

A philosophical analysis might be useful to try to understand an individual expression of existence and place. Some philosophies, perhaps all, start with the notion of what it is to be human, even those that start with existence and being start with human existence and human being. But, why not start with what it means to be or to live? A more comprehensive declaration might be: 'I live here and therefore I am in place and an part of it and know it.'

The fundamental characteristic of analysis is taking apart. Taking apart living beings or working cycles can result in their death. However, more general kinds of analysis rely on good observations to allow deductions about internal workings. Goethe rejected destructive analysis; his approach was passive attentiveness. Certainly many forms of analysis can approach this level of sensitivity.

1.6.1. *Linguistic & Historical Analysis*

In order to make sense of the sheer multiplicity and complexity of their environments, human beings create abstractions. An abstraction is an idea created to refer to all objects that have certain characteristics in common, e.g., all birds. Abstractions can be generalized, e.g., all things that fly, but at each outer level, the objects have less in common; thus flying things include insects, mammals, reptiles, seeds, and spores.

Human beings also classify and label their abstractions. The systems of classification are reflexive and pragmatic; that is, they refer to the classifier as well as to the object, and they are guidelines for how to think about, treat, and relate to an object, according to S. I. Hayakawa. As soon as a classification is no longer useful, people stop using it and become receptive to a better classification. This gives a historical and linguistic caste to analysis, which can be used to differentiate between abstractions and classes.

1.6.2. *Systems Analysis*

To understand the workings of complex systems, systems thinking is useful, with its concepts of feedback and emergence. Complex adaptive systems display emergent behavior. As an example of emergence, slime molds can form a community without a pacemaker cell that determines when the cells need to combine. It seems that self-organization is bottoms-up. Emergent systems are rule-governed, though; slime molds explore by adhering to low level rules. Individuals coordinate work, even if they cannot assess the global situation. Emergent systems are local; individual molds "think" locally and act locally. Random action serves to explore local space. Individuals pay attention to their neighbors, and patterns emerge from local activity. Simple behavior seems to work, with local feedback, and more sophisticated behavior "trickles up" to approximate a global perception. Mathematical analysis measures those behaviors with standards and converts them to numbers in an abstract space.

A system is a set of things—such as people, cells, or molecules—according to Donella Meadows, interconnected in such a way that they produce their own pattern of behavior over time. The system may be buffeted, constricted, triggered, or driven by outside forces, but the response is characteristic of the system. A system is not just a collection. It is inter-connected set of elements that is coherently organized to achieve something. Systems can be distinguished into three parts: Elements, connections, and function (purpose). Functions are harder to see than connections, and connections are harder to see than elements. A system is a way to explain part of the universe and to deal with complex behavior. Mario Bunge defines a system as a complex object, every part of which is connected with other parts of the object in such a way that the whole possesses emergent properties that the parts lack. The parts of the unit keep its structure and function stable, despite changes and disturbances in its surroundings. Intrasystem bonds are stronger than intersystem bonds. If they were not, the system would fall apart.

Mario Bunge makes a strict dichotomy between formal and concrete systems. A concrete system is composed of concrete things linked together by real physical, chemical, and biological ties. Cells, wolf packs, and nongovernmental organizations (NGOs) are concrete systems. Many concrete systems are open and self-regulated. Many are closed and artificial. Every concrete thing has properties. Some properties can be known easily; others can be revealed by research. A list of known properties describes the state of the system with finite quantities. A qualitative description of properties describes qualities of the system. Quantitative descriptions measure quantitative variables. The state of a system, definite and objective, can be conceptualized with theories and models. All concrete systems change as reality unfolds. That is, the properties of systems change through time. Some change, such as growth or decay, is quantitative. Other change is qualitative, resulting in breakdown or formation of the entire system. Conceptual systems can be linked by logical relations.

Regularities in systems are patterns. Patterns can be seen in things or even cultures. For instance, laws of genetics are natural patterns; human customs are artificial patterns. Where the natural ratios of females to males are altered by female infanticide or other human action, the pattern is semi-natural.

Bunge distinguishes four kinds of real patterns: Laws, trends, correlations, and rules. A law is a stable pattern inherent in things; it is discovered; laws like gravity are boundless; biological laws are bounded. After the "Big Bang," gravity was bounded. There are examples of social or cultural laws: "The inertia of a social system is directly proportional to the number of components and inversely proportional to its cohesiveness;" or: "Higher culture does not emerge in society until the basic needs of some of its members have been satisfied."

A trend is a temporary pattern, such as the globalization of capital or fertility. Trends can be reversed. A correlation, usually statistical, is a covariation of two properties, e.g., a correlation of sickness and education—but this correlation is problematic, due to fact that educated are wealthier and report their sickness more than the poor; the real correlation could be reversed. A better correlation is: Single-species forest stands can be correlated with standing armies in the northern hemisphere; standing armies were thought to be characteristic of people in tougher climates, which also encourage large stands of trees. A rule (or norm) is a social convention set up by people, in force in a social system. The analysis of patterns is the strength of systems analysis. The systems lens allows us to ask 'what-if' questions.

1.6.2.1. Physical & Ecological Systems

Physical systems can be composed of atoms, molecules or organisms. All concrete systems change as reality unfolds, that is, the properties of systems change through time. Some change, such as growth or decay, is quantitative. Other change is qualitative, resulting in the breakdown or formation of the entire system.

In biological systems such as organisms, ecosystems, or the biosphere, most parameters must stay under control within a narrow range around a certain optimal level under certain environmental conditions. The deviation of the optimal value of the controlled parameter can result from the changes in internal and external environments. A change of some of the environmental conditions may also require change of that range to change for the system to function. The value of the parameter to maintain is recorded by a reception system and conveyed to a regulation module by an information channel. Biological systems contain many types of regulatory circuits, both positive and negative. As in other contexts, positive and negative do not imply that consequences of the feedback are good or bad. A negative feedback loop is one that tends to slow down a process, while the positive tends to accelerate it.

Every real thing possesses properties, which can be distinguished in several ways, for instance as intrinsic or relational. An intrinsic property is one that a thing possesses regardless of relationships with other things; a population of wolves is an intrinsic property (although it can be affected by other things). The emigration of wolves, young males, is a relational property, that is, it occurs due to the relation to other things.

Table 1621-1. Systems (Concrete Living)

Properties	Local	Global	Differences
Whole	Subwhole	Whole	Level
Open	Self-making	Self-regulating	Bottom up
Feedback	interconnect		Scale (for albedo etc.)
Complexity	interdependence		Maintain
Emergence	+	+	Required (vs. ubiquitous)
Adaptive	To local		
Hierarchical	Simple focus		Frame top
Boundary	Bounds	Bounds	Kind
Lifetime	Ephemeral successional	Maintain	Length of time
Nonlinear	Motions	Constrained	

An ecological system is a material thing. Unlike an animal, which is a tangible thing, an ecosystem is an imperceptible thing, like an electron, which is also material and real. Real things are changeable objects. Of course, as human beings, we can create conceptual objects, such as theories, which do not change (unless we think about changing them). Other conceptual things are written words, which can stand for ideas. Material processes can produce conceptual objects, which are embedded in the processes, which are embedded in webs of processes.

Processes encounter each other in a functioning web of an ecosystem (with tangible and diffuse surfaces). Lynn Margulis qualifies her definition of an ecosystem: The smallest unit capable of recycling the elements of its membership. For example, organic carbon can be expired, fixed, reacted, or transformed. This is done through the physiological activities of the members of the system, through breathing, enzymes, or some other way. Margulis states that elements recycle faster within ecosystems than between them.

Global (or emergent) properties are possessed by wholes, regardless of components, as in "the territory of the wolf pack is eighty thousand hectares." This property emerges with the system and disappears if the system breaks down. It is an emergent property if no component of the system has it individually. History, structure and stability are emergent properties of an ecosystem. Role and scarcity are not. An ecosystem has emergent properties that are different from the sum of community interactions. They also affect biogeochemical cycles.

1.6.2.2. Cultural Systems

Any large system, such as an ecosystem or culture, is a high-order, multiple-loop, nonlinear, feedback system. In such a system, feedback loops are the basic structural elements. Each loop is a circular path of interaction between several elements. Like many systems, cultural systems need energy for their maintenance. Social systems have all sizes and degrees of complexity. Governments are more complex than families. Systems can also be constituents of larger systems, like nations are part of the United Nations (UN).

People and their food are kept in dynamic balance by feedback loops, where the system is maintained by information. Information is fed back into system and causes change, sensing and reacting to environmental conditions. A positive change in a system leading to one direction can cause the system to change in the same direction, which can disrupt stability. For example, population growth can be positive, especially with algae and eutrophic lakes, but, also with human populations. More food is needed; then, more people are needed to grow the food. And, possibly, the economy needs more people to eat the increase in food. The population birth rate is positive, although the death rate is negative; other factors can be positive or negative including cultural things related to marriage.

Like humans, mammals are bound by biological requirements that must be met if a population is to survive. These functional requirements are rather minimal for humans, however, being only food, clothing, shelter, and reproduction. Other requirements, such as respect, comfort, and self-fulfillment, depend on cultural systems. Culture is a codification of reality that increases human fitness in an uncertain environment, through expressions, which can be preserved and transmitted through generations through language. Different languages program events differently, therefore no culture or belief system can be considered entirely apart from language, or language entirely apart from place. A cultural system surrounds the network of human interactions with raw materials, forms of life and other humans.

The systems approach to culture allows us to consider specific dimensions of culture, from kinship to intellectual, economic and political culture. Cultural systems have all sizes and degrees of complexity. Governments are more complex than families. Systems can also be constituents of systems, like the United Nations (UN). Culture evolves. Cultural evolution depends on two things: How fast useful changes arise and how fast they spread.

1.6.2.3. Systems Models

The word 'systems' has been used as a model of a field. As models, systems are useful; a system is a way to explain part of the universe. A basic definition of a system is a complex unit (in space-time or the STEM field) whose components keep its structure and function stable, despite changes and disturbances in its surroundings. Systems can deal with complex behavior. There are two theories of systems. General systems theory is a holistic description of the hierarchical order of nature as a complex of open systems of increasing complexity. J. C. Smuts described holism as the "whole is more than the sum of the parts." Ludwig von Bertalanffy, Arthur Koestler, Ervin Laszlo developed the foundations of systems philosophy and applied it to numerous spheres of inquiry. Cybernetics is a theory of self-regulation of those systems through deviation counteracting (negative) and deviation-amplifying (positive) feedback cycles. Norbert Wiener generalized cybernetics, feedback, and information beyond computer technology. Shannon and Weaver's information theory and John von Neumann's game theory are also fundamental contributions. Systems can be concrete or artificial, according to Irvin Laszlo. Systems can be described as physical, biological, ecological, and cultural.

Systems theory analyzes events, processes, and patterns in the world. An event is a change of state. An event is described by two points, an initial and final state. A process is a sequence of states, also called a history. A process creates a path, described by a trajectory of states. A process is described as evolutionary if it involves emergence and the creation of new things, as in general speciation; to be a species, however, the novelty has to reproduce, multiply or diffuse.

The concept of state precedes the concept of process, although a process can define a state. In complex systems the past contributes to or constrains the state, from the magnetic hysteresis of ferromagnets to organisms, which have memory. Changes in systems can be recorded, but only partially and not continuously. The discontinuous recording is digital. The process is continuous, but the measurement is discontinuous.

1.6.2.4. Several Properties of Systems

Systems have properties, in fact, much the same as fields: Process (speed, flow), Autopoesis (self-making, wholeness participation form shape boundary), Differentiation, Integration,

Constancy, and Development. Systems can also be analyzed in terms of operation: Motion (turning), using the special terms of systems theory, such as feedback, connectivity (coupling or linking), and triggers (energy amplification). In fact, systems can serve as models of the field, ecological groupings (ecosystems), and human complexes. Some of these properties are critical when trying to design large ecological or cultural systems, especially limits, feedback and connectivity.

1.6.2.4.1. Limits

What is a Limit? The word limit comes from the Latin word for boundary or frontier. It means at least three things: Boundary, utmost extent, either the largest or the smallest, and restriction. Thus, a limit is a point or line where something must end, or a boundary beyond which something ceases to be or to be possible. A limit may be unknown or invisible.

Motion results in limitation. Limitation is the principle by which the many can come to existence out of the one, the unlimited of Anaximander. All forms—universes—are possible, and any particular one is mutable; but the laws relating all forms are the same in any universe. Is this sameness independent? This sameness leads to mathematics, a conceptual reality independent of how the universe appears. Limit is the essence of forms and patterns, which are defined by their limits. As Pythagoras is thought to have said: "Limit gives form to the formless." A limit is defined as a boundary or the utmost extent, smallest or largest.

Mathematically, the concept of a limit is fairly simple: It is something that is approached, but not reached, as some value descends to zero. In specific applications the limit is usually a number, that is, the rate of photosynthesis or the carrying capacity. Every finite system has limits. Systems have a limit in size, which may be determined by structure or the speed of turnover of components. Systems may have limits in terms of the number of connections that can be maintained with other systems. Any physical system, with multiple inputs and outputs, is surrounded by layers of limits. There are always limits to growth. A limit can be a threshold to a new system that may be less stable. If not system imposed, limits may be self-imposed or cultural imposed.

Spencer Brown states that a universe comes into being when a space is severed or taken apart. The skin of an organism cuts off the inside from the outside, as does a circle from a plane. The act of severance is remembered as our first attempt to distinguish different things in a world where boundaries can be drawn anywhere we please. We define order by making boundaries based on the perception of particles or of wave size. The universe cannot be distinguished from how we act on it, according to Brown.

Systems rarely have real definite linear boundaries. Connections ruin the neatness. The lesson of boundaries is hard for systems thinkers. There is no single legitimate boundary to draw around a system. We artificially create boundaries to simplify our problems and insulate our sanity. We need to be flexible to find appropriate boundaries to think about new problems and make models for understanding new problems.

1.6.2.4.2. Feedback

Feedback is part of the science of cybernetics. When Norbert Wiener, Arturo Rosenblueth, and Julian Bigelow needed a name for their new discipline, they adapted a Greek word cybernetics, meaning 'the art of steering' to evoke the rich interaction of goals, predictions, actions, feedback, and response in systems of all kinds (the term 'governor' derives from the same root, after N. Wiener 1948). Early applications in the control of physical systems, such as aiming artillery, designing electrical circuits, and maneuvering simple robots, clarified the fundamental roles of these concepts in engineering.

A control system usually has input and output to the system; when the output of the system is fed back into the system as part of its input, it is called the 'feedback.' In cybernetics and control theory, feedback is a process whereby some proportion of the output signal of a system is passed (fed back) to the input. This is often used to control the dynamic behavior of the system.

There are two kinds of feedback, positive and negative. With positive feedback a system responds to the perturbation in the same direction as the perturbation (or deviation-amplifying, or error-amplifying, or cumulative causation, or destabilizing, or centrifugal). In contrast, a system that responds to the perturbation in the opposite direction is called a negative feedback system (or Deviation-reducing, or stabilizing, or centripetal). The term 'positive' means responding to the same direction as the perturbation whereas 'negative' means responding to the opposite direction. A system in which there is positive feedback to any change in its current state is said to be in an unstable equilibrium, whereas one with negative feedback is said to be in a stable equilibrium.

The end result of a positive feedback is often amplifying and 'explosive,' that is, a small perturbation will result in big changes. The feedback is deviation-amplifying (sometimes referred to as error-amplifying). This feedback, in turn, will drive the system even further away from its own original set point, thus amplifying the original perturbation signal, and eventually become explosive because the amplification often grows exponentially (with the first order positive feedback), or even hyperbolically (with the second order positive feedback). Indeed, chemical and nuclear fission based explosives offer an excellent physical demonstration of positive feedback. Bombarding fissile material with neutrons causes it to emit even more neutrons, which in turn affect the material. The greater the mass of fissile material, the larger the amplification, resulting in greater feedback. If the amplification is great enough, the process accelerates until the fissile material is spent or dispersed by the resulting explosion.

Feedback is usually bipolar—that is, positive and negative—in natural environments, which, in their diversity, furnish synergic and antagonistic responses to the output of any system. Bipolar feedback is present in many natural and human systems. Feedback is information directed back into system that causes a change. If there is too much positive feedback, then the system can overshoot an equilibrium. The concept of overshoot allows human consumption to increase at the same time that ecological capacity is shrinking. There is no contradiction, just a form of madness. Both are happening currently.

There is no direct, immediate feedback to counter or correct overshoot apparently. The wastes become global, that is CO_2 can leave auto exhausts despite the concentration in the atmosphere. There is feedback, but it is delayed for a long time by the size, flexibility and redundancy of the system. Meadows notes that a person who makes a decision based on feedback cannot change the behavior of the system that drove that feedback. Decisions only affect future behavior. There will always be delays, since nothing can react instantaneously.

1.6.2.4.3. Connection

A system is defined as a complex object, every part of which is connected with other parts of the object in such a way that the whole possesses emergent properties that the parts lack. Systems are considered to be wholes where the internal connections are stronger than the external connections. There is no whole system without an interconnection of its parts, and there is no whole system without an environment. Everything is connected in a system; removal of any part alters the dynamics of the system. Connectivity is related to the size and density of a system. To connect means to bind together, from the Latin words, or to link or couple. A connection is the relationship or association. A complete connection is regarded as a circuit, which becomes required for certain patterns and their continuation. A cycle is movement through a circuit.

The organization of an ecosystem is described by diversity and connectivity, among other characteristics. Every local system is connected to others some degree, but usually the degrees are very weak, through gravity, global chemical cycles, or just through patterns. Too little connection between parts and the system may fall apart. Too strong a connection between parts, and the system may not be able to renew itself through processes like metalysis, which requires that connections be broken and reformed. Too little connectivity and species are too independent; they can die (no food or prey). Too much connectivity and each species has to compete with all species; there is little flexibility to change. So it seems that connectivity must have regions of operation. Overconnection can be compared to power grid connections and failures. If overconnected, all can fail; if underconnected, many local areas fail; at a mid-range there can be power transfers.

In an ecosystem, individuals and species are connected in terms of quality and quantity of connections. The connectivity of a component of a system is a measure of the number of direct connections between it and the rest of the system. Connectance is a percentage of the number of connections through predation or exploitation as a percentage of the total number of possible interconnections. A system is more connected if the absolute number increases and the percentage and strength increases.

In an ecosystem, functional connectivity is measured by the potential for movement, dispersal and interchange between populations of species, especially those subject to fragmentation. Connectivity in an ecosystem is a function of the mobility of a species, that is, its dispersal characteristics, its autecological characteristics, such as food or shelter requirements, the structural characteristics of the landscape, such as spatial patterns, the distance between patches, the presence of barriers to movement, such as highways or rivers, predation patterns, and disturbance patterns, including human interference. An ecosystem also has to be connected to global cycles, such as phosphorus. To preserve connectivity under uncertain conditions, it is necessary to maintain large areas of natural habitat in a larger matrix, as well as maintain natural connections and to minimize artificial connections, such as roads.

By being overconnected a system can lose stability. Overconnectedness creates a lag in signals. Agricultural fields are good examples of overconnectedness. This is even more true with corn than with wheat due to the changes on the species from domestication. Overconnection causes greater lag times. But, in a city, feed bins damp food variations so the lag is less important. Underconnection can also create instability. Parts go their own way and there is no communication between them. If a system is underconnected, it may become unstable or lose resilience, because there is no pathway to allow a signal between significant parts.

The connectedness of the subsystems or systems is necessary to understand some concepts, such as diversity. Therefore, it cannot be related well to stability. In addition, for each direct species to species relationship, additional bundles or sequences of indirect interconnected relationships spiral outwards logarithmically into the ecosystem and the biosphere as a whole. Indigenous peoples have names for many of these relationships, usually in verb form, such as pollinating, germinating, and shading. They understand and respect the importance and intelligence of species connectedness.

At a critical point of connection, there is a phase change from a disconnected phase to a connected phase. This allows nodes to interact or communicate. That is, one location has the potential to affect another. Otherwise local events would only be felt locally. Global connectivity arrives in a sudden jump. Experience is local, in a neighborhood. But, there may be global connections. Disconnected networks can prevent global cascades. What is the trade-off between local stability and global connectivity? A global connection can be overwhelmed if it replaces local to local connections.

1.6.2.4.4. Triggers

A trigger is a mechanism that activates an immediate change in scale or a release of information or energy. A trigger can have connections to a global system. A trigger often creates positive feedback. A signal is a sign that may be recognized or transmitted, often to initiate some action, such as communication. A triggering signal can cause dramatic amplification (See Section 1.7.3.2.6 for a further discussion).

1.6.3. *Ecological Analysis & Synthesis*

Traditional science attempts to control a situation by thought or experiment. An experiment limits degrees of freedom in a formal manner; it screens out quality. The gain from an experiment is an 'if/then' structure, resulting in mathematically specific functions. Portions of the universe are placed behind glass in a laboratory world. As science cuts the connection to direct observation, it becomes blind to outer world. The formality of science makes statements about the outer world tautological. This is a problem with quanta, species fitness and psychological needs. Colloquially, trying to control ecosystems is like trying to control your spouse in marriage or your machines in a factory; they will not be controlled well or completely.

Modern science does not capture the state of being that experiences symbols. It does not value rare experiences of direct observation or even experiences that yield to clear description. Symbolic understanding could improve the quality of human observation. The modern biological view of man is the result of man's loss of a symbolic understanding of nature, according to J. Needleman. The symbols are only understood in a different state of consciousness. Early teachings are often judged as picturesque or insane. John Ruskin regarded the false perception of smiling fields and somber mountains as a pathetic fallacy. Emotional responses should be tied to the plain facts of nature; but Goethe knew that there were no plain facts. Now our perception is crippled by an apathetic fallacy—we see nature as dead. The pathetic fallacy is a fallacy only to apathetic, who are victims of the apathetic fallacy.

Unfortunately, the language from a mechanical world view dominates even ecologists and politicians. This world view impoverishes humans by claiming all consciousness for humanity. It claims that nature offers no joy, or love, or peace, or certitude. Emphasis on the evil of nature creates a gap between humans and their universe. In contemporary cosmology, there is no room for the intrinsic worth of nature. But science undermines the scientific cosmology and provides the elements of its alternative: wholeness and relatedness. The scientific cosmology is deficient; it is limited to a certain category of facts. But modern science is undermining the scientific world view. Quantum mechanics has altered the picture of the universe, making it mysterious again.

The scientific method is a formal extension to traditional ecological knowledge. The modern scientific method of knowledge can be differentiated (in italics) from the traditional (in roman type) in a series of modified statements.

Many scientists, such as M. Polanyi, A. Maslow and A. Koestler, have questioned the adequacy of scientific method and theory, with the intention of eliminating its reductionism and broadening its sensibilities. Science can enlarge its capacity for corrective self-awareness by becoming binocular.

Table 163-1. Methods of Knowledge

1. Observe phenomenon and record facts.
2. Analyze the phenomenon into components, *using scientific ideas. Create standards.*
3. Experiment with the phenomenon and facts to see what happens. Experiment, in a controlled way. Logically isolate the phenomenon. *Measure the components of phenomenon, before and after manipulation, allowing the results to be quantified.*
4. Make guesses and generalizations about future conditions, using the logic of deduction. *Frame a hypothesis, a statement about relationships that can be shown to be untrue.*
5. Formulate rules from the generalizations about the phenomenon. Such rules describe the behavior of the world. *Formulate laws from the generalizations about the phenomenon. Such laws describe the behavior of a natural system.*
6. Describe the rules using words. Put the words in stories. Describe the laws verbally *and mathematically using numbers. Express in stories and equations.*
7. Develop ideas, metaphors, models and myths to understand and explain phenomenon. *Develop a theory to predict new phenomenon. Theories can lead to new conclusions and sometimes altered perspectives about phenomenon. A scientific theory is a statement that postulates an ordered relationship among natural phenomenon and explains some aspect of the world. It allows one to ask certain kinds of questions, some as specific hypotheses. A theory cannot be tested by hypotheses.*
8. Act accordingly.

Ecological analysis has to be one of the largest forms of analysis. It includes limits-analysis as well as footprint analysis. The ecological footprint method offers a way of calculating the ecological capacity, or stock of natural assets, in terms of area, similar to Eugene Odum's method in hectares per person or to ghost acreage calculations by William Catton. The ecological footprint method (Wackernagel and Rees, 1996) offers a way of calculating the ecological capacity, or stock of natural assets, in terms of area, in hectares per person (similar to Eugene Odum's method) and comparing demand for energy and resources with the regenerative capacity of the biosphere to provide such goods and services. The method has been successful and some nations, cities, and businesses are already more efficient as a result of considering their impacts.

Ecological Footprint analysis adds up the area of different land types, such as forests, fishing grounds, cropland, pasture, that are required to support humans' activities. This is done by normalizing the biological productivity of the different land types in a common unit, such as the global hectare or global acre in the United States. One global hectare (gha) is equal to one hectare of land with world average productivity for all land types. The supply of ecosystem service is called 'biocapacity;' the human demand for these services is the 'ecological footprint.' By comparing supply (biocapacity) with demand (footprint), Ecological Footprint analysis provides a metric for indicating unsustainability. Globally, as of 2002, there exist 1.8 gha per capita of biocapacity available to support human activities, and not setting aside any biocapacity for nonhuman species. In 2002, the most recent year for which data are available, the global human economy used 2.2 gha per capita, resulting in an ecological overshoot of about 20 percent. This means that it takes the Earth about 15 months to regenerate what humans use in 12 months. This condition of global ecological overshoot cannot continue indefinitely. A catastrophic ecological failure may occur if the demands for "natural capital" are not reduced to a level that the biosphere can provide on an annual basis.

Footprint analysis can be subdivided itself, into local analysis or ghost acreage analysis and trade analysis. The geographer Georg Borgstrom in 1961 named "ghost acreage" as the acreage that the average citizen occupies that it is out of sight. The footprint adds up all the real and ghost acreage that a country uses, as part of the footprint. That is, the imports are added and the exports are subtracted.

Although the footprint is a good metaphor, there seem to be many things left out. Footprints not only have an area, but also a depth, not just a simple size, but also a shape. Footprints can be immediately gone or persist for a long time. Wastes are far more critical than resources, as regards ecological cycles. Many uses are not included, so the efficiency calculation may not be accurate. As Rees and Wackernagel admit, the "bio-productive areas that are really needed to maintain today's usage of nature's benefits are most probably larger" than the ones in the calculation. Ecological analysis would force us to look at the obvious— generating nonmarketable use values occupies the center of every culture because it provides a satisfactory life to its members.

1.6.3.1. Binocular Science

Analytic science has reached its limits. Data and information developed by hard studies have undercut the paradigms that guided their investigation. The compartmentalization of scientific fields has exposed the complex connections of the subjects. Science does not need to be based on logical positivism and reductionism, though these have allowed great, although insensitive, changes. A. N. Whitehead thought that what had been missing during the formation of science was a sense of relatedness. Early science saw the world as mechanism; modern biology is seeing it as resembling an organism. Organismic trends can be seen in sciences, from relativity and gestalt psychology to ecology.

Reliance on physical explanation impoverishes the complexity of ecological reality. Morowitz's portrayal of each living thing as a dissipative structure is reductive and distorting. Ilya Prigogine defined a dissipative structure as one of two types of organization, whose order is governed by amplified fluctuations; his examples include walls and slime molds. However, Prigogine misuses the concepts of order and complexity by making them dependent on random events; furthermore, his concept of stability assumes a reversibility of biological time (the result of its basis in quantum mechanics). Dissipative structures are more applicable to pans of boiling water than to black bears. The reduction of ecological patterns to dissipative structures ignores observed behaviors like communication and intention.

In describing a general concept of nature, Prigogine emphasizes process over structure. He assumes that energy flow is more primary than matter, that energy is a more fundamental reality than discrete entities. Although an organism may be characterized as a `configuration of energy,' that is an artifact of the quantum perspective (and, perhaps, of the desire for an absolute reality). There are philosophers and ecologists, such as Ramon Margalef, who consider `information' to be more basic than energy or matter and more in-line with patterning. Thus, organisms are reduced to information in a cybernetic perspective. These perspectives are useful to an understanding of complex behavior, but sometimes ecologists and philosophers simply take over the vocabulary of a paradigmic trend, new physics or information theory, and apply it uncritically to the epistemology of the older paradigm. Thus, 'efficient cause' can disguise itself as a `genetic program,' according to Margalef.

Like quantum physics, in biology and ecology, the observer is within the theory in a very fundamental way. By the act of observing the observer influences the outcome of a phenomenon, as Wheeler (1987) says, taking part in the construction of physical reality. David Bohm suggests that the Heisenberg indeterminacy may be because the past history of a particle is not taken into account in predicting its behavior. Nature is unpredictable. Its fuzziness is due to indeterminate (to us) states.

There was a paradigm change in metaphor from machine to organic system that undermined atomism and the old animism alike in developmental biology. The modern notion of organicism can be traced to the foundation of Taoism. Things are what they are and act upon one another by virtue of their position within a system of patterns. The tao was the great pattern, a field of force in the physical and spiritual world. This organic conception was carried to Europe by the Jesuits in the eighteenth century and had a profound influence on the philosopher Leibnitz. Leibnitz influenced Morgan, Smuts, Whitehead, Needham, and Bertalanffy. For Bertalanffy organicism was necessary to accomplish three specific jobs in biology: appreciation of wholeness (regulation), organization (hierarchy and level laws), and dynamics (process, behavior of open systems). Certain psychologists (e.g., A. Maslow) and philosophers (e.g., M. Merleau-Ponty) have preferred qualitative description to quantitative analysis. The blending of scientific objectivity with the sensuous and intuitive capacities of the mind is called hierarchical integration by Maslow (1968). He suggested science should strive for comprehension above clarity.

Sciences could use synthetic as well as analytic approaches to their subjects. Synthetic branches would be concerned with providing coherent pictures of the realms of study, as exemplified by the application of General Systems Theory to agriculture and farm animal welfare (M. W. Fox, 1983). Ecology could become a unifying science, including the whole of human experience, and permitting science an ethical dimension.

1.6.3.2. Ecological Science & Perspective

Ecology is one of the oldest disciplines, rather than one of the youngest. Early in their development, human beings realized the value of recognizing edible plants and animals and their interrelationships, and they built up traditions of knowledge. This practical ecology has been obvious for a long time, but because it is subtle and complex, it is not easy to quantify, and its development as a science is recent. This practical ecology enabled some cultures to achieve long-term stability in a natural environment; it also embodied teleological and holistic concepts expressed in qualitative terms rather than in mathematical forms. As a young science, these practical concepts were formulated as various principles: wholeness, the relationship of complexity and stability, succession and climax states, and the balance of nature. As ecology became more quantitative, that is, mathematical and reductionistic, its methods and topics more reflected the old physics. The new ecology, according to C.H. Waddington, places emphasis on the discreteness of individual genes, the randomness and nonrelational nature of the process of mutation, and the unimportance of the experience and reaction of an organism to its environment. Old ecological principles were rejected. D. Simberloff, for instance, even argues that the ecosystem model of A.G. Tansley is only another way of formulating the 'balance of nature' and must be rejected. Similarly, holism is unacceptable to the new ecology and is replaced with an individualistic view.

The new ecology is a central dimension in biology and overlaps many specialties. It deals with different levels of a hierarchy, focusing on organisms, populations, communities, and ecosystems, but with attention to genetics as well as to geological and evolutionary events. Although ecology is a scientific newcomer, it is certainly a foundational science. Any science that studies those basic relationships is foundational.

If the old ecology were truly linked to the new physics, at least in the terms of field and wholeness, it would have a much different flavor than that presented by several scientists. For example, the field has a historical character; a whole is self-making. The implications of these attributes are neglected. Ecological principles are not the same as physical laws. Physical operations necessarily apply to biological systems, and ecological theory must be consistent with physical laws. But biological systems exhibit unique regularities that are not reducible to lower levels of activity or understanding. Although the ideas of ecology to some extent parallel the ideas of physics, ecology lacks the laws and constants of physics. Physics lacks the high-level predictability of ecology. The ideas of ecology and physics can benefit from cross-fertilization, but ecological ideas are not reducible to physical ones.

Table 1632-1. Sequences of Ecology

Name	Common Name	Characteristics
Old Ecology	Natural History	Historical Observation and traditional knowledge
Modern Ecology	Science	Physics, chemistry, energy budgets, discreteness, randomness
New Ecology	Quantitative and Mathematical Science	Quantum physics, information theory, systems, cybernetics, thermodynamics
New Old Ecology	Holistic Science	Field (historical), holism, autopoesis, construction, relational, nonreductive

The environment cannot be separated from what organisms are and what they do. Lewontin states (1983): "The organism and the environment are not actually separately determined. The environment is not a structure imposed on living beings from the outside but is in fact a creation of those beings." Ecology deals with the relationships of organisms to environments. It is not a reductive discipline, and not amenable to easy quantification. Ecology itself cannot be studied, only its components and relations; it is therefore a perspective, a way of seeing. It is a perspective of the human situation in its interconnection. Ecology includes all events.

An ecosystem is a complex system that interacts with four large global fields—atmosphere, lithosphere, hydrosphere, and biosphere—and their cycles. The properties and behavior of a complex system are determined by its internal organization as well as its relations with environment. There are two fundamental modes of behavior: (1) Maintenance, based on negative feedback loops and characterized by stability, and (2) Change, based on positive feedback loops and characterized by growth or decline. The two modes can create a typical series of behavioral patterns, from stagnation to rhythmic regulation.

Any large system, such as ecosystem or city, is a high-order, multiple-loop, nonlinear feedback system. In the system feedback loops are the basic structural elements. Each loop is a circular path of interaction between several elements. Ecological analysis forces us to look at the obvious—generating nonmarketable use values occupies the center of every culture because it provides a satisfactory life to its members. These values are represented in a unique image of the world for a culture.

The interconnectedness of the world means, according to Susan Oyama, that there is no intelligible distinction between inherited and acquired characteristics (thus making defense of Lamarck unnecessary). Oyama states: "What is required for evolutionary change is not genetically encoded as opposed to acquired traits, but functioning development systems: ecologically embedded genomes." C. S. Pierce suggested that Lamarckian evolution is evolution by force of habit.

Assumptions of a new paradigm (based on Ho and Fox 1988) are that form and variation are not arbitrary or random. Processes that generate form and variation at every level occur before natural selection is said to act; evolution can be understood in terms of this process, more than in terms of maximum fitness (exemplified in protobiotic evolution, molecular genetics). The emphasis is on integration, transcending disciplinary boundaries. In protobiotic evolution, physical and chemical processes are responsible for molecular selection and for the fitness of the environment for life. In the generation of organic forms, physical and chemical processes provide organizational principles that coordinate detailed biological mechanisms, including viscoelastic changes of the cytoskeleton and expression of different genes. The new paradigm will have new metaphors (Table 1632-2):

Table 1632-2. New Paradigms for the New Old Ecology

A process view (A. N. Whitehead 1920, M-W. Ho 1988)
Holism (J.C. Smuts 1926, A. Koestler 1958)
A field concept (C. H. Waddington 1962, S. Goodwin 1988)
Self-organization and constructionism (Francisco Varela 1982, S.W. Fox 1988)
Reciprocally constrained construction (Russell D. Gray, 1988)

In a process view, organisms are dynamic structures that are immanent and simultaneous with the process, rather than a consequence of natural selection of past random mutations. Like quantum physics, in ecology, the observer is within the theory in a very fundamental way. By the act of observing the observer influences the outcome of a phenomenon, taking part in the construction of physical reality.

In trying to synthesize the evolutionary theory of Darwin and relativistic physics of Einstein in a frame of holism, J. C. Smuts presented the whole as a powerful organizing principle inherent in nature. Koestler extended this to a system theoretical model of self-regulating, open hierarchical order (SOHO); he proposed the term holon to designate the Janus-faced entities on intermediate levels of any hierarchy, which can be described either as wholes or parts, depending on the frame of reference above or below. A holarchy of holons replaces the notion of a hierarchy of parts in a whole.

The field concept for development emphasized dynamic transformation (form as organized spatiotemporal domain), in contrast with the particulate concept of an organism. The field is understood in terms of group dynamics rather than selective advantage or cost.

Self-organization replaces the idea of natural selection. Any organisms or system is the history of the maintenance of its identity through continuous self-making, or autopoesis. The system develops through a continuous dance of autonomy and control; autonomy represents generation, internal definition, internal regulation, and self-assertion, whereas control represents consumption, instruction, assertion of other identity, and external definition.

Reciprocally-constrained construction replaces the concept of adaptation. The organism and environment are co-implicative, co-defining, and co-constructing. A process of self-assembly, where the self is the organism/environment system.

A holon is any stable subwhole in a hierarchy that displays rule-governed behavior and structural Gestalt constancy, to paraphrase Koestler. The rules lend order and stability, as well as flexibility, to a system. Wholes are mutually defining, but also self-making.

Nature is a self-making system; species and organisms are self-making. The ontology of any living system is the history of the maintenance of its identity through continuous self-making, or autopoesis. The evolutionary stability of the subassemblies—organs, organisms, species—is reflected by the degree of autonomy (self-government) each has, according to F. Varela. The system develops through a continuous dance of autonomy and control; autonomy represents generation, internal definition, internal regulation, and self-assertion, whereas control represents consumption, instruction, assertion of other identity, and external definition. Furthermore, the holistic nature of the STEM field eliminates the unsatisfactory notion of the priority of relationships to beings or of wholes to components.

Ecology deals with the relationships of organisms to environments. It is not a reductive discipline, and not readily amenable to quantification. Even scientific ecology is an integrative discipline that extends beyond the bounds of science. In a way, ecology is an amphibious discipline, with the authority of science and the force of moral knowledge. Ecology, studied through its components and relations, is a perspective, a way of "seeing," according to Paul Shepard. It is a perspective of the human situation held in a web of interconnections. For Paul Sears, ecology is a "subversive subject." Ecology is normative and sensible. Ecology also offers a "sacramental vision" of nature. Ecology is radical— from the Latin word meaning "rooted"—and forms part of a new metaphor that is more appropriate to the unity and interrelatedness of the earth. Ecology is part of a movement of consciousness, concerned with equality, diversity, health, with humane methods, and with a holopoetic cosmology, and ecology affects them simultaneously. Radical ecology offers a new perspective of humanity in the total field of nature and defines balanced relationships with ultrahuman beings and species. Radical ecology addresses the determination of separate wilderness areas necessary for a healthy ecosphere, and an optimum human population, based on net ecosystem productivities and modified by appropriate technologies within ecological and cultural restraints. It urges local, self-reliant cultures with adaptive cosmologies and natural values in wild ecosystems.

1.6.3.2.1. Patient Practice

There are ways of dealing with the earth that are not scientific or technological; they are aesthetic or ethical. They are not incompatible with a whole science. In his field theory of being, Heidegger stated that a concern with being implies a patient existence, a willingness to lie in wait for an image to produce itself (without imposing meaning).

Goethe's natural philosophy incorporates a world view described as organic dialectics; its method is described by contemplative nonintervention and the primacy of the qualitative. Organic Dialectics is a phenomenological science. Goethe was aware of the pervasive pattern of process, of transformation, unity in process. This unity in process was referred to as morphology by Goethe. Evolutionary theory is built on morphology. Goethe's conception of evolution merged selectivity with destiny and external pressure with internal thrust. The two methods are described as follows.

1.6.3.2.1.1. *Contemplative Nonintervention*: An observerless and valueless science may be contrasted with the Goethean ideal of contemplative nonintervention. Eddington had asked if advanced equipment does not tell us how nature can be made to behave. Goethe felt that the human being was the best apparatus (perhaps to a foolish extreme). Later Eddington raised the same issue, that experiment might only tell us how nature can be made to behave. And Heisenberg expressed the exasperation that scientific investigation showed us nature exposed to our methods of inquiry. Goethe rejected domineering analysis; his radically different approach was passive attentiveness. An observer tried to get into the flow of phenomena; by observing patiently and receptively, an astonishing insight could cap years of patient watching. This was the attitude of relaxed attention, a receptive state of creation. Knowledge came of itself, in quantum leaps, a gestalt was perceived.

1.6.3.2.1.2. *Primacy of the Qualitative*: An exact sensory imagination is needed to midwife the deep-down phenomenon. Qualities must be evaluated; a script of qualities to be read for meaning. Sensory data were qualities savored, not measurements taken. Starting with qualities, the "exact sensory imagination" could midwife the deep phenomenon.

Goethe attempted to use his method to produce an organic and morphological world-history. This methodology is part of nature. It reflects nature. Rilke wonders whether or not all the dynamics of nature, including those of human society, are hieroglyphics of the methodology of thinking. Early science saw the world as mechanism; modern biology is seeing it as resembling an organism; perhaps it will be seen as spirit or as a composite of all. Consciousness research chooses paradigms of nature that are mental, not mechanical (in the investigations of Tart, LeShan, Pribram, Krippner, White, Kamiya, and others).

Maslow, in "Towards a Humanistic Biology," saw the good specimen as chooser for the whole species. The organism may be seen as having biological wisdom; it can be trusted as autonomous, self-governing and self-choosing. To examine organisms, and nature in general, we must shift to a taoistic approach: Asking rather than telling; observing rather than manipulating; receptive and passive, not active and forceful; "nonintruding," (Ibid, p. 18.) and noncontrolling. It stresses noninterfering observation rather than controlling manipulation; it is receptive rather than forceful. This is part of the paradox of duality; it is detached yet concerned; free yet committed; and independent yet responsible. Classical objectivity may be contrasted with taoist, which is another path to objectivity with greater perception. Loving perception provides kinds of knowledge not available to nonlovers; this is especially true in ethological literature. Maslow cites his own work with monkeys. Lorenz, Tinbergen, Schaller, Van Lowick-Goodall, and Fox have found it to be true. This is the way a good psychotherapist, teacher, scientist, parent, or friend functions.

1.6.3.2.2. Soul Science

Science sometimes undercuts its neutral, objective approach, and allows conclusions to be reversed. For instance, one result of Elton's food chain was the realization that the bottom link—plants—is the most important part. The use of energy flows in ecosystems resembles descriptions of yogic meditations (compare the idea of enantiodromia). Nature is an extremely sensitive nexus of means and ends. Nature is a feeling system. We need a new animism to approach nature. This animism would allow us to behave "as if" nature was intelligent and sensitive.

What is necessary is not a primitive animism or a single-vision science, but a scientific animism, to replace scientific humanism and to understand our animalistic nature and use it as the foundation for a sound human ecology, philosophy and psychology. Anima is from the Latin word for soul. Animism could be an inquiry that would carefully and appreciatively consider the animal aspects of ourselves and how we understand and empathize with other living organisms. A scientific animism would consider the relations of humans to vegetation and the human attitudes toward ecotypes, like open plains and dense forests; it would consider the need for sacred places, and open, quiet or wild landscapes; and, it would consider territoriality, aggression, and the aesthetic reaction to the wonder and beauty of life.

A scientific animism would be concerned with far more than the anatomy and taxonomy of animals; it would be concerned with the mutual training between human and nonhuman animals (the emotive bond); and, it would be concerned with the need for touch and phylogenetic possibilities of animal empathy—dogs, for instance, exhibit strong physiological changes when they are petted and human blood pressure drops. It feels good to be touched.

A scientific animism also needs to understand the meaning of being human, to go below cultural or social explanations of love and alienation. It may, as a genuine science, forget the analysis and lose itself in the ecstasy of the phenomenon that it seeks to explain. Science functions best when we understand so well that we no longer need it.

The human mind shares nature's intent—producing experience—since the mind fits nature. The whole ecology is related to furthering the greater pattern of the universe itself.

The universe is a regenerative system, of which we are a part. Theodore Roszak asked if ecology could approach the sacramental vision of nature. He characterized ecology for its sensibility; it is holistic, receptive, trustful, aesthetic, and intuitive; furthermore, it is judgmental. This holistic attitude also flows through some professional ecologists, from Clements to Odum. Ecology must be capable of assimilating moral principle and visionary experience to be a science of the whole planet.

1.6.3.2.3. Being Other: Animism and Conservation Totemism

Deep ecology and its source, ecosophy, could make a basis for a new animism. The new animism would be expressed in specific societies for plants and animals. Totemism could be used for protection and conservation of nonhuman species.

Enlightened modern hunters often form associations, such as Ducks Unlimited, to protect animals and their habitats from the accidental threats of modern civilization. A good hunter learns to think like her prey and to recognize its signs and requirements. Hunting traditions have taboos against taking too many prey or vulnerable individuals.

In the past, hunters revered their prey. Reverence for life is an ancient attitude, recently revived as a principle by Albert Schweitzer, who distilled it from his examination of religions. Most philosophies are not adequate to deal with nature and ecological relationships; many religions are too narrow to consider nature as more than a dominion. Ethical thought had been developing since prehuman history, Schweitzer said, and it culminated in the principle of reverence for life. When humanity extends its concern to relationships with all life, and intelligence operates on the will to live, then the reverence for life arises. True reverence for life makes no distinction between higher and lower forms. If we were to act so, distinguishing between pests and pets, we must do so in the sorrow of the recognition that we are killing. True reverence for life entails reverence for death, since life and death are inseparable. No pattern can survive death, when death is the destruction of individual patterns. No one would mourn the content, which is even more evanescent. All life is sacred, but this can never be a reason for not killing, because that is how lives are sustained. Since life is of the utmost importance to the living, it should only be taken in sorrow, used and shaped with respect, and experienced with awe, for underneath it is still unfathomable mystery. Lao Tse and Confucius taught universal reverence and nonviolence as well. These ideas also have been stressed in modern times by Vivekananda, Tagore and Gandhi, and by St. Francis and St. Thomas, R. W. Emerson, J. Maritain, and Paul Tillich.

People have identified with animals for thousands of years, tracing their own ancestral lineages from them. In totemic religions, typical of foraging groups, each group is symbolized by a particular plant, animal, object or phenomena (the word 'totem' is from the Algonquin word meaning relative). Clans are descent groups from these chosen or fictional ancestors. The first totem often came in a vision; it was used to coordinate out-marriages between groups. A totem, as defined by J.G. Frazer, identifies an intimate and special relation between humans and every member of the class.

For instance, the Arunta tribe in Australia is divided into totemic bands. The "witchetty grubs" had about forty people living near Alice Springs. These bands owned territory, for example, the hare wallaby and carpet boa clans owned part of Ayers Rock. Traditionally, the bands went from one favored locality to the next, and at a central gathering place, had sacred ceremonies. All people, totems and natural phenomena are assigned to moieties. Moieties divided people into two groups within an estate group: Half of society might be black, kangaroo, acacia tree, and goanna lizard, and the other half, white, emu, gum tree, and rainbow serpent. Such divisions made it simpler for marriage and unions. Spouses would be drawn from opposite moieties. The rules of a totem also constituted good conservation practices: No hunting in sacred sites; prohibitions against hurting totemic animals; and, compassion to animals. Totem animals (or objects or phenomena) could be contacted in Dreamtime (every-when) when the invisible side of life was made visible.

In another instance, the Kwakwakawakw people in the Pacific Northwest of North America shared their ancestry with each other on a crest pole (sometimes called totem or princess poles). They believed that each person had a special totem or guardian spirit, who bestowed some of her powers. The totem could be an animal, plant, substance, or event, such as lightning or cloud. These poles were commissioned by the head of the family to display status and created by artists working within traditional styles of representation; The poles were cut and carved with axes, adzes and chisels, then colored with dyes from hemlock, cedar, alder, and other sources. Their completion was often celebrated by a potlatch ceremony. Animal ancestors included wolves, bears, whales, seals, sea slugs, barnacles, as well as mythical beings such as Thunderbird or Seamonster. At first, poles were kept exclusively inside houses, but later outside poles intrigued visiting sailors and businessmen, who sometimes took them from seasonal villages.

The poles were not worshiped, but embodied beliefs about social realities, including descent, inheritance, power, privilege, and social worth. The poles were meaningful because they were the chosen visible expression of the history of the family, which had to be publically recounted and witnessed. The construction of the poles was accompanied by the appropriate rituals, since each pole was owned by a supernatural being. The artist presented the faces of beings only, so the stories had to be retold by the owners. The poles, especially later model poles, were also used for funerals and as teaching devices, as children and spouses had to be trained in the totem. Totems, more than being good to eat or use, were chosen because they were good to think with, according to Claude Levi-Strauss. They were metaphors for people and things, as well as for trees, which conveyed history, strength, growth, and longevity. The chief of the tribe was the post of the world, as his ancestors were the root of the tribe. Honors, an important form of wealth with goods, were displayed in a totem.

Totemism is a good way to have animals be represented in human society and law. It is more selective than pantheism or animism, and the totemist provides a commitment to preserve the subject through the bond of brotherhood. Totemism teaches an invisible whatness about the self, since it is an analogy between a social system and the natural world, in a religious context. People are parts of society, while people, animals and plants are parts of ecological system. In this way, empathy is accentuated.

Paul Shepard, and independently Michael W. Fox, proposed a human association devoted to each species of wild animal, so that every creature on earth would have a human constituency. Groups of people would form voluntary leagues dedicated to single species and their habitats and ecosystems, which are crucial parts. A feeling for participation in ecosystems is necessary. Leagues would have the ability to foster concern for nature, with a tendency to preserve it. A totemic disposition would increase a "primitive" awareness of the earth. There may be secondary or tertiary totems that overlap with other groups or leagues. Social networks could be extended into the ultrahuman world by imagining that life and obligations were shared. Kinship could be reflected in totems.

Is there a way to replace cultures, economic systems, or bioregions with the alliances of totems? I myself am aligned with black bears, lichen and lightning and ask you to join me in the Black Bear spirit clan founded in Coeur d'Alene. This is not just a duty, it is a fun! Look for us on www.beartotem.net

1.7. *Threading Systems with Physical & Metaphysical Themes*

Certain themes operate throughout any discussion of the design of human places, cultures and their environments. These themes can be used to compare different places, cultures and environments. This section offers some of the basic themes that appear again and again in different levels of design. These themes include: Limits, maxima, optima, scale, place, patterns, models, feedback, quantity, connection, stability, interaction, symbiosis, exploitation, interference, difference, diversity, repetition, revolutions, cycles, change, and succession. The discussion expands many of the themes, which are discussed as characteristics of systems.

1.7.1. *Physical & Ecological Limits*
Our modern cosmology, based on machine metaphors and the principles of plenitude, gets in trouble because it does not understand how basic the concept of limits is to the physical universe, to life, to ecosystems, and to human constructs, such as cities and economics. Limits are important at all levels, starting with the physical.

1.7.1.1. What are Local & Global Physical & Biological Limits?
We have learned that there are physical limits to many physical phenomena. We cannot measure light going faster than a maximum speed. We cannot reach, even in theory, a lowest temperature of absolute zero. We see that certain masses of hydrogen cannot exist without exhibiting qualitative changes in form or temperature.

When it comes to biological phenomena, we notice that groupings of individuals cannot exceed a certain number before the group fissions. With human beings, we almost never have a permanent family group of over 150 individuals. Individuals themselves rarely remember over 9 items at one time. These limits have to do with the characteristics of patterns that make up individual things and entities. The limits can be described in terms of the number and strengths of connections, as well as of limits of freedom.

1.7.1.1.1. Limits of Freedom & Order
The limits of the universe, like the speed of light or the quantum of a field, put limits on freedom; freedom is defined here as the absence of restraint or confinement, or the state of being free from rules or patterns. But, limits cannot be complete, any more than freedom can. The universe as a whole seems to work with fifty percent reliability. Existence is already half determined; freedom and necessity seem to be balanced at about fifty percent. This can be compared with redundancy in information theory. The predictability of particular events within a larger aggregate of events is referred to as redundancy in information theory.

Random order is open; it has many degrees of freedom. Differences can be related to degrees of freedom. An order with more differences has more degrees of freedom. The macro-order limits the degrees of freedom in underlying orders. Movement could not occur without a free order or disorder. There could not be any activity at all within a complete order—perhaps this is what Plato had in mind with perfect forms. This implies that the universe is partially ordered and partially disordered. The distinction between order and disorder disappears when the domain of both is the universe. Freedom, as meant by social or political, also applies (see later discussion). Life and art require a balance of rules and freedom.

Balance means an equality of basic processes, such as integration and disintegration. It means that things are being built up as they are being torn down. It means that natural processes such as motion and rest, turning and returning, are approximately equal. Being in balance is like a form of maturity, while growth is usually a characteristic of immaturity. Being out of balance for too long leads to death. Of course, nothing can be in balance indefinitely, or for very long, depending on its size and complexity. Balance is an ideal for humans. Too much novelty is stressful; it is too new to understand. Too much of sameness is stultifying. Too much conflict is stressful, but too much peace is boring. Perhaps the fifty-percent rule applies to balance, and to the ideas of maximum and minimum. Maybe 50% of life is spent in balance. Maybe 50% is the amount of error?

1.7.1.1.2. Physical & Ecological Connections

A system is defined as a complex object, every part of which is connected with other parts of the object in such a way that the whole possesses emergent properties that the parts lack. Systems are considered to be wholes where the internal connections are stronger than the external connections. There is no whole system without an interconnection of its parts, and there is no whole system without an environment. Everything is connected in a system; removal of any part alters the dynamics of the system. Connectivity is related to the size and density of a system. To connect means to bind together, from the Latin words, or to link or couple. A connection is the relationship or association. A complete connection is regarded as a circuit, which becomes required for certain patterns and their continuation. A cycle is movement through a circuit.

The organization of an ecosystem is described by diversity and connectivity, among other characteristics. Every local system is connected to some degree, but usually the degrees are very weak, through gravity, global chemical cycles, or just thought patterns (see Section 1.6.2.3.3 for part of this discussion). By being overconnected a system can lose stability. Overconnectedness creates a lag in signals. Agricultural fields are good examples of overconnectedness. This is even more true with corn than with wheat due to the changes on the species from domestication. Overconnection causes greater lag times. But, in a city, feed bins damp food variations so the lag is less important. Under-connection can also create instability. Parts go their own way and there is no communication between them. If a system is underconnected, it may become unstable or lose resilience, because there is no pathway to allow a signal between significant parts.

The concept of diversity does not take into consideration the connectedness of the subsystems or systems. Therefore, it cannot be related well to stability. In addition, for each direct species to species relationship, additional bundles or sequences of indirect interconnected relationships spiral outwards logarithmically into the ecosystem and the biosphere as a whole. Indigenous peoples have names for many of these relationships, usually in verb form, such as pollinating, germinating, and shading. They understand and respect the importance and intelligence of species connectedness.

1.7.1.2. Biological Limits & Carrying Capacity

Life involves a vast number of interacting structures. Living consists of complex behaviors whose limits are defined by rules of order that can be empirically described. Biological order is built on physical and chemical orders. That is why life is limited to such a narrow range of conditions, as regards temperature, pressure and the composition of air or water. And, that is why the most complex orders are vulnerable to changes in their substrates; energetic radiation can alter and destroy an individual, a small change in climate can destroy crops and human civilizations. Complex orders always depend on simple orders. Where a planet entirely of algae is conceivable, one inhabited by only rats and humans is not. The earth is suitable for life because of three kinds of limits: The solar radiation that has stayed within certain limits for four billion years; the biogeochemical cycles of oxygen, water and other nutrients have stayed within certain limits; and, the constancy of the environment, constant enough for organic evolution, but variable enough for natural selection to be challenged.

1.7.1.2.1. Organismic Limits

Ecosystems and organisms depend on a complex set of conditions, which can be limiting factors. Organisms are affected by the quantity and variability of materials, if they require a minimum of them. Organisms are also affected by their own limits of tolerance to those materials. Life is limited by elements and physical factors, such as light or water; it is limited by too little of an element, such as phosphorus—this is Liebig's law, or by too much of an element, such as salt—this is Shelford's law of tolerance. Every tree species has a lower and upper critical temperature, which limits their growth. Low temperatures prevent trees from absorbing the moisture needed for transpiration. High temperatures result in excessive losses of moisture.

The life-image limits the goals of an organism. Each animal is a participant in a field of existence. Using its senses, each participant creates an image of nature, or world (*umwelt*, life-world, is the term used by von Uexkull), from the sensations that are meaningful to it, and which limit it.

The behavior of mammals is controlled and population regulated through the use of space. Most populations, furthermore, regulate their density well below the limits of the food supply, often by as much as 50-70 percent. Territoriality can be correlated inversely with trophic levels and productivity. But populations can also be limited by the specificity of prey or plant source, size of prey or plant populations, predators, or natural events or catastrophes. The preference for a certain prey, or taste, can be a limit for a predator. Wolves in the Arctic disperse with the migration of the caribou. According to David Klein, they prefer Caribou to other often more easily obtained species, such as mice. This preference reduces their hunting efficiency, however.

Competition limits the number of species in a niche (the competitive exclusion principle). Garrett Hardin states the competitive exclusion principle as: complete competitors cannot coexist. Niches must be different for species. Krebs states that the fundamental niche of a species has an "infinite number of dimensions," making a complete determination impossible. Territory can limit breeding percentage. In some birds, less that 30% may breed. In wolves, only 60 percent.

Whenever there are too many replicating units for space and resources, whether genes, organelles, individuals, families, cultures, or species, some persist and some fail. Self-organization and co-construction of the organisms and environments allow selection to act on various levels at different time scales. More than one unit is selected. No one unit is the key. System complexity is a limit for rogue species, since all have the same building blocks.

1.7.1.2.2. System Limits: Carrying Capacity

The system itself has limits. It has a maximum biological load that can be carried indefinitely; this is the carrying capacity. This carrying capacity is usually considered to be the maximum population sustainable on a long-term basis of renewable and nonrenewable resources.

Calculating a carrying capacity for many mammals is relatively simple. For instance, the number of caribou that could be supported on the North Slope of Alaska is basically determined by a long-term average of primary productivity, that is, food for browse, according to David Klein. Calculating an optimum population for human beings, based on caribou as a food source, is also relatively simple.

Carrying capacity can be defined basically in terms of energy. For instance, an adult human requires 2300 kilocalories just in food. For heat and clothing, more calories would be required; for transportation and shelter still more; and, for luxuries, many more. By 1990 the average American was using 2,300,000 kc per day, almost 1000 times the minimum.

For humans, this carrying capacity must include domesticates, as human equivalents, since many domesticates compete for protein consumption. Carrying capacity calculations often just consider food energy, but all needs—clothing, shelter, transportation, information generation, aesthetic satisfaction—must be included. This introduces cultural elements into consideration, so human carrying capacity must be considered as cultural carrying capacity.

Cultural carrying capacity involves many more variables, such as luxuries, aesthetic space, the use of technology, the implications of images of place, and the idea of an optimum. Since an optimum is always less than a maximum, according to Eugene Odum, the carrying capacity would be reduced by as much as half. Many mammals adjust their numbers below a maximum capacity, especially when the variability of the system is considered. Wolves underutilize their resources, as do most mammalian predators, perhaps since the same level of resources are not always available every year. Because people use culture to adapt to the earth, a figure for carrying capacity has to be variable or inconstant, to reflect the annual changes in productivity.

Furthermore, the optimum carrying capacity decreases as the per capita use of energy and resources increases. Technology could expand the carrying capacity to some extent, with more efficient use and resource substitution, but it could reduce the capacity with unforeseen effects, from the use of pesticides, for example. The optimum could be reduced more to reflect the possibilities of catastrophes; perhaps it has to be the lowest possible number to meet the worst conditions in a satisfactory way.

1.7.1.2.3. System Limits: Cultural Connections

For many peoples, such as the Luiseno of California or the Huichol, their metaphorical image, the Tree of the World connects all worlds, all the layers of the cosmos, the netherworld and heaven. Modern cosmologies are unconnected to changes in their context. The tools will have to be used in new ways in a new framework, perhaps with topology and holograms as metaphors. Topology provides the mathematical model for processes. A hologram provides a model for connected wholeness.

New images can bring people, even poor industrial people, into better contact with nature, through connectedness. But we are connected by more than images. We are already connected by existence and feeling, even if we are no longer aware of it. Human life is linked in an intricate web of connections. Of course, images let people rank the connections, as a function of survival or style.

Cultural patterns may increase or decrease connections. Villages lower connectivity to the landscape by increasing patches and corridors. In city suburbs, corridors and networks increase, patches increase, but connections decrease as does productivity. Species diversity may increase, but the species often have no connections to resident species. Suburbs are the result of other patterns, such as the supply of rural land, cheap gasoline, and a network of highways.

Some industrial artifacts become overconnected to the habits of people. Perhaps one reason for cultural collapse is the number of tight overconnections, as sometimes happens in ecosystems. This is a problem with modern globalization, in the 1990s, as well as with simpler global connections that began in the 1300s.

1.7.2. *Scale of Matter & Energy*

This ground—the world moving, turning, processing, rising, and intending—is the source of everything. Matter and energy, molecules and elements, combinations and societies, are the basis for living, and form constructs on various scales.

1.7.2.1. Materials

Materials on the earth include elements, compounds, mixtures, and combinations. The atmosphere is composed of elements and compounds, such as nitrogen and carbon dioxide. The ocean is composed of hydrogen hydroxide (water, of course), with many dissolved elements, such as carbon dioxide, gold, and sodium chloride (salt, of course). Materials can be extracted from places in the landscape: Metal ore from rock, wood from trees, oil from tar sands, or stone from formations.

Organisms themselves can be described as material patterns in space (as well as energy patterns in time). Plants concentrate materials and energy in their form. Insects and animals eat plants, although significant amounts of plant material end up in the soil. Organisms can change size and grow, but as the scale changes, the organisms changes. A skeleton, for instance, changes shape if it has to bear more weight. Additive change also scales up, but it is limited by the skeleton. At some point the qualitative change requires a new skeleton.

For example, an endoskeleton, such as that of a typical insect, protects the vital inner tube that has organs and supply lines. This can accommodate many kinds of external forms, although the skeleton is relatively static. In fact, an organism with an endoskeleton cannot grow beyond a specific size, due to the limitations of the skeleton with regard to structural strength and heat exchange. To be stronger at larger sizes the skeleton has to become internal support. Internal skeletons also have size limits, although some of these can be exceeded with a change in life style, for instance, water dwelling.

Scale has another positive effect for larger organisms: The ability to eat lower quality food, and an increase in longevity. Thus, if an animal is 10 percent larger, it may expect to have a lifespan 6 percent longer. A biological approach recognizes that as the scale changes the pattern changes also. That is, the body has to adapt its shape to get larger. This approach also puts growth and development into an organic pattern.

All living and nonliving beings participate in some kind of material system. A material ecosystem is an adaptively organized system of living beings and elements that recycle important elements within the system (after Lynn Margulis's definition); it is the smallest unit that recycles biologically important elements, a natural system where the exchange of materials between living and nonliving parts follows circular paths. Cultural materials also become the source for new materials.

Cultures have been unique programs, using local materials and ingenuity, for satisfying basic material and spiritual needs. Cultures are recognizable in the residue pattern after the flood of goods and ideas. Groups of people eat plant matter or animal flesh to get energy to survive and work on their needs for survival. They use materials for clothing and shelter, rocks for tools, and goods for status. Art materials were taken from what exists in place: For the Kwakiutl, cedar was begged from trees for straight boards, canoes and posts. Materials can limit how a culture expresses itself in tools. But, materials can also be traded in raw form or finished form, that is, materials flow between cultures. Materials have also been used to judge status in a culture.

In hunting and gathering cultures, materials had to be carried with the band, or left behind for seasonal use. The materials of human societies have been increasing steadily for forty thousand years, limiting regular mobility, at least until transportation became capable of hauling the material goods; horses and wagons lead to trains and trucks. One wonders if accumulation was one reason for advances in transportation. The number of artifacts seems to increase as the populations increase. The number of artifacts seems to increase as the complexity of a culture increases. Materials can be used to express power.

1.7.2.2. Energy

The sun provides energy that drives global cycles and ecosystem development, although the earth and the moon provide energy that they are dissipating from their formation. Geological formations and plants store energy that can be used by living beings. Human beings can release stored energy at hundreds of times greater flow and many hundreds of times faster than it can be fixed. Human civilizations all together have a very high energy use, perhaps thirteen times the use of all other mammals on the planet.

Human cultures, in addition to the energy put out by individuals, have various degrees of access to different kinds of bound energy, such as fire or coal. Energy is a useful tool of analysis of kinds of culture. A culture can be defined by its levels of energy use. Traditional economies are based on solar energy and natural productivities. Human energy is used for building and foraging for game and plants. Fire, as the release of organically-bound energy, is used for heating and cooking. Sometimes, animals or human slaves are captured and used for work. Production is kin-oriented and reciprocal.

Carbon products are the largest source of energy. The earth is also a source of energy from tides and hot springs. The discovery of other forms of energy, such as fossil fuels, lead to technologies that could leverage them. Technologies require energy. The industrial revolution increased the quantity of energy, but decreased the variety of energy resources. The three main uses of energy for industrial civilization are motor traffic, heat for buildings, and manufacturing. Manufacturing is a heavy consumer; it includes: Pulp and paper; primary metals (for cars); nonmetallic products (for plastics or shoes); and chemicals.

The invention of ways to use nuclear energy increased the energy available, but has been limited by problems with the disposal of heat and long-lived by-products ('by-products,' like 'side-effects,' are equal things, just not expected or desired). Nuclear fusion promises to proved cleaner energy, but with a different set of limits, such as containment and safety.

There are other kinds of energy that it might be possible to use, such as gravity, or the rotation of the planet. But, at the planetary level, we are at the beginning of the Kardashev scale, a general method of classifying how technologically advanced a civilization is, using planetary, solar, or galactic energy.

1.7.2.3. Cycles

In normal stochastic processes, things bunch together or move apart. We can identify cycles of these processes. These cycles are not just repetitions; they spiral and take up the old processes—who says life does not learn unconsciously from history? There are many kinds of natural cycles. For instance, the climates cycle as a result of planetary motion. Every being has a life cycle; every ecosystem has a development cycle.

In a mature forest, for instance, very little material actually leaves the forest. It is held in local cycles. Materials cycle above and below ground, between the atmosphere and trees, between trees and insects, and between squirrels and fungus. Chris Maser is fond of saying that most of the cycling is invisible because it is underground or in the air. Many cycles can be investigated through ecosystem analysis, where energy and materials are traced through transfers through compartments in an ecosystem. Virtually every material cycles through the forest, including, but not limited to: Nitrogen, carbon, phosphorus, potassium, calcium, sulfur, magnesium, and water. Nutrient cycling involves many of these materials. Nutrient cycles change with the succession of a forest. A variety of fluxes and accumulation rates of several nutrients are needed to maintain a growth rate of twenty cubic meters per hectare per year in Corsican pine in Scotland. Half the nutrients for new growth are drawn from senescing needles and twigs.

There are some material cycles between other ecosystems or in larger patterns around the planet. Some of the cycles are daily; some are seasonal or annual; others are years or decades; a few, like the carbon cycle, are century or millennia long. Human beings do not pay much attention to very long cycles or to those we perceive as not affecting us. These cycles, applied to cultures, allow for their growth and development, expansion and contraction as cultures. They can influence the efficiency of a culture.

1.7.2.4. Scale

Size is that quality of a thing that determines how much space it occupies. And, because space is integral with a Space-Time-Energy-Mass (STEM) dimensionality, it cannot be separated from the other three. The scale of any one changes the properties of the other three; so, for instance a significant mass of hydrogen can change the development of helium and the production of energy in a star. Space-time at large scales can take on different shapes, which we can correlate with gravity.

There are barriers to the size of nuclear matter, of stars, and as well as of planets. Things divide naturally. They make barriers or walls. The proper size of things depends on their function, according to D'Arcy Thompson. The scale of a thing depends on its place in nature. Kohr states the principle that stability and soundness adhere to bodies that are relatively small—this is his theory of small cell architecture. Dynamic balance is self-regulatory, Kohr states, because of "the coexistence of countless mobile little parts of which no one is ever allowed to accumulate enough mass to disturb the harmony of the whole."

1.7.2.4.1. Ecosystem Scale

Scale has several meanings in ecology. Basically, it has to do with the level of measurement in a STEM context. The scale of the system defined, for instance a mangrove forest ecosystem, depends on the scale of the phenomenon being addressed. Ariel Lugo and his associates found that mangrove forests are in phase with the frequency of hurricanes, although hurricanes may not influence the life histories of short-lived organisms as much as seasonal cycles. The forest ecosystem is a large-scale pattern of millions of minute events. Microbes, for instance, are affected by short-term cycles of precipitation and temperature. The overall extent of forests is often determined by large-scale weather patterns and environmental changes. Watching the history of the earth from an orbiting macroscope, over millions of years, it might be possible to see forests move across continents like the shadows of clouds. The possibility of extreme change is rarely considered in wilderness design.

Problems in management arise where applications that work on a small scale are expanded to large scales, without thought for the difference or changes in patterns. For instance, it is well known that Douglas-fir is shade intolerant and grows best in openings that get light. Rather than simply remove single trees or small groups of trees, and release or plant the fir in the openings, industrial forestry applies the treatment to the entire landscape with large clearcuts, which alter the other conditions that fir requires—some shade, water, protection from browsing, associated species. For the forest in which Douglas-fir grow, increasing the cuts to increase the potential for the fir has the unintended effect of drying out the soil, as well as destroying the infrastructure of plants, animals, and fungi, that the forest requires to continue. Things change when scale changes. Changes in scale put pressures on human and natural systems. The current economic style is too great, fast and reckless for ecological systems to absorb its impacts. The scale of things is an independent problem that can ruin the best intentions of policy.

Other forest-cutting systems, such as high-grading or thinning from below, may also have unintended scale effects. High-grading is a form of group selection, in which the best and largest trees are removed. The practice was not overly destructive on a small scale, with many of the matriarchs and patriarchs remaining, but on a large scale, all old, large heritage trees were removed. Thinning from below may be a good idea in a stand, but like high-grading, it is biological selection—the gene pool is altered as the small and suppressed trees, which may be genetically stronger, are removed.

An inappropriate time scale, its shortness and urgency, is a cause of many of these problems in industrial forestry and ecosystem management. Although forests are considered renewable resources, they are slowly renewable, requiring hundreds or thousands of years to renew from catastrophic disturbance; this time is far longer than any economic plans and nonrenewable on a human life scale. This has important implications on sustainability.

One solution to many of these problems is a reduction in scale for everything from forest use to management units, with local controls and local use primary. Temporal scales, however, should be expanded with long-term research and management. Ecological management must observe the proper scales, especially in terms of size, age, and patterns.

1.7.2.4.2. Human Scale

Human institutions are also subject to effects of scale. At the level of families and bands, which are often extended families, behavior is understood and exchange is reciprocal. At the level of tribe or chiefdom, conflict resolution cannot be done without everyone knowing everyone else, and it cannot be left to members who are strangers.

Nations become organized on territorial or organizational models, rather than on kinship. Nations can include more than one ethnic group or language. Nations have advantages over smaller groups because of specialization (especially of soldiers), technology (especially of weapons) and population size. Nations arose in Mesopotamia, China, Nile, Indus, Mexico, Andes, and West Africa. Chiefdoms were formed in even more places, including Amazonia and Polynesia. Larger amalgamated nations only arise through conquest or from duress for protection against other nations. What leads to this complexity? Population growth and its attendant processes, including intensive food production and resource use, resulting in public works programs, long distance trade, and specialization.

Each place has a temporal and spatial order, with a physical scale and rhythm. Small-scale cultures were egalitarian. But, there is a qualitative change when scale increases. The villages became ranked. The shift in size to cities changed the scale of food collection and trade. For instance, time is perceived at a faster rate in cities than in rural settings.

Economic systems all seem to work on a small scale, and to work less well on larger scales, when the market and the production become too large. The problems increase and more is taken up by the problems. Many kinds of products do not add to the wealth of people. L. Kohr identifies three kinds of such products related specifically to growth: (1) Power commodities, enhance the standard of society without contributing to material welfare of the members of society; these include tanks, bombs, and government services. (2) Growth products of a second category, density commodities, are necessary due to the increase in population, but do not add to individual happiness. These are traffic lights and hospitals, and fire losses. Growth requires growth products to address the problems of growth. (3) Growth products of a third category, progress commodities, are improvements made necessary by previous improvements (anti-aircraft guns need better ammunition) or unwanted tie-in products such as license plates or television repair.

To work in a large society, economics needs to be redistributive as well as reciprocal, that is, a central authority redistributes excess goods to individuals and groups with deficits. Economics was useful as the harmonious distribution of wealth among people. But, there are problems outside small-cell scales. As governments and corporations grow larger, they tend to dedicate themselves to self-preservation instead of fulfilling their designated purposes. More and more of the resources of corporations, for instance, are diverted to the infrastructures, such as new buildings, status, advertising, and employee rewards. Instead of representing and regulating people and institutions, larger governments simply take over the unwanted functions of corporations and communities. The scale itself requires changes in resource use and organization.

Kohr's principle argument against large-scale economies is the law of diminishing productivity. Variables added to fixed units eventually add less to total product than a preceding one. Increase in size or power does not produce an increase in satisfaction or productivity. Kohr points out that Marx failed to link misery to the scale of economics rather than to the system. The problem is with overgrowth more than style, which is why socialism based on overgrowth looks the same as capitalism overgrown.

Small-cell theory, Kohr says, is a fundamental principle of health. We are happiest in smallest units, such a family or county. The pattern has to be small cells under a federal unit larger than the largest cell. That allows harmony and manageability by ensuring a physical and numerical balance. Then, central authority can be weak.

1.7.3. *Fields*

The universe at large, with its clusters of galaxies, clouds, stars, and planets, extends itself through space and time. Parmenides held that space was a plenum. On the other hand, Leukippus conceived of space as emptiness. The physicist David Bohm combines both ideas in the concept of a field. He describes the universe as a field with waves of infinite size. The universe is permeated by septillions of waves at all times. A wave is an integral pattern of the physical continuity of a particle. Waves shape the field, that is, they excite the field. The field is an invisible, nondetectible source from which elementary particles draw order and energy. There is no place for both field and matter, "field being the only reality," according to Einstein. The field here and now depends on the field in the immediate neighborhood at a time just past. Excitement is generated with a temporal dimension. Humans generate their own waves that are added to the infinite variety coursing the universe.

The field concept was originally introduced by Michael Faraday into studies of electricity and magnetism, and was expanded by Clifford. The field concept is central to the unification of theories of light, electricity, and magnetism. Einstein extended Clifford's ideas of the field in his theories of relativity. In large systems, time acquires a new meaning associated with irreversibility. The classical mechanical concept of time results from simplifications. Einstein induced that time is relative to the frame of reference. Space-time is an ensemble of occasions and places, held together by duration. The future and past are tied together by duration. To the largest duration—the universe—all time is present. Time is not empty or abstract. According to Einstein, every change of coordinate systems mixes space and time in a mathematically defined way.

In his Special Theory of Relativity, Einstein included gravity in the picture, making space-time curved. But as massive bodies have greater gravities and mass can be converted to energy, all four perspectives form a field that can alter each component. Through the principle of equivalence of gravitation and acceleration and through the use of a symbol that mathematically described the local rate of 'turning' of the curvilinear coordinates, Einstein was able to relate curvilinear order and measure to the gravitational field. Both the electromagnetic field and the gravitational field can be understood as aspects of curvature. The particle affects the field only in its own locality. Particles move along paths that are intrinsic to curvature of space-time field independent of coordinate frame used for measurements (but not from the act of measuring). These paths are called geodesics and are the shortest distances between two points. Einstein's 1907 principle of the local equivalence of gravitational and propulsive accelerations (geometrodynamic law) linked the two currents of thought from Riemann ("geometry is part of physics") and Mach ("inertia is influenced by mass elsewhere").

The physicist John Wheeler regards curved geometry itself as the building material of the universe. Gravity can be regarded as slow curvature; the electromagnetic field is rippled with different curvatures, and the particle is a knotted up region. Wheeler hypothesizes that a pregeometry is necessary because geometry fails to explain some events such as gravitational collapse. If one regards geometry as an abstraction from a moving Space-Time-Energy-Mass field (STEM), the pregeometry is the STEM field, the ground of being. The STEM matrix is a cosmic, transformable field. The STEM field is primary. Time and space are secondary to the field that contains them. No one component is ontologically subordinate to another. The fundament is an immense multidimensional ground, from which orders are projected into a unified field. The STEM field is a meso-field in the universe; it vanishes at both ends as a knot dissolves into the identity of rope after being analyzed. Particles dissolve into identity with the universe. The field is a paradox; localities are part of it but do not communicate instanta-neously, by light. The speed of light is still a limit for most communication, but not for the behavior of the field as a whole. Space is filled with local relationships. The limits of the array are nondefinable. The STEM field has general properties, such as motion and autopoesis.

1.7.3.1. Properties of the Field

A property is an attribute proper to a thing or characteristic quality common to all members of a class (by comparison, characteristics are qualities that distinguish unique individuals, systems, or patterns. Gregory Bateson calls them differences that make a difference). The properties of a field are shaped by the historical operation of the field itself. This means that these properties are reflected in different levels of organization.

Mario Bunge distinguishes three kinds of collective properties: Aggregate, structural, and global. Aggregate properties are often statistical, as in the average age of the wolf pack is 3.9 years. Structural properties can be possessed by individuals or groups on the basis of their relations to others, e.g., high-tail is the daughter of nick-ear. And global (or emergent) properties are possessed by wholes, regardless of components, as in "the territory of the wolf pack is eighty thousand hectares." This property emerges with the system and submerges if the system breaks down. It is an emergent property if no component of the system has it in-dividually. History, structure and stability are emergent properties of an ecosystem. Role and scarcity are not. The basic properties of a field, any field, are: Motion (process), autopoesis (self-making), differentiation, integration, constancy, and development.

1.7.3.1.1. Motion as a Property of a Field

Nature, for Alfred North Whitehead, consists of patterns whose movement is essential to their being. These patterns are analyzed into events (or occasions). An event is a change of state. An event is described by two points, an initial and final state. A process is a sequence of states (also called history). Process is the dynamic change in an unfolding flow. Process is a fundamental feature of nature. A process creates a path, described by a trajectory of states. A process is described as evolutionary if it involves emergence and the creation of new things (general speciation; to be a species, however, the novelty has to reproduce itself or multiply or diffuse). The concept of state precedes the concept of process, although process can define state. In complex systems the past contributes to or constrains the state (from the magnetic hysteresis of ferromagnets to organisms, which have memory).

The activities and movements of an organism are united into the being of the organism. The organism is a dynamic structure that is immanent and simultaneous with the process, rather than just a consequence of the selection of individuals modified by mutation. Beyond being merely relations of relations, organisms are pulsations of process, natural units of fact. For example, since subatomic particles are part of a field, and they are internally related within the field; they cannot exist without the field. The field provides the order required for producing individual actual occasions. The coming-to-be of organisms, i.e., process, is a fundamental feature of reality. The organism undergoes a process of evolution in which it produces new forms in itself.

1.7.3.1.2. Autopoesis as a Property of a Field

The field is autopoetic, that is, self-making (from the Greek words auto and *poiesis*), according to Francisco Varela and his colleagues. Autopoesis refers to the dynamic self-producing and self-maintaining activities of living beings. The tenets of autopoesis are presented in six principles: (1) Identity: Identifiable components are organized internally with structural boundaries; (2) Integrity: The self is a single, whole, dynamic functional system; (3) Self-boundedness: The boundary is produced by the system; (4) Self-maintenance: The boundary and components are produced by the functioning of system; (5) External supply of materials: Elements, such as carbon or water, are obtained beyond the boundary; and (6) External energy supply: Light or chemical energy from beyond the boundary is converted into organic bond energy.

In an autopoetic framework, every being is embedded in a world and observed by an embedded observer. The material components of life move through physiological processes. Autotrophs, such as bacteria, algae, and green plants, convert energy into organic compounds; heterotrophs reconvert autotrophs into heterotroph flesh. The unit of evolution at any level is a network capable of a rich repertoire of self-organizing configurations. Reactions at the chemical, organism, or ecosystem level, when combined, produce unexpected positive results from the sum of single reactions. An ecosystem has emergent properties that are different from the sum of community interactions. They also affect biogeochemical cycles.

1.7.3.1.3. Differentiation as a Property of a Field

The field universe is a vast order of individual events, discrete unfoldings within limits, following certain laws, but in unique patterns. Patterns of complexity shade and grade into one another endlessly. A large family of similar unstable particles can been classified phenomenologically in a complex set of interrelated orders. It appears to many physicists as if the beginnings of new orders of natural laws are being revealed, in which particles would be like flower designs on a carpet pattern, while something unknown corresponds to the structure of the cords of the carpet. In fact, orders reform in the 'carpet,' as John Wheeler suggests. Analyzing the world as if it were made of particles would be similar to analyzing the carpet as if it were made of flowers: it would give some results and have some predictability, but the metaphor would limit better understanding. Wheeler states that patterns in the 'foam' are seen as subnuclear particles. The foam can be seen as a metaphor for the self-reconstruction of the universe from standard parts to novel structures, through a process of metalysis, that is, the break-down of parts to form new structures.

Order is measured through relationships. That which cannot be related to something else, i.e., to a curve mathematically, is referred to as disorder. In fact, Bohm himself states that order is basically a set of similar differences. Differences can be related to degrees of freedom. An order with more differences has more degrees of freedom. The macro-order limits degrees of freedom in underlying orders. Bohm contends that order is more fundamental than relationships and classes; order is logically prior to relationships. But this seems to be a misconception. Order is measured through relationships. The brain registers differences and similarities, leading to perception of order, and this is related to other orders stored in memories (perhaps in a holographic manner). Nature is often ordered only by our understanding. Bohm sees that there are two kinds of differences: Constitutive determine the essence of order; distinctive determine how one order is distinguished from another. But are the two really different? Consider a series of points, where a curve can be drawn several ways, using some different and some same points. The constitutive becomes distinctive.

1.7.3.1.4. Integration (& Participation) as a Property of a Field

John Wheeler believes that since law, field and substance all exist after the theoretical big bang, the universe owes its existence to trillions of acts of registration. The phenomenon comes through elementary acts of participation by different observers. Wheeler asks if the universe might not be brought into being by the participation. Quantum mechanics strikes down the concept of the neutral observer; participation is vital. Wheeler noted that "To observe the electron even, the experimenter must shatter the glass—must reach in with instruments." The quantum principle destroys the observer behind glass (*in vitro*). The universe is not the same after measurement; the observer becomes a participator. Einstein said that no event can be postulated without the presence of an observer; but no observer can see the whole system; and anything can be an observer or participant. A world without participants is impossible. The observer is part of a natural or social community. No system can exist without observers. Fortunately, nature contains many eyes. There are no lone observers. The autonomy of our social system goes beyond our individual autonomy; the knower is the observer-community. Humanity participates in the natural world, so human history is part of natural history. All beings participate in relationships that make up their worlds. Full participation removes the barriers of language or thought; "otherness" pours through.

How tightly is the universe integrated? From one perspective, the universe seems to be nonlocal, with instantaneous action-at-a-distance and unbroken wholeness (connection is extended to the entire universe); from another, it seems quite local. Nonlocal influences cannot be argued then, since no information is proved to be transmitted between devices. The theoretical cross-correlation of separated events does not occur in a local frame. It requires a global frame, which can only be used over a long time. The failure of the principle of locality would mean that there are no separate parts in the universe, but we know that nothing is entirely separate anyway. The conclusion that experimentally separate parts are correlated forever must consider the "parts" as eternal things, not as phenomenal processes. In fact, the particles are connected by the field, not by superluminal signals or any other kind. In a holonomic field, the resolution of any one perspective of the whole is so weak that it has no predictive value. Information is not transferred because the field is in-form-ed. An adequate model must provide for unified understanding of all of nature.

1.7.3.1.5. Constancy as a Property of a Field
Bishop Berkeley proposed a bootstrap principle when he argued that the inertia of any body is determined by the distribution and masses of all other bodies in universe. Ernst Mach's principle repeated the same idea, demanding the closure of the universe, to make it finite and bounded. The mechanical properties of space are determined completely by matter only in a space-bounded universe.

In the bootstrap principle, the universe is what it is because it is consistent with itself; we are not free to sort out accidental properties and distribute them with various values among different universes. No properties of the universe, however, are fundamental; they follow from other properties in a web of interrelationships. Geoffrey Chew's bootstrap model considers hadrons as temporarily stable configurations that result from an interaction of processes. They may transform themselves into other particles. The bootstrap relation takes the form: "Universe=subatomic-particles," where '=' means 'equivalent to.' Each particle represents a facet of the universe and not just a small part. All electrons have the same charge because they represent a single aspect or perspective of the universe. Each particle is only an abstraction of a relatively invariant form of movement in the whole field of the universe. The proton and neutron are just two different points of view of the nucleon; the sigma has three different points of view. Both nucleon and sigma particle may be just different points of view of the same particle, related by an extended isospin symmetry. Specificity can be a property of the path, what Whitehead a 'concrescence,' and what Waddington calls a 'chreod.' Yet, if the universe is open-ended, then the consistency of the whole can never be proved logically.

1.7.3.1.6. Development as a Property of a Field
The loss from motion creates a sediment, which is history. Sedimenting is irreversible and gives direction to time. The sediment of the past is a given and different for each present. The ceaseless activity of being in history creates newness. Novelty is born from the womb ('hystera') of change. History is the result of this hysteresis process. The universe restructures itself constantly at a more complex level. The evolution of the universe is a history of the unfolding of differentiated order or complexity. Building up emphasizes structure and levels of hierarchy by joining systems from the bottom up. David Bohm states that we have been conditioned to the belief that higher orders of nature are determined by the lower order of the mechanical motion of particles. It is impossible to exclude the contrary assumption that high order features of natural laws are as fundamental as those referring to atomic movements.

By comparison, unfolding implies an interweaving of processes structured at different levels. Evolution acts in sense of simultaneous and interdependent structuration of microworlds and macroworlds. Complexity emerges from the interpenetration of the processes of differentiation and integration, processes running simultaneously from top and bottom and shaping the hierarchy from both sides. Patterns of complexity shade and grade into one another endlessly. Reciprocal causal processes can increase structure, differentiation and complexity in natural systems, according to M. Maruyama.

1.7.3.1.7. Properties Across Levels
The properties of the field are modified by the operation of the field. Different properties emerge at different levels. For instance, stability at the ecosystem level is replaced by loyalty at

the social and psychological level. It is possible to trace the changes in characteristics from the field to the ecosystem and then to place and society. All levels can be addressed as systems.

Table 17317-1. Contrasted Properties of Different Levels of Patterns

— Nature —		— Culture —		— Design —	
Field	*Ecosystems*	*Place*	*Culture*	*Good Places*	*Good Society*
Process	Course	Dynamicism	Conduct	Action	Method
Autopoesis	Self-making	Identity	Wholeness	Individuality	Extension
Differentiation	Diversity	Uniqueness	Flexibility	Richness	Variety
Integration	Construction	Investment	Adaptation	Conviviality	Cooperation
Constancy	Stability	Regularity	Endurance	Consistency	Loyalty
Development	Productivity	Renewal	Vitality	Health	Harmony

Animation and ecological value changes the differentiation of the field to the openness of the ecosystem. The addition of communication and cultural values to that characteristic of the ecosystem results in the richness of place. And, the addition of social values and awareness of a place leads to variety in that society. These six properties of a field can be related to equivalent properties in different aspects or parts of the field, from ecosystems to human societies.

1.7.3.2. *Field Changes*
Field changes result from the operation of the field. Some relevant changes include growth, breakdown, emergence, patterns, trigger, scale, and history.

1.7.3.2.1. Growth (Anabolic Change)
To grow is to come into being or to increase in size or in some other quantity such as mass or energy. The universe and its field grow from some unknown impulse. We have observed that the universe has been growing, as has our local solar system. Everything seems to grow. Gravity encourages accumulation of mass and energy in black holes. Everything tries to expand, to survive within serious constraints or limits or to exceed them. Why is that? Some metaphysical urge? Growth seems to be a primary thing, yet limited, from aggregation to organic size. A living body, for instance, produces many more cells than are needed, and then it corrects the excess by signaling some cells to die. A body also produces many more sperms or eggs than are ever used. In some cases, many more offspring are produced than ever can mature to reproduce themselves. Life seems geared to cycles of overproduction and correction.

The first growth of an organism is to fill out its form within the limits of its genetic possibilities. The most obvious aspect of form is its constancy in spite of growth. In some cases, growth is necessary to sustain the form. The growth of one part must affect that of all. Growth serves as a mechanism of evolutionary adaptation, by carrying out genetic instructions in an environment; but growth is also conservative and stabilizing rather than innovative and reorganizing. Growth and development are homeorhetic; in a homeorhetic process, the flow is constant, not a stationary state. Flow processes follow fixed trajectories, called chreods. Growth, from fertilization, embryo states, birth and maturity, represents a homeorhetic process following more or less fixed chreods, programmed genetically and conditioned environmentally. Chreods must be like electron paths, multiply probabilistic.

In the organic world, growth is healthy only when the rate of change is decelerative in the long run; cancer and population are constant or accelerative. Some economists once assumed that growth is limited to one of two kinds of growth, steady or logarithmic (arithmetic or geometric), but that is palpably untrue. Three particular causes of growth can be distinguished: Additive, an accumulation of more; replicative, an accumulation of more through reproduction; and mutualistic, where all agents change structure, as in meiosis.

1.7.3.2.2. Breakdown (Catabolic Change)

John Wheeler states that patterns in the foam are seen as subnuclear particles. The foam is more like the fabric of a dream. Wheeler states that patterns in the foam are seen as subnuclear particles. The foam is a metaphor for the self-reconstruction of the universe from standard parts to novel structures, through a process of metalysis, that is, the break-down of discrete units to form new structures. The process of renewal can be called metalysis (means 'loosening change,' from the Greek *meta* and *lys*). Living order can also be defined in terms of the influence of whole over the parts. Disorder at the level of a molecule can reflect the higher level of the order of a cell. The machinery of a cell is not a permanent fixture, but is disassembled and rebuilt periodically, according to specific patterns and in harmony with its functioning in its environment. Weiss shows that parts of a cell are constantly changing, growing, dying, breaking up and recombining, but under control at a cellular level. This is metalysis.

Metalysis is a technical term for the process of dedifferentiation in biology. It can also be used to describe physical processes like quantum foam or social processes like institutional revolution. It is used here to describe biological processes, such as metamorphosis, where changes are often so extreme that the organism constructs many specialized parts of the adult from cells set aside in the embryo and that are nonfunctional in the juvenile; these cells are imaginal disks, which are controlled by juvenile hormones. These rules work well with natural processes that operate in ecosystems. There is an operation of metalysis with nations as well. Through isolation and identity, people divide into cultures and nations. Nations have unified and disunified. The process of dedifferentiation is also forwarded by catastrophes.

1.7.3.2.3. Development (Complex Change)

Development means the introduction of an innovation. The evolution of matter, for instance, proceeds through a spiral process, as exemplified in the carbon cycle in stars, where a carbon nucleus captures four protons and emits them as an alpha particle at the end of the process. In some cases physical behavior depends on past history. Magnetic hysteresis is one case. Hysteresis is not confined to magnetism—structural deformation and colloidal behavior also may depend on past history. Louis de Broglie, and later David Bohm, think that the Heisenberg indeterminacy may be the result of not taking into account the past history of an elementary particle in predicting its behavior. Perhaps some of the problems attendant in scientific inquiry arise from the inability to take into consideration the complex history of a particle. Nature is unpredictable. Its fuzziness is due to indeterminate states—indeterminate at the human level, where we have limited perspectives in a local system.

The development of life since the Cambrian era displays a diversity of forms in an expansion of life into places that can only be described as self-realization, since it is far more active than the passive adaptation of self-preservation. This development is reciprocally constrained construction. The organism and environment are co-implicative, co-defining, and co-constructing. They engage in a process of self-assembly, where the self is the organism/environment system. A biological organism grows to maturity, which is a stopping point for size. The organism continues to develop, however, experiencing and learning the environmental complexities through mating and then to the end of life. Development may include growth at some early stages, but development refers to the continued change after growth has stopped, but continuing for the lifespan (and dedevelopment after death).

1.7.3.2.4. Emergence (Novel Change)

Complex adaptive systems display emergent behavior. As an example of emergence, slime molds can form a community without a pacemaker cell that determines when the cells need to combine. It seems that self organization is bottoms-up. Emergent systems are rule-governed, though; slime molds explore by adhering to low level rules. Individuals coordinate work, even if they cannot assess the global situation. Emergent systems are local; individual molds "think" locally and act locally. Random action serves to explore local space. Individuals pay attention to their neighbors, and patterns emerge from local activity. Simple behavior seems to work, with local feedback, and more sophisticated behavior "trickles up" to approximate a global perception.

Emergent properties (or global properties) are possessed by a whole, regardless of its components, as in "the territory of the wolf pack is eighty thousand hectares." This property emerges with the system. It is an emergent property if no component of the system has it individually. History, structure and stability are emergent properties of an ecosystem. Role and scarcity are not. The emergent properties of organisms include discreteness (physical), genetic homogeneity, recognizable subsystems, the coordination of parts, irritability (response to a stimulus), and reproduction (genetic consistency). Of course ecological design should consider these properties as well.

1.7.3.2.5. Pattern (Constant Change)

A pattern is an arrangement of form of elements. Process applied to components yields pattern. Nature is composed of patterns. Patterns are not still. A circular pattern through time can be recognized as a spiral (the earth's orbit for example). The pattern should allow for surprises and discontinuities; it can do this if it is flexible. Regularities in systems are patterns. Patterns can be seen in things or even cultures. For instance, laws of genetics are natural patterns; human customs are artificial patterns. Where the natural ratios of females to males are altered by female infanticide or other action, the pattern is semi-natural.

Paul Shepard describes living natural "objects" in terms of events that constitute a "field pattern." Relations are not prior to objects; they arise together. The wasp and the yucca coevolve; they are not co-linked by prior relations. Furthermore, a specimen is more than the sum of its species' relationships to an environment; it is an intentional being that, with other members of the species, can create niches, as well as adapt to them. Because the STEM field produces life, the qualities of life cannot be separated from its physical qualities. While it is true that living subjects are at a different level of description than events in field patterns, they should not be treated as ontologically subordinate. All of the aspects of the field have equal status. The ecosystem model attempts to be a field theory, but is limited by its parentage, thermodynamics, and has been rejected by some new practitioners.

Just as the new physics has transformed the mechanical picture by placing atoms in a field that accounts for the qualitative emergence of properties from simple quantities, the new ecology has placed living "objects" in a field. This field determines the limits of any ecological field of activity, and no field of ecological activity can be described without taking the physical field into account. The field is living and intentional, as well as physical, and an ecosystem model must address intention or other emergent properties.

1.7.3.2.6. Triggering (Explosive Change)

Mario Bunge distinguishes two types of causal mechanisms in systems: Energy transfer, and a triggering signal, where the energy is far disproportionate to its effect. The triggering signals are small changes in energy that cause large changes in the physical or cultural output of energy. With a trigger, information guides energy use. The trigger mechanism applies to simple phenomena, such as explosives, or to complex biological and cultural phenomena. For instance, there does not seem to be a single trigger for El Nino. One trigger has to do with water overflow and then back flow from the western pacific. Others may have to do with sunspots, which would reduce radiation, or volcanic eruptions. Many of these atmospheric, oceanic and geophysical triggers may converge. Nineteenth-century famines may be correlated with ENSO events that influenced China, Indonesia, Brazil, and southeast Africa.

In a way, culture is like a trigger. It allows less information to activate the system. And, this is the only way to increase information handling, when the system is larger and more complex—of course, this is what stereotypes and metaphors do. Leaders or chiefs can give signals that channel energy into large projects. Population density was controlled by the traditional approaches to resources. In archaic societies, cooperation and consensus, as opposed to competition and individual exaltation, permitted planning to remain informal. Population growth triggered competition and conflict, which lead to positive feedback of the thing that caused the stress, that is population growth and conflict. Art can be a trigger for change (or explosive decompression, to continue the weapon metaphor). The metaphor itself is compressed information that can be uncompressed by the receiver.

1.7.3.2.7. History (Irreversible Change)

History is a sequence of states. In complex systems, the earlier state, also called the 'past,' contributes to or constrains a subsequent state, such as the magnetic hysteresis of ferromagnets or the behavior of organisms having a form of memory. The ceaseless activity of being in history creates newness. Novelty is born from the womb (*hystera*) of change. History is the result of this hysteresis process. Motion, Version, Scension—all together form the ontological spiral of being/environment. The motion of the field creates a turning, which leaves a history, out of which patterns arise. Scension is a precession of motions, resulting in directional change—evolution. The transformation is a historical expression.

The evolution of matter, for instance, proceeds through a spiral process, as exemplified in the carbon cycle in stars, where a carbon nucleus captures four protons and emits them as an alpha particle at the end of the process. In some cases physical behavior depends on past history. Magnetic hysteresis is one case. Hysteresis is not confined to magnetism—structural deformation and colloidal behavior also may depend on past history. Louis de Broglie, and later Bohm, think that the Heisenberg indeterminacy may be the result of not taking into account the past history of an elementary particle in predicting its behavior. Perhaps some of the problems attendant in scientific inquiry arise from the inability to take into consideration the complex history of a particle (compare Heisenberg's uncertainty principle). Nature is unpredictable. Its fuzziness is due to indeterminate states—indeterminate to us, who have limited perspectives in a local system.

1.7.4. *Living Fields*

Field theories in physics provide for continuity, in space-time, for instance. Charles Sherrington developed a field theory of subjective space. He distinguished between exteroceptive, interoceptive and proprioceptive fields; the exteroceptive receptive field was coextensive with the body surface and richer in receptors. These elements comprise a biological explanation of subjectivity. The subjective field is characterized by a broadening of lived time. This extension includes the lengthening of responses.

The subjective field of self is parallel to Kurt Lewin's concept. The self is a creature committed to a specific association, a genome plus place; the fitting of self to setting is partially definitive of both. Place/person/act are as indivisible as the physical STEM field. Lewin thought that the psychological fields joining the personality to its life-space, the immediate environment, and life-space to a larger environment, were strong enough to alter objective facts, that is, to make them normative.

Lewin was willing to study the objective factors that were potential determinants of life-space. The organism's own world is usually left out of consideration by Psychology. Ludwig Binswanger broke the life-world into three interrelated modes for human beings: umwelt, mitwelt and eigenwelt. The first refers to world of natural objects (environment); it is approximately equal to von Uexkull's idea of umwelt. The second to the human alone, the interrelationships of humans; the third is one's self-world (=thought world). The same division could be made for animals, but would be more difficult to describe.

Lewin's life space combines the umwelt and life-world (of Edmund Husserl). Topology provides the mathematical model for Lewin's representation of psychological processes. Topology is a geometry in which spatial relationships are represented in a strictly nonmetrical manner. Since topology had no directional concepts, Lewin invented a qualitative geometry called hodological space, represented by vectors. He used two-dimensional planar maps to represent life spaces (now, recent researchers use a linear graph, on which an indefinite number of points can be mutually interconnected; and asymmetrical relationships).

Lived space requires a three-dimensional representation. It may be horizontal or vertical. A vertical aspect could include the high of a drug user and mountain climber; it could mean the loss of horizon and the horizontal. The horizontal dimension requires reciprocity; it is an intersubjective domain. Older cosmologies balanced the horizontal and vertical, but unbalanced cosmologies exhibit problems. New mathematical treatments of fields tend to be three-dimensional. Rene Thom's catastrophe theory and Conrad Waddington's epigenetic landscape are two such theories. The epigenetic landscape field can be used to explain why chickens rarely walk around a fence to get food. Is it empirical? Could it be measured by intensity of brain waves in a chicken? The chicken's need chreod is deeper than the path of a cognitive chreod around the fence. So the chicken can only go toward the food. If the need chreod is too deep, it cannot explore first with its cognitive chreod, where a less hungry chicken could. In humans, this explains why necessity cannot be the mother of invention; the broader landscape of leisure is needed.

1.7.4.1. Field of Person

Our direct intuitions of nature tell us that it is bottomlessly strange; it is alien, yet beautiful, and ruthless, yet gentle. It has innumerable modes of being that are not human modes of being. It is always mysteriously impersonal, unconscious, and immoral; sometimes we interpret it as hostile or sinister. Although we can intuit the interdependence of nature, we sometimes mistakenly conclude that our skin is the boundary to ourselves. We extend tentacles of personality to other things and people. As Whitehead pointed out, everything prehends everything else. The human skin is a delicate interpenetration, like a pond surface, according to Paul Shepard. The skin's interpenetration links a living field to a larger field.

What is impaired in the absence of a rich ecology is the individual's knowledge of himself, not only as a person, but as a member of a species. Harold Searle's thesis is that the environment constitutes one of the most important ingredients of human psychological existence. There is within the individual a sense of relatedness to total environment or field.

A deep relationship with a place is as necessary as one with other people. Without such a relationship existence loses much of its significance. A range of experiences can spring from a place, from depression to the peak experiences described by Maslow. Then there is the opposite feeling, the dullness of place, where everything becomes oppressive and life becomes tedious. Drudgery is part of commitment to place; it is acceptance of restrictions. The word nostalgia was coined by Johannis Hofer (1678) to describe an illness characterized by insomnia, palpitations, stupor, fever, and persistent thought of home. The disease could result in death. Thus far, the sense of place cannot be gleaned from an analysis of the nervous system. Yet a place shapes the nervous system, somehow, until it becomes a necessary part of health.

1.7.4.2. Field of Place

The making of places from undifferentiated wilderness is the living ordering of the world into places. A place changes qualitatively; it becomes structured. Natural complexity decreases as the human increases, although the two are not mutually exclusive. Places are ecosystems intimately associated with living beings. Fitness is achieved after progressive reciprocal adaptations; it requires a stability of relationships between societies and the place.

The specificity of place is important. The earth extrudes itself into particular plants and animals; flexes mountains; and sweats weather. It is being homogenized by industry. The spirit of each place may remain to influence successive generations.

Ecology relates to at-homeness; a dwelling in neighborhood of its source. A. Portmann observed that insects and animals displayed a powerful attachment to places; that it was best understood as home. The attachment to a place is rootedness. What does it mean to be at-home? The fundamental ambiguity of existence is that humans have different capacities for feelings and awareness. Some feel strongly about a place or home; others never do. We need only to reciprocate and be available; to be and let be; reaching out and being ready to receive the other—any being. We need not feel homeless. But home is also an enclosure, a place for protection and privacy.

A place is a part of the environment claimed by feeling. Emotion binds together motion and perception. Emotion can transcend distance. Emotion creates an 'in-place'. A place must be found and made; it does not exist independently. We discover that there is an ecology of the spirit, which must be trusted like all life support systems, to find its way to balance and compensation. We need to trust a place. Home requires rootedness, at-easeness, and regeneration. Von Uexkull described the importance of rootedness in his concept of lived-world (umwelt). Feeling at home is a state of awareness; losing the feeling may cause crises.

Being inside is knowing where; it is safety, cosmos, enclosure. Inside and outside, like a dialectic, can always be reversed. Empathy is willingness to be open to significance, to know and respect symbols of place. A house is a place that provides shelter; it answers social needs; it is a repository of memories, a field of care. For the private self the house is a world; for public selves, the world is a town or civic center.

Neil Evernden asked what it does to a person to see his place, his context, destroyed? What does it do? It disrupts the field. It creates violence and apathy. Modern populations are rootless, moving about from city to city. We are suffering from a placelessness, which arises from our style of efficiency and proclamation of mass values. The environment of few significant places becomes a flatscape. It is turned into uses. John Fowles states: "We shall never fully understand nature (or ourselves) and certainly never respect it, until we dissociate the wild from the notion of usability—however innocent and harmless the use." Fowles continues, "This uselessness lies at the root of our hatred and indifference toward it."

1.8. *Education: Fun with Words Numbers Links & Images*

An effective education need not be bound by the conceptual and economic limits assumed by most institutions. A minimum education may train students for an economic role in society, but a good education teaches them how to enjoy living among other human beings in an ultrahuman nature and to perpetuate a good society. Poets like Wordsworth and Auden recommended that broad training in science and technology was necessary for poetic knowledge, which is part of a good education. Novalis considered that the study of the external world, through science, was only the first, half-way, step to full human consciousness. The second step was introspection, the contemplation of the self. Subject areas in traditional institutions concentrate on one step or the other. Any student can achieve both steps, leading to a complete education, with time and inclination. A complete education requires intense effort, discipline, patience, and a tolerance for failure. Elite institutions, with their richness of culture, can offer more potential than the pedestrian paper mills, but are often limited by social fashions and finances. An ecological education could offer greater benefits, with a teacher "leading" a pupil to an education (which is derived from the Latin word meaning 'to lead'). In his educational theory, Plotinus went still further and laid down a triple organization of education, requiring a social education, a personal and self-revealing education, and a synergetic one that would permit a perspective of the whole of human existence. Only institutions that integrate education within a balanced society can achieve this triple objective. By encouraging students who are already working outside academic walls, ecological institutions can foster this necessary kind of synergetic education.

The most valuable qualities of an ecological education are personal contact, which allows noncompetitive constructive criticism, and flexibility, which allows the educational process to fit an established and meaningful life-style. Ecological education lets one learn how to feel and live, as well as to think. As Aristotle recognized long ago, experience is necessary for thinking. The very cloisteredness of traditional university education works against it in this respect. The university fails to teach communal responsibility, self-reliance, and physical work—those qualities most dear to R. W. Emerson. Education must embrace three concepts, according to Schiller: Liberation, play, and community. Liberation is freedom from the limits of identity; play is imaginative experience; and community is the supporting matrix of life. This ideal is most closely approached in already established communities (not the artificial and involuted, temporary university dormitories) and when play and freedom are not limited by arbitrary rules and economic goals.

Ecological institutions provide for education within the larger community, in the larger context of work and recreation. Perhaps this attention to context accounts for the success of business training institutions, such as hamburger university or insurance institute. Relevant institutions stand a better chance of surviving social changes than those built on self-perpetuating administrative interests and alumni sports empires, whose vast buildings and grounds might better be turned into shelters for the poor or enclaves of employment, to produce real goods for society as part of their programs.

All human beings need a life that is protected and ordered, loving companions and contact with the wild universe. But, one cannot pretend that this little world is not part of a larger one riddled with hunger and fear—one which everyone's actions affect. The beliefs people hold are worthwhile only if they enact them in a larger world, ethically (which comes from the Sanskrit word meaning 'doing together'). Very few institutions concern themselves with the scope of ethics; that is left to the student.

Although it is pretentious to assume the responsibility for a class or culture, sincere action begins with personal responsibility and responsiveness within a smaller system. Unpopular questions, ethical expressions, and confrontations all have a price. It may be only silence or deprecation; it may be expulsion, imprisonment, or death. Everyone needs the courage to question society and express their beliefs and findings without worrying about the cost. For only through sincere, studied expression or example can anyone hope to influence the consciousness of others. And only by surrounding society with a new field of consciousness—not by attacking it—can any transformation occur. When human beings cooperate spontaneously, because they understand what it means to be human, because they understand how to treat their places on earth, only then can education be considered successful and, perhaps, lead to a more peaceful and humane world.

Garret Hardin points out that education is not just literacy, and that literacy, the ability to understand what words really mean, is not enough anyway. It needs to be supplemented by "numeracy," the ability to quantify information and interpret it intelligently— computers, remember, use numbers for everything—and, on another level, by "ecolacy," the ability to take into account the effects of complex interactions of systems over time, for understanding of the complexity of the world, that things are interconnected and affect one another. Together, these are three major filters against folly that citizens can use against the blindness, short sightedness, and sheer idiocy that experts disguise as eloquence or expertise.

1.8.1. *Literacy* (Words)

Literacy is the quality of being literate. Specifically, it is the ability to read a short passage and answer questions about what was read. The word comes from the Latin word meaning letter or later 'writing.' Being literate is being characterized by learning, cultured, educated. A person who is educated has been "lead" from ignorance, out of the self in other words, by fostering the growth and expansion of knowledge through a course of formal study. Knowledge is a condition of knowing, an acquaintance with theoretical or practical understanding. There are no limits on what can be known. But, most knowledge is concerned with survival first. It is important to know what plants to eat, where to find shelter, how to make clothing. This was, and still is, the most basic level of literacy.

But, literacy occurs on more than one level. Beyond familiarity with the terminology and applications, computer literacy is competence at solving problems using a computer, or even the professional design of new machines. Complete computer literacy is the knowledge of how the computer operates, as well as its design, manufacture, and programming—few have that complete kind of literacy. Classical literacy can survive the collapse of printers and newspapers. But, could computer literacy survive the collapse of computers? Perhaps this question hinges on the definition of literacy.

Gandhi put literacy in a similar perspective, "Literacy is not the end of education, not even the beginning, it is only one of the means." Certainly, computers can be valuable for certain aspects of education, but we must not forget what function the computer is assisting. In other words, computer use should not displace the skills themselves. Education should include a core of mathematics as a liberal education always has. And poetry and narratives should still be memorized, as well as written or examined on a computer.

Literacy, as the skill in the written and spoken language, enables readers to draw on the wisdom (and foolishness) of human beings distant in space and time. Hardin notes, in a discussion of the sins of the literate, that language has two functions beyond communication: "To promote thought and to prevent it." The second function is why literacy has to be accompanied by the ability to think critically.

1.8.2. *Numeracy* (Numbers)

Numeracy involves the ability to measure and to interpret quantities, proportions and rates. Hardin warns that human beings have learned how to use literacy to hide numbers and the need for numerate analysis. He draws attention to the problems created by always thinking solely in terms of dichotomies, e.g., safe vs. unsafe or pure vs. impure, rather than in terms of relative risks and benefits. James Lovelock has also noted our inability to assess risks mathematically. The quantitative analysis that is so important in science, technology, business, and government is dismissed with indifference. In a complex, rapidly changing society, understanding quantities, ratios, rates and duration of time is crucial. Numeracy has limitations, also—the conclusions of an accurate mathematical analysis are only as good as its premises.

We have already had fun with words in section 1.3.2. We can have fun with links and connections throughout the remainder of this project. There are many interesting numbers that play a part in the design of the planet. The Universe has 100 billion galaxies in the universe; there are 100 billion stars per galaxy. The Sun is a fusion reactor; its mass is 330,000 times earth, and the volume is 1.3 million times earth. Its core density is 300 billion earth atmospheres and central temperature is 15 million degrees C. The surface temperature is 5500 degrees C. Hydrogen gets converted to helium at 600 million tons per second. The size of the sun (864,000 miles diameter) is a balance between gravity and gas pressure. The sun has increased its output by 25% in the past several billion years.

The sun is orbited by eight planets, including earth. Earth it is part of double planet system with the moon. Its diameter is 8000 miles. Its elliptical orbit is off 8 degrees from a circular orbit of sun. It tilts on its axis, but the moon stabilizes the tilt. The surface is 197 million square miles, 29% of which is land (and 11% of that is arable). Temperatures range from –127F to 136F. Mountains are limited in height by the thickness of the crust (not a problem with Hawaii, which rests below the crust).

At the beginning of the last ice age, 100,000 years ago, human numbers may have dropped to 10,000 adults or merely 600 adults. In the Upper Paleolithic, human numbers rose to 500,000, then 10,000 years ago (before agriculture) to 6,000,000, and now 6,600,000,000. The changes in density were also dramatic: 100,000 YBP, there was 1 person/12,500 km^2 or 0.00001 person per square kilometer. In the Upper Paleolithic, the numbers were 1/255 km^2 or 0.08. At 10,000 YBP 1/25 km^2 and at 5,000 8/25 km^2 and at 2,000 YBP 42/km^2 and 300 years ago 160/km^2 and now 1013 people per square kilometer.

Human populations changed the character of many ecosystems. In 1979, rainforest was disappearing at a rate of 300 acres per hour. In 2000, 1500 acres an hour became desert. Brazil lost 25,000 square kilometers of jungle in 2002. Indonesia loses almost as much. Forest worldwide is lost at 0.4 hectares per second (= 1 acre). M. Williams calculates that over 7×10^6 km^2 of dense forest and $3.13. \times 10^6$ km^2 of open forest have been eliminated since post-glacial times.

In 1986 Peter Vitousek calculated 40% of all plant energy was used or wasted or destroyed. Around 1980, human demands for plant materials was equal to the regenerative capacity of the earth. In 1999, it exceeded it by 20 percent

For the first time in human history, in the year 2000, as many people lived in large communities (over 20,000) as in small communities (under 20,000). By 2001, the number of people living in cities reached 56 percent. Sixty percent of Earth's inhabitants are expected to live in cities by 2030, according to the United Nations—the same year global carbon dioxide emissions are expected to increase by almost two-thirds of what they are today. Intense interactions of people in larger concentrations also produce more information. Information in the entire world is estimated at 2000 petabytes (petabyte = billion megabytes or a quadrillion books of 170 pages) by K. Kelly. What was it in 12,000 BC or 0 BC or 1400 by the way? The econosphere may be as big as the biosphere, according to Kelly. In 1998 dollars he estimated the global infrastructure is worth 4 quadrillion dollars—perhaps an overestimate. Mary Daily estimated ecosystem services of the earth at 33 trillion. This is doubtless an underestimate. If you include the worth of the capital that is producing those services, at some nominal interest, such as 5 percent, then the value of the living earth is over 3 quadrillion dollars, which still might be an underestimate.

1.8.3. *Ecolacy* (Links)

Ecolacy was once achieved by studying natural history, the plants and animals that surrounded every human group. Ecolacy is the ability to ask and answer the question: 'And then what?' This would allow the effects of the interactions of systems over time can be taken into account.

Scientists have been extremely successful using reductionistic methodology on every problem, breaking problems down into their components and studying the properties of these components and their interactions. This has led to the ascendancy of mechanical science—thinking that one can do just one thing. Garrett Hardin stated the important ecological understanding of ecolacy as: 'We can never do merely one thing." This statement is now known as Hardin's Law, and it means that there are always wanted and unwanted effects, products and wastes.

Reality is composed of causal chains of events rather than single effects. Events are embedded in causal networks and are produced by multiple causes and have multiple effects, each of which triggers a causal chain of future events. Hardin contends that since we cannot do just one thing we must always ask and answer that question: "And then what?" We have to try to ascertain the benefits and costs of proposed courses of action on both the individual, social, and ecological levels. The ecological systems way of thinking employs scientific theories and knowledge to study the interlocking processes characterized by many reciprocal cause effect pathways. The ecological systems way of thinking must become an integral part of the thinking of the well educated person if we are to adequately control technology and human actions.

1.8.4. *Discussion of Filters*

David Hargreaves judges that our current educational institutions resemble a curious mix of a factory, an asylum and a prison. This command and control model creates pressure on overloaded people. Ivan Illich's proposal to deschool society means creating learning webs by using existing technologies and spaces, such as town halls. Learners would connect with their peers in new contexts to learn. Learning would not be funneled through one teacher.

The world is too complex for our minds, suggested G. P. Marsh and many later thinkers. So, our minds have to filter out what is less important. We filter data, arguments, emotions, and information. The filter allows a total picture of the whole with relatively little information. Of course that picture might be wrong. But, it is clear, as a result of filtering and thinking. Other human activities act as filters, also, especially culture.

In education and communication, noise is a problem. We are flooded with information, and much of that noise. We have to filter out as much noise as possible and much of the information. But, many cultural filtering systems—William Thackera mentions peer reviews—are collapsing.

The industrial system developed like a jigsaw puzzle with an unknown design. With only a few pieces the pattern is unknown, notes Charles Taylor. Partly, however, that is because we are making the pattern, generating it, as we go along, with bricolage. Our efforts at aggregation and filtering result in a form. As we generate it, more of the pieces fill in and a pattern emerges. But, if the pattern does not fit the environment, it has to change.

Garrett Hardin contends that most of the major controversies of our time can be understood as the result of the participants relying too much on any single one of these three filters. No one filter by itself is adequate for understanding reality and predicting the consequences of our actions. There is, however, a filter that has gone unmentioned, in its ability to concentrate or distort human perception. That filter is the image.

1.8.5. *Imagacy* (Images)

When a hierarchical order cannot adapt its function to an actual situation, then there is a synchronous change in the vertical order. The change in order is discontinuous, although the function is continuous. Since the details are not abstracted to a higher level, one cannot predict when the revolutionary change will take place. In general, revolution is discontinuous and synchronous, whereas evolution first appears to be continuous and diachronous. In the synchronic mode, form is complete as soon as it appears; in the diachronic, the form is slowly elaborated. It would appear that complex processes, like evolution, actually use both modes.

Hierarchical thinking is limited in cybernetics. The concept of control might better be stated as kind of reciprocity or mutual causality. Furthermore, in an open system, the environment, structure, program, and feedback all govern the system in concert. Hazel Henderson states that only the system can model or manage the system, but this is not entirely true. Images can model the system in miniature.

An image is an imitation or representation of something. It can also be a symbol or type, a metaphor or concept. An image can stand for something else, for instance the image of a dove is often used as a symbol for peace. In the etymological sense a symbol is something 'thrown together,' as a problem is something 'thrown forward.' Unlike an image, a symbol often represents some other thing, process or quality. Symbols can be processed by 'Analytical Engines' or computers. These machines have been used for metaphors of the brain, which also processes symbols, that is, the operation of both parts of the metaphor, brain=computer, can be described in terms of algorithms, or mathematical rules for manipulating symbols.

An image of course can be used in a variety of ways. It can reference similarity, correlation or a formal linkage, according to the categories of association developed by C.S. Peirce: icon, index, or symbol. Icons have a similarity with an object, as when a landscape painting depicts a landscape. An index is when an image is causally linked with something else, as when a wolf's howl is related to location or emotion. A symbol has a social convention that establishes the relationship linking an image and a thing, as the Coat of Arms of the Woulfe family symbolizes the family line. Physical connection is not necessary for any of these modes. Peirce suggested that the difference between the modes of reference could be understood in terms of hierarchical levels of interpretation. Symbolic relationships are composed of relationships between sets of indices, and indices are composed of iconic relationships between sets of icons. Understanding a higher level means decomposing it to a lower-order form. Understanding the logic of images, as well as their power, is important to their use in communicating and design.

An image, especially a cosmology or an image of the world, models the system in miniature. From the image, which can be a paradigm or mind-set, a whole system can arise, with unique goals, rules, parameters, and structures. A cosmology includes a mythology constructed as a poetic system. Joseph Campbell states "Mythology—and therefore civilization—is a poetic, supernormal image." Mythologies and religions can be understood as great poems. When recognized as such, they point through individual things and events to the ubiquity of a presence that is whole in each. This is what P.B. Shelley meant when he wrote that poets are the 'unacknowledged legislators of the world'—not that they pass laws or prophesize the future, but that they generate images for the future, and these images can be articulated into goals to influence our actions. Bad images, from indifferent poets, can relate to severe cultural problems, as when popular Italian poets romanticized the violence and hatred of Fascism. The mechanistic images of science, from Shelley to the Fascists, determined much of the violent conquest of nature.

Kenneth Bounding notes that the image as a cognitive construct of the world has several aspects: Spatial, temporal, personal, relational, value, and affection (emotional) for each individual. The total sum of individual images is a world of interrelated constructs. This parallels the experiences of other living beings. Using its senses, each participant creates an image of nature, or world—umwelt, life-world, is the term used by Jakob von Uexkull)—from the sensations that are meaningful to it, and which limit it. Simple beings, such as bacteria, make a relatively simple image, whereas more complex beings, such as apes and humans, forge more complex images.

In fact, human beings design complex images for a variety of purposes, including to make a profit or to persuade others to join a group. In this, design reflects the economic and political bents of humans. In a hyperactive marketplace, design responds with sophisticated images and crass intentions. In the city, design responds with an architectural icon, the skyscraper, for expressing power, status, success, and victory over limits and the environment—however temporarily.

In the discussions of design, the importances and weaknesses of images will be explored. Images will be added to the complex ecology of people, projects, tools, and social structures in an open living system. William Thackera wondered if such an ecology could be designed. Perhaps it can.

Figure 1-1. Late Harvest at Altazor Forest

2.0. Approaching Local Regional & Planetary Systems with Design

How can we approach an ecosystem as a whole? Scientifically? Technologically? Aesthetically? The planet is relatively old and large, with a unique history and many dynamic ecosystems. There are ways of dealing with systems that are not scientific or technological; they are aesthetic or ethical. To examine something that is larger than our logic or perception, we need a different approach. To examine nature in general, we must shift to a taoistic approach, asking rather than telling, and observing rather than manipulating. We need to be receptive and passive, rather than active and forceful, and nonintruding and noncontrolling, instead of pushy and manipulative. A taoistic approach stresses noninterfering observation rather than controlling manipulation; it is receptive rather than forceful. This is part of the paradox of duality; it is detached yet concerned, free yet committed, and independent yet responsible. Perhaps the taoist is another path to objectivity with greater perception.

We are skilled at iconic perception and aesthetic perception, but seem unable to use a double perception. Aesthetic perception offers a reassuring vision, which interprets or identifies nature—we need a naive vision, which surprises, shocks, fascinates or seduces the senses, which awakens desire and stirs the imagination, and which furnishes a feeling of the invisible. Iconic perception offers a cognitive view of the planet that reduces the planet to the limits of logic—we need a holistic view, compounded with the unconscious and feeling.

Nature is an extremely sensitive nexus of means and ends. Nature is a feeling system. We need a new animism to approach nature, the understanding that perception is part of feeling. This animism would allow us to behave "as if" nature was intelligent and sensitive, and as if we were loving and wise. Loving perception provides kinds of knowledge not available to nonlovers; this is especially true in ethological literature. Abraham Maslow cites his own work with monkeys. K. Lorenz, N. Tinbergen, G. Schaller, J. Van Lowick-Goodall, and M. W. Fox have found it to be true. This is the way a good psychotherapist, teacher, scientist, parent, or friend functions. And that has to define our approach to the planet.

We can apply designs to ecosystems on many levels. We are doing so accidentally now. But, the designs have to consider the limits of systems, as well as limits to the human enterprises. We might start by designing ecosystems, artificial ones on a small scale, for the express response to problems caused by inappropriate scale, interference and waste disposal.

2.1. *Human Design & Systems*

Humans have designed many things, from tools to trade routes. And, we have changed many different ecosystems to be more productive for us. But, we have not attempted to design humanity, human systems or ecosystem patterns.

2.1.1. *Why Design?*
Why should we need to design ecosystems? An ecosystem is self-making, self-managing, and self-renewing. What can human design add to that? As a young, exploring innovative species, we have added diversity and interest to other living forms. We became hosts for many new parasites and types of bacteria. Our feeding patterns encouraged other species, which did not taste as good to us, to compete successfully. Other animals and plants found us useful to live near. As an older, confident, successful species, we started to convert ecosystems to what we wanted. Now that we are self-satisfied and arrogant, we find that we are influencing every

natural system, taking what we need from some ecosystems, enhancing a few, misusing others, and interfering with the rest; we are creating patterns that we cannot control. So, what can we do? We can control ourselves, we can control our influences, and we can abstain from trying to control every variable in a system and every system on the planet. That may be the concern of the design of human limits.

There is no guarantee that nature can provide humans with everything they want. Recognizing the lack of guarantee is simply recognizing that nature is wild and we must come to terms with nonhuman beings and processes. It is not enough to arrange trees in rows to maximize future harvests; it is not enough to preserve small areas of old-growth without natural disturbances. We must pay attention to the processes that make up the habitat, for example, the role of herbivores on trimming vegetation (and diversifying it by predation). The design of an ecosystem and its management must ensure that the processes operate to maintain a dynamic state. Furthermore, the context must be conserved. The ecosystem, however, cannot be considered outside of the context of the entire landscape, including human images and institutions.

We need good ecological designs to restore the balance between human needs and natural processes. Ecological designs focus on whole communities that work in the same self-sustaining and self-limiting ways as nature. By consciously creating meaningful order, we can develop ways of producing widespread community wealth while positioning the community for a long, sustainable future in a healthy environment.

What are we considering when we talk about the design of regions? Should we build new islands or paths? Restore species or guilds that have dropped out or become extinct? People are often discussing solutions to local problems, such as deforestation, that would entail massive planting programs in some places. These technical aspects may also be the concern of a design of physical alterations.

2.1.2. *What is the Local or Global?*

What does the word local mean? A local system, for instance, has unique characteristics that emerge from smaller local interactions. What does it mean to be local? What does the word 'global' mean? In the Old English Dictionary, it meant 'spherical' referring to the earth. By 1945, the word was used by the United Nations to describe political action, referring to actions to commonize costs. The United Nations used it thus. Marshall McLuhan and other used it in reference to a global village. Then, it began to mean ubiquitous, or everywhere. Garrett Hardin contrasts the ubiquity of potholes around the world with the specific meaning of a property of the planet earth, such as the atmosphere or ocean, perhaps life. Global things are quite different than things that appear globally. The global system, for instance, has unique characteristics that emerge from local interactions. What does it mean to be global or local? These terms are terms of scale, used to describe a field.

2.1.2.1. The Field

The concept of the field allows many things to be related internally and connected by scale. These operations of the field scale to other aspects of the field need to be addressed by designs on various levels. These properties of the field are important to design because they must be considered if the design is to be stable and useful. Local design has to address these properties to be successful as design. Local design has to consider the emergence of new characteristics and the qualities of patterns, especially those related to connectedness and scale.

Local system properties are a set of constant constraints on local systems. These differences will have important affects on local ecological design. Cultural systems develop patterns (especially as related to design): Paths, Nodes, Networks (hierarchy), Fractals (Explosions Implosions, traps), Spirals, and Cycles; and these can create problems with feedback to local design (where there is not enough feedback or none).

2.1.2.2. Local & Global Scale Differences

Differences in scale are often called local or global differences (in systems theory). Concepts of scale apply to physical and biological fields. Local fields or events are those separated from other fields or events and exert minimal influence on the others. That is, the internal connections inside a local field are stronger than the connections between local fields.

Local systems create regional and global systems; at the same time local systems emerge from a regional and global system, which has characteristics that emerge from the interplay of local systems. Regional and global systems have properties that no local systems have, such as an overall climate or global biogeochemical cycles. James Lovelock suggests that the interaction between hardwood forests and softwood forests may act as a global regulator of oxygen for the entire planet. Many things, such as human poverty and species extinctions, only seem global because they are happening in many local systems at the same time, and may effect regional cycles.

Local fields have characteristics that distinguish them from the regional fields. Local fields can be characterized by: Separation, limited influence, disjunctiveness (the future is not like the past), reflection of local conditions, contribution to the regional, and territoriality. Separation refers to location in the regional field (space-time), as well as having different make-up and different environment. The local is independent of other local fields. Local properties may not determine the variables in other local fields. Having a limited influence means that there is often a weak connection with other local systems, and sometimes little affect on the global. A local field has an immediate effect on only a small fraction of the totality of parts of a regional field. Ross Ashby uses the example of a chemical reaction of silver nitrate and sodium chloride, producing AgCl, but not influencing the regional system. A local system reflects local conditions. Of course, local systems develop historically. In this case local fields have similarities with regional fields.

Table 2122-1. Local Regional & Global Properties

Local	Regional / Global
Separation	Connection
Limited influence on other locals	Affect on all locals
Reflection of regional influence	Constraint on local
Disjunctiveness (future different)	Internal temporal organization
Contribution to regional	Independent of any local
Limited by regional	Emergent properties from local

Local fields can be described as: Emergent, with new properties; internally organized; constraining the local, affecting all locals, and independence from any local. In general, regional systems emerge from local systems; they have unique properties that local systems do not have. The properties are not repeated independently in local systems. Often the properties are too interconnected to be independent. Ashby uses the example of chemical and biological reactions that form protoplasm. Properties are emergent if they are not predictable from the analysis of local systems. For example, hydrogen and oxygen are gases; under certain circumstances, they combine to form a fluid, water. For another example, the tasteless gases hydrogen and oxygen can combine with carbon, in certain biological processes, and form a sweet, solid sugar. In a third example, amino acids, which can form as a result of volcanism in an early atmosphere of the earth, can be combined in living processes to make proto-plasm, which is self-making; this new property is a surprise. The new emergent properties are protected from factors that could reverse or destroy them. Furthermore, they are not reversible due to historical development. The regional can organize local systems in internal relations (and have relations with an external environment). The regional puts constraints on the local; that is, it limits the action of local systems. The regional has connections to all local systems, even those that have no connections to other local systems. Regional change affects all local systems. But, a regional system is independent of any local system; the local may have minimal influence on the regional. Regional systems develop historically. As regional systems become larger, when the range of size of part to the whole is larger, the properties of the whole are more likely to be very different than the local.

The regional level has many more kinds of feedback, as it consists of many more sub-systems. Furthermore, the rate of feedback is expected to be much slower. This means longer times and longer lag times or delays. Laws defining constraints on the whole cannot be de-rived from the laws of parts. Regional systems properties are a set of constant constraints on local systems. These differences will have important affects on regional ecological design.

Switching our attention from a focus to the larger frame usually involves a change in scale. Scale has several meanings. Basically, as it is used here, it has to do with the level of measurement in a space/time energy/mass context. For instance, in forestry the measure-ments of leaf litter by foresters can be made at several scales: Single tree, stand, annual mea-surements, or stand life measurements. Processes that are unimportant at a small scale might be vital at a large scale. Too much litter by one tree might be unimportant, but too much litter in one stand in one year, for instance, might suppress a soil cycle; too little litter over a century might interfere with several regional or global atmospheric cycles (for more forestry examples of scale in design, and delays of feedback, see Chapter 3.3).

2.1.3. *Why Start with the Local or Global?*

Because an ecosystem is a unique whole, local design has to address the uniqueness of the whole, the special emergent properties of the whole. For example, it makes no sense to dis-cuss the amount of CO_2 in the atmosphere without paying attention to the carbon, oxygen and water cycles—or in fact, any cycles, ecosystems, or organisms that impinge on carbon dioxide—or in fact the deep history of the atmosphere and planet, as well as the structure and function of the atmosphere.

All the layers and processes within the whole are subject to restraints and limits on the whole system. The whole is not equal to its levels; it is more than the sum of the levels, to sharpen an old saw, so to speak. The local system moves and develops as a whole, subject to regional and global constraints. The whole system maintains its identity through continuous self-making, or through autopoesis in the words of Francisco Varela. The system develops through a continuous dance of autonomy and control, according to Varela. By reducing the study of the planet to the chemistry and physics of a subwhole, or holon, science can lose sight of larger emergent patterns.

The framework of our thought has to be concerned with the whole idea of ecosystems bound up in living patterns. A coherent, flexible framework can provide the context for the problems of atmospheric warming or extinction pulses. The organization of our thought requires the vision of the whole. Our efforts have to be directed to the good of the whole system.

2.2. *Holecological History & Life*

Places and life have a long history on the planet. They are influenced not only by the geology of the planet, but by the moon and solar system. The history of the planet very much shaped the history of life, until life itself began to change the atmosphere and the ocean. Variations allowed by the electric force enable rare events to occur; life is a rare event. Life is a rare, but possibly a certain event. Given definite conditions and processes, it is certain to arise. Rare events are fundamental to the universe. Living beings all use the same 20 amino acids, but arrange them individually according to a species blueprint, from which the protein is built. Even the least complex protein has 100 acids in a chain—20 to the 100th power possible proteins. But most are never tried. If one combination occurred every four seconds over four billion years, only 10 to the 60 power possibilities would have been tried. Even figuring trials on all the estimated planets in the universe (10 to the 20th power), only 10 to the 80th power would have been tried. The enormous information capacity of biological molecules makes production by chance improbable. A random process could not have produced even a single organic molecule. But once one rare event occurs, another becomes possible. Humans have difficulty observing events on this time scale. Rare events are not studied scientifically. Rare events are also revolutionary, and may be very important. History is the study of rare events.

2.2.1. *Geological Cycles & Climate* (Moon to Elements)

The geological level contains many resources for living beings. It also provides a stable platform, at least in terms of the lifetimes for most living beings. The reasons for this relative stability reside in the composition of the planet and in the double-planet influence of the moon on the earth's wobble and tides. The moon not only stabilizes the tilt of the earth as the two revolve around each other, but it provides energy and challenges to organisms in the ocean.

The amount of light hitting the atmosphere in the short-term is two gram-calories per square centimeter per minute—this is the solar constant (a gram-calorie is the amount of heat required to raise the temperature of one gram of water one degree Centigrade at fifteen degrees Centigrade; this amount is relatively small, so Kilocalories is used as a more convenient measure), which varies minutely. This radiation is reflected or attenuated as it hits the atmosphere—on a clear day at noon, perhaps two-thirds of it reaches the surface. Light also provides a suite of challenges for living organisms.

The ecosphere acts as one system in which energy from the sun is cycled. The functioning biomass is integrated by feedback responses to extract the maximum of energy and still maintain a balance. Most of the solar energy is used for maintenance by the biosphere, which indicates the biosphere's high degree of ecological maturity. Chemical equilibrium is a global regulator. The steady effect of light, the availability of oxygen, the thickness of soil, and the area and depth of the oceans is almost a steady state, or rather a steady flow, and that results in a good mean temperature for the planet. Living fossils demonstrate a continuously available environment for long periods; horseshoe crabs show that the sea has been relatively unchanged for the past 200 million years; algae goes back to microfossils that are over 3 billion years old. Elements frequently associated with life—calcium, phosphorus, iron, and sulfur—are preferentially concentrated and deposited by living organisms.

The single most important system factor is climate, and for humans especially, the fluctuation of the climate over the past 10,000 years. Climate determines the distribution of water. We depend on water, and on its collection in aquifers, for drinking and irrigation.

Because the climate is variable, the ecosystem is ever changing, and humans have unforeseen effects on the system, we have to consider water use and recycling. Yet, the climate is roughly predictable and stable, whereas the weather seems unstable and unpredictable.

2.2.2. *Holecological History*

Water and gas in the atmosphere seem to have resulted from volcanic outgassing after the formation of the planet, although bolide collisions may have contributed significant quantities of mineral atoms and water molecules. The original atmosphere of cosmic gases—neon, argon, and xenon—may have been lost as the planet heated up (although argon is still well-represented). The new constituents—carbon, nitrogen, oxygen, hydrogen, and sulfur—came from inside the earth. These changes may have created a reducing atmosphere. Free oxygen may not have been abundant until after life developed.

The surface of the planet was exposed to ultraviolet, until an ozone layer formed. Silicon, oxygen, magnesium, and aluminum in stone have a fantastic history. Continental plates turned over the crust, bringing mineral deposits to the surface, burying others. The plates also pushed land, weather and environment. Each of the continents is unique and takes a part of the cycling of elements.

Variations allowed by the electric force enable rare events to occur. Life is a rare event, although given definite conditions and processes, it is certain to arise. Rare events are fundamental to the universe. In a long evolutionary process rare events are very significant. Events can be divided into regular, like heartbeats or solar revolutions that occur in cyclic succession, and irregular, like earthquakes or mutations. The rarity of irregular events can be measured by a period of probability. A devastating earthquake that is expected about every 1000 years has a probability of 1/12,000 in any month. It would not be wise to situate a nuclear waste dump, with material having a half-life of 5,000 years or longer, in such a place. Over long enough periods of time, many improbable events are certain to occur. If the creation of self-replicating molecule has an annual probability of one in a billion, it is fairly certain to occur in a billion years.

Life depends on chemical and electrical events. The balance between fidelity and change depended on electrical forces between atoms in living creatures being neither too weak nor too strong; too sloppy and everything would perish of genetic mutation or too exact and all would remain at level of microbes. The electrical force strikes balance only in certain materials at a certain temperature range. Heredity is embedded in a chemical code. Plants are basically chemical systems. An animal is an electronic system dominating a chemical one. The level of the animal is judged by the complexity of the system.

The steady effect of light, the availability of oxygen, the thickness of soil, and the area and depth of the oceans is almost a steady state, as is the diversity of living species and the mean temperature of the planet. Living fossils demonstrate a continuously available environment for long periods; the survival of horseshoe crabs indicates that the sea has been unchanged for 200 million years; algae go back to microfossils 3.2 billion years old. The amount of material that has passed through living organisms since Cambrian times, roughly a billion years, is of the same order of magnitude as the earth's crust. Assuming carbon is fixed each year by, the total mass recycled of carbon alone exceeds the mass of the crust. Total coal and oil deposited in lithosphere are estimated at a percentage of the crust.

Elements frequently associated with life—calcium, phosphorus, iron, and sulfur—are preferentially concentrated and deposited. Life has created the present soils and atmosphere out of the original rocks and gases. Autotrophs remove high energy CO_2, H_2O and traces of other compounds and elements from the environment and convert them into low energy compounds. Detritus accumulates when autotrophs, with a smaller number of heterotrophs, die and are not immediately consumed or recycled; these ordered compounds form deposits. Marine organisms produce deposits of oil (possibly in a reducing atmosphere); forests leave coal. Microorganisms may have selectively deposited iron oxide in a shallow sea to form the Mesabe iron range. Ecomass has been steadily growing since Cambrian times, thereby increasing the order of the earth's surface. The earth breathes. Berkner suggests that there may have been cycles of oxygen production and carbon dioxide consumption depending on relative abundances of plant and animal life. In natural systems, balance is maintained when everything changes.

2.2.2.1. Deep History of the Current System

The current planetary system rose during the Cambrian explosion, about 600 million years ago; less than 10% of the history of the planet. The evolution of eukaryotic cell allowed explosive radiation. Life has been relatively quiet since then, as it was before. The Cambrian explosion could have been predictable outcome of a process set in motion far earlier. Much life rose during the initial rapid diversification, but much died out during stable times. The one burst must have filled up the oceans. Since then evolution has basically recycled the basic designs. S.J. Gould claims that the mystery of the Permian extinction can be solved by an ecological theory relating organic diversity to habitable area. In short, the Pangaea coalescence caused the area of shallow seas to shrink and deepen drastically, causing extinctions.

By 400 million years ago, in the early Devonian period, primitive plants covered much of the surface of land; in order to compete better for light some plants developed strong stems and vessels to support their leaves above other plants. These plants quickly, in an evolutionary timeframe anyway, attained heights of 5-6 centimeters (or several inches). Partly as a result of competition and of predation by early terrestrial arthropods—scorpions, spiders, and centipedes—these primitive plants competed and diversified into club mosses, horsetails, and ferns. They also grew taller, which challenged insects to climb to get to the nutritious tissues, and eventually to glide or fly between leaves.

After 50 million years of evolving, and now in the Carboniferous period, club mosses could grow to 52 meters high (170 ft.), while horsetails could top 15 meters (50 ft.) and ferns 21 meters (70 ft., with a 24-inch diameter!). The giant "seed ferns" were neither truly ferns or trees or flowering plants. In terms of height and complexity, these forests were as dramatic as any modern tropical forest, although most of these species are extinct and little is known about their relationships.

These first forests were composed of such trees in dense jungles over swampy ground. The tree crowns were formed by spreading branches covered with green spiny leaves; limbs and trunk were coated with brown scales. Although insects—beetles, flies, lice, dragonflies, and cockroaches—and amphibians were evolving throughout this period, there were no: flowers, fruits, nuts, pollen, seeds, nectar, frogs, reptiles, birds, or mammals. The jungle was relatively noiseless and monotone. For over 100 million years (345-225 million years YBP—years before present), while dinosaurs and winged insects were evolving, these seed ferns covered vast expanses of land. As the continents broke up further from one large plate, Pangaea, these forests formed the basis for most coal fields found around the earth. This break-up caused a massive drying of the land.

About the time glaciers started appearing on several of the continents, 225 million years ago, cone-bearing plants, conifers similar to modern pines and spruces, supplanted (no pun intended) the seed ferns. Conifers could colonize dry ground, especially because of the innovations of pollen and seed, whereas ferns had to live near water and mate with their immediate neighbors. Pollen in clouds can travel hundreds of miles, offering greater variety for mating. The seed ferns had naked seeds not enclosed in fruit and were, according to Lynn Margulis, sensitive to cold. The conifers had a greater tolerance to cold, even subzero temperatures, because the seeds were wrapped, even though they are labeled gymnosperms, 'naked seeds'). The fungal root networks allowed them to ingest phosphorus and nitrogen and to expand into higher elevations and colder latitudes. The first conifers were the main diet of vegetarian dinosaurs.

After roughly another 100 million years, plants with flowers descended from the same plants that produced the seed ferns. Within a mere nine million years, flowering plants had colonized most of the land areas, coevolving with the first mammals, warm-blooded egg-layers and small marsupials. Coevolution of life forms is far more important than previously thought. Mammalian interest in these plants as food probably leads to their rapid dissemination. Flowering plants (or angiosperms, 'clothed seeds') had fruits and seeds to protect the embryos from being made into animal flesh. Margulis suggests that plants have been seducing animals for millions of years, tricking us into helping them to move and forcing us into more complex patterns of behavior. Angiosperms developed concurrently with mountain building, such as the Alps and Himalayas, in the Miocene. Coniferous forests began to diminish in the Miocene as grasslands spread and deciduous trees appeared. Poplar and plane trees were established by the early Cretaceous. Deciduous trees had evolved with many more plant, insect, and animal partners.

The glacial-interglacial alternation during the last million years has a periodicity of about 100,00 years, with interglacial durations of 10,000-12,500 years. The current interglacial time began about 11,000 years ago, after glaciers covered vast areas of the northern hemisphere. The advance and retreat of ice sheets and climactic changes have been linked with large migrations of plants and animals, and more recently peoples. During the last glacial stage in the Pleistocene, both coniferous and deciduous woodlands were forced southward, often to isolated refuges. The return of the forests by the Holocene, our current geological epoch, is one of the great stories of natural history, according to Neil Roberts. Even in the past 10,000 years forests have expanded and contracted with environmental change, for instance, the vast dawn redwood forests north of the present Arctic circle died out in the Eocene; long thought extinct, fragments remained in China, preserved on monastery grounds, to be discovered again a few decades ago. The shape of forests depends on a suite of physical conditions that vary from continent to continent. Weather patterns on North America, for instance, have created, in Paul Colinvaux's term, "nation states" of trees that surprised the first European naturalists because they were so different from the European forests. Ecohistory should consider human dispersions, adaptations, invasions of wild areas, perceptual problems, and language and mythology as tools of cognition.

2.2.2.2. What is Life?

There are several approaches to defining life: Intuitive, functional, structural, synthetic, or abstract. The intuitive definition assumes that we can recognize living beings: A pine marten is aggressively alive; salt is not. One functional definition states: A living being affects a local decrease in entropy. This definition is not exclusive, however, since crystals also decrease entropy. A sample structural definition is: A living being is made of cells. This is inadequate, also, since dead organisms are made of cells.

A synthetic definition combines the latter two: A living organisms is made of cells, grows and reproduces, and causes a local decrease in entropy. But, the synthetic definition leaves out computers, as well as viruses.

A molecular biological definition of life does not need to mention cells: Life is a structural replication of enzymes based on a close (but imperfect) reproduction of nucleic acid molecules. By this definition, a virus is alive, even though it uses the host's enzymes. Enzymes are tools, like hammers or computers. Life can be characterized by the possession of at least one nucleic acid capable of replication.

Nucleic acid is the basis of organic life, but it is not the only means of carrying information for a self-replicative pattern. Consider this final, more abstract definition, that could include the possibility of computer life: A living being is complex, reactive, and able to maintain itself under changing conditions; it has a stable structure that carries information on its pattern, and it can reproduce its pattern in a separate being.

The classical criteria of life are: irritability, the ability to react to change through motion; metabolism, the ability to compensate for energy loss; variability and selection, the ability to enlarge the amount of information available; memory, the ability to keep the information immediate; and, self-reproduction, the ability to preserve biological information after individual death. In ordinary human experience, it is not difficult to distinguish living forms from nonliving. From a cosmological or ecological perspective, living organisms interpenetrate deeply into nonliving forms and the earth. Individual organisms are woven into a complex fabric. Once living beings were formed on earth, their populations must have exploded to use every source of energy available. The fossil record indicates that three billion years ago the earth was inhabited only by blue-green algae. From a simple base, complex forms arose.

In the evolution of life, archaebacterium combined with swimming bacterium and then combined with green bacterium. The first protists were anaerobic, allergic to oxygen. They lived in muds, rocks, and pools, where oxygen is absent. Five kingdoms are: (1) bacteria, (2) protoctists, (3) fungi, (4) plants, and (5) animals. However, the first kingdom could be subdivided into Archaea and bacteria, on the basis of comparisons of ribosomal RNA by Carl Woese.

Viruses have no metabolism and cannot live outside cells. Probably, they evolved from bacteria. Viruses spread genes among bacteria, trees and humans. Viruses are sources of evolutionary variation. If an ecosystem is disrupted, viruses can outgrow their resources, like any living being. No virus can live outside a cell. No DNA can replicate itself outside of a cell. It requires RNA (an older version?) and enzyme proteins of a cell to replicate. Cells are membrane-bounded, self-replicating things.

How did life continue? Life involves a vast number of interacting structures. Living consists of complex behaviors whose limits are defined by rules of order that can be empirically described. The earth is suitable for life because of three kinds of limits: (1) solar radiation has stayed within certain limits for 4 billion years; (2) the biogeochemical cycles of oxygen, carbon, nitrogen, phosphorus, sulfur, water have stayed within certain limits; and (3) the environment has been constant enough for organic evolution, but variable enough for natural selection to be challenged. Other forces are also stable, such as gravity or nuclear interactions. Life is limited by elements and physical factors (light, water, gas, salt); too little of an element limits life (Liebig's law); too much limits (Shelford's law of tolerance). Life is also challenged by energy, gravity, and the moon's behavior.

2.3. *What is Natural Design?*

Poets and biologists often look at patterns and events in nature and express appreciation at how well nature designs things, from eyes to sunsets. But, we have to ask, is it design?

2.3.1. *Does Nature Design?*

Does nature design? Does life design living beings? What about self-designing systems? Does nature really design forests or streams? Perhaps we should say that nature is self-organized? The self is a scaled phenomena. The small self is organism-centered. The larger self is part of a self-organizing web. But, self-organization is not self-design. Are living beings designed? Nature obviously produces complex beings over long periods of time, and it does so in a way that has been referred to as design by people for several hundred years at least. It seems possible that motion, over time, at various scales, results in greater complexity of elements and arrangements, automatically resulting in living beings. This process takes place in a changing biophysical environment, as a form of ecological play, complete with system economies and with living values. The process of nature is described in those four categories.

More complex kinds of patterns, such as the circular pattern, suggested by the myth of eternal recurrence, depends on regular repetition, as with the seasons. It allowed disintegration to be replaced by regeneration. The helical pattern is innovatively cyclic, and the cycle is additive and seems to reach a higher stage with each repetition. Finally, there are nonlinear patterns. These seem to have more surprises due to the acceptance of complex nonequilibrium systems. They may evolve in a definite direction, but can do so by leaps and turns.

Of course perception is a large part of patterns. And, we perceive the direction as being towards more complexity and more integration, until we have a global society, coordinated on several levels, within a more complex biosphere.

2.3.2. *Natural Design*

The patterns of nature form due to the characteristics of nature. These characteristics can be grouped according to their human use, from the surrounding biophysical environment to act of valuing.

2.3.2.1. Characteristics of the Surrounding Biophysical Environment

There are some characteristics that we can say that nature has, as a result of its historical motions and patterns. This set ranges from patterning to interaction limiting (a kind of principle of least effort applied to physical process).

2.3.2.1.1. *Self-patterning.* A pattern is a regularly array of similar units. The units do not have to be exactly the same shape or size, and the regularity does not have to be perfect. Many of the laws of physics are nondeterministic laws (or stochastic) and they influence natural and human systems. Our ignorance of them lets us get caught by surprises. These laws create patterns in space, as well as in human history. There are simple kinds of patterns. A linear pattern tends to be interpreted as progress or regress—this is the dominant concept of unending progress in modern history. Patterns have a form, sometimes repetition, and sometimes regularity, but each of these is caused by some limiting factor. As Paul Weiss noted, the patterns of organic nature are a combination of order and diversity; order involves constraint while diversity requires freedom for difference. Some scientists describe organisms as being configured by energy through time, but organisms are material patterns in space as well. Regularities in systems are patterns. Paul Shepard describes living natural "objects" in terms of events that constitute a "field pattern."

2.3.2.1.2. *Creating forms.* History creates unique patterns, especially in ecosystems. Each ecosystem is unique in its form, information, and in its dynamics and history. In creation and recreation, things form and unform, reform and unform, then form again. Creativity is the process of recombination into forms.

2.3.2.1.3. *Self-making.* Nature is autopoetic, that is, self-making (from the Greek words *auto* and *poiesis*), according to Francisco Varela and colleagues. Autopoesis refers to the dynamic self-producing and self-maintaining activities of living beings. That means that organisms have identity, integrity, self-boundedness, and use external supplies of energy and materials for their own maintenance.

2.3.2.1.4. *Wholeness.* Energy, information, matter, life are part of inseparable whole, with no formal theory linking them. Relativity, quantum theory, and entropy imply individual wholeness, in which the analysis of something into distinct parts is no longer relevant to understanding the whole. Relativity and quantum theory both imply the need to regard the world as an undivided whole, not as a spectral oneness, but as a diversity of the whole from many perspectives. Wholes are mutually defining, but also self-defining or self-making. Nature is a whole system, as well as a self-making system; species and organisms are self-making. Wholeness is related to health— indeed, the word 'whole' comes from the Indo-European root *kailo*, which is also the root for the words health and holy.

2.3.2.1.5. *Systemic.* A system is a way to explain part of the universe and to deal with complex behavior, using the concepts of feedback and emergence. Systems theory analyzes events, processes, and patterns in the world. An event is a change of state that is described by two points, an initial and final state. A process is a sequence of states, also called a history. A process creates a path, described by a trajectory of states. A process is described as evolutionary if it involves emergence and the creation of new things, as in general speciation; to be a species, however, the novelty has to reproduce, multiply or diffuse. Systems have a limit in size that may be determined by structure or the speed of turnover of components. Systems may have limits in terms of the number of connections that can be maintained with other systems. Systems are considered to be wholes where the internal connections are stronger than the external connections. Connectivity is related to the size and density of a system. There is no whole system without an interconnection of its parts, and there is no whole system without an environment. Everything is connected in a system; removal of any part alters the dynamics of the system.

2.3.2.1.6. *Processing Patterns.* Nature, for A. N. Whitehead, consists of patterns whose movement is essential to their being. These patterns are analyzed into events or occasions. Process is dynamic change in an unfolding flow. Process is the fundamental feature of nature. Life has patterns as well, that are shaped by rules and physical limits. The first thing is that every living being creates an inside and walls out the outside, but not completely. Living beings tend to have a few themes that can be expressed in many variations. Living beings tend to optimize their shapes and sizes, depending on the environment. Living beings are opportunistic, that is they take advantage of new resources that may result from physical or biological processes, for instance tides or other forms of life. The act of living includes errors in form and information. Living beings build from the bottom up, even complex ones repeating older patterns of development.

2.3.2.1.7. *Development/Change & Filtering/Sorting.* Patterns change and develop. Evolution can be described as an unfolding of patterns that are filtered by a changing environment. These patterns can be analyzed into events. Biological reproduction is prolific, but is limited by processes such as predation, by material and energy cycles, and by the ability to find solutions to filtering processes. Francisco Varela analyzes the evolutionary process as satisficing rather than optimizing; that is, a suboptimal solution is adequate to continue living; striving for an optimum or maximum does not pay off in terms of investment of effort. Wallace described evolution as a conservation process, similar to the centrifugal governor of a steam engine. Darwin used 'natural selection' as a metaphor for evolution. Evolution is then the changing theme of the conversation between species and environments. Evolution may be interpreted as the development of new channels for communication. A process is evolutionary if it involves emergence and the creation of new things. The process of evolution at all levels is the formation of varying sets, and the differential elimination of some sets and survival of others. Consequences include directional change at all levels, an increase of organized diversity in the universe, and the evolution of wholes maintained by interlevel feedback in a relatively isolated and stable environment.

2.3.2.1.8. *Place generating.* A place is generated by living beings as they exist, bounded by their environment. That boundedness gives a place its identity and integrity, and allows it to maintain itself in a definite form and process. Places are patterns of things and webs of relations that can be understood by observing and participating. Like any "thing" or system, a place is extended in space and bounded, so it has an inside and outside. As part of a geographic approach, most places can be located on the earth. Like any finite thing, a place is a unique set embedded in a larger system; it is a local thing connected to other places by dynamic processes and cycles. Places have meaning for the species that have modified them by living there for generations. For humans, places have even more of a psychological dimension that reflects the personal investment of energy and emotion in a place.

2.3.2.1.9. *Place responsive.* Place can be described as a field that orders itself. The types of ordering depend on the local characteristics of place. Plants and animals respond to place by creating their own world, according to their complexity and needs. An organism is inseparably related to its habitat, the place where it lives. Individuals are related to places by basic physical factors. Place provides limits to which individuals have to adapt. Places and organisms shape each other through feedback. Insects and animals displayed a powerful attachment to places, as Adolf Portmann observed; this attachment can be best understood as home. Several metaphors have been used to describe the human place on earth: Storehouse, property, or a spaceship. But the earth is not a spaceship or storehouse; it is home.

2.3.2.1.10. *Dynamic nonequilibria.* Like any finite thing, a place is a unique set embedded in a larger system; it is a local thing connected to other places by dynamic processes and cycles. Individuals are part of complex large dynamic systems. Most of the solar energy captured is used for maintenance by the living forms on earth; this is one indicator of the biosphere's high degree of ecological maturity. The organism is a dynamic structure that is immanent and simultaneous with the process, rather than just a consequence of the selection of individuals modified by mutation. The organism is what it does. Dynamic change is the result of a process unfolding in an advancing flow through place. Process, the coming-to-be of organisms, is a fundamental feature of reality. The process of nature is not merely rhythmic change, but it is a creative advance, producing new forms in place.

2.3.2.1.11. *Interaction limiting.* To survive, an ecosystem depends on the interactions and balance of many variables, most of which are not well understood. In any system the possible number of interactions is limited by the specific variables of the system, with its complement of living forms and resources. While it is meaningful to speak of an optimum diversity, as the result of limits and the interaction of many factors, a maximum diversity may never be reached in any system. Maximum order would result in a static universe, whereas a maximum diversity would create a nonordering chaos. As Paul Weiss noted, the patterns of organic nature are a combination of order and diversity; order involves constraint while diversity requires freedom for difference.

2.3.2.2. Developing Ecologically

Biophysical systems develop into ecological systems, which can be described in terms from aggregating to actual existing.

2.3.2.2.1. *Nature-based*. Nature in general (certainly prehuman nature) is nature-based, that is, it does not seem to rely on any external factors, other than physical constrains and sunlight.

2.3.2.2.2. *Aggregating*. Aggregation is usually the first step in the self-assembly of individuals, from cells, and groups, from individuals. The word 'individual' is a metaphor. As such it refers to aggregates (sand), functional units (populations), and autonomous, self-regulating beings (a wolf). The world is a sloppy place; to describe it, it is better to build a model with a high redundancy of subsystems, each of which is coarse, but the aggregate of which gives good adjustment to environmental events. Claude Levi-Strauss described such a process as bricolage, fitting the bits together, identifying impressions of life as sets and forming them into systems. This is what nature does with organisms and niches. An organism is an intentional being that, with other members of species, can create niches, that is custom places, as well as adapt to them. M. Krebs states that the fundamental niche of a species has an "infinite number of dimensions," making a complete determination impossible. Those species that enlarge their niche also enlarge the ecology as a whole. Species integrate themselves into the system over time, using previously unused productivity or waste in developing their special niche. Each ecosystem, however, sets the limits and determines the style and complexity of its individual species.

2.3.2.2.3. *Trial & Error process/Novelty testing*. Animals learn through a kind of empirical testing from trial and error (for humans, books are a kind of redundancy of trial and error learning, or as Buckminster Fuller said, 'trial and error error error.'). Evolution is trial and error process of learning that takes place on an immense timescale of billions of years. All learning contributes to the evolution of a larger pattern. The activities of two communicators combine to make the universe of the observer more ordered and redundant. After all, the goal of life is experience, not efficiency; and redundancy promotes experience as well as stability. Natural processes that seem destructive are cyclic and preservative also. Ecosystems that seem inefficient and wasteful are many times extremely redundant, and therefore stable and flexible.

2.3.2.2.4. *Trial generating*. Natural environments test organisms with challenges. This generates trials of the kinds of organisms and can result in success or failure if the challenges are survived or not. Nature generates trials of new forms, not only on a large time scale, but on the physical scale of the planet. Most trials result in failure. Even success applied to new conditions can result in failure. Failure from success can be tragic at the human level.

2.3.2.2.5. *Novelty generation*. Ecosystems with living beings change and turn. The sediment of the past is a given and different for each present. The ceaseless activity of being creates newness, that is, novelty is born from the womb (*hystera*) of change. History is the result of this hysteretic process.

Change maintains the openness of an ecosystem, and allows novelty in the system. Individuals also use various patches and areas in one or more systems; they create paths between these areas. Their activity can change the character of these areas. The organism and environment are co-implicative, co-defining, and co-constructing. They engage in a process of self-assembly, where the complete self is the organism-environment system. The process of construction involves a self-presentation offering new symbiotic relations and novelty. Novelty always enters with environmental change, which serves to maintain the openness of the system. The "strategy" of ecosystem development is increased control of, or homeorhesis with, the physical environment and novelty—probably to protect itself from perturbations.

A process is described as evolutionary if it involves emergence and the creation of new things, as in general speciation; to be a species, however, the novelty has to reproduce, multiply or diffuse. Balance is an ideal for humans: Too much novelty is stressful, it is too new to understand, and too much of the sameness is stultifying; too much conflict is stressful, but too much peace is boring.

2.3.2.2.6. *Reciprocal interactions*. Ordering of the world makes places from wilderness. A place changes qualitatively; it becomes structured. Fitness is achieved after slow, progressive, reciprocal adaptations; it requires a stability of relationships between societies and place. The nodes in tree or net represent elements in a system. Their links suggest interconnections.

These are three general levels of interaction: Individuals, species and systems. Animals and plants, algae, bacteria, fungi, live together in ecosystems. In an ecosystem, individuals and species are connected to some degree, in terms of quality and quantity of connections. There are three basic kinds of interactions: Neutral, negative, and positive. Living together involves many kinds of interactions, from competition and conflict to cooperation and mutualism. Interactions may be reciprocal or complementary. They may dominate or control. Interactions are multidimensional. A wolf, for instance, may howl to communicate, or to restore proximity with a mate, or for simple pleasure. Many animals, such as wolves and caribou, develop together over time, adapting to each other's strategies. Paul Ehrlich and Peter Raven refer to this mutual adaptation as coevolution. Coevolving systems never completely adapt. Coevolving increases the quality of the environment.

2.3.2.2.7. *Bottom-up organization*. Complex adaptive systems display emergent behavior. As an example of emergence, slime molds can form a community without a pacemaker cell that determines when the cells need to combine. It seems that self-organization is bottoms-up. Emergent systems are rule-governed, though; slime molds explore by adhering to low level rules. Individuals coordinate work, even if they cannot assess the global situation. Emergent systems are local; individual molds "think" locally and act locally. Random action serves to explore local space. Individuals pay attention to their neighbors, however, and patterns emerge from local activity. Simple behavior seems to work, with local feedback, and more sophisticated behavior "trickles up" to approximate a global organization.

Unfolding implies the interweaving of processes of structuring at different levels. Evolution acts in the sense of a simultaneous and interdependent structuration of a micro and macroworld. Complexity emerges from the interpenetration of processes of differentiation and integration, where the processes run simultaneously from top and bottom, and shape hierarchy from both sides. The predator serves as a top-down limit for prey species, keeping them healthy by altering their behavior, removing young sick and old individuals. Of course, in the system, there are bottom-up limits also, as plants change their chemistries to attract or avoid predators of their own.

2.3.2.2.8. *Physical restraints*. The limits of the universe, like the speed of light or the quantum of a field, put limits on freedom; freedom is defined here as the absence of restraint or confinement, or the state of being free from rules or patterns. But, limits cannot be complete, any more than freedom can. Individuals adjust to the restraints of an environment, but their activities change the environment, putting new constrains on other beings or systems. The process of nature is not merely rhythmic change, it is creative play, transcending restraints and resulting in a creative advance, in A. N. Whitehead's words, producing new forms everywhere.

2.3.2.2.9. *Natural process.* The goal of processes is continuity, as living beings emerge from previous states. Species and places depend on other species and natural processes. Natural processes, such as fire, wind, or species explosions, operate freely, even as they alter the functioning of the system. Natural processes that seem destructive may preserve larger patterns. Natural processes are their own purpose and constitute their own value. Natural physical and living processes sustain each other. Natural processes take on significance of their own without reference to humanity.

2.3.2.2.10. *Actual Existing.* In fact, nothing exists by itself, that is, not in relation to other characteristics, activities, or systems. Nature takes in, or conceives, things as parts and wholes. Wholes and parts do not exist absolutely. John Wheeler's alternate worlds are really the umwelts, the existence worlds of particles. Particles choose their paths from probabilities. The universe owes its existence to trillions of acts of registration by beings, according to Wheeler. For example, since individual things or organisms are part of a field, they are internally related within the field, and they cannot exist without the field. The field, or physical society, provides the order required for producing individuals. Individuals are connected in place by infolding movement. Each animal is a participant in a field of existence. Living beings synthesize the parts and find meaning in living there, and in doing so revitalize a place. Karl Popper conjectures that besides the theory of the hostile environment (passive Darwinism), there is a complementary theory of the friendly environment. Many organisms are active explorers, searching for new, friendly environments. Conrad Waddington said that the "general anagenesis of evolution is towards what may be crudely called richness of experience." The goal of all creatures is to come into the fullness of existence.

2.3.2.3. Economizing

Nature makes an economy of the interactions of beings in patterns. This economy can be described in terms ranging from productivity to gigatrends.

2.3.2.3.1. *Productivity* (Or creative waste or natural capital). Nature is productive, that is, plants living in ecosystems transform solar energy into food energy. Systems of forms produce the materials and energy needed for others. The amounts produced by each system are limited by temperature, moisture, and a variety of other parameters. Resources and living beings in patterns are the natural capital of systems; productivity is the interest of the system, and productivity that is not use by the system or any system is the interest that is not available (although, as creative waste, it may be used later in other systems).

2.3.2.3.2. *Interrelational.* The expression of life points the mind toward the broadest meaning. It points, not toward truth, but towards an ever-enlarging relational field. It points to the frame, in fact. That which is the frame must be ambiguous. Frames themselves must be used as metaphors. For human ecology, Arne Naess rejects the image of man-in-environment for the relational, total-field image. He characterizes organisms as knots in the biospherical net, a field with intrinsic relations. The relationship with other beings becomes part of the basic constitution of a being. Relational qualities can be understood better by a morphogenetic logic, which can be characterized as relational, qualitative, symbiotic, heterogenistic, reciprocally causal, and interactionist.

2.3.2.3.3. *Nested.* Ecosystems are scalable and can be nested within others. Every thing could be considered as a set of nested ecosystems, from a body to the planet. Each nested system acts as a subwhole at a lower level and a whole at a higher (or outer) level. This nestedness allows linkages between scales.

2.3.2.3.4. *Interdependent.* As with an organism, the various parts of nature are so interdependent that nothing can be abstracted without altering the identity of it or the whole. Whole things are mutually defining, but also self-defining. Any living system is the history of the maintenance of its identity through continuous self-making or autopoesis. By comparison, unfolding implies the interweaving of processes of structuring at different levels. Evolution acts in the sense of a simultaneous and interdependent structuration of a micro and macroworld. The elegance of a fish pond and the delicacy of its stable disequilibrium are the outcome of a long evolution of interdependence. But, no one has ever made a complete census of even so simple a system as a pond.

Many local patterns flow together through time interdependently, sharing materials. The death of one pattern sometimes leads to the death of other patterns. None of the bodies or systems are completely independent or completely bounded; they are interdependent and open systems. A body or system is only maintained by a flow of energy and materials from its surrounding environment.

2.3.2.3.5. *Solar-life driven.* All of the energy on earth comes from the original formation of the universe, solar system, sun, and the planet, although the transformative energy of the sun directly drives most biogeochemical cycles as well as plant life. The traditional economies of animals and humans are based on solar energy and natural productivities. The earth is suitable for life because of three kinds of limits: the solar radiation that has stayed within certain limits for four billion years; the biogeochemical cycles of oxygen, carbon, nitrogen, phosphorus, sulfur, water, and other elements have stayed within certain limits; and, the constancy of the environment, which is constant enough for organic evolution, but variable enough for natural selection to be challenged.

The planet started as a mixture of chemical elements (over a hundred, including hydrogen, helium, and nitrogen) circulating over an active geology, driven by solar energy. The ecosphere acts as one system in which energy from the sun is cycled. The functioning biomass is integrated by feedback responses to extract enough energy and still maintain a balance. Most of the solar energy is used for maintenance by the biosphere. The regular effect of light, the availability of oxygen, the thickness of soil, and the area and depth of the oceans is almost a steady state (or homeorhetic state), that is, a constant pattern from the steady flow of energy.

2.3.2.3.6. *Aesthetic creation.* Living in place orders experience, which is an aesthetic function. As animals perceive nature, they use pieces of things as food, nutrients, resources, or tools—at least wolves, termites, and chimpanzees do. Most animals and maybe plants in a species share similar aesthetic preferences. For human beings, most prefer natural scenes similar to an African savanna: Rolling expanses of grasslands with clumps of trees in a warm but dry climate. For a long time, that environment dominated created images. This may be why humans value walking in the woods or observing the production of art.

2.3.2.3.7. *Waste generating.* As a mature ecological system becomes more efficient, it supports a larger biomass with the same amount of energy. The food chains become more web like (dominated by detritus chains as opposed to linear grazing). Mineral cycles become closed and the nutrient exchange between organisms and the environment slows. All systems generate some waste, from degraded energy to unused productivity.

Waste is defined as the materials or energy not used by a single system. Waste is a strangely inappropriate category; the waste of one system is almost always a potential resource for the next system. Even the waste of the sun, light, is a resource for plants on earth. Renewal of a system is limited when too much of the system is lost through waste. In nature, other systems develop that can use the waste of 'up-stream' systems. These become partnerships when the first system can also use wastes from the second, as a result of larger cycles. In economic terms, waste can be part of the capital for a subsequent system (in space or time).

For example, coal formed as the result of slow cycles that produced waste vegetation (not consumed by insects or animals and returned to the system—however, this waste is capital in modern industrial systems).

2.3.2.3.8. *Free play/change.* Change occurs as the result of the free play of elements in a system. Change does not seem to have a purpose or function. Nature has no cost or efficiency considerations, although more complex forms may be more efficient—actually the system, an ecosystem, may be more efficient as it matures.

2.3.2.3.9. *Gigatrends.* Things change, sometimes rapidly, sometimes slowly. Rapid change may be considered a revolution or a catastrophe; slower changes as sometimes referred to as trends or fate. A number of large, long-term ecological trends are evident, for instance, long periods of increasing or decreasing temperatures. The long-term increase in complexity that emerges from historical life activities, as well as the increase in cooperation between members of different species, either partnerships or symbioses. Ecosystems build up information. There are at least three different interacting channels of information in an ecosystem: The genetic, in replicable individuals; an ecological based on interaction between cohabiting species, expressed in changes in their numbers; and the cultural, transmitted through individual learning based on experience. Feedback within the interaction of species is expensive memory with little storage capacity. Whenever succession starts again, after a volcanic eruption, for instance, old information of interactions has not been saved. Genetic memory has a larger capacity and is long-term. Cultural memory has a large potential capacity.

2.3.2.4. Valuing

Living beings value their living. They value the things they need to keep living. They modify themselves and their environment, as a result of living.

2.3.2.4.1. *Living.* A response is a reaction to a challenge (as an event or condition). This is not the same as a responsibility that involves an obligation or accounting. During the evolutionary development of living responses to environmental challenges, living beings tested many kinds of designs for their tools, places, and living habits. These initial designs often worked well in situations of limited resources or limited power.

2.3.4.2. *Fitting.* Animals and plants try to fit into places in the environment to use what it offers to survive. That is their strategy. Patterns have a form, sometimes repetition, and sometimes regularity, but each of these is caused by some limiting factor. Fitting the pattern to patterns of the environment can lead to both continuity and predictability, and both of these are needed to adapt living activities to natural limits. Being fit is the ability to function and reproduce under normal environmental conditions. Fitness can be measured quantitatively by testing. Fitness builds up in an ecosystem as it matures. Selection at the organismic level is selection of the fit. The levels of selection must balance, so that life is not too fit or too unfit. Evolutionary fitness cannot keep increasing. In some cases it decreases with time (E. Haeckel's observations on senescence supports this idea). Perhaps species are self-limiting in fitness.

2.3.2.4.3. *Harmony-building long-term.* Harmony is related to wholeness— indeed, the word 'whole' comes from the Indo-European root *kailo*, which is the root word for harmony, health and holy. Living order can also be defined in terms of the influence of the whole over the parts. Disorder at the level of a molecule can reflect the higher level of the order of a cell. The machinery of a cell is not a permanent fixture, but is disassembled and rebuilt periodically, according to specific patterns and in harmony with the functioning of the cell in its environment, that is, the organ or body. Paul Weiss shows that parts of a cell are constantly changing, growing, dying, breaking up and recombining, but under control at a cellular level.

Harmony can contain discordant notes and themes and weave them into a rhythm.

2.3.2.4.4. *Existence Value.* In the larger view, evolution is value-free. Creation and destruction, as well as beauty and ugliness, are expressed in one complex pathway. At the same time, a reversal of values associated with an evolutionary, or ecocentric, perspective supports the concept of intrinsic value. Values usually encode information having survival or prestige importance. Perhaps the most valuable thing is living time. The experience of life—aesthetics—is also valuable. Each being has intrinsic value as a perspective, a unique packet of in-form-ation and experience. Bees have bee value; wolves have wolf value. Wolves are not efficient at binding nitrogen; neither are humans. Lichen are poor predators, but they break apart rock better than bighorn sheep. Living beings also have value to others; according to Eugene Odum, every being that is part of the food chain has value to many others. Dung has value to a dung beetle, mice to a coyote.

Individual beings may have not only self-value and other-value, but ecosystem value, as when mycorrhizal fungi fix nitrogen and support cycles and systems. Values are not always hierarchical or consistently ordered, however. Some things have value for the species (or ecosystem or culture) that may not be apparent to or wanted by the individual—in the sense that the predator contributes to the diversity and health of the prey. Some things of value to the group are harmful to the individual. Ecological value includes predictable changes to the system. In this sense, the everglades in Florida depend on hurricanes as much as Ponderosa pine forests in Oregon depend on occasional fires.

2.3.2.4.5. *Learnable.* Through living, an individual learns the seasons, the foods, and the values of the group. They learn what they need for living. Ecosystems 'learn' the rhythms of patterns of disturbance or change. The ecosystem 'learns' the changes, e.g., seasons, of the environment. Any system formed by reproducing and interacting organisms must develop an assemblage in which production of entropy per unit of information is minimized. It is a general property of some systems that acquired information is used to close the door to further inflow. A mature system needs less information, since it works toward preservation rather than growth or expansion. The limit of maturity allows maximum variability between systems with slight external differences, like temperature. Ecosystems consist of different pre-fabricated pieces: species. Since the supply of species is limited, succession becomes asymptotic, that is, it leads to maturity (or in the old terminology, a climax).

2.3.2.4.6. *Adjusting.* Living beings make the places they live in, or rather they adjust themselves to fit the places and then remake places by adjusting them to fit themselves. They fit into a place and make images of a place that also fit, that is, they adjust themselves as well. Species are place-making species. Adaptation is not simply beings adjusting themselves to place or adjusting the place to them. Each is a constraint on the other. Adaptation is a process of making fit by adjusting to circumstances, environmental or cultural. Here it means fitting into an ecosystem, within established cycles and functions. Once a living being adjusts, it learns to exclude information that may not be relevant to living. It is adaptation that improves the chances of survival for a living being. Life does not adapt to a passive prior environment, it produces and modifies its surroundings.

2.3.2.4.7. *Formation.* Formation is the interaction of a physical and an intentional process. Physical or living things take forms. The making of form is informing. Ramon Margalef considers "information" to be more basic than energy or matter and more in line with the concept of patterning. Information is not transferred because the field itself is in-form-ed.

2.3.2.4.8. *Chaotic/Free.* Nature is chaotic and unpredictable. Nature may treat species well, or not. We should recognize that nature will provide opportunities for us, but also that it may remove them through the occasional violence of its chaotic systems. There are chaotic events, plagues and random frenzies in every system. An ecosystem is defined as a biotic community and its nonliving environment, functioning together. An ecosystem is a self-organizing, chaotic system with emergent properties. unique from those of individuals, species, or communities. Ecosystems develop in time. That is, an ecosystem develops by a reasonably orderly, directional process that involves changes in structure that result from community modification of the physical environment. Although the physical environment imposes limits and sometimes determines patterns and rates of change, the community controls the development of the system. The recognition of possibilities can excite and inspire living beings.

2.3.3. *Principles of Nature?*

Are there principles of nature? Can we say that explosiveness is a principle? Or intensity? We often describe design principles as being based on basic principles of nature, such as the principle of change. It might be more accurate to consider change a property of nature, that can then become a design principle or scientific principle. Nature is in flux, culture is in flux, everything is. The climate will change, the shorelines will change. Human understanding and behavior is changing. But, nature is also self-regulating—this is the principle of self-regulation. Nature has evolved to maintain its stability in the face of many kinds of disturbances from planetesimal impacts to changes in atmospheric composition. Nature will continue to regulate itself even as human beings make dramatic changes. The danger is not so much that nature will collapse, as that humanity will lose those things that it values the most, from cool air to wilderness.

Nature seems to flow, which is necessary to the functioning of organisms as well as to the biosphere. Flow can only be realized through structure, however, a cell or an ecosystem. It is the problem of free flow or division by membranes. At each level of a natural system, from cell to biosphere, the units involved do more exchanging internally than externally with other units at the same level. Flow and division must be in balance.

Individuals in nature exist through separation, by walls, barriers or membranes, which means nonflow or limited flow. Barriers are necessary to maintain form and integrity of individuals as well as of ecosystems. By removing some barriers, such as releasing carbon that has been locked up for eons, we unbalance natural cycles. Even though ecosystems exist within large ecoregions, which exist within the biosphere, they need closure to maintain their integrity. At the level of local ecosystems, there need to be fewer closures, so that there is a flow of genetic information within a species as well as between species. Human activities block the flow at local ecosystems, with asphalt and wires, yet increase the flow between large regions with ships and airplanes—mostly the flow of pests and domestic species.

Nature exhibits error, or play, which permits diversity. So many things are thrown at the flow of life. There is not just a little error. Half of everything seems to be error: Billions of pollen grains, billions of eggs, millions of species. Transmission is not flawless or efficient, but it is generative of difference and diversity. The diversity of the current ecological world evolved through the breakup of Pangaea, which provided the distances and barriers to isolate species. Many other properties of nature, such as least action or complexification, can become the basis of principles.

2.3.4. *Summary*

Nature produces many complex patterns. Living beings create further complex patterns, but this is done without conscious design, without plans. Natural ordering of the world makes places from spaces. A place changes qualitatively; it becomes structured. Some change, such as growth or decay, is quantitative. Other change is qualitative, resulting in breakdown or formation of the entire system. Life is also a property immanent in an organization of molecules; and language emerges in a higher level of organization. New qualities that emerge at every step are unpredictable on the basis of the past; natural reality is creative. Nature is more like an artist than an engineer in this sense, that is, nature provides novel patterns that can be duplicated or reformed by living beings within nature.

The scale of nature is fundamentally different than the scope of any design. The time scale of evolution is measured in billions of years. The size scale of evolution is measured in trillions of living individuals. These scales permit any speed of operation of selection through reproduction and genetic mutation. Nature does not design at all in the human sense, limited by a single lifetime or even group longevity. Nature makes patterns that change, on many levels, over an immense scale.

2.4. *What is Traditional Human Design?*

Human beings have an ability, shared with many species, to alter their environments. They are able to reform natural forms and flows with their hands or with tools that have been made for that purpose. They are able to design the tools, as well as other larger objects and patterns, such as landscapes.

2.4.1. *What is Traditional Design?*
The word 'design' comes from the Latin word for "to mark" and means 'marking off a pattern.' Tracing lines can be inclusive or exclusive, but if the line is more like a membrane then it permits certain sizes and shapes of intrusion. The verb form of the word 'design' means to make a pattern or plan, or to intend for a purpose. A design is a purposive plan, which itself is a diagram in two dimensions. Design is a human project in which, as Oliver Lucas says, "visual and physical parts are assembled in order to achieve a specific end result." A design can also be an orderly arrangement of parts in an overall pattern. Designers make visible creations from plans, using the resources available.

Most design is concerned with products, from magazines, clothing and toys to houses and buildings. Our designs are often shaped by the impulses of technological feasibility, assembled randomly in neighborhoods and cities. Of course, cities are the results of complex adaptive behaviors in response to environmental conditions. Cities are not designed as whole systems, with a few possible exceptions, such as Paolo Soleri. Landscapes around houses and buildings are often designed. But, large landscapes are not considered or designed. They are made up of individual patterns, such as farming and manufacturing. Sometimes the landscape is charming; other times it is not. The landscape is a common property that can be understood and integrated by perception. The loss of design is the inability to imagine, shape and build things that enhance life and safety; it is the inability to respond to changing circumstances. Too much of a world is overdesigned by professionals, according to Victor Papanek, with the result that design simultaneously ignores things that need to be design and creates further dangers with many designs.

2.4.2. *What is the Difference between Natural & Human Design?*
Design is creation for reproduction. Design is intelligent reproduction. Life is experience that continues by reproduction. Life is experiencing and reproducing forms to continue experiencing. Design is the production of forms for experiencing and reproducing (See Table 2421-1).

2.4.2.1. The Biophysical Environment of Design

The environment provides a framework for design, but it also limits some of the possibilities of design with resources and cycles.

2.4.2.1.1. Patterning. The word 'design' itself means 'marking off a pattern.' The verb form of the word means to make a pattern, so the concept of patterning is built in to the idea of design. The human creation of designs extends a natural process.

2.4.2.1.2. Conscious intention. Design is a conscious and intuitive effort to impose significant order, according to Victor Papanek. Design implies intent, and intent implies foreknowledge, as Anatol Rapoport notes. Van der Ryn states that 'Design is the intentional shaping of matter, energy and process to meet a need or desire.'

2.4.2.1.3. Making/Generating. Papanek also states that design is a process for making meaningful order (as well as objects). Design makes things into a new pattern.

2.4.2.1.4. Focus. Focus is the ability, as Erich Jantsch says, to organize one's total experience towards a purpose. The focus must include rational experience and technical substance. The intent of a designer focuses on the object, however, not on the process or the fitness.

2.4.2.1.5. Feedback accepting. Anatol Rapoport notes that improvisation is an important part of design. Being open to feedback allows design to function without complete rational knowledge, as long as we recognize that the design has negative and positive consequences. This is a way to make improvements and have design adapt to change.

2.4.2.1.6. Objective/object-oriented. Design is used for shaping and producing functional industrial and artistic things. The word 'thing' goes back to Old English words that mean 'object' or event—the word might have been used originally as a metaphor for some event under certain conditions. Nature is regarded most often by design as a set of challenges or a storehouse of resources than as a mutually created matrix.

2.4.2.1.7. Growth-permitting. Design can operate without regard to expanding or contracting conditions, that is, it does not pay attention to the over-all growth of the number of objects or to their relationships in an economy.

2.4.2.1.8. Place independent. Although design can take inspiration from the uniqueness of a place, it does not depend on that place. It can take inspiration from anywhere and combine it without any reference to place. Formal development of design is more recently concerned with an assembly line model—simple, isolated, efficient, and easy to maintain—than with the uniqueness of a place. As a result of indifferent designs, we become remote from, and indifferent to, the system that supports us. We acquire unrealistic images of the world and harmful values, and then make bad decisions based upon them.

2.4.2.1.9. Principle limited. Design is often based on principles of form and style. In fact, designers are expected to follow recognized design principles. But, these principles do not incorporate principles of ecology or properties of nature to use as a foundation.

2.4.2.1.10. Equilibrium. Design is concerned with balance and balance is conceived as a state of equilibrium between two or more elements. Sometimes, design depends on a static environment.

2.4.2.1.11. Maximizing/Optimizing. Design is obsessed with maximizing beauty or optimizing usefulness. Yet, it does so without attention to any social or environmental constraints on a maximum.

Table 2421-1. Contrast between Biophysical Environment of Design.

Natural Creation	*Traditional Design*
Self-patterning	Patterning
Creating forms	Conscious intention/form
Autopoetic	Making/Generation
Whole	Focus
Systemic	Feedback accepting
Processing patterns	Objective/object-oriented
Development/Filtering/Sorting	Growth-permitting
Place generating	Place independent/isolation
Place responsive	Principle limited
Dynamic equilibria	Equilibrium
Interaction limiting	Maximizing/Optimizing

2.4.2.2. The Ecocultural Level of Design

At this level, design is influenced by the environment and shaped by the images and demands of a culture.

2.4.2.2.1. *Culture-based.* Designers, like others, are raised in a single culture and acquire the unique attitudes and values of that culture. Many of their designs are more successful in their culture than in other cultures. This means they also imbibe the weaknesses and peculiarities of one culture. Design limited to shaping and producing industrial functional and artistic objects unconsciously supports a division of artifact and nature. In fact, nature becomes more of an artifice, when we make meadows and lakes, and restore rivers and forests. Most approaches to design are dualistic (wilderness-industry) or triune (wilderness-design-industry).

2.4.2.2.2. *Bricolaging.* Design puts pieces together to see if they fit. The pieces may be unrelated or related. They may fit well or not. The assembly is sometimes just an aggregation that falls apart or unravels quickly.

2.4.2.2.3. *Conscious Trial & Error process.* Unlike natural processes, design is almost always a conscious process of trial and error. Papanek states that design is a future-oriented, trial and error process for making meaningful order. This kind of experimentation does produce things where the pieces may fit within a new object.

2.4.2.2.4. *Learning from failure.* Designers often have an acceptance of failure and the ability to learn. Designs are improved in this way. Design as an art shares the same relationship with failure as 'science,' where knowledge is increased through the failure of experiments more than from success.

2.4.2.2.5. *Novelty emphasis.* Novelty is attractive to many people. Many designs aim for novelty to be successful. People recognize, value and seek novelty for itself. Thomas Birch and John B. Cobb Jr consider that the richness of the world and the freshness of a living response are matters of novelty.

2.4.2.2.6. *Rules/Geometric grids.* In some designs, grids are used as a frame that supports certain elements, but also limits the freedom of possibility. Using grids has been basic for many kinds of design, from graphic to urban design. Movements that tried to sweep aside all rules and start from zero, like the Bauhaus School, were not as successful as they wanted to be.

2.4.2.2.7. *Top down direction.* Design tends to impose patterns from the top down.

The direction is orchestrated from the top. That has advantages for individuals and teams, especially when it comes to control of all aspects of a design, but it requires knowledge and control of all details and levels.

2.4.2.2.8. *Creative Play.* Design is goal-directed play, states Papanek. The creative play of design often starts with creating a line or boundary. Play is the method of learning for most juvenile animals and a means of enjoyment for many adult animals. For humans, play is imaginative experience, entered into freely. Much human activity is play, in place in a community. Even science and philosophy are forms of play, as attempts to solve the puzzles of existence. The object of the senses is life, the object of reason is form, and the object of play is living form—called beauty in the widest sense. Aesthetic play, like physical play, requires order and control.

2.4.2.2.9. *Goal-directed.* A goal is an end to which effort is directed. Papanek states that design is goal-directed play. Design requires understanding the process as well as the tools and goals.

2.4.2.2.10. *Conceptual.* A concept is an idea or generalized idea of a thing (as well as a unifying idea or theme). Designers often start with the concept before the particulars. The conceptual design may not always translate well into the physical.

Table 2422-1. Comparison of Ecocultural Levels

Natural Creation	*Traditional Design*
Nature-based	Culture-based
Aggregating	Bricolaging
Trial & Error process Novelty testing	Conscious Trial & Error process
Trial generating	Learning from failure
Novelty generation	Novelty emphasis
Reciprocal interactions	Rules/Geometric grids
Bottom-up organization	Top down direction & control
Physical restraints	Creative Play
Natural process	Goal-directed
Actual Existing	Conceptual

2.4.2.3. The Economies of Design

At this level, the designer has to consider the commitment of time and expense in relation to the product.

2.4.2.3.1. *Capital-using.* Larger designs rely on the input of capital from some source. Finances have to be borrowed or granted in order to start a design. This requires trust or advances.

2.4.2.3.2. *Constructive.* Designs are constructive, in the sense that elements are fitted together systematically to build an object. The object is put together from a variety of parts at a certain scale.

2.4.2.3.3. *Auto separation.* The design is usually conceived as an independent object that is addressed separately from other objects or their contexts. It does not have to fit within a definite pattern nor does it have to participate in a network of ecological relationships.

2.4.2.3.4. *Independent destructive.* Because design is such a limited system, however, waste is produced and sent to other systems and may be destructive to other systems. The design itself may explicitly wreck other designs or systems.

2.4.2.3.5. *Energy use*. Van der Ryn repeats that design is "the intentional shaping of matter, energy and process to meet a need or desire." Design requires energy to make its forms. Design may use energy efficiently. It is, however, usually unconscious of the energy source or of the costs of producing and using that energy. Most of the energy used in current designs is derived from fossil fuels.

2.4.2.3.6. *Technological tool use.* Design is an art that uses technology to produce its object. Technical proficiency is an aspect of design. In fact, some design is treated as a technical problem, to be solved through an understanding of tools and technology. Design may overemphasize technological considerations, which can shift or shape the actual design. Design must include technical substance and rational experience, but also systematic organizational and cultural themes.

2.4.2.3.7. *Waste production.* Design seems to be unconscious of waste, also, not only the waste from energy and the acquisition of materials, but the waste from the use of the designed object and its premature disposal. Although waste can often be used by another system, many designed objects cannot be deconstructed or reused without great care and expense. The waste is independent of waste-incorporating cycles.

2.4.2.3.8. *Cost consideration.* The true or actual cost of an object is rarely considered by design. This includes the energy and material costs as well as the psychological and social costs of poor or thoughtless designs. Many modern designs are driven by market considerations, such as profit or turnover, and are created for specific temporal markets at the lowest possible cost.

2.4.2.3.9. *Fashion & Style.* Thoughtless design is susceptible to fads and styles. Designing superfast cars for slow, crowded highways (or for rare circular tracks somewhere) is an example of an expensive, wasteful fad. Many designs are created exclusively to fit an ephemeral style or fad.

Table 2423-1. Comparison of Economies Differences

Natural Creation	*Traditional Design*
Productivity	Capital-using
Interrelational	Constructive
Nested	Auto separation
Interdependent	Independent/destructive
Solar-life driven	Energy use (any, fossil)
Aesthetic creation	Technological tool use
Waste generating	Waste production, Efficiency
Free play/change	Cost consideration, Market-oriented
Gigatrends	Fashion & Style

2.4.2.4. Values of Design

At this level of design, value is incorporated into traditional design, although the value may be limited to immediate economic value or restricted aesthetic value.

2.4.2.4.1. *Responsibility* to product, consumer, and community. Perhaps the first responsibility for design dealt with effective usefulness—either the digging stick worked or it did not work. Later, designers discovered a responsibility to the consumer, for the safety of the materials and design, and later, to the community, that a product not harm others.

2.4.2.4.2. *Planning.* A design is a purposive plan, which itself is a diagram in two or three dimensions. The plan was concerned with the orderly arrangement of parts in an overall pattern. Designers can be thought of as people who make visible creations from plans, using the resources available, according to Erich Jantsch. Specific designs based on plans can help people visualize the consequences of policies. Large-scale designs represent a synthesis of all planning studies, according to Frederick Steiner. Site design is the application of planning process to a specific parcel of land, according to Kevin Lynch.

2.4.2.4.3. *Meaningful order.* Meaning in design seems to refer to usefulness. Victor Papanek states that design is a process for making meaningful order. And what makes it meaningful is that it enhances not just survivability but human life, as it fits into the needs and expressions of a culture. The lifetime of a designed object is not considered in terms of its context or extended use. In fact, the object tends to be short–term and tied to economic time frames, which tend to be much shorter, at 2-5 years, than the artifactual use—digging sticks are still being used after at least 40,000 years.

2.4.2.4.4. *Goodness concepts.* We judge the results of our designs as good or not good. The word 'good' is derived from the old English word meaning 'suitable' or 'pleasing.' Design is usually considered good when it is pleasing and functional. David Holmgren suggests that good design depends on a free and harmonious relationship to nature and people, generated through continuous reciprocal interaction.

2.4.2.4.5. *Teachable.* Traditionally, design is taught from master to apprentice or mentor to student. A designer acquires knowledge and technique from practice and imitation. History and principles may be learned at that time. Recently, design started to address specific problems, where students model possible solutions and faculty evaluate their work. The evaluation focuses on the object, not on the process or its fitness. It may address questions of goodness or meaning, but it rarely tries to fit the surrounding culture or ecology.

2.4.2.4.6. *Exclusive.* By its nature in marking off boundaries, design is exclusive of those things considered irrelevant or uninteresting. Many things are considered to have no immediate impact, regardless of subtle or long-term influences. Other considerations are excluded as being irrelevant or too expensive.

2.4.2.4.7. *Information.* Designers need information to create designs. Information is the presentation of the forms of nature before they is distorted into facts or data. The information considered necessary for a design is often a small subset of the information needed for a design to fit in a particular context.

2.4.2.4.8. *Exciting.* Design is exciting. It is exciting to design. The product develops out of materials and energy, as well as human effort and inspiration, and can contain many surprises. Design should be exciting, and excitement can come from novelty, beauty, or many other sources.

Table 2424-1. Comparison of Values Levels

Natural Creation	Traditional Design
Living	Responsibility to product, consumer, community
Fitting	Planning
Harmony-build/long-term	Meaningful order/short-term
Existence value	Goodness concepts
Learnable	Teachable
Adjusting	Exclusive
Formation	Information
Chaotic/Free	Exciting

2.4.3. *Design Components Elements & Variations*

There are basic geometric elements of any design, from 3 (volume) to 2 (plane), 1 (line), and 0 (point) dimensions. These elements can vary in numerous ways, by number, position, direction, size, shape, interval, texture, color, and temporal. Furthermore, the elements can be organized into groups by nearness, similarity, and difference (diversity). Then they can be combined into whole structures by principles of rhythm, balance, and finally sensory force. The design of any dimension can affect all dimensions.

2.4.3.1. Elements

There are basic geometric elements of any design, from the 3-dimensional (volume) to 2 (plane), 1 (line), and 0 (point) dimensions.

2.4.3.1.1. *Point.* In geometry a point is an element having no size, shape or extension. As an element in design, a point has extension as a detail or striking feature, for example as the center of a book cover or single treetop in an urban park. A point may be a value. Points may be anywhere.

2.4.3.1.2. *Line.* A line is a mark between two points (containing an indefinite number of points) or the path of a moving point, having length but no breadth, straight or curved. It is sometimes mathematically a limit. A line may simply be where two shapes meet. In design, a line may be a border between colors.

2.4.3.1.3. *Plane* (lines). A plane is a surface that contains every line that joins two points on the surface. The plane may be horizontal, vertical, or diagonal. A flat surface, such as a desert or grassland may form a plane.

2.4.3.1.4. *Volume.* A volume is the amount of space occupying three dimensions. It can be created by the interaction of points, lines, and planes. A stream or volcanic island has a volume.

2.4.3.2. Variations

Mathematically, a variation is the manner in which two or more quantities change relative to one another. In design, it is the degree or process of change in appearance. Thus, there is great variation between a concrete walkway and flower garden.

2.4.3.2.1. *Number.* A number is a symbol showing how many items are in a series or collection. Numbers measure quantitative change, the source of all qualitative change.

2.4.3.2.2. *Position.* A position is a location or place where a thing is. It is also a description of earlier elements, such as lines in the environment. Diagonal lines are most pleasing to most observers; lines at right angles to the contour are rarely pleasing because the landscape is broadly horizontal; geometrical shapes look artificial (even when they are natural); natural shapes, perhaps fractal, are considered more natural and interesting. Fractals, or self-similar structures, allow larger units to have the structure of smaller units. This applies to coastlines, mountains, and art, as well as to political structures. The diagonal lines, as opposed to horizontal and vertical, of hills is both dynamic and pleasing to human senses.

2.4.3.2.3. *Direction.* A direction is a point towards which an element faces or a line along which something moves. A horizontal direction may be associated with tranquility, whereas a vertical may suggest formality. In the design of ecosystems, up is preferred to sideways or down. Diagonal lines give the impression of energy and movement, which is quite true as the hills are still being shaped geologically by erosion and wind. The eyes of travelers are drawn down one slope and up the next, or along the series of slopes. As people respond to one element and then another, the elements are perceived as parts of the whole. The sinuous path of the road through the corridor draws the eye toward the end of the corridor. The skyline is made more interesting by the shape of the hills.

2.4.3.2.4. *Size*. Size is the quality of an element that determines how much space it occupies. It is the extent or magnitude of the element. The size of an element determines the perceived size and relation of other elements. For instance, a large talus slope on the side of a small mountain may look larger than it is.

2.4.3.2.5. *Shape*. Shape is the quality of a thing that depends on the relative position of all points on the surface, or the spatial form characteristic of a thing. Shapes are one of the first things one notes in a design. For instance, a complete inventory of elements in an Oregon creek starts with the shapes of the features in the area. The large volumes are rounded and natural hills—even the agricultural evidence is almost natural, that is, from the roadside, not the air, the fields appear not to be squares, triangles, or circles; a small number of geometric shapes exist in the buildings by the road, but because of their scale are not too intrusive. Although the road itself has been flattened, it is not perfectly straight and does not conflict badly with the curving planes of the hillsides.

Shape to some extent determines how we see our surroundings. Shapes dominate other design factors, so appropriate shapes are critical. Proper scale or diversity cannot save a design if shape is wrong; the mind can pick up incongruities and artificial geometric qualities. Suitable shapes are vital for the unity of the landscape. As stated in by the English Forest Authority: "The perception of shape is influenced by overall proportions, viewing position and direction, and the nature of the external boundary edge."

2.4.3.2.6. *Interval*. An interval is the space or time between two things. The relation between objects in space or time. For instance, the distance between streams is greater than the distance between draws on the slope.

2.4.3.2.7. *Texture*. Texture is the surface quality of a shape or volume. The graininess of the system, the structure of an ecosystem, or quality as a result of the interval of elements. The texture of an old growth forest is rougher and more interesting than a tree plantation.

2.4.3.2.8. *Color*. Color is the property of the reception of different wavelengths of light. Color is distinguished by the qualities of hue (e.g., red or yellow), brightness (or lightness for pigmented surfaces), which may result in value, and saturation (the degree of intensity of a hue). Colors are vivid qualities in objects or ecosystems. There is an indefinite number in nature, although some ecosystems, such as forests, have smaller palettes.

2.4.3.3. Organization of Variations

Variations can be organization within patterns, which may be across many scales of time, space and embeddedness.

2.4.3.3.1. *Time*. Time is a duration in which things happen. Time is a period or measurable interval between two events, or a period of existence. Temporal patterns can be classed into four groups: circular, the eternal return, disintegration, reintegration, as of forests, nothing new; spiral, that is, a circle under stress, a cycle under change, something new; linear, the straight line of industry, progress to an end point, such as heaven; and, nonlinear, the line combined with chaos, ending in extinction or creation.

2.4.3.3.2. *Groups*. A group is an aggregation or a number of elements forming a unit, regardless of whether they have common characteristics or similar connotations. Elements can be grouped by likeness, patterns or weights.

2.4.3.3.3. *Nearness*. Nearness refers to the spatial or temporal closeness of elements. Nearness implies an intimacy of elements.

2.4.3.3.4. *Similarity*. Similarity is a state of likeness or resemblance between elements. Elements may be almost identical in shape, but of different size or position, for instance, an old understory tree may have the same shape as a dominant, but only be one-fourth as tall.

2.4.3.3.5. *Density*. Density is the quality of being compact or having a quantity per unit. Density can be related to closeness or connectivity. Blades of grass in a prairies may be perceived as dense from a low perspective.

2.4.3.3.6. *Diversity* (difference). Diversity is the number of differences in a framework. Increased diversity also has the effect of reducing scale, so adding diversity can be used to do reduce the scale. A high level of diversity is acceptable if one element is clearly dominant or if the differences cannot be recognized from a distance.

2.4.3.4. Organization into Structures

Patterns can appear in structures, during and after formation.

2.4.3.4.1. *Structure* (Paths, Patches). Structure is the manner of organization of elements in space or time; it can also be thought of as the arrangement of the parts of a whole. This is especially so for ecosystems, as whole units; a tropical forest ecosystem, for instance, has a typical structure of six levels, from the herbal layer to rising dominant trees.

2.4.3.4.2. *Rhythm*. Rhythm is a flow characterized by the regular recurrence of elements in space or time. Ridges on the side of a mountain range provide a rhythm in space. (Biologically, rhythm is the periodic occurrence of specific physiologic changes in organisms in response to geophysical factors, as when deciduous trees lose their leaves with low light levels and cooling temperatures. The units of time may range from milliseconds to thousands of years, although most will seem to be daily, seasonal, or long (in 2-25 years). Rhythm generates interest.

2.4.3.4.3. *Tension*. Tension is the balancing of elements in opposition or the interaction of elements without resolution. Tension connotes a state of strain, stress, or force in a pattern (sometimes exerting force against resistance). A large clearcut area may create tension in a mature forest landscape.

2.4.3.4.4. *Balance*. Balance is a state of equilibrium between two or more elements. It is also related to the harmony of elements in a system or the state of the system. It can also be defined as the stable movement of elements around a center (or attractor) in a dynamic equilibrium. In forest design, a larger meadow is needed to balance a dense woods, due to differences in perceived masses.

2.4.4. *The Principles of Traditional Design*

In addition to the basic elements, structure and variations, design can be presented through a number of principles. Principles are fundamental rules or laws, based on the characteristics of objects or systems, that we can use to create images or models to meet stated objectives, that is, the goals towards which our actions are directed, such as a functional beautiful bicycle or a comfortable inspiring city park. Principles unify our images. Select principles are introduced briefly to show the depth and breadth of design.

The principles presented are derived from the typical characteristics of design objects. Characteristics are qualities that distinguish unique individuals, systems, or patterns—Gregory Bateson refers to characteristics as differences that make a difference. From these principles, standards for design activities can be established. Standards are models or examples of quality or value, established by authority or mutual consent, which can be repeated as procedures.

The principles identified form the basic axiomatic truths of design. They represent the basic assumptions of the human world that guide the practice of design. These principles affect the arrangement of objects, elements or components within a composition. These principles, not exhaustive by any measure, state that any composition has to display: Proportion, Rhythm, Emphasis, Balance, and Unity.

2.4.4.1. The composition has to exhibit *Proportion*. Proportion is the comparison of the dimensions or distribution of forms in a composition. It is the relationship in scale between one element and another, or between a whole object and one of the parts. Differing proportions can suggest different kinds of balance or symmetry, or relationships between scale and proximity. Elements of a larger scale can dominate the composition, presenting more visual weight or altering perspective.

2.4.4.2. The composition has to have *Rhythm*. Rhythm is the repetition or alternation of elements, often with defined intervals between them. Rhythm can create a sense of movement or progression; it can establish pattern and texture. There are many different kinds of rhythm, such as regular, flowing, or progressive, which are defined by the feeling evoked.

2.4.4.3. In most compositions, there is a *dominance* of one or more elements. This may emphasize elements that are in closer proximity, are larger or are centered.

2.4.4.4. The composition must be in *Balance*. Balance is an equilibrium that results from looking at images and judging them against our ideas of physical structure (such as mass, gravity or the sides of a page). It is the arrangement of the objects in a given design as it relates to their visual weight within a composition. Balance usually comes in two forms: symmetrical (also formal) and asymmetrical (also informal), depending on whether the weight of composition is evenly distributed around a central axis or not, and whether the axis divides identical forms or not.

2.4.4.5. The composition has to have *Unity* (or be in harmony). The concept of unity describes relationships between the parts and the whole of a composition. Unity, the relational aspects of a design, provides a sense of wholeness to the composition. Without unity, a design may fall apart. Gestalt theories of perception have contributed to understanding of the organization of information into categories. In a harmonious design, the contradictions, conflicts, influences, and emphases are recognized as a whole composition. The elements are no longer independent things; they are constrained by the unity or harmony.

2.4.5. *Applied Design Stages*
Traditional design proceeds in four stages.

2.4.5.1. *Review the function* or goal. What is the composition supposed to inspire or do? Is it to be a public or private composition?

2.4.5.2. *Find appropriate materials.* Some materials are better than others for certain purposes. Understanding the purpose, as well as the potential materials that can be used, insures a better design.

2.4.5.3. *Create outline* or plan of design, using elements and principles to guide the design.

2.4.5.4. *Build the object.* If a single object, it is given or sold. If it can be reproduced or mass-produced, it usually is.

2.4.6. *Three Levels of Traditional Design*
The traditional design process considers three levels: Components, products, and the community.

2.4.6.1. *Material Components.* Many cultures have found, released and used materials for tools and designs. The materials themselves were often the thing itself.

2.4.6.2. *Object/Products.* Often, materials were melted, woven, or combined into specific objects as tools or chairs.

2.4.6.3. *Social context* in Community. The objects that were made were usually integrated into an order described by a culture, based on knowledge and techniques handed down through generations. These things acquired social value even when their utilitarian value diminished. Chairs, for instance, beyond being useful for sitting, can grant authority to the person sitting in it.

2.4.7. *Evolution of Design*

Charles Darwin and Alfred Wallace described a theory of evolution that was based on natural selection. Through many forms of interactions, from competition to cooperation, members of different species exploited each other and the environment, changing the whole structure and process, and generating conditions of diversity and complexity that made the process effective in a changing and developing planet. The success of a species was measured by its fecundity, which allowed the actualization of more possibilities. Some species were able to colonize harsh habitats, others mild ones. A few species, such as lichen or human, were able to thrive in almost every environment. Species adapt to the environment and then adapt to the environment as it is changed by them. The progressive development of wholes is evolution. Evolution can form a harmony over many decades or millennia. At all levels evolution includes freedom of action as well as interdependence.

Evolution uses different levels: The physical was replaced by the chemical, which has generally been replaced by the electronic. Plants are generally chemical; electronics becomes important with motion for processing information. But electrical systems outran their design (hunting) in humans. Although humans have created artificial systems that are faster than their own organisms. Perhaps a new quantum level (like culture?).

James Lovelock thinks of Gaia as a collective to perpetuate life as a planetary phenomenon. Richard Dawkins rejects Gaia as the desire to believe that evolution works for the good of all. But, Dawkins is mistaken, since evolution does not work for anything. Obviously life has continued as a planetary phenomenon, but that does not mean life or the planet have any goals beyond survival, much less to be optimally productive.

Evolution does not have goals. But, design has to be based on goals and images. The goals have to be comprehensive at a larger scale. Like several other species, humans were able to use tools to adapt to challenging situations. They exploited materials and patterns to develop tools, clothing and culture to accelerate their adaptation. The sophistication of early designs seemed to be correlated with the difficulties of the environment. Design was limited by the energy and materials available in the local ecosystem. They were integrated into the ideas of the culture.

Culture, with its new way of transmitting information, involves a mix of: Trial and error learning, social learning through observation and imitation, and finally teaching. By this definition, animals as well as humans can have culture. Some classic criteria of life require self-reproduction, to preserve biological information, and variability and selection, to enlarge an individual information store. Memory keeps the information immediate. Culture is an emergent property of the evolution of speech and tool-making; economic and political structures, with their ideas of disparity and wealth accumulation, emerge from culture. We can think of agriculture as an information system that can overcome some limits of local environments. Cities are another kind of adaptation to uncertain environments.

Tools and technology have evolved also, as human culture has evolved. According to Radovan Richta, technology is evolving through three stages, from tools to machines to automation. Tools, from plows to microscopes, augmented physical labor or senses. A powered machine, such as a water mill, tractor or computer is a complex, often powered tool that replaces all physical effort but allows human control. An automation is a machine that uses an automatic algorithm to replace human control; automation can be observed in digital watches and automobile assembly factories.

It looks like a fourth autonomous stage is developing, where the tools are capable of self-direction (and perhaps rudimentary intelligence and consciousness). Possibly robots and living cities may decide how to achieve harmonious relations with human or wild systems. At the same time a fifth stage may be developing, hybrid systems, in which humans become additional components in a larger whole.

After each fundamental type is introduced, however, it continues to be used widely, thus simple tools, such as digging stick, or a machine, such as a cotton gin, are still being used with automated telephone interchanges. The overall use of technology displays several gigatrends (refer to Section 1.5): Increased efficiency, increased control over external conditions, and increased complexity of technology nests.

T.M. Lenton, K.G. Caldeira et al. suggest that humans have managed to escape the constraint of adaptive genetic selection (the Red Queen Hypothesis, drawn from Lewis Carroll's character in Alice in Wonderland, who has to run to stay in place; constant rates of speciation and extinction lead to general stability). The escape is based not just on the advantages of reproductive sex in resisting parasites, but the advantages in culture for escaping environmental constraints. The general hypothesis proposes that evolution within a species must keep pace with environmental selection or the species will go extinct. Like other species humanity altered its habitat. Unlike most species, humanity seems to have escaped the constraints of a single habitat. The evolution of tools may have destabilized our levels of exploitation. We can alter all ecosystems, even those where we do not want to dwell. The alterations are so widespread and fast that they themselves exert selection pressures similar to catastrophic interference events, such as meteor strikes or ice ages. The alterations are not limited to single habitats but extend to biomes and planetary biogeochemical cycles, such as nitrogen and carbon. Lewis Mumford suggests that at larger scales and with centralization, technology becomes another tool of authoritarian decision-making.

Design adjusted to increased knowledge of materials and possibilities. It adjusted to the environment. It started to adjust to new limits, to issues of safety, to new techniques and technologies, and now to flows and life cycles. Some tools, however, may work against biological fitness. David Berman suggests that the slot machine is an ingenious piece of industrial design that assists in the triumph of the greedy over the vulnerable. The design magnifies a combination of subtle weaknesses in the human decision-making schema. The schema works quite well for mammals in the wild, but it is short-circuited by the a low grasp of probability.

Evolution seemingly did not prepare the human species to exceed certain limits. Humans as a species have figured out how to rearrange resources so that this drive could be regularly over fulfilled. The result is that we are caught up in seeking, as a society, more than simply enjoying. Has this has to do with our many generations as hunter-gatherers? Berman notes that design can affect how you behave and think; he asks designers if they really want to spend the best years of their careers spreading misery and feelings of inadequacy, or encouraging addiction, or destroying hopes and dreams. Does this happen because the profession has no code of ethics?

2.4.7.1. Expanding Ethics for Design Responsibility & Safety

Traditional societies usually had a strong moral compass. People sought fulfillment and balance, some more mystical or secular than others. In modern societies, once people tracked the dangers of a particular design to the designers, design as a profession started to address matters of safety in their products. Designers gradually became responsible for using safe materials and safe configurations, mostly because it was good business not to be attacked or sued for designing flammable night clothes or bedding.

Berman notes that design can affect how you behave and think and then he asked designers if they really want to spend the best lives of the best years of their career are spreading misery and feelings of inadequacy, or encouraging addiction, or destroying dreams.

Howard Gardner points out that the unit of human thought is the symbol. It could also be the metaphor of course simple. Through design many symbols are presented to deceive. Gardner suggests that symbols are the building blocks of reason and the memory. Berman suggests that it is unethical to insert misleading symbols into our environment. He suggests that as our civilization matures we will recognize that visual communicators manufactured misleading memories, and that these visualize can be as dangerous as hot steel.

As we know advertising is capable of many manipulations of symbols that compel people to consume too much. David Berman suggests that legislation to prohibit lying with imagery might help. Berman suggests that design needs to be more ethical and express its professional behavior, which could inoculate culture from the downside of global velocity or rather the velocity of globalization. The greatest threat to human future is consumption or overconsumption. Design fuels mass overconsumption. Victor Papanek noted that designers were trained to aid and abet corporations whose policies are relatively unenlightened as regards aesthetically pleasing, well made, ecologically responsible products. Often the corporations litigated to avoid their legal responsibilities, since enforcement and judicial agencies were so weak. Berman suggests that designers have more power than they realize and could short-circuit this.

Human civilization cannot afford another major mistake, such as agricultural conversion, overconsumption or mass pollutants. We can leave a better legacy by using our best ideas and not by copying our chromosomes. So the question is why do we consume so much? Why do we consume some things and not others? Overconsumption is a very destructive pattern and the pattern is defined by deception and lies. Our entire society now focuses on spending, for psychological reasons as well as economic ones. The banks do not help either with their irrational campaigns to push second mortgages at higher interest rates. Tim O'Reilly suggested that more than real estate bubble, we need to be concerned with an inflated reality bubble.

According to Papanek, the good life depended on satisfying four basic social desires: Conviviality, religion, artistic/intellectual growth, and politics. In response to ethical dilemmas, designers need to learn applied design ethics, so that when they perform deeply satisfying responsible work, and expose their clients to social and environmental needs, they might not feel guilty doing or earning less. Creating new approaches and professions is endorsed by Papanek and Fuller.

Cultural ethics are usually restricted to some members in a local ecosystem; such ethics are assembled inductively, from experiences from living in specific places. Philosophical ethics have been traditionally restricted to the human species and human situations. The areas of concern of ethics are not broad enough; their foundations are not deep enough. Philosophical ethics is the ethics of the human species living alone, without wild animals or plants in modified ecosystems.

Extensions of ethics are developed in response to problems that arise from increasing knowledge. Science has phenomenally increased our knowledge of physical and biological processes. It has now become the basis of our moral code, but it cannot very long be a science divorced from feeling and art if that code is to help us survive. To do this science requires aesthetic perception as well as disciplined thinking and feeling. As there is a rational component to ethical judgments, so there is an intuitive and emotional one, also.

The extension of ethics to animals and land is an ecological necessity with the history of human pandomination. This extended ecological ethics defines a social conduct that is a mode of cooperation and, ultimately, symbiosis. Aldo Leopold argued that voluntary limitations of freedom are necessary in a complex world of which we remain incredibly ignorant. An ecological ethics suggests that humans avoid tampering with complex evolved systems, not because they are good, but because they are the basis of life. Ecological ethics is situational because ecology is the study of changing systems. It is pluralistic, as Stone notes, because of the variety of entities involved. The morality of the act is determined by the current state of the system. Adaptive modes should conform to ecological patterns. An ecological ethics is based on attributes of ecosystems and human compliance with ecological laws. The aim of an ethic must be harmonious with the whole population of living beings.

An ecological ethics is a set of rules for living together with other beings. It is based on ecological knowledge, grounded "in the breadth of being," in Hans Jonas' words, founded on principles discovered in existence. An ethics based on ecological knowledge places human behavior in vital social and biological communities in nature. The frame of reference of ethics is enlarged, as Albert Schweitzer predicted, leading to appropriate behaviors in a larger context, through reverence for life.

Most sets of ethics make the rules easy to follow. They emphasize the differences (relativism) or similarities (absolutism) of human beings only; or of the individual or the group; or of good feeling, reason, or desire. We must develop specific rules to live with other species, more formal than isolated cultures like the Campa and more comprehensive than modern cultures like the French or German. Ethics has to confront the individual, embedded in a community, located in a bioregion, on earth. And, the rules really are not as easy as human systems have presented. Schweitzer made them too difficult, with a constant valuing, but neither are they that difficult. An ecological ethics can be detailed only on a local level—even when it uses a global strategy.

An ecological ethics is not distorted by human needs and wants when it argues for the preservation of animals and habitats themselves, because they are as they are, independent communities on which we rely for all levels of 'services.' Because of the uncertainty of human actions, ethics has to encompass the far past and distant future. No one knew that when DDT killed mosquitoes, it would concentrate in the food chain to kill birds. Values are time dependent, and ecological time can be very long indeed. The futures we invent are viable only if compatible with constraints imposed by evolutionary past. An ethics that requires a long-range responsibility also requires a new humility, since technological power exceeds the ability to foresee its consequences. An ecological ethics recognizes the moral obligation to leave the world habitable for future generations. Ecological design ethics has to do more. Berman believes that the future of the world is now "our common design project."

It has to show that ethical designs have immediate and long-term values. It also has to recognize that speed of response is essential.

2.4.7.2. Emergency Design: Starting with Urgent Needs

Is there an emergency? Certainly there is if you consider the extinction of species, a thousand-times higher than background, or the premature deaths of millions of people every years. Certainly if we consider our civilized debt-loads and how little food and fuel reserves we have, or if you look at the natural devastations from natural events and poorly planned wars. It is an emergency.

And, it is a global emergency, caused by an emerging global system of industrial capitalism that is being adopted by most of the major cultures. Unfortunately the goals of the system are growth and profit, at the expense of the health of natural and human systems.

What emergencies? Species and ecological communities are dying out. Millions of humans are ill, unemployed or homeless. Industrial civilization seems to be on a path to collapse. In the case of industrial nations, we have been embracing excess for many generations, so that we are crippled by stress and sickness. Perhaps all we need is a diet. The essence of a diet is to restore one to health, by restricting unhealthy consumption or changing patterns of behavior. As societies and cultures may also be guilty of this kind of behavior, so they need to put themselves on a diet. Archaic and agricultural nations have been strapped by historical inequities and unfair trading. Their challenge is to avoid simply repeating the same errors and consequences in the rush to acquire minimum standards and wealth. The solutions for all nations include trying new ways of balance for self-reliance, paying attention to cultural and physical catastrophes, and striving for better equity. Because of the extent of our overuse and misuse of ecosystems, and their effects on natural processes, systems are collapsing. As the catastrophes are large, the emergency responses have to be large, also.

How should we act? Should we declare war on other people or on the environment or the planet? Usually the goal of a war is to destroy soldiers, civilians, towns, fields, resources, or all of those. Wars are more than battles and destruction of course. Mobilizing for war usually gives more power to the government, to enlist people, to create special laws for sharing or retooling industry to weapons. Maybe is we mobilized for war, but only declared our intention to destroy bad patterns of behavior: The relentless conversions of habitats, the constantly growing consumerism, our addictions to cheap fossil energy, or desires to cheat the poor for more personal gain.

Maybe war will not work. Science has not been normative enough to say what we should do. Technology is too busy growing to illuminate any one possible direction. Governments remain dedicated to preserving the status quo of greed and acquisition in unfair ways. Maybe design would work, since good design or ecological design focuses on whole projects, the entire psycho-socio-ecological network of civilization.

The nature of an emergency requires everyone to drop their normal activities and normal behaviors and to respond to a catastrophe. The catastrophe is usually quite evident, a wall of fire or a massive surge of water that will destroy or has already destroyed homes and people, as well as insects and birds, plants and animals, and their habitats. We seem reluctant to give the causes of these catastrophes the status of real emergencies, partly because the catastrophes seem like natural events, such as a warming trend, and partly because they are related to our industrial habits, which provide us with necessities, as well as with comforts and luxuries.

The planetary emergency exists due to a series of slow, long, large, invisible catastrophes that are resulting from the normal wildness and uncertainty of planetary conditions and the human modification of and interference with planetary cycles and diversity.

The problems may be regional and global, but they can be addressed on a local scale, decentral and human scale. A locality can have the authority, the power to take responsibility and make decisions, for global problems that impinge on the local. It may not solve the global problem, but it will affect it, especially if other local communities exercise their authority.

An ecological or global ecological design project could address every aspect of civilization, adding goals, resizing formal, corporate or cultural efforts to survive and prosper. Good design can make it easier to act on a local level, for instance, by providing human-powered transport for all countries. Pedal power can trump gasoline, even those that get 100 mpg. By offering simple, good, lower-carbon stoves. By combining functions to improve social use; Papanek suggests putting washing machines (or Laundromats) in playgrounds, so mothers or fathers could socialize and work while watching their children. Papanek notes that the easiest way to save resources and energy, and cut waste, is to use less. Conserve. But, we can also produce designs for survival.

Let us make it a year of celebration (Emergency), where we start no new products, no new house starts, until the old ones are renovated or resettled. No new profits for individuals or corporations (all given to the government for redistribution to the neediest). Volunteer couples could put off having children for a year.

Using less seems to be connected with economic health. In a dramatically changing world, with most of us fearing change, design needs flexibility with a synthetic approach. It needs to be integrated, comprehensive, and anticipatory, as Fuller and Papanek urge. We have to direct our representatives into global emergency actions, and participate ourselves. This is another function of design, to change human behavior.

2.4.8. *Summary*

As a profession, design is a special self-consciously separate, economic discipline that adds value to human artifacts, and in fact distinguishes its objects from lower-quality goods. It is the separation of specialists from the vernacular design of self-sufficient people. Design is a historical process of the development of a profession. Of course, it can be regarded as more. Victor Papanek characterizes design as the "primary underlying matrix of life" and states, "All men are designers."

The basic elements and variations apply to all kinds of design. But, as traditional design expands to ecological design and to global design, the number of groups and principles has to increase to reflect the larger scales and increasing connections. Heroic design and extravagance in life is needed. It is not contradictory or antithetical to frugal lifestyles or to restoring a healthy environment. Life is exuberant; energy is used, lives are lived and used, not saved. Life is the accumulation of individual experiences that cannot be saved, stored, or owned. Design is a way of expressing conscious life.

2.5. *What is Ecological Design?*

What is ecological design? Benign Design? How is it different from traditional design? Is it different enough to justify more words about it? This is an academic question. Ecological design is, in Nancy Jack Todd's words: "design for human settlements and infrastructures that incorporates principles inherent in the natural world …" David Orr defines ecological design as: "the careful meshing of human purposes with the larger patterns and flows of the natural world and the study of those patterns and flows to inform human actions." Let us take these ideas as a working definition and expand on them, by comparing ecological design to traditional design and 'natural' creations.

2.5.1. *Definition of Ecological Design*
The word 'design' comes from the Latin word for "to mark" and means 'marking off a pattern.' Marking creates a line or boundary that is inclusive or exclusive, but, if the line is more like a membrane, then it can permit certain sizes, shapes and regularities of intrusions. In ecological design boundaries must be designed as membranes.

Traditional design is a human project in which, as Oliver Lucas says, "visual and physical parts are assembled in order to achieve a specific end result." But, the end result has to be placed in context. As Lewis Mumford pointed out: All thinking must now be ecological. Ecological design has to consider the context and the long-term implications of any designed object. Van der Ryn has qualified design as ecological by stating that it "is any form of design that minimizes environmentally destructive inputs by integrating itself in living processes."

Ecological design has two meanings in this discussion. The first regards the connection of design to its ecological context, while the second refers to the design of ecosystems. In its large sense, ecological design is the creative modification of ecosystems to repair or enhance their ability at self-organization and maintenance of their complexity and diversity. Diversity, as in biological diversity, means species richness, different age and size classes in a population, and genetic differences in a species, as well as kinds of habitats present in an ecosystem and the kinds of communities occupying the habitats; and the kinds of ecological processes that maintain habitats; and the variety and richness of the planet's genetic heritage. Ecosystems that are designed so are healthy. A definition of health is the condition of being sound in body or well-being.

2.5.2. *Differences between Traditional Design & Ecological Design*
The four topics from the design matrix— Biophysical systems, ecocultural play, economies, and values —are expanded to highlight the differences between traditional design and ecological design. The characteristics of natural systems are also repeated with both kinds of design.

2.5.2.1. Physical & Biological Environment of Ecological Design
At this level of ecological design, the object is related to the environment, with all its relationships, constraints and details.

2.5.2.1.1. *Repatterning.* The physical world is a patterning, a flowing whose constituent functions are interacting fields of force. Patterns of complexity shade and grade into one another endlessly. Their history allows things to express minor differences that become larger differences as the history of a pattern unrolls. Differentiation is the discrete unfolding of things, within limits, in a place. The ecosystem is generated by the ebb and flow of energy, substances, individuals, and species across a suitable landscape, and design should fit the flow.

2.5.2.1.2. *Ecological restraints.* Ecological design does not have to merge technology with nature, as Van der Ryn and others suggest. For buildings, technology is an important component, but for the shape of rivers or ecosystems, it may play a minor role. Ecological design has to respect nature and natural limits, while integrating human patterns into natural processes.

Connection is important factor related to restraints. Too little isolation can lead to the weaker or smaller system being overwhelmed; this is similar to overconnection. Overconnected designs and processes are not as flexible under changing circumstances. Designs that are underconnected may be too independent to have mutual restraints.

2.5.2.1.3. *Remaking/Regeneration.* Ecological design stresses regeneration (especially in the sense that Robert Rodale uses the word, sharing meaning with restoration and renewal). Ecological design has to fit the design into systems that remake themselves over time. Ecological design is also the creative modification of ecosystems to repair or enhance their ability at self-organization and the maintenance of their complexity and diversity. Like John and Nancy Todd, Van der Ryn suggests that seeding a design with a diversity of elements makes environmental processes more diverse. Certainly this is true with depauperate systems, although there is a risk of exotic species displacing native species under such circumstances, when both are introduced simultaneously.

2.5.2.1.4. *Peripheral frame/system.* Most designers, especially architects and city planners focus on instrumentalities and institutions. Industrial designers seem to focus on human wants and relations. Ecological design needs to pull back from a focus. It needs to approach sideways, searching for those edge effect perceptions of the whole. It is not quite right to say that ecological design is ecocentric, because it is actually concerned with the frame and not the focus, the periphery and not the center. It is acceptable to focus on a design but the frame has to always be a part of any design.

2.5.2.1.5. *Feedback driven.* Places and organisms shape each other through feedback. There are two fundamental modes of behavior: (a) Maintenance, based on negative feedback loops and characterized by stability, and (b) Change, based on positive feedback loops and characterized by growth or decline. The two modes almost always work simultaneously to create a typical series of behavioral patterns, from stagnation to rhythmic regulation. Attention should be directed toward dynamic equilibrium of ecosystems as the subjects of design. And design has to relate outputs—material and social technics—to positive and negative feedback. This kind of design has to be more comprehensive and consciously planned.

Nature is expressed by complex adaptive systems with nonlinearities, feedback loops, and thresholds (Holling, 1973). By ignoring such dynamics, the ecological design cannot indicate possible ecological consequences of overshoot. It is also important to keep in mind the scale and acceleration of changes. Any large system, such as ecosystem or city, is a high-order, multiple-loop, nonlinear feedback system. In the system feedback loops are the basic structural elements. Each loop is a circular path of interaction between several elements. Ecological analysis forces us to look at the obvious—generating nonmarketable use values occupies the center of every culture because it provides a satisfactory life to its members. Improvisation is an important part of design. Open to feedback, we can design without complete rational knowledge, as long as we recognize the process. It is the only way to achieve balance. To use a simple analogy from gymnastics, we cannot plan every arm motion on a balance beam, although we can design a routine; arm motions are reactions to small changes in activity, which cannot be predicted ahead of time.

2.5.2.1.6. *Dimensional process relationships.* Paul Shepard and others have written that relationships are as real as the objects that result from them. The science of ecology attends the overall pattern of relationships, beyond the details. A specimen is more than the sum of its species' relationships to an environment; it is an intentional being that, with other members of the species, can create niches, as well as adapt to them. The relationships of humans with plants and animals have changed dramatically in the past 12,000 years, since the beginning of domestication and landscape conversion. The increase of humans and the destruction of plants and animals have unbalanced the relationships. Ecological design has to reroute relationships into a better balance.

2.5.2.1.7. *Development after initial growth.* In organisms and systems, growth serves an immediate function of allowing the system to reach an effective size. After growth stops, the system continues to develop, or mature. In economics, as Herman Daly and others have shown, there is no necessary association between development and growth. A community is forced to accept an upper limit, beyond which it cannot grow any further. Further growth results in destruction or disruption of itself and nature. There is another distinction between growth and development. The ecological social approach (or a redistributive environmental strategy) to development makes it irrelevant to discuss global limits to growth. Local limits are far more significant to the majority of population.

2.5.2.1.8. *Place specific.* Ecological design can fit society within an ecological perspective by following the principles of ecology and applying them to the properties of good places. The approach has to be tradition-based and partially self-conscious. It has to be responsive to the genius of place, as well as to human meaning derived from the existential and phenomenological significance of the place. The concept of place incorporates physical, biological, and cultural dimensions. Land use should be matched to the limits of the land. The task of ecological design is to create land use and structures adapted to place. Design should respond to the deep structure of place, the fundamental geological and ecological processes that form the landscape.

There have been attempts to define and design places. Edward Relph outlined the essence of place and then the disappearance of variety that results in placelessness. Christopher Alexander decomposed environmental objects and activities into their constitutive elements, to be reconstructed into designs that could fit local places. These formal solutions can improve strategies for design and can provide a matrix for the making of places. But, they should not assume that human variables can be manipulated to achieve a predicted response. And, they cannot ignore psychological or cultural patterns.

2.5.2.1.9. *Fitness-oriented.* Fitness is the ability to function under normal environmental conditions. Fitness of the components builds up in an ecosystem as it matures. Selection at the organismic level is selection of the fit. The levels of selection must balance, so that life is not too fit or too unfit. Evolutionary fitness does not always keep increasing. In some cases it decreases with time. Fitness is achieved after progressive reciprocal adaptations; it requires a stability of relationships between societies and the place.

Ecological design has to incorporate the idea of fitness to be effective. Ecological designs are modes for conveying ecological sense; they are less concerned with survival than the survival value of a good fit between agents of life. Balance is needed between self-restraint and self-expression, between self-protection and self-restriction. It is not self-expression or self-restraint, but both in a satisfactory fit. Fitness attunes us to limits.

2.5.2.1.10. *Nonequilibrium incorporating.* Ecological organization is a nonequilibral state, where order is governed by amplified fluctuations. Organization is also related to the harmony of elements in a system or the state of the system. It can also be defined as the stable movement of elements around a center (or attractor) in a dynamic process. Design has to address the continuous shifting and change of systems that may be stable for thousands of years or only weeks.

2.5.2.1.11. *Optimizing/Satisficing*. Individual people are concerned with having maximum freedom or producing maximum values, and in fact, the idea of unlimited value tends to encourage the effort. Values may be indefinite, but they may not be maximal or infinite. Like the principle of limited good, there may be a principle of limited value. Even aesthetic appreciation requires limits, to avoid having our over-appreciation overwhelm the values evident in nature.

A value in nature, such as diversity, may seem to have an optimum. While it is meaningful to speak of an optimum diversity, as the result of limits and the interaction of many factors, a maximum diversity may never be reached. As Paul Weiss noted, the patterns of organic nature are a combination of order and diversity; order involves constraint while diversity requires freedom for difference. A maximum order would result in a static universe, where a maximum freedom would create a nonordering chaos. What is a minimum, optimum or maximum size of a city design? Science might identify a few minima or maxima but ecological design should aim at what is optimal or satisficing (satisfactory).

Table 2521-1. Comparison of Natural, Traditional and Ecological Design of Biophysical systems

Natural Creation	Traditional Design	Ecological Design
Self-patterning	Patterning	Repatterning
Creating forms	Conscious intention/form	Ecological restraints Experiential
Autopoetic	Making/Generation	Remaking Regeneration
Whole	Focus	Peripheral frame/system
Systemic	Feedback accepting	Feedback driven
Processing patterns	Objective/object-oriented	Dimensional process
Development/change Filtering/Sorting	Growth-permitting	Development after initial growth
Place generating	Place independent/isolation	Place specific
Place responsive	Principle limited	Fitness-oriented
Dynamic equilibria	Equilibrium	Disequilibrium incorporated
Interaction limiting	Maximizing/Optimizing	Optimizing/Satisficing

2.5.2.2. Ecocultural Play of Ecological design
At the ecocultural level of ecological design, designers have to contend with the diversity of human cultural values and behaviors.

2.5.2.2.1. *Nature & culture-based.* Ecological design extends the consideration of cultural value to the values and limits of nature. Design that is nature-based and culture-based looks different than one that singles out just one. It has to fit the local ecosystems and constraints of the landscape, as well as the values and experiences of a culture. A better approach to dualism and division would show how these interweave so extensively that it would be hard to just use boxes or circles as metaphors. The dualism of nature and culture can be resolved with an understanding of the recent (60,000 years at least) coevolution into domiture (a neologism encompassing the previous nature/culture divide). Domiture has to include human reason and human emotion, rather than simply putting them on opposite columns in a table, as with order and chaos, higher and lower, and linear and cyclic.

Nature evokes feelings of beauty and terror, joy and sadness—we can stare at the wild or feel pleased from the experience of nature. Nature is fun. It invites play. Ecology can be fun. It invites play. No matter how tiresome or frustrating ecology as a science can become, there is always the potential to enjoy it. Who cares if nature is not natural anymore? This is only human words and ideas. Nature is healthy, and that lets us be healthy. Who cares if culture is claiming dominance? Hurricanes, tsunamis and other events can act as agents of cultural humility.

2.5.2.2.2. *Niche-assembling.* An organism is more than the sum of its species' relationships to an environment; it is an intentional being that, with other members of the species, can create niches, as well as adapt to them. Those species that enlarge their niche also enlarge the system as a whole. Species integrate themselves into the system over time, using previously unused productivity or waste in developing their special niche. Each ecosystem, however, sets the limits and determines the style and complexity of its individual species. Species diversity increases because there are more possibilities for making niches in the increased structural variation. Within species there is more genetic variation. Less energy leaves the system, because it is bound up into maintenance of the structure.

Ecological design can create new niches and increase the diversity of a system; it can modify niches and restore natural systems, but it has to do so with caution. Up to a point, niches can be enlarged or increased, but with so many humans, it creates the danger of supplanting other species entirely.

2.5.2.2.3. *Ecological redundancy.* The universe as a whole seems to work with fifty percent reliability; existence is already half determined. Freedom and necessity seem to be balanced at about fifty percent—perhaps this is comparable with redundancy in information theory, where the predictability of particular events within a larger aggregate of events is technically referred to as redundancy.

Some forms of industrial design build with adequate physical redundancy to allow objects, such as airplanes and bridges, to fail partially. Ecological design has to incorporate appropriate levels of redundancy to allow the system to be relatively stable and flexible. Ecological redundancy has value to an ecological system.

2.5.2.2.4. *Failure incorporation.* Ecological Design, like science and art, can learn from previous failures and add that knowledge to future designs. Many unconscious designs of large-scale efforts, such as agriculture or urbanization, have begun to fail, but the lessons have not been added to a redesign effort. Ecological design can approach these patterns with a holistic perspective. Design as usual, like business as usual, could precipitate a catastrophic ecological failure if smaller failures are not corrected and if demands for "natural capital" are not reduced to a level that the biosphere can provide on an annual basis. Of course these failures can occur in endless combinations.

2.5.2.2.5. *Historical tradition.* Design is experiential and historical. As it proceeds, it not only changes parts of the environment, but also human reality in its total context. The approach of ecological design has to be tradition-based and self-conscious. It has to be responsive to the genius of place, as well as to human meaning derived from the existential significance of the place. The concept of place incorporates physical, biological, and cultural dimensions. The solutions cannot, and do not need to, be precise; they can be unfinished and ambiguous. They can be fuzzy. They do not need to guarantee rootedness or workability, but they can identify limits. They can provide possibilities through a matrix that allows tradition and richness. They can provide direction from the understanding of the parts. For instance, they can understand how to make a fertile kind of soil, as a metaphor, where things can live in and develop. Living beings synthesize the parts and find meaning in living there, and in doing so revitalize a place. Design can learn from the history and successes of many lives and cultures to create high-quality designs.

2.5.2.2.6. *Quality emphasis.* Quality can arise as a break in quantity, that is, it can emerge from simple quantity. In an ecosystem, individuals and species are connected to some degree, in terms of the quality and quantity of their connections. By emphasizing quality, ecological design contributes to long-term products and patterns. Design can make models of quality environments. By embracing ideas from ecology, design participates in a movement of consciousness, concerned with equality, diversity, and health, as well as with humane methods, and a holopoetic cosmology—and is affected by them simultaneously.

2.5.2.2.7. *Bottom-up action with top-down restraints.* Complex adaptive systems display emergent behavior. As an example of emergence, slime molds can form a community without a pacemaker cell that determines when the cells need to combine. It seems that self-organization is bottoms-up. Emergent systems are rule-governed, though; slime molds explore by adhering to low level rules. Individuals coordinate work, even if they cannot assess the global situation. Emergent systems are local; individual molds "think" locally and act locally. Random action serves to explore local space. Individuals pay attention to their neighbors, however, and patterns emerge from local activity.

Simple behavior seems to work, with local feedback, and more sophisticated behavior "trickles up" to approximate a global perception. As design becomes more complex, individual design can be aided by the emergence of social creativity. For a social, bottom-up design to work, there has to be sufficient inflow. The cultural basis may have to be slightly redesigned to allow top-down restraints and forms to shape the whole design.

2.5.2.2.8. *Consequences attention.* One challenge for ecological design is to avoid simply repeating the same errors and consequences in the rush to acquire minimum standards and wealth for most people. To survive, an ecosystem depends on the interactions and balance of many variables, most of which are not well understood. In agriculture or forestry, we try to maximize one of those variables. When that happens, the balance or harmony is altered, and although it may take decades or centuries for the consequences to be known, the system is affected. The decline of Rome and many other civilizations demonstrates that ignorance of ecology can have important consequences. Some cultures ignore the long-range ecological consequences of drainage, irrigation or overexploitation, and these cultures may decline and die. But many archaic cultures display a form of fitness and limitation. Some, such as the Tukano Indians, try for adaptation before domination, according to Ricardo Reichel-Dolmatoff. Ecological design can learn from cultural history to foresee as many consequences as possible.

2.5.2.2.9. *Permanence sustaining.* Permanence is important element in the idea of place and dwelling. In English the term for dwelling means 'to stay.' This is the symbolic opposite of moving or changing. It means to 'withstand time.' Dwelling resists and persists. Attachment leads to valuing a place. In terms of ecological design, this can bring about a change from unsustainable activities to sustainable, since movement and waste would be reduced. The benefits of settlement are economic, ecological and spiritual. Economic because everyone will have to learn to live within the limits of photosynthesis and watersheds. The ecological benefits of rootedness are that people will take care of their place if they realize they are going to be there for a thousand years. Having a place means that the inhabitant has stock in it and participates in its unfolding, through planting and caring. Detailed understanding of plants in a locale allow gathering of food and medicine. People in place acquire a sense of community, nonhuman and human; shared set of values and concerns; health and spiritual benefit.

2.5.2.2.10. *Perceptive.* The use of a metaphor alters the perception of the secondary as well as the primary systems of reference. Using metaphors, ecological design can shift our perception of humanity and the earth, from the center of the universe to participants in a multidimensional process. No culture is the center, or the evolutionary survivor of other cultures. By enlarging human perception, ecological design can create designs that are appropriate for individual cultures or for the common needs of many cultures.

Table 2522-1. Comparison of Natural, Traditional & Ecological Design of Ecocultural Play

Natural Creation	*Traditional Design*	*Ecological Design*
Nature-based	Culture-based	Nature & culture-based Connecting
Aggregating	Bricolaging	Niche-assembling/ connecting
Trial & Error Novelty testing	Conscious Trial & Error process	Ecological redundancy
Trial generating	Learning from failure	Failure incorporation
Novelty generation	Novelty emphasis	Historical tradition
Reciprocal interactions	Rules/Geometric grids	Quality emphasis
Bottom-up organization	Top down direction Control of actions	Bottom up action with top down restraints
Physical restraints	Creative Play	Consequences attention
Natural process	Goal-directed	Permanence sustaining
Actual Existing	Conceptual	Perceptive

2.5.2.3. Ecological Economies of Ecological Design

By considering an enlarged concept of economy, in which there are no real externalities, ecological design can produce forms that fit well within the larger ecological regional and global systems.

2.5.2.3.1. *Natural capital-preserving.* Production forms need to be redesigned to incorporate renewable energy forms and longer-used materials, that is, fewer materials that can be recycled by the system driven by plentiful alternate energy sources. Using renewable energy, and fossil fuels at one six-thousandth the current rate, would allow the energy capital of the system to remain intact. Keeping materials in the system longer allows the natural capital of the minerals or the overburden—which is usually a functioning ecosystem—to be saved, also.

2.5.2.3.2. *Adaptive process.* A material ecosystem is an adaptively organized system of living beings and elements that recycle important elements within the system (after Lynn Margulis's definition); it is the smallest unit that recycles biologically important elements through circular paths. 'Culture' is a word that refers to a form of human expression that is adaptive to the environment of nature. Culture has been a useful adaptation to environmental challenges, such as cold, drought, and rapid change. Design can use purposive policies that mobilize the adaptive capacities in these dimensions of culture and nature. Without certain knowledge, however, we can make adaptive decisions based on partial knowledge; this is adaptive design, which is very much like adaptive management.

2.5.2.3.3. *Integrative.* Most approaches to design and production are dualistic (wilderness-industry) or triune (wilderness-design-industry). Ken Yeang states that ecodesign is the biointegration of artificial systems to natural systems. The ecosystem is the level of integration and the unit of organization undergoing a directional development, according to Eugene

Odum. Energy is bound into organic material, measurable as productivity. It is not landscape strictly—it is domiture, which includes all agriculture, wilderness, and other forms. Ecological design can make connections visible and restore the interconnectedness of the systems, especially ecological and human social systems.

2.5.2.3.4. *Dependent constructive minimally destructive.* Ecological design minimizes the waste produced by designing for reuse. By using more benign arrangements of elements and materials, design can minimize damage from unnecessary or toxic waste. Ecological design depends on the coconstruction of artificial and natural systems moving under constraints from each other.

2.5.2.3.5. *Appropriate energy use.* Ecological design tries to determine the source of energy and to use energy from renewable sources, such as solar energy and wind power. But, it also has to be cognizant of the scale and appropriateness of these sources and try to use small-scale, distributed sources. When possible, power generation could be integrated into the design. With efficient homes and electric transport, all of the energy needs of some areas, the Pacific Northwest for instance, could be met by established water power, which is estimated to be half of the potential.

2.5.2.3.6. *Appropriate technology.* Van der Ryn states that ecological design is about merging nature and technology. Certainly technology is emergent from human activity, which is emergent from "nature," but ecological design is also concerned about separating technology from nontechnical species. Advanced technology is not always necessary. To be sure, it is wonderful, but it should not be the driving force for ecological design. Conservation and fitness are far more important. Since 'Natural services' cannot be replaced by technology, either economically or completely, large areas are needed to be left wild. However, due to our extreme, inconsiderate interference, we need to design a protective framework, for that "nature." Artists have certainly agonized over the dehumanizing effects of technology on natural landscapes and the human spirit. Perhaps art can help shape an emerging ecological consciousness (or a stage for disillusionment and collapse).

2.5.2.3.7. *Waste integration/Ecological Efficiency.* Other designers, including Nancy and John Todd, and William McDonough, suggest that design should create no waste by following natural systems. But, every group, and many systems, create waste; often it is just used as a resource by a system down-stream. Although in the evolutionary past, and possibly in a few systems now, so much waste was produced that it was incorporated into geological processes as carbon forms or other metallic forms, like iron waste. Waste could be drastically reduced by designing the process to use common reusable materials. Much unavoidable waste from production can be used in other processes. Some waste may have to be separated and buried, that is reintroduced to geological processes.

2.5.2.3.8. *Cost Constraints.* What is the total cost of any design? Not just material, but ecological and psychological costs? It may include pollution and loss of community, with the attendant loss of trust. Cost constraints should be based partly on physical constraints. Ecological accounting includes all the costs that impact consideration. Costs to ecological systems should also be calculated and minimized. Longer time-lines make a difference, also.

2.5.2.3.9. *Size/scale constraints/Quality consideration.* Although fads are part of biological and human life, ecological design considers the material and scale constraints that will discriminate against wasteful or stupid fads. The second law of thermodynamics is a real constraint. It guarantees that cost is always a consideration with design. Gravity also limits extents in design. Limits reduce options. Constraints and limits, however, can be sources of creativity rather than just hindrances.

Table 25239-1. Comparison of Natural, Traditional & Ecological Design
Use of Ecological Economics

Nature Creation	Traditional Design	Ecological Design
Productivity	Capital-using	Natural capital-using
Interrelational	Constructive	Adaptive process
Nested	Auto separation	Integrative Identity?
Interdependent	Independent/destructive	Dependent constructive minimally destructive
Solar-life driven	Energy use (any, fossil)	Appropriate energy use
Aesthetic creation	Technological tool use	Appropriate technology
Waste generating	Waste production Efficiency	Waste integration Ecological efficiency
Free play/change	Cost consideration Market-oriented	Cost constraints based on physical constraints
Natural Teratrends	Fashion fads Style (e.g., more speed)	Size/scale constraints/Quality consideration

2.5.2.4. Values of Ecological Design

Ecological design treats value differently, partly because it is holistic, and partly due to the increase in the number of values.

2.5.2.4.1. *Responsibility to living community & environments.* Ecological design takes responsibility for the state of the ecosystems in which design projects, at any level, are embedded. This responsibility includes caring not to disturb the operation and cycles of ecosystems by adding unbearable amounts of waste, energy or materials. The ecological responsibility for the production of goods is important. Choosing to use designs for yourself involves knowing an ecology, knowing a politics (being astute), and protecting what is valued. People in wealthier nations could cultivate voluntary simplicity, living in a way that is outwardly simple and inwardly rich, by consuming less, and thus release more rare goods to other people.

2.5.2.4.2. *Strategies.* Ecological design uses strategies to shape the overall plans. These strategies consider ecosystem use and exploitation, as well as long-term effects on the human realm. An ecological strategy defines a larger end—one that includes the welfare of other species and ecosystems, as well as the welfare of individuals and other people, over a longer period of time.

2.5.2.4.3. *Flexible order.* Order must be more than meaningful to humans. It has to consider time scales beyond human, on evolutionary living order times. The order has to be flexible to accommodate changing conditions and changing human meaning. Flexibility allows for more options under different constraints. This flexibility allows designs to be useful for longer times under different cultural values.

2.5.2.4.4. *Fitness concepts.* Fitness is treated as a value by ecological design. In an organic world, fitness is defined as the continued free interplay of energies and structures without dissolution. Perhaps, as a working definition, we can just use 'harmony,' that is, a good fit into the environment. Harmony as an ecological concept includes the health of people, other life forms and the living systems.

2.5.2.4.5. *Limitable Cautious.* David Orr's characteristics of ecosustainability include the idea of nature summarizing the code of evolution as a model. Nature uses limits; ecological design can treat limits cautiously. Many things 'stand in the way' of designing a sustainable livable environment. Lack of knowledge is one thing. Another is the focused self-interest of an institution or a culture. Another is individual self-interest, which may be broadened by culture or narrowed. The goals of a collective society may not reflect the goals or intentions of individuals. Our imagination is limited, our abilities are limited, and our vision is limited. In our excitement to design good things for ourselves, we have to be careful not to dispose of a tremendous base of traditional ecological knowledge (and we might avoid movements, like the Bauhaus School, with its slogan of "Start from Zero" and desire to sweep aside all rules). Relearning is the process of investigating past designs without rejecting them. Strong past designs created a meaningful human order. The modern order of design has to recognize, respect, fit, and work with environmental conditions. Of course, some people use design to try to force the environment to fit human desires or to try to escape from its limits, but designers need to be cautious about displaying too much power.

2.5.2.4.6. *Inclusive.* Ecological design has to be inclusive, not only of objects and ecosystems, but of a long temporal order. Anything that impacts the ecosystem health has to be considered in design. Although a design has a boundary, ecological design considers the boundary more like a large enveloping membrane that allows flow through the design.

2.5.2.4.7. *Knowledge.* Design has to consider innate knowledge and the spectrum of ignorance, rather than just rational knowledge, which is just part of a large system of knowing. Designers need more than sensitivity and ambition. They need an understanding of human experience and an understanding of a balance of nature and culture. They need the ability, as Erich Jantsch says, to organize their total experience towards a purpose. Design is presented as a counterpoint to analysis, but analysis is necessary in design, before it can begin a synthesis. To design products with minimal environmental impact, we need to have a basic understanding of ecosystem impacts and the interactions between them.

2.5.2.4.8. *Inspiring.* Design can be inspiring, as well as exciting. It can inspire almost everyone to try to design objects and forms to make things more interesting, as well as safer and more useful. Design can capture the excitement of the ecosystem context to provide inspiration for creators, as well as for users and viewers.

Table 2524-1. Comparison of Natural, Traditional & Ecological Design in Values

Natural Creation	Traditional Design	Ecological Design
Living	Responsibility to product, consumer, community	Responsibility to living community & environment
Fitting	Planning	Strategies
Harmony-building long-term	Meaningful order short-term	Flexible order Living order times
Existence value	Goodness concepts	Fitness concepts
Learnable	Teachable	Limitable Cautious
Adjusting	Exclusive	Inclusive
Formation	Information	Knowledge
Chaotic/Free	Exciting	Inspiring

2.5.3. *Ecological Design Components, Elements & Variations*

Ecological Design builds on the basic geometric elements of traditional design, from 3 (volume) to 2 (plane), to 1 (line), and to 0 (point) dimensions. These elements can vary in numerous ways, by number, position, direction, size, shape, interval, texture, color, and temporal. Furthermore, the elements can be organized into groups by nearness, similarity, and difference (diversity). Then they can be combined into whole structures by principles of rhythm, balance, and finally sensory force.

2.5.3.1. Elements
Ecological design has the same elements—point, line, plane, volume—as traditional design (see Section 2.4.3.1.), but they are understood in an ecological context foremost, in treetops, promontories, and mountain ridges, in the outline of a forest edge or a volcanic island, the surface of a grassland or the path of a shark.

2.5.3.2. Variations
Ecological design shares the same variations—number, position, direction, size, shape, interval, texture, color—as traditional design (see Section 2.4.3.2.), but there are many more occasions: Sand blowout and a forested tertiary dune, the diagonal lines of hills, the path of a corridor, the framing of a skyline by the shape of hills, a talus slope on the side of a mountain, the volumes in and around an Oregon creek, the shape of an agricultural field squared by a road, the texture of an old-growth forest compared to a young tree plantation, or the palettes of a coral reef.

2.5.3.3. Organization of Variations
Ecological design shares the same organizations of variations—time, groups, nearness, similarity, density, diversity—as traditional design (see Section 2.4.3.3.).
However, time in ecological design is ecological time, which ranges from the lifetimes of bacteria to the development of landscapes over tens of thousands of years or the lifetimes of species in millions of years. Trees in the temperate northern hemisphere tend to grow in stands. Nearness for ecological design can range from the interpenetration of endophytes to a general community grouping in a desert mesquite landscape. Blades of grass in a prairie may be perceived as dense from a low perspective. With fewer elements, a Japanese garden of the same area as a Dutch Tulip garden may look smaller.

2.5.3.4. Organization into Structures
Ecological design shares the same organization into structures—structures, rhythm, tension, balance—as traditional design (see Section 2.4.3.4.). The structure may be more complex for ecosystems, as when a tropical forest ecosystem has a typical structure (of interrelated parts) of six levels, from the herbal layer to rising emergent trees. Rhythm in ecological design has units of time that may range from milliseconds to thousands of years, although most will seem to be daily, seasonal, or long (in 2-25 years). A large clearcut area could create tension in a mature forest landscape because it would rarely develop under natural conditions. In forest design, a larger meadow could balance a dense woods, due to differences in perceived masses (especially if it reflected differences in soils or topography).

2.5.3.5. Concluding Thought: Ecosystem Complexity & Wicked Problems
There are not many examples that give the interests of the ecosystems independence or priority, perhaps because the project would be too large or complex, or perhaps because there is no funding for it. Eric Higgs thinks that this may be a wicked problem for ecological design. Rittel Horst, in 1960s, defined a wicked problem as a kind of social problem that is formulated with confusing information, too many clients, and confusing values. Imagine how wicked ecological problems are when they include all those nonhuman perspectives and values. Problem definition is subjective based on a point of view (or two or three). The solution is participatory design, which can reduce domination by one point of view. Design is political; there is no way of envisioning other approaches.

Complexity is not the only variable that can make a problem wicked. The scale is a problem if we think we can destroy the planet by burning fossil fuels. And, it is a problem if we think that nothing we can do will affect the planet.

Orientation is another variable that can influence the wickedness of problems. Design can be object oriented, like a building, and not seem wicked. But, design can be pattern-oriented, like planning for minimized polarized light surfaces on buildings in sensitive eco-systems, and then it pulls in scale and complexity—we are finding that we cannot avoid the things that make designs wicked, and that denying them does not make design good.

But, design can work things out in an artistic way. There are ways in which design can be subversive; the magazine *Adbusters* is an example. Perhaps that might solve the wicked problem situation of design. Being subversive may turn over and expose wicked design problems. That artistic way is a wild way of thinking and meshes or can mesh with restoration design better than a simple technical approach. Ecological design requires an ecological perspective, systems understanding, participation, and standards of knowledge. However, it does not require a degree or certificate, which means that fewer people, educated or self-educated, will be ecological designers.

2.5.4. *Characteristics of Ecological Design*
The characteristics of ecological design have to address the properties of ecosystems and places. The courses of the design have to imitate the processes of systems and places. The wholeness of the design has to address the ideas of spirit of place and sensory force. After the processes are identified, they have to be related to the patterns. Design may have to try short-cuts, unless it can afford to take ecosystem time, which is often many human lifetimes.

2.5.4.1. Imitating Courses
Design has to identify patterns of movement and construction, then it has to imitate the processes and patterns, modifying them to include more things of human interest and need. This is more difficult than copying a shape or a structure. Fortunately, imitation is a human strength, even if the recognition of complex, long-term, moving patterns is not. Courses can be thought about using topological and mathematical models representing a four-dimensional landscape.

2.5.4.2. Extending Identity & Wholeness
Ecological design has to capture the integrity of a place, or to restore it. Unity is a fundamental objective of landscape design. Unity is the way the elements, including shape and scale, of a landscape are combined. It is sometimes related to simplicity and character.

For instance, visually, a forest ecosystem usually dominates the landscape. From a distance, even-aged forests have much the same impact, in terms of color, shape, and scale, as uneven-aged forests. Diversity becomes more important visually at a smaller scale. Natural forms of the forest are unified with the landscape because the margins are very uneven, and open space in the forest is part of the mosaic caused by birth and death of individual or groups of trees. Whole means containing all of the elements to be complete in itself or being a system. In Arthur Koestler's idea, things are wholes related in nested levels.

2.5.4.3. Enhancing Diversity
Because the operation of the universe tends to change systems, the design of a place should be open to the types of processes that could destroy the design, as well as enhance it. Furthermore, flexibility, defined as the unused capacity for change, can be designed into the system.

The parts of a system have to maintain the potential for many possible behaviors that could flow from any other part of the system. Openness and flexibility are characteristics of healthy ecosystems, and they can be considered and enhanced by design.

Play is an activity that can lead to relaxation, or practice for more serious things relating to survival, such as food and mating. In another sense it means to move freely within limits. As an activity, play can increase the openness and flexibility in design.

2.5.4.4. Participating in Coconstruction

Designs are limited by the real biological constraints of ecosystem processes and biogeochemical cycles. We must know the constraints in order to create a healthy design. The design has to work within the constraints of the ecosystem. Rather than emphasize static equilibrium, the design should emphasize heterogeneity and adjustment to disturbance, within a dynamic nonequilibrium.

People have needs, animals and plants have needs, the site has constraints, and these things can be married into a good pattern. Design is the participation in the process of the ecosystem as a harmonious system, with mutually restrained conflicts and constrained influences. In fact, harmony is a constraint of the whole system. An ecological design is a form of co-constrained construction, where the organisms, environment and designs are co-implicative, co-defining, and co-constructing. They all engage in a process of self-assembly, within the whole system.

2.5.4.5. Investing in Stability & Constancy

People often judge the health or wholeness of ecosystems, or the goodness of designs, by how they look. Traditional design has emphasized visual results above all else. Ecological design, however, achieves the same results by paying attention to the structure and function of the ecosystem first, before sketches out an ideal form. Design has been concerned for centuries with making domesticated landscapes out of wild ones. Now, design is addressing the opposite problem: How to preserve or provide the conditions for wild ecosystems so that they are stable and healthy within the processes shaping and affecting them.

Signs of ecosystem health include the homeorhesis of the system, that is, the stable, directional flow of the system capable of resistance, resilience, and accommodation. The design for an ecosystem describes the system in a comprehensive interdisciplinary approach, using dynamic concepts such as constancy and stability.

2.5.4.6. Stimulating Productivity & Health

Most products of an ecosystem are produced, consumed and recycled within the ecosystem. Humans need to minimize the external inputs to systems in the form of energy and exotic substances. The community must be restored to health. This means balancing human needs with bird or fish needs in a sustainable pattern.

Ecological design is the creative modification of ecosystems to repair or enhance their ability at self-organization and maintenance of their complexity and diversity. Health is the overall ability of a system to maintain itself under a normal range of environmental conditions. Ecosystem health is one of the goals of design. The goal, of course, is not an end point that can be reached once, but is rather a continual striving. Responsibility is the condition of being accountable for actions or obligations. Here it means undertaking all aspects of a design, regardless of the expected level of success. These six characteristics have to be related to other levels by ecological design.

Table 2546-1. Properties at Different Levels.

| | — Nature — | | — Culture — | | — Design — |
Field	Ecosystems	Place	Culture	Good Places	Good Society
Motion Process	Course	Dynamic Change	Conduct	Action	Method
Autopoesis	Identity	Self-making	Wholeness	Individuality	Self-extension
Differentiation	Diversity	Uniqueness	Flexibility	Richness	Variety
Integration	Constrained Construction	Integration Investment	Adaptation	Conviviality	Cooperation
Constancy	Stability	Regularity	Endurance	Consistency	Loyalty
Development	Productivity	Renewal	Vitality	Health	Harmony

For instance productivity of the ecosystem level promotes vitality in cultures and harmony in good societies, which are embedded in human places, which are embedded in ecosystems. And, diversity at the ecosystem level promotes flexibility at the cultural level and variety in good societies.

2.5.5. *Principles of Ecological Design*

Principles reflect the differences between local and global systems. Principles cannot be reduced to other principles, because they emerge from earlier ones and they expand into new forms. This expansion is why we cannot fit them into the old can of old principles. These design principles are based on basic principles of nature, such as the principle of change. Nature is in flux, culture is in flux, everything is in flux. The climate will change, the shorelines will change. Human understanding and behavior is changing. But, nature is also self-regulating. This is the principle of self-regulation. Nature has evolved to maintain its stability in the face of many kinds of disturbances from planetesimal impacts to changes in atmospheric composition. Nature will continue to regulate itself even as human beings make dramatic changes. The danger is not so much that nature will collapse, but that humanity will.

2.5.5.1. Ecological Principles

Principles of ecology, such as stability, have to contribute to the design of landscapes because those landscapes have derivative principles, or maybe emergent principles. Design has to incorporate those principles because it is in place, in an ecosystem, in a culture. Some cultural principles, such as sacredness, are actually emergent from ecological principles, such as the principle of pattern. It expresses the values of the locale and the culture. Principles must be flexible to mirror the flexibility of open systems; flexibility is provided by diversity in fact.

    2.5.5.1.1. *The principle of flow.* Nature is in flux. It changes; it flows. The principle of flow is necessary to the functioning of organisms as well as the biosphere. Flow can only be realized through structure, however, cell or ecosystem. Another principle is the principle of flow, which is necessary to the functioning of organisms as well as to the biosphere. Flow can only be realized through structure, however, cell or ecosystem. It is the problem of free flow or division by membranes. At each level of a natural system, from cell to biosphere, the units involved do more exchanging internally than externally with other units at the same level. Flow and division must be in balance. The diversity of the current ecological world evolved through the breakup of Pangaea, which provided the distances and barriers to isolate species.

    2.5.5.1.2. *The principle of separation.* The principle of separation, by walls, barriers or membranes—nonflow or limited flow, is crucial to design. Barriers are necessary to maintain form and integrity of individuals as well as of ecosystems. By removing some barriers, such as releasing carbon that has been locked up for eons, we unbalance natural cycles. Even though landscapes exist within large ecoregions, which exist within the biosphere, they need closure to maintain their integrity. At the level of local ecosystems, there need to be fewer closures, so that there is a flow of genetic information within a species as well as between species. Human activities have the effect of blocking the flow at local ecosystems, with asphalt and wires, yet increasing the flow between large regions with ships and airplanes—mostly the flow of pests and domestic species. Relate closure to freedom with the 50% rule.

2.5.5.1.3. *The principle of growth.* Everything seems to grow, from the universe to the human population and its activities. Urbanization attracts people to live together in larger populations. Everything tries to expand, beyond serious constraints or limits if it can, or if, not to survive within them. Cells produce many more daughter cells than can be used. Many species produce more offspring than can ever live. This principle seems to be limited by the principle of pattern, which provides limits to growth, and the principle of separation.

2.5.5.1.4. *The principle of self-regulation.* Nature is self-regulating. It has evolved to maintain its stability in the face of many kinds of disturbances from planetesimal impacts to changes in atmospheric composition. It does not need to be regulated by external factors or an internal regulator—the regulation emerges from the complexity of patterns.

2.5.5.1.5. The principle of error (or play). Error permits diversity. So many things are thrown at the flow of life. There is not just a little error. Half of everything seems to be error: billions of pollen grains, billions of eggs, millions of species. Transmission is not flawless or efficient, but it is generative of difference and diversity. The diversity of the current ecological world evolved through the breakup of Pangaea, which provided the distances and barriers to isolate species.

2.5.5.1.6. *The principle of pattern.* A pattern is an arrangement of form of elements. Process applied to components yields pattern. Nature is composed of patterns. Organisms have characteristic patterns, such as the branching of trees or the cloud forms of tree crowns. Lichens have lobes, wood grain under stress has spirals. The cracks in tree barks form nets. Patterns are not still. A circular pattern through time can be recognized as a spiral (the earth's orbit for example). The pattern should allow for surprises and discontinuities; it can do this if it is flexible. The design of ecosystems is vulnerable to surprises because nature is chaotic (unpredictable) and science itself is uncertain (by definition) about patterns of change in ecosystems.

These principles seem to be interactive, each forming part of the environment for the others. These principles result in a definite set of standards and behaviors that can guide ecological design. For instance:

- Work within the scale and duration of an ecosystem. Processes like succession can be assisted, slowed, or speeded up, but not skipped or ignored.
- An ecosystem is designed by its limits, time, scale, and complexity.
- Only details and exceptions form the ecosystem; only generalities are important—so you have to live with contradictions.
- Make the smallest number of changes, in case they are not effective and have to be modified. Nature seems to follow its own principle of least effort.
- Adapt process of change to the site (you fit the ecosystem, not other way), so that you do not have to put as much effort into control.
- Seek the best use for the products of the system; everything in the system is a resource for something; many can be directed to human use, but not all or most of them should be.

- Extend the life of things through cycling, then return the waste to the ecosystem; things can be recycled indefinitely, as in an old growth cycle.
- Preserve the components, structure, and function of an ecosystem. The redundancy of a system will allow the system to continue to be self-making.
- Understand the patterns and connections. Make sure they are not broken. Try to fit within their limits.

2.5.5.2. Ecological Design Principles

Ecological design principles emerge from knowledge of ecological principles. They must combine the constraints of ecology with the ascendant sensitivity of design.

2.5.5.2.1. Spirit of Place

The spirit of each place is unique. Place is not just location; it is the total sum of objects in the landscape combined into a unique whole. The identity of place often leads to human identity, thus people call themselves by their place names. The more unique a place the stronger the emotional attachment of the inhabitants. Every place has certain characteristics that enforce the spirit of place, for instance, a strong definition of place or indicators of great age (trees or rocks), or where a place distills the essence of larger landscapes. A sense of wildness and water also contribute greatly to the spirit of place.

Each place expresses a unique combination of elements, including contrasts, dramatic features, and the presence of water. Design can work to be consistent with the recognized spirit of place. If the design recognizes this aspect of the landscape, it may be stimulated by spirit and it may further enhance it—what it should not do is degrade it. Forest design can emphasize some features above others. The goals of good designs include: To relink people with genius of their places, to revivify image and identity with places, and to develop and maintain the identity of places.

2.5.5.2.2. Sensory Force

All the elements of design can be combined in an image. Every organism creates an image of its place from what is meaningful to it. This image is what fits the organism to its place. Suckers and caddisworms have simple images; coyotes and humans have more complex ones. Kenneth Boulding notes that the image as a cognitive construct of the world has several aspects: Spatial, temporal, personal, relational, value, and affectional (emotional) for each individual. Cognition is an active relationship that is creatively shaped by the participants. Participation is not an option by the way—every scientist or inhabitant becomes part of the system of observation. The total sum of individual images is a world. Some of the images we impose on nature result from idealized notions of pastoralism or technological futures. Thus landscapes abound in nostalgic or consumptive trends on many levels of explication—some are iconic, some invisible. We originally perceive the landscape symbolically, but the landscape has other functional dimensions that increase according to use.

Visual force is a psychological interpretation of perceived power in a landscape. As a principle, it is embodied in psychology, art, graphic design, and architecture. The human mind responds to visual force in predictable and dynamic ways, for instance, visual forces in landscapes draw the eye down convex slopes and up concave ones—the strength depending on the scale and irregularity of the landform. The effect of a landscape is not completely visual, however. Smell, sound, touch, and even taste play a large part of our appreciation of forests. Crawling, which is highly recommended by Gary Snyder, climbing, listening, and tasting things, such as soil or lichen, can expand our perception of other aspects of the forest.

2.5.5.2.3. Precautionary Principle

We should adopt a precautionary principle, which asserts that, if harm is threatened, and if there is uncertainty about the seriousness of the harm, then precautionary actions must be taken. Since the 1970s, in fact, this principle has been incorporated into Swedish and German environmental laws. This principle means that not doing something, "benign neglect," becomes a valid management option. Carl Walters suggests that inaction is an inappropriate alternative to gambling as a result of confusion, but inaction is a very appropriate alternative in the face of confusion—when in doubt about an ecosystem, we should not interfere. As appropriate alternatives, both inaction and action can be suggested by theory.

Traditional planning asks "what is most likely to happen," where uncertainty planning asks "what has already happened to create the future," in terms of long-term trends and demographics, according to Peter Drucker. Traditional approaches to uncertainty management are probabilistic, focusing on uncertainty as a measure of randomness or of confidence.

2.5.5.2.4. Scale Principles

Scale means basically a larger size or time. There is more mass and more momentum. There are more connections. This can mean lower interunit communications or connections get weaker as they are farther apart. There is a longer response time with size. Collapse may be absorbed by the connections, but sudden collapse becomes more likely. The interior/exterior system ratios change. A larger surface area is required to dissipate heat or pollution. This is why giant ants are not possible; they would cook in their exoskeleton. The breakdown would occur immediately if something was made larger suddenly.

Principles must be flexible to mirror the flexibility of open systems; flexibility is provided by diversity and redundancy. These principles also result in a further qualification of standards and behaviors. Some actions based on these principles might be listed. For instance: Contribute to excitement and to life (being); Act from the bottom, restrain from the top (and act locally in parallel, within the constraints from the top and blueprint); Cultivate feedback; Develop rather than grow; Use a sideways approach, rather than focus on details; Work with errors, learn from, and allow them; Work with and allow disequilibrium; Structure the design to allow change; and, Pursue the satisfactory (saticficing in H. Simon's words) instead of a perfect maximum or even optimum.

2.5.6. *Applied Ecological Design Practices As Lessons from Nature*

Design can imitate nature on many levels, from structure and process to landscapes. We can imitate the structure of mature forests by planting on every level of the forest hierarchy, from canopy to below ground. We can use native species. We can imitate the process of forests by allowing birds, bats, and other animals the opportunity to distribute seeds and energy to other areas or to prey on "pests." We can create microclimates within the landscape that may shift the landscape in new directions. Planting trees, for instance, allows new species to become established under their protection. Ecosystem health is one of the goals of design. Although the goal is not an end point that can be reached once, but is rather a continual striving.

The landscape provides its own metaphor for design. The landscape is a unique individual, a community, a dynamic system of interacting patterns—the human pattern is a part of it now and should be preserved as part of the whole pattern, but not necessarily as the only pattern or a completely dominant one. Most products of an ecosystem are produced and consumed and recycled within the ecosystem. Humans need to minimize the external inputs in the form of energy and exotic substances. The community must be restored to health. This means balancing human needs with birds or fish needs in a stable pattern. Each element in a pattern relates to others and to the whole.

Natural patterns can suggest a number of regularities to understand for the design of ecosystems, according to Michael Soule. Well-distributed species are less at risk than concentrated ones. Large blocks of habitat are safer from species extinctions. Blocks close together are more effective than those far apart. Contiguous blocks are better than fragmented blocks. Interconnected blocks are better than isolated blocks. Corridors can make functionally larger blocks functionally. Roadless blocks are better. Human disturbances similar to natural ones are more likely to be resisted or accommodated.

These rules work well with natural processes that operate in ecosystems, such as metalysis (building up and breaking down), animal movement, interelement flows, human interactions, and shifting mosaics. For instance, forest fragmentation can be reduced through the design of forested areas, taking into account the genetic diversity of the trees, catastrophic conditions, minimum viable populations, corridors, and edge effects. The survival of organisms usually depends on one of two factors in the web of relations. These factors can be modified by design.

Wild landscapes are affected by climate, soils, interactions, and disturbances. Domestic landscapes are affected by these and by land use as well. The greatest changes have been brought about by the destruction and creation of forests. With the predominance of artificial forests, it is important to consider the qualities of naturalness in the landscape. Forests are expected to meet the needs of society by producing timber, creating wildlife habitats, and providing recreational opportunities for people. But, forests are also expected to look natural.

The English Forestry Commission's guidelines (1994) to principles and practical applications of forest design may be of use. They represent an established standard. They do not cover every aspect of landscape design or all of the details of design techniques, however, nor do all of them apply to wild forests. The guidelines indicate what to look out for, and which situations may need special attention. Forest landscape design is a complex subject, as are forestry and ecology.

The values of the land and forest must be most carefully assessed. The characteristic qualities must be identified and measured for uniqueness. Comprehensive landscape plans should be required when planting or extensive felling is planned on a large scale. The patterns established at these times may persist for many years or centuries. Good design maybe able to resolve conflicts between characteristic qualities of the landscape and the changes from use. Also, the design should last as long as possible and should be self-sustaining.

2.5.6.1. Levels of Ecological Design
As a design process includes the planning of ecosystems, this means a fourth and fifth level of design— notice that numbers lower than number four, all are properties of traditional applications of design: (5) Regions, landscapes, and watersheds (as well as airsheds), (4) The community, for instance forest, city, or corporation, (3) systems, such as ecosystems, traffic, or industrial areas, (2) products, as habitats, houses, roads, or plant sites, and (1) components, such as trees, fungus, bats, tools, rooms, cars, or land. Many problems occur at level three—more are to be expected at the new levels four and five.

All levels of design need to be addressed, from the conceptual to the political, and are involved in all stages of the process. This involves new challenges for ecological design, which has to:

6. Relate a project to its global context (sixth level of design). This is the level of global

cycles relating to renewal of living processes and resources from long and short-term global cycles.

5. Relate a project to its regional context (fifth level of design). This is the level of regional cycles relating to precipitation and waste recycling.

4. Relate a project to its ecosystem context (fourth level of design); be concerned as much with cultural survival, justice, and wilderness preservation as with efficiency and aesthetics.

3. Consider the whole perspective (ecocentric, perhaps); the proper vision is of the whole community in which we dwell. Apply ecological concepts, such as networks and carrying capacity.

2. Make designs that are anticipatory, flexible, pluralistic, polyvalent, and polytechnic. Make open guidelines for long-term decisions.

1. Essentially, work backwards from values and goals, and from the bottom up and inside out, drawing designs from the genius of place.

0. Participate in place, care for all inhabitants, and assume responsibility for the designs.

One of the shortcomings of design has been the lack of consideration of the higher levels, which can result in specific problems with safety or recycling.

2.5.6.2. Stages of Ecological Design

Ecological designs can be applied in eight stages. These stages should be typical of all designs:

- First, Decide to design; Use a design management matrix, review the situation; Evaluate ecological history of the site, observing patterns of movement; Account for interconnectedness, population change, land use, building and development, boundaries, limits, and life; Conduct ecological and functional analyses.

- Then Inventory, record all of the resources, from physical resources to cultural resources. Survey the area and create base maps, from geological to zoological maps.

- Next, evaluate the interactions in terms of impacts, needs, goals, and limits. Assess the whole system. Delineate and balance components.

- Create a series of plans, from the site plans to value plans. Integrate with biosystem services. Kinds of integration: horizontal, Vertical, temporal. Consider aesthetic-perception factors.

- Start to design, which is a community process requiring the participation of all people, including the elderly, handicapped, and poor, as well those ultrahuman beings who cannot voice their concerns. Synthesize simulations and models (conceptual, capability, and suitability). Make another series of plans, from landscape plans to policy plans, within a master design.

- Implement the design together. Use appropriate measures and techniques, emphasizing native species over an adequate time period to ensure the stable processes of transformation. Create new connections, Reduce unwanted effects, such as road effects or heat island effects. Integrate into context. Optimize passive mode integrate water, biomass, materials flows. Conserve water and energy

- Maintain and manage design. Provide services for continuity and management. Monitor and improve constantly. Monitor, Reassess.

- Prepare for undesign and disintegration.

Ecological design must work within the components, structure, and function of ecosystems. Unless it does, it will not be long lasting or satisfactory. Because ecological design has to work with wild ecosystems, whose aspects are often ambiguous, fuzzy, changing, and general, design has to be able to work with these aspects. Furthermore, the design has to work within the constraints of the ecosystem.

Ecological design takes far more time than graphic or automobile design, due to the complexity, size and longevity of its subject. A number of factors have to be carefully assessed before design work starts. Wild ecosystems require a lot of observation before activities can take place. Wild ecosystems can be highly reactive to change. Some people value different character of such ecosystems than others.

2.5.7. *Applying Ecological Designs to Places*

All design elements are related psychologically by designers, as focus or frame, as contrast or uniformity, as dominant or recessive, or in a number of other pairs. Good ecological design means not violating any of the aforementioned principles and ideas.

Design can improve the results of bad practices. Bad forest harvesting practices, for instance, often result in geometric wastelands. Good design can correct reliance on straight lines, parallel lines, right angles, and perfect symmetry. In cutting or planting forests to improve natural appearance, a number of things have to be considered, including the age of the forest, windthrow, the width of corridors, and the minimum size of the habitat.

2.5.7.1. Place-Based Ecological Design

Geology, climate, disturbance, and stability all produce diversity. Landscape diversity is linked to ecological diversity, which depends on diversity of the substrate. Different ecosystems introduce diversity into a landscape, but different ecosystems often can look similar. Excessive diversity can lead to confusion in a landscape design. Increased diversity also has the effect of reducing scale, so adding diversity can be used to reduce the scale. A high level of diversity is acceptable if one element is clearly dominant or if the differences cannot be recognized from a distance.

Psychologists have recognized the need for diversity for people's quality of life and emotional well being. Ecological diversity in the forest has been reduced by human activities, such as planting or grazing. The overall landscape diversity of many forests has not fared as badly, due to the addition of human artifacts, which can increase the diversity. An increase in ecological diversity would lead to an increase in diversity of the landscape, however. It would also tend to reduce the scale, but this would not be a problem in large forests.

Process applied to components yields pattern. Nature is composed of patterns. Organisms have characteristic patterns, such as the branching of trees or the cloud forms of tree crowns. Lichens have lobes; wood grain under stress has spirals. The cracks in tree bark form nets. Patterns are not still. A circular pattern through time can be recognized as a spiral (the earth's orbit for example). The pattern should allow for surprises and discontinuities; it can do this if it is flexible. The design of forests is vulnerable to surprises because nature is chaotic (unpredictable) and science itself is uncertain about patterns of change in forests.

2.5.7.2. Ecological Design of Buildings

Design cannot be separated from social and political questions. Designs, as Winston Churchill recognized, especially buildings, steer our experiences and actions. According to Langdon Winner, design includes the deliberately chosen, enduring forms of both material and intangible entities that affect human relations. Ecological design is a field that aims to recalibrate what humans do in the world according to how the world works as a biophysical system and a cultural entity. Design in this sense is a large concept having to do as much with politics and ethics as with buildings and technology.

Green buildings are more sustainable, energy-efficient buildings. In the United States, buildings account for 36 percent of total energy use, 65 percent of electricity consumption, 30 percent of greenhouse-gas emissions, and 30 percent of waste output. Making new buildings that use recycled, nontoxic materials, recycle their own waste and derive multiple benefits from the renewable natural resources around them, such as the sun, wind, rain, not only reduces their environmental impact, but also creates indoor environments that are healthier and are more inviting places to live and work.

Adapting existing buildings has advantages. The creative reuse of existing buildings offers the multiple benefits of "harvesting" already built structures, reducing the demand on our overflowing landfills and allowing people to maintain a stronger connection with the history and development of a city. Changing buildings leads to rethinking transportation. Efficient, convenient public transit can be used to get within and between neighborhoods, reducing congestion, pollution and the use of fossil fuels. Modifying roads, especially in neighborhoods to slow or redirect traffic, could allow more walking and bicycling.

Design could integrate the city into its environment like a living organism in an ecosystem, although in this case the city is an extension of living human organisms. The city could generate its own energy with solar and wind power. The effects of climate could be modified with building shapes and vegetation. Design could influence human connections and diversity, with good psychological shapes and connections. Design could also reduce the scale to a more human scale, requiring less management and rules.

2.5.7.3. Ecological Design of Tools and Things

The ecological design of tools considers tools in every aspect as an extension of the human mind. Tools and things are extensions of human ingenuity. The design examines the effects, short-term and long-term, of the tools, and attempts to balance the trade-offs. Tools have environmental effects, as well as physical, biological, and cultural effects on their inventors and users. Tools have different levels of involvement, as well as levels of intensity.

2.5.8. *Ecological Design Management*

Ecological design is not finished with the design. The system may need to be managed as a result of the design. Noninterference matrix management is proposed as a technique for design management. An ecosystem exists as part of matrix that many interacting elements. Any activity in the matrix can have some effect on these elements. The whole matrix needs to be managed with the ecosystem in mind.

Understanding of the principles of ecology can lead to better management. One critical message of ecology is that if we diminish variety in the natural world, we debase its—and our own—stability and wholeness. Many ecosystems have been simplified and degraded. Perhaps we do not have sufficient knowledge to manage a complex landscape because it is too complex to understand scientifically. But we can understand the pattern and drive it in a healthy direction with minimal intervention. We must do all that we can to restore its richness and the natural processes that created the richness.

2.5.8.1. Principles of Ecological Design Management

Management, whether it is archaic, traditional, industrial, or ecological, should adhere to basic principles.

- Holistic Practice. Forestry practice should be holistic since it affects the whole forest. You should harvest in the context of planning, measuring, monitoring, protection, and restoration. Manage for preservation, reservation, protection, and use. Create a practical plan, based on the forest (especially riparian areas). Emphasize interconnectedness over separate structures or operations. Sustain the forest before any yield.

- Retention. Retain the structures and processes (which produce the complexity and diversity) of the forest, including legacies and special areas. Soil is an important structure. Preserve the diversity of microsites. Preserve all components, structures, and functions. Protect and maintain diversity/preserve the patterns. Design your uses or harvests to maintain connectivity, thereby minimizing fragmentation.

- Noninterference. Do not interfere with the health and stability of the forest. Do not block flows of air or water. Water Flow. As the medium for life, water is used and transported by organisms and systems. Water is necessary in any organic energy conversion. Air flow. As the medium for gaseous and water cycles, air is essential for most life. Do not block flows of nutrients, materials, or animals and plants.

- Least Effort (also a principle of economics and cybernetics). Encourage natural regeneration; let the forest do as much work as possible—it has millions of years of practice in some cases. Make the fewest cuts, go the least distance. Broaden the base of use of materials from the forest; avoid concentrating on timber only.

- Respect and maintain natural disturbance regimes through time and space in order to maintain forest landscape patterns. Natural disturbances, from the death of individual trees to large fires or windstorms, are responsible for critical composition, structures, and ecosystem functioning necessary to maintain fully functioning forests. For example, the death of an individual tree sets off a process of change, beginning with a standing snag that provides habitat for cavity-nesting birds and ends with a fully decayed fallen tree that serves as Nature's water storage and filtration system. At a landscape level, natural disturbances, large and small, are responsible for diversifying habitat patterns and, therefore, maintaining a natural diversity of plants and animals. Natural disturbance regimes are also critical to the maintenance of soil nutrient cycling and adequate levels of soil nutrients. Protecting, maintaining, and, where necessary, restoring natural disturbance regimes provides for natural composition, structures, and functioning at the forest landscape and stand levels. Respect traditional practices in place-based cultures.

- Managerial Flexibility. Be flexible—keep your options open. Principles must be flexible to mirror the flexibility of open systems; flexibility is provided by diversity in fact. Small is more flexible (and beautiful as Schumacher says); small mills, small operations, small businesses are easier to run and easier to change with conditions or trends.

- Work with the changing ecosystem. Succession can be assisted, slowed, or speeded up, but not skipped or ignored. Only details and exceptions form the forest; only generalities are important—so you have to live with contradictions. Make the smallest number of changes. Adapt process of change to the site (you fit the forest, not other way). Keep acquiring understanding and knowledge. Manage personally. Probably no one will know that forest as well as you do (especially if you live in it).

- Appropriate Scale & Time. Scale management to forest size. Approach the forest as in a partnership; do things to benefit the forest as well as yourself. Give back to the system. Work with forest time; do not try to do everything in the industrial schedule. Be as slow as you want or need to be. Attempt to set up a transgenerational land tenure system to accommodate forest time, which spans human generations.

- Profitability. Make your business profitable (by profit, I mean simply the excess of returns over expenditures—thus your health and happiness might be considered a return from caring for the forest). Adjust your input to expected goals. If your goal is to restore the forest to health and enjoy it, then the profit may be spiritual. Seek best use for products; everything in the forest is a resource for something; many can be directed to human use.

- Keep wild ecosystems wild by minimizing control and interference. Preserve the structures and functions of a forest. Preserve minimum viable populations of species. Create standards of forest classifications.
- Accept Management Limits. Management is limited, by costs mostly, to regulating animal and plant populations in a system, rather than climate, geology, or water, soil, mineral cycles. Manipulating plants and animals to the environment is more likely to succeed than the opposite. Management limits are determined by the ecological characteristics of the resources, before any economic, technological, or political limits are applied. The use must be limited by the ecology before monetary considerations. For legal stability, find the best combination of ownership, trusts, easements, and plans to ensure that the forest will be cared for in the long-term.

2.5.8.2. Noninterference Matrix Management

A noninterference approach to ecosystem management, using the preceding principles, allows the system to take its own course (the essence of a taoist way). Therefore, once the temporary constructs were in place, whether planting or cutting or any other manipulation, the system would be allowed to develop without further interference.

In nature, noninterference means 'letting be.' Noninterference matrix management (NMM) is not indifference, which is diffuse. It is caring. Noninterference will not lead to chaos, poverty, and stagnation. The technocratic vision strives for "life under control," but the forest is self-managing, productive, efficient, and orderly. We need to practice the rule of noninterference so that all beings can enhance themselves. Noninterference can be derived from nonviolence, or from taoist nondoing. This attitude would entail using what is necessary, exploiting parts of some ecosystems, changing a place to fit human aspirations, and killing plants and animals for sustenance. But it would also mean limiting humanity and its technological effects, limiting human use to local impacts, and letting other beings live without interference. It is not necessary to dominate or terraform the ecosystem completely to save it. Noninterference matrix management weaves people back into the fabric that supports them and in a sense makes them subject to the constraints of ecosystem processes.

NMM would manage the system with minimum subsidies; manage activities that could upset equilibrium; manage sustainable conditions; align human activities with natural processes; work with system instead of attacking it; and, restore context. According to Garrett Hardin, many of the ideas necessary to fitting humanity into the pattern of nature are known but not yet popular. For instance, exponential population growth, or economic growth, cannot be maintained very long. Human communities cannot grow 4 percent per year without disastrous consequences to the infrastructure and the quality of life. Growth cannot be continued because the landscape is limited, in terms of productivity, energy, and resilience. Thus, we need to fit our population into the limits of the landscape, although some limits can be expanded by technology or by lowered expectations. The carrying capacity of the area is not only a function of the limits of the community, it is equal to the number of people multiplied by the level of comfort, the quality of life style. Having more energy and space means having fewer people.

The ecosystem may be too complex to design. How do we design an ecosystem, such as a complex, self-making, self-sustaining wild forest? We could identify the parts and functions and try to duplicate them, but that would prohibitively expensive and time-consuming. Management has to recognize the limits of design. Limits of ecological design include the fact that ecosystems are wild, and we have no real control over them. The scale of ecosystems is often too large to manage everything. The longevity of ecosystems is too long, and we will never complete the design in human lifetimes. The costs may be prohibitive—indeed, we have depended on the free goods of ecosystems for economic advantages, which would disappear if we had to rebuild the system. Other human limitations apply to our ability to see and understand any ecosystem.

2.5.8.3. Uncertainty & Management

Carl Walters discusses how we recognize and measure uncertainty. He presents material on decision theory. He also distinguishes between three kinds of uncertainty in natural systems:

1. Regular disturbances over time generate unpredictable and uncontrollable changes. Certainly this is true, but we have found that regular disturbances are also responsible for creating diversity in ecosystems; furthermore, regular disturbances "habituate" the ecosystems, so that often the system requires the disturbance to continue. Certainly, as he suggests, monitoring is necessary. The uncertainty of these disturbances is usually just in the exact timing and scale, not whether they will occur or not.

2. Statistical uncertainty about parameter values of functional responses, e.g., production rates as a function of stock size, can lead to problems with exploitation. He suggests the solution to this is simply assigning probabilities to parameter values. Probability is mathematical way of describing "random events," such as coin tosses. Nature, of course, is not random, even at a quantum level. Probability is also a degree of belief, according to Savage (1954). Probability theory centers on the notion of uncertainty as it applies to independent or conditional propositions. There are at least two kinds of probability: Personalistic, which depends on personal beliefs and confidence, used in Bayesian contexts, and Objective, which deals with repeatable events. Probability, in the face of some history, becomes conditional probability to reflect events that have already occurred. In nature, unfortunately, we are not always sure which events have already occurred and which have not.

3. Basic structural uncertainty about what variables to consider can lead to indecision. Incomplete structural representation implies time-varying parameters and surprises involving changes. Surprises are usually the result of incomplete knowledge, but also can be from the emergent properties of systems.

These three kinds of uncertainty are not that different, being basically levels of ignorance. There are other levels and kinds of uncertainty. There are levels of uncertainty, from the quantum to the human. Elementary processes at the quantum level are not subject to a precise description in time and space. Predictions about location and velocity are just statements of probability—this is Heisenberg's uncertainty principle. The effect of this principle on epistemology is that our exact interpretation has to be abandoned. There are uncertainties at higher levels of organization as well; for instance, the hysteresis of some magnetic solids determines their subsequent behavior—without knowledge of initial conditions and all past events, it is not possible to predict present or future behavior. Furthermore, the higher up in levels of organization, the more kinds of uncertainty there are. There is genetic uncertainty, as well as environmental and social.

There are also many kinds of uncertainty, as suggested by Max Black. There is fundamental uncertainty in the thing/event/pattern, as well as in the channels (noise, meaning) of relationship. At the human level, uncertainty can be distinguished between ignorance and conflicting knowledge.

Ignorance can further be divided into kinds: vagueness (indeterminate knowledge),probability (confidence in partial knowledge), incompleteness (of knowledge, missing elements), irrelevance (place in a pattern is unknown), and fuzziness (with overlapping interpretations).

Conflicting knowledge also can be broken down into kinds: anomaly (incongruity of knowledge, simple error), ambiguity (alternative interpretations of meaning), inconsistency (simultaneous untruth), equivocation (knowledge is constant and inconstant, true and untrue), and belief (confidence in subjective knowledge, taboo).

There are probably other components of ignorance, as well as other ways that they can be combined into a formal typography. There are many different kinds of ignorance, maybe more than kinds of knowledge, and that these determine our approach to the practice of making places. If we think of knowledge as an expanding sphere in a space of ignorance, then as the sphere grows, the surface area in contact with ignorance also grows.

Walters stresses that managers have to live with uncertainty, and then makes the great connection that management decisions are essentially gambles. Gambling is a profession that acknowledges the operation of chance and makes conclusions in the absence of facts—few people are successful at it. This is an important admission, that we do not have facts to base our actions on, that nature is a stochastic process, and that ecosystems always changing. Furthermore, we do not know for sure what effects our actions will have on the forest, which used to live so long. If we do not act responsibly, we are gambling that the cost is more than damaging the planet. If we act responsibly we are gambling also. Successful gambling suggests that the proper attitudes for gambling with nature are awareness, humility and courage, not arrogance, fear and maximum use.

Although he suggests that inaction is an inappropriate alternative to gambling as a result of confusion, in fact, inaction is a very appropriate alternative in the face of confusion—when in doubt about an ecosystem, perhaps we should not interfere.

Walters suggests that managers should embrace uncertainty rather than always strive for "certainty-equivalent" policies. That is, they should try to define a set of possible outcomes, e.g., models, consistent with experience, rather than make a single best prediction. Managers have to live with uncertainty, but they can make intelligent gambles, based on their ability to work within real ecological and physical limits. Of course, they should not gamble with all their resources at once.

2.5.9. *Ecological Design Limits*

People often judge the health or wholeness of ecosystems by how they look, as a function of richness, despite the fact that ecosystems have to fit into extreme climates and places. Traditional design has emphasized visual results above all else. Ecological design, however, achieves the same results by paying attention to the structure and function of the system first. Design has been concerned for centuries with making domesticated landscapes out of wild ones. Now, design must address the opposite problem: how to preserve or provide the conditions for wild ecosystems.

Design must address the common good, that is, the good of the entire ambihuman community; it can do so by: Promoting the well-being of all individuals in larger community, deciding what is preferable, attempting to regulate and anticipate all effects, encouraging convivial activity, recognizing links and dependencies, mediating the relation between technology and community, and alleviating some of the problems of industrial society.

Designs provide a framework for natural and artificial processes to work in. The patterns in design are echoes of patterns in nature. Good designs learn to embrace error and failure, so necessary in open systems. Most ecological designs of ecosystems will not be restorations, because of the uncertainty about the kinds and associations of native vegetation. Furthermore, humans are now a large part, although not yet an integral part, of the system; therefore it could not be restored to a premodern or prehuman state, even if we knew the proper or historical state. This design is not the biotechnological design of a new ecosystem, either; we cannot accurately control and predict ecological events in most ecosystems. However, we can steer some of the events in a known direction—known because we have historical records of the system, although not complete. We can also reduce those human activities that we know alter the conditions of the forest, such as overcutting and pesticide use.

2.5.9.1. Ecological Balance

Although ecological design attempts to restore some kind of balance, the balance does not exclude human activity. Rather, it integrates it into the larger community. A moderate number of human impacts can be absorbed by the system—too many destroy the systems capacity for self-maintenance. The design should be open to evolution and to human technological and social development. The design should be based on a model of ecosystem functions, considering diversity, complexity, and the maintenance of natural process—natural here meaning a self-sustaining system composed of elements now lost through human disturbance.

How do large-scale processes influence design? That is, how can design be flexible and open enough to cope with change? How does design accommodate those processes? The processes provide constraints on the designs, which when implemented, force some constraints on the processes themselves. The question is how to limit the latter constraints so that the processes are not fundamental shifted into a new regime. For example, a dam is located depending on waterflow, canyon walls permeability, location, rainfall, and other factors. After being built, however, the dam changes rainfall, erosion, waterflow, species groups, and other factors, which may reduce the rain upstream that reduces water behind the dam.

The role of designers is to optimize or satisfy the fitness of people with their environments. To fit cultural goals to ecological characteristics and limits. It is adaptive creativity, not just for the current technology, but because it needs to adapt to the technological and natural environments. An ecological design involves designers and people in reshaping and recreating a self-sustaining community. Individual resources are limited. The relationships to strive for here are community relationships. Furthermore, there are limits for human manipulation of other communities. Total control has limits, also. We should not aim to try to control ecosystems. We have to trust that natural processes are self-correcting and organizing.

Ecological design is the design of communities. We design places as organic wholes to promote the well-being of individuals and the common good. The immediate goals of design are to reverse degradation and reclaim places for communities, but also to work to increase public awareness of the interdependence of communities, to create environmental quality, and to transform public values by generating new metaphors for living. Unlike what Robert Bailey and others imply or say, local design cannot ignore culture, politics, and economics. Design has to commit to limits, constraints, and optima (physical, biological, psychological, and social). Ecological design can restore the interconnectedness of the systems, especially ecological and human social systems.

2.5.9.2. Ecological Perspectives

According to Victor Papanek, an ecological perspective can change design: With a greater emphasis on quality and permanence. So products can be more timeless and age gracefully. By questioning the ultimate consequences of a new product, regardless of profit or prestige.

By adding new products for new approaches or new professions; monitors, scrubbers, converters. By understanding that all design has ecological and cultural consequences that need to be discussed and evaluated. And, by having a greater concern for and understanding of nature, preserving and restoring the health of the environment. Ecological design importantly has to do with ecological scale, which explains Papanek's conviction as a designer that: Nothing big ever works! Not corporations, building, schools, or bureaucracies. Ecological design has to fit designs into the scale of their context, so that they harmonize with it. Ecological design fuses art and ecology from the work on forests and rivers to agriculture and buildings.

On the other hand, our tendency to redesign nature, rather than to tolerate and cherish, is dangerous; it leads to "curing" abnormal people and "solving" weed problems. We do not respect the wild, complex side of human nature. So we need ecological designs that seem like 'dedesigns.' Ecological design does not have to merge technology with nature, as Van der Ryn and others define it. For buildings, technology is an important component, but for the shape of ecosystems, it may play a minor role, of necessity; we may not have the knowledge or the ability to completely control technological replacements for natural services. Ecological design has to respect nature and natural limits, while integrating human patterns into natural processes. The role of designers is to optimize the fitness of people with their environments, to fit cultural goals to ecological characteristics and limits. Ecological design is adaptive creativity, not just for the current technological fix, but because it needs to adapt to the technological and natural environments.

An ecological design is the creation of a clear vision of the ecosystem that is aesthetic, useful, and self-sustaining. Ecological design fuses art and ecology from the work on forests and rivers to agriculture and buildings. Some of the relationships can be captured by maps and drawings, but not the dynamic four-dimensional qualities of the system itself, which can only be understood by dwelling there for years. Nevertheless, a simulation of the view from foot or airplane is more compelling than a recital of the statistics or species lists.

The goal of ecological design is not to restore a damaged wild system to some vague prehuman state—it is to revitalize and reinhabit the wild ecosystems. We do not want to live in the dead bones of a mechanistic failure. We want to live in a healthy environment with aesthetic appeal—aesthetic appeal is a requirement for human health. Every system has physical, biological, economic, and political characteristics. The ecological design, planning, and management for a forest, for example, describes the system in a comprehensive interdisciplinary approach, using dynamic concepts such as feedback and stability, recognizing limits to change and sustainability with different levels and scales of structure and function in an anticipatory, flexible planning approach, recognizing human and nonhuman goals, and incorporating personal and institutional interests.

Ecological design is the design of whole communities. We design places as organic wholes to promote the well-being of individuals and the common good. The immediate goals of design are to reverse degradation and reclaim places for communities, but also to work to increase public awareness of the interdependence of communities, to create environmental quality, and to transform public values by generating new metaphors for living. The long-term goals require a wild, heroic design and extravagance. It is not contradictory or antithetical to living frugal lifestyles or to restoring a healthy environment to create ambitious, heroic designs—this needs to be repeated often. Life is exuberant; energy is used, lives are lived and used, not saved. Life is the accumulation of individual experiences that cannot be saved, stored, or owned. This local ecological design is what brings all the problems into focus—or rather into context and out of focus. We have to creep up on the problems sideways like a crab.

2.6. *What is Global Ecological Design?*

Global Ecological design is not the additive design of ubiquitous factors such as roads or popular products. Nor is it the design of common structures or symbols, such as cities or money. Although it has to address truly global phenomena, such as the atmosphere or ocean, there is not the knowledge or power to change or reconstruct those structures and processes. So, it has to do mostly with fitting human activities within the framework of the planet. It can minimize the impact of human activities on the global anatomy of the planet, and it can minimize the impact of the planet on human constructs.

2.6.1. *Design at the Level of the Planet*
Global Ecology itself has to be broader that regional ecology, than community ecology, which is broader than human community ecology, which is broader than human ecology. Global ecological design is a human process that considers the emergent and unique factors of the planet. We have to change the framework so that we can shape new approaches. We no longer have an external point of view. We are inseparable from the environment and each other, but we can still differentiate and develop new designs.

2.6.2. *How is Global Design different from Ecological Design?*
The major difference between ecological design and global ecological design is scale. The scale of the planet requires changes in design. In some cases, design simply permits global processes through benign neglect. In others it has to provide the constraints for local and regional design. Four areas of differences—biophysical systems, ecocultural play, economies, and values—are detailed between these two levels of design.

2.6.2.1. *Physical & Biological Environment of Global Eco-Design*
At the biophysical level, global design has to address many more systems as well as the new emergent properties of the globe.

    2.6.2.1.1. *Repatterning.* The patterns of concern here are global patterns that are limited by climate and regular catastrophes. This includes the global cycles of elements and well as global distributions of species and communities.

    2.6.2.1.2. *System emergent.* As the result of the local creation of forms and ecological constraints on systems, large-scale patterns emerge. But, these patterns depend on the continued operation of local processes, as well as on new global cycles. Design has to consider all scale levels as well as the planetary system itself.

    2.6.2.1.3. *Growth/Development.* The autopoetic process is not just about generating forms, it is about degenerating and regenerating, where the two anabolic and catabolic process run up and down simultaneously. It is about making, demaking and remaking on a planetary scale.

    2.6.2.1.4. *Whole system.* The frame contains the local systems that may be the focus of local design, but the whole system also needs to be considered by local or regional design. Global design has to pull back from a focus and consider the aspects of the entire frame. The whole system has unique properties that have to be given opportunities to continue the process at the global scale.

2.6.2.1.5. Feedback cultivating. The ecosphere acts as one system in which energy from the sun is cycled. The functioning biomass is integrated by feedback responses to extract the maximum of energy and still maintain a balance. Most of the solar energy is used for maintenance by the biosphere, which is an indicator of the biosphere's high degree of ecological maturity. Lovelock hypothesizes that every element in the system is related in a feedback network to every other element, from the very strong to very weak. Cycles that do not operate with the right kind of feedback function as traps. Thus, on an elemental level, phosphorus becomes trapped in an ocean sink, and can only be recycled by long geological processes on a global level. On a cultural level, some organismic behaviors, for instance specific harvests of marine organisms by human activities, may unbalance a complete food chain, which would change rates of photosynthesis, which could affect planetary albedo. Global design has to consider many kinds of traps.

2.6.2.1.6. Diversity allowing. One of the important factors in diversity is locality. Differences emerge from communities and species developing in isolation. So, isolation must be an important part of global design. Global designs must erect or allow barriers to human-assisted migration of exotic or invasive species to allow diversity to develop as it has. Combining many organisms at random may increase diversity briefly, but it may decrease it in a new settled-out system. On the other hand, human additions to some systems, such as reintroducing camelids to North America, could increase diversity in the long term.

2.6.2.1.7. Evolution. Evolution is the process of development. Although it applies to the survival of individual species, it also applies to their communities and ecosystems, and possibly to the entire planet. Landscapes evolve as well, even at the continental scale. As species compositions shift the landscapes shift also, and as they shift the planet changes.

2.6.2.1.8. Place framework. With global design, we want a framework that permits natural processes and places to form and operate freely. Place generates design through participation. The framework of place comes with a long history of changes and successful patterns that can inform and limits designs. Global design considers places simultaneously.

2.6.2.1.9. Place-supporting. Although ecological design is place-specific, global design has to be context-aware. The framework for making and maintaining places has to be healthy and productive.

2.6.2.1.10. Change permissive. The matrix of nature occurs in a dynamic disequilibrium. The framework of global design has to permit the processes to continue, in order to have designs on the local level.

2.6.2.1.11. Possibilizing. Living beings in local systems may maximize or optimize their populations, depending on their environment and qualities, such as territoriality. They may settle at satisfactory levels of population or feeding. Global design needs to keep many sets of possibilities open, rather than striving for explicit structures with optima or maxima.

2.6.2.2. Ecocultural Play of Global Design

At the level of ecocultural play, global design considers the sum of cultures and evolution acting on planetary patterns.

2.6.2.2.1. Planet-based. More than being nature-based or culture-based, or nature and culture-based, global design has to consider the entire planet, so that processes of natures and cultures may be isolated or limited when necessary. The health of the planet, in all its history and variety, is considered first.

Table 2621-1. Comparisons of Biophysical Levels of Design

Natural Creation	Traditional Design	Ecological Design	Global Ecodesign
Self-patterning	Patterning	Repatterning	Repatterning
Creating forms	Conscious intention/form	Ecological restraints Experiential	System emergent
Autopoetic	Making/Generation	Remaking Regeneration	Growing/developing Degen-regenerating
Complete	Focus	Peripheral frame/system	Whole system
Systematic	Feedback accepting	Feedback driven	Feedback cultivating
Processing patterns	Objective/object-oriented	Dimensional process	Diversity allowing
Development/change Filtering/Sorting	Growth-permitting	Development after initial growth	Evolution
Place generating	Place independent /isolation	Place specific	Place framework / context
Place responsive	Principle limited	Fitness-oriented	Place-supporting
Dynamic equilibria	Equilibrium	Disequilibrium incorporated	Change permissive
Interaction limiting	Maximizing/ Optimizing	Optimizing/Satisficing	Possibilizing

2.6.2.2.2. Nesting. Nature deals with aggregates. Ecological design may assemble species to restore an ecosystem; it may connect the niches and behaviors of species. Global design needs to nest the systems and species in the global framework.

2.6.2.2.3. Cyclic linking. The trial and error process of design works well with traditional design and even some ecological designs, where species are assembled in restorations according to their functions and allowed to sort out over time. Global design is concerned with making sure that adapted local species are not replaced with faster-growing or tougher exotics over large areas. Furthermore, species have to be linked with local and global cycles.

2.6.2.2.4. Failure allowance. Ecological design incorporates the failures of previous designs into newer designs. Global design has to be sure to allow failures to continue to occur within the framework of the designs. The failures are only of concern if a critical level is reached or if the failure is at the cellular level of life.

2.6.2.2.5. Novelty from history. Global design has to consider novelty as well as tradition and historical processes. In fact, it has to allow the processes of novelty and the history of traditions to both operate on a larger scale.

2.6.2.2.6. Reciprocal Quality breaks are important. In a system of global design, quality can be submerged by quantity. This can be a limit on the lifetime of products and patterns, as things break down and are reformed.

2.6.2.2.7. Chaotic development within restraints. Although local systems are formed bottom-up, global systems provide the context and restraints that limit not only global cycles and patterns, but local ones as well. At the global level, a designer can assess the overall situation and tune some of the limits and patterns. Trickle-up and trickle-down processes can be identified ad monitored to make sure that both processes continue.

2.6.2.2.8. Courageous neglect. Although traditional design and ecological design may be very attentive to forms and their consequences, global design has to display a courageous neglect, when it comes to allowing natural processes to work. That may mean that not every charismatic species should be saved in zoos or gene banks. Not every favorite ecosystem should necessarily be restored. Unique patterns of plants and animals may come and go rhythmically, without human design trying to freeze them in one configuration.

2.6.2.2.9. Health emphatic. At the global level, the overall health of the system, with degenerative and regenerative processes continuing, is more important than sustainable permanent patterns. Health is a measure of the harmony of the degenerative and regenerative processes, which may recycle dwelling places, especially those badly sited in earthquake or tsunami zones. Designer should take care to create good places, but on the global level especially, they have to be aware of short or long-term destructive processes at work. Health is related to stress, both good stress and bad stress. Stress may be related to the rate of change for the system, in addition to loss or gain of components or changes in structure. Health is the overall ability of a system to maintain itself under a normal range of environmental conditions. Obviously, a pioneer community may change the conditions to favor a new level of the system with new components. The global level may require a balance of mature and young systems.

2.6.2.2.10. Receptive. Simple behavior seems to work, with local feedback, and more sophisticated behavior "trickles up" to approximate a global perception. A global design level has to be aware of local centers and their places in a global context. Receptive in this sense means incorporating patterns that may not be conceived or perceived.

Table 2622-1. Comparisons of Ecocultural Levels of Design

Natural Creation	Traditional Design	Ecological Design	Global Ecodesign
Nature-based	Culture-based	Nature & culture-based connecting	Planet-based
Aggregating	Bricolaging	Niche-assembling/ connecting	Nesting
Trial & Error process Novelty testing	Conscious Trial & Error process	Ecological redundancy	Cyclic linking
Trial generating	Learning from failure	Failure incorporation	Failure allowance
Novelty generation	Novelty emphasis	Historical tradition	Novelty from history
Reciprocal interactions	Rules/Geometric grids	Quality emphasis	Reciprocal
Bottom-up organization	Top down direction Control of actions	Bottom up action with top down restraints	Chaotic development within restraints
Physical restraints	Creative Play	Consequences attention	Courageous neglect
Natural process	Goal-directed	Permanence sustaining	Health emphatic
Actual Existing	Conceptual	Perceptive	Receptive

2.6.2.3. *Ecological Economics of Global Ecological Design*

In terms of ecological economics, global design has to consider economics as the basic exchange of energy and elements at all levels.

2.6.2.3.1. Capital-creating. Ecological design is concerned with using natural capital at renewable rates, with a minimum use of the productive accumulated "interest." Global design considers the capital-creating process on several time scales. Global design has to take care that those long-term processes are allowed to continue, even while we use renewable energies and materials.

2.6.2.3.2. Symbiotic. At a global level it is easier to identify and appreciate the advantages of systems living together for mutual benefits. These benefits extend to the global level. Although integrated, it seems that many of the partnerships become obligatory.

2.6.2.3.3. Reintegrative. All levels are integrated by global cycles. But, the patterns weave and reweave so much that global design has to constantly reintegrate the pieces. It can do this by following a monitoring program that is concerned with regular changes in patterns. Tools and designs are important extensions of the human mind. Their purpose is to foster and assist survival, not to make it more difficult. Tools and designs can be made appropriate to environmental limits and cultural preferences, both of which are often ignored by industrial approaches, by integrating them into regional and global patterns.

2.6.2.3.4. Constructively indifferent. Where ecological design is dependently constructive and minimally destructive, global design has to appear indifferent to the destructive parts of the process. The destructive is a necessary part of the cycle, in balance with a constructive process. Both are necessary; neither can be overemphasized. Small-scale operations in general, are less harmful to the environment than large-scale operations. But, we are pulled by our idea of global economics to make large-scale decisions beyond our immediate comprehension. Wendell Berry suggests that we are not smart enough or conscious enough to work on a global scale. This kind of design would assume that and scale down things to understood limits

2.6.2.3.5. Energy scales. More than appropriate energy use, global design has to balance all kinds of energy use. For example, energy use on a local scale, for instance capturing tidal energy in bays, might have negative consequences on a global scale, perhaps affecting local life styles in the bays or even the rotation of the planet. We have to ask if wind farm energy on a global scale will affect the quality and kinds of winds on a local level, as well as populations of avian and mammal species.

2.6.2.3.6. Minimal technology. Appropriate technology in ecological design is crucial. We cannot be sure what appropriate design is on a global scale. At the very least, technology on a global scale implies the ability to predict, control, and respond to the technology and its consequences. For instance, placing millions of small mirrored surfaces in orbit to reduce the solar constant might have effects on many species, as well as unforeseen impacts on ecosystem surface that have developed with expected light levels. Van der Ryn states that ecological design is about merging nature and technology. Certainly technology is emergent from human activity which is emergent from "nature" but global design also has to consider separating technology from nontechnical species.

2.6.2.3.7. Waste sorting. The adaptive ecological design applied to a global level has to be sure to channel the unavoidable waste of local systems into the global cycles that reuse it. This means limiting waste and eliminating unnecessary kinds of waste that come from lack of design or bad designs.

2.6.2.3.8. No costing. Ecological design has to be cost effective at local levels. At the level of the planet, however, cost is not as important as effective survival. When we approach global design, it is because we have already thrown global cycles out of balance, and we have interrupted or interfered with local systems crucial to global functioning. In most cases design at this level is more about control of human interference than about replacing global functions. Losses from accidents and diseases can be reduced by preparedness. Losses from earth and climate changes can be reduced, also, with preparedness for 'normal' events, such as hurricanes, earthquakes, and droughts. Design can also be used to reduce impacts from these events; for instance, by denying building permits on floodplains. The losses from some events, such as droughts resulting from El Nino, can be ameliorated, by having surplus food and supplies stockpiled.

2.6.2.3.9. Multiscalar trends & Interweaving. Scale constraints are equally important at the global level, but perhaps it is more about minimizing the scale of designs. The patterns of designs can be interwoven at a global scale rather than trying to replace processes with specifically global ones. Scale-linking is important. Scale-linking in nature occurs from energy and matter flows (which can be represented by fractal geometry). Global biogeochemical cycles link organisms across many levels of scale, from photosynthesis chemistry to the planet surface. This has holistic ramifications, as any action at any scale impacts others at other scales.

Table 2623-1. Comparisons of Economic Levels of Design

Natural Creation	Traditional Design	Ecological Design	Global Ecodesign
Productivity	Capital-using	Natural capital-using	Capital creating
Interrelational	Constructive	Adaptive process	Symbiotic
Nested	Auto separation	Integrative	Reintegrative
Interdependent	Independent/ destructive	Dependent constructive minimally destructive	Constructively indifferent
Solar-life driven	Energy use (fossil)	Appropriate energy use	Energy scales
Aesthetic creation	Technological tool use	Appropriate technology	Minimal technology
Waste generating	Waste production	Waste integration	Waste sorting
Free play/change	Cost consideration Market-oriented	Cost constraints based on physical constraints	No costing
Gigatrends	Fashion fads Style	Size/scale constraints	Multiscalar trends

2.6.2.4. *Values of Global Ecological Design*

At the level of global design, values are expanded from human and species to those useful for the planet.

2.6.2.4.1. Responsibility to not interfere with living cycles. Ecological design has to be responsible for the local environment and community. At a global level, responsibility has to be concerned with not interfering with natural cycles and landscape processes. In fact, it might be practically impossible to be responsible at the scale of the planet. Much of design, again, becomes concerned with limiting and directing human influences, rather than replacing and controlling natural cycles.

2.6.2.4.2. Whole goals. At the local level, we can design strategies to accomplish our ends. At the global level, we have to consider global goals and impacts before implementing those strategies.

2.6.2.4.3. Whole order geological time. Global ecological design is concerned with much larger frames of space and time. The recycling of elements globally, for instance, should consider at least a 10,000-year time frame for many compartments and pools, and millions of years for geological shifts. Although there is quite a bit of flexibility for local orders, at the global scale, it would be better to keep flows in dynamic equilibria. Long-term catastrophes teach long-term thinking, Large catastrophes teach large thinking.

2.6.2.4.4. Free motion interchange. The goodness of a traditional design can be related to any destruction it may cause. Ecological design considers how the design fits into the environmental context. At the global level the design has to permit the free interchange of elements and patterns, before any human design is applied to the pattern.

2.6.2.4.5. Respectful. As the precautionary principle needs to be followed in ecological design, a principle of respect needs to guide interactions at the global level of design. Perhaps even a principle of nonintervention should guide our thoughts of changing global cycles. By promoting the understanding of the inadequacies of bad characteristics and bad designs, eutopias can stop the plague of uniformity and paucity. Through an understanding of the consequences of human ambitions and actions, eutopias can avoid many of the evils that result from a civilization on technical autopilot.

Table 2624-1. Comparisons of Value Levels of Design

Natural Creation	Traditional Design	Ecological Design	Global Ecodesign
Living	Responsibility to product, consumer	Responsibility to living community & environs	Responsibility to not interfere with cycles
Fitting	Planning	Strategies	Whole goals
Harmony-build long-term	Meaningful order short-term	Flexible order Living order times	Whole order Geological time
Existence value	Goodness concepts	Fitness concepts	Free motion inter-change
Learnable	Teachable	Limitable Cautious	Respectful
Adjusting	Exclusive	Inclusive	Binding/spiritual
Formation	Information	Knowledge Ecology	Wisdom/harmony
Chaotic/Free	Exciting	Inspiring	Dangerous

2.6.2.4.6. Binding. Design needs to be inclusive at every scale, but at the global scale it can also bind us to the planet spiritually. James Lovelock suggests that regarding the planet as a living entity may suggest proper ways to attempt changes that could affect global cycles. Is this what religion does, entreating us to ask rather than tell? We have lost a meaningful frame of reference. Our fabulous cyborgs, computer communications, and genetic engineering coexist with burning poverty, urban violence and ethical erosion. Few of us can bear to look at the total situation. Of course, no one wants this other dimension, the slums and the pollution, but they are emergent characteristics of an unplanned, free society—just as a leader may sincerely not want a war, but may be helpless to stop the mechanisms that trigger a war. The social roles that determine the social action that could correct these sad things do not seem to exist. But, can they be designed?

2.6.2.4.7. Wisdom. Knowledge, and the understanding of its limits, is important for ecological design. At the global level, the right application of knowledge, at the right time and in the appropriate way, is necessary. This is one definition of wisdom by Jonas Salk.

2.6.2.4.8. Dangerous. Nature is chaotic, exciting and inspiring, and these emotions add content to ecological designs. Global ecological design accepts that nature, and the planet, is dangerous also. Although design can allow us to prepare better for natural catastrophes, through siting or not, through fit designs or not, design cannot guarantee complete safety.

2.6.3. *Global Ecological Design Components Elements & Variations*

Global Ecological Design builds on the basic geometric elements of traditional design, which can vary in numerous ways, by number, position, direction, size, shape, interval, texture, color, and temporal. Furthermore, the elements can be organized into groups by nearness, similarity, and difference (diversity). Then they can be combined into whole structures by principles of rhythm, balance, and finally sensory force.

2.6.3.1. Elements

Global ecological design has the same elements—point, line, plane, volume—as traditional design (see Section 2.4.3.1.), but they are understood in a global ecological context foremost, in the movement of continents, the submerging of a peninsula, the volume of the atmosphere, or the spreading of the oceans.

2.6.3.2. Variations

Global ecological design shares the same variations—number, position, direction, size, shape, interval, texture, color—as traditional design (see Section 2.4.3.2.), but there are many more occasions: In global ecological design it is the degree or process of change in appearance. Thus, there is great variation between a seabed and a cloud formation. At a global level, the spin of the planet pushes air and water patterns 'eastward.' The moon looks larger than it is when it is near the horizon. For instance, the continents have unique shapes that effect air and water flow and quantities, as well as animal and plant movements; animals on the Eurasian continent can move along latitudes that have similar climates much easier than those on the American continents, which are oriented north-south. On the global level, the effect of human activities on the shapes of landforms and seaforms is less evident. For instance, storms in the Caribbean form at certain times of the year, depending on dust storms and water temperature near Africa. The larger palettes of regions seem more uniform than the local variations in a rainforest, for instance.

2.6.3.3. Organization of Variations

Global ecological design shares the same organizations of variations—time, groups, nearness, similarity, density, diversity—as traditional design (see Section 2.4.3.3.). Global time is planet time, measured in billions of years, and covering dramatic changes, from a volcanic planet, to a rustball planet, slimeball planet, snowball planet, and possibly a desert planet. Global ecological design has to consider much larger timescales for design patterns and events. Large islands seem to appear on the east side of continents, as a grouping. Nearness for global ecological design refers mostly to landforms. Cloud pattern density can be related to land elevation and wind currents. On a global level, diversity seems to decrease due to the scale, as elements become more difficult to perceive individually.

2.6.3.4. Organization into Structures

Global ecological design shares the same organization into structures—structures, rhythm, tension, balance—as traditional design (see Section 2.4.3.4.). On a global scale, biomes form that reflect similarities of temperature and precipitation. Many global rhythms are influenced by the tilt of the earth and its orbit around the sun, as well as its movements in the plane of the planets and through interstellar dust lanes. The dramatically large clearcuts in the Brazilian state of Rondonia create tension in the mature tropical forest landscape, as well as with global chemical cycles and rainfall patterns. In global ecological design, many human landscapes can balance wild landscapes.

2.6.3.5. A Thought on New Scales of Design

Global ecological design extends the scales of design from the traditional and ecological to the planetary level, with associated changes in perspectives, for example, in the following sequence:

Designing objects (in a product-centered environment)

Designing interacting objects and human subjects (user-centered)

Designing systems of interacting objects (practice-centered)

Designing systems of interacting objects within dynamics of living subjects

Designing systems of interacting objects within systems of interacting objects (multi-centered)

Designing systems of interacting objects within ecosystems of interacting objects, species and nonliving things (ecologically conscious).

Designing systems of interacting objects within ecosystems of interacting objects, species and nonliving things within a global system (noncentered, globally conscious)

2.6.4. *Characteristics of Global Ecological Design*

The characteristics of global ecological design are similar to those of ecological design. Due to differences in scale, they have to address the characteristics of ecosystems, places, biomes, continents, and the planet and its environment. The courses of the global ecological design have to imitate the processes of systems and places. The wholeness of the design has to address the ideas of spirit of place, and sensory force. After the processes are identified, they have to be related to the patterns. Design may have to try shortcuts, unless it can afford to accommodate global time, which is longer than the lifetime of species or continents.

2.6.4.1. Method of Imitating Courses

Design has to identify patterns of movement and construction, and then it has to imitate the processes and patterns, modifying them to include the things of human interest and need. This is more difficult than copying a shape or a structure. Fortunately, imitation is a human strength, even if the recognition of complex, long-term, moving patterns is not. Courses can be thought of using topological models representing a four-dimensional landscape.

2.6.4.2. Self-extension of Identity & Wholeness

Global ecological design has to capture the integrity of a set of places, or restore them as a unit. Unity is a fundamental objective of landscape design. Unity is the way the elements, including shape and scale, of a landscape are combined. For instance, visually, a forest ecosystem usually dominates the landscape. From a distance, even-aged forests have much the same impact, in terms of color, shape, and scale, as uneven-aged forests. Diversity becomes more important visually at a smaller scale. Natural forms of the forest are unified with the landscape because the margins are very uneven, and open space in the forest is part of the mosaic caused by birth and death of individual or groups of trees.

2.6.4.3. Variety Openness & Flexibility

Because the operation of the universe tends to change systems, the design of a place should be open to the types of processes that could destroy the design. Furthermore, flexibility, defined as the unused capacity for change, can be designed into the system. The parts of a system have to maintain the potential for all possible behaviors that could flow from any other part of the system. Openness and flexibility are characteristics of healthy ecosystems, and they can be considered and enhanced by design.

2.6.4.4. Cooperation in Coconstruction

Designs are limited by the real biological constraints of ecosystem processes and biogeochemical cycles. We must know the constraints in order to create a healthy design. The design has to work within the constraints of the ecosystem. Rather than emphasize static equilibrium, the design should emphasize heterogeneity and learn to adjust to disturbance.

People have needs, animals and plants have needs, the site does have constraints, and these things can be married into a good pattern. Harmony is a constraint of the whole system. Design is the participation in the process of the ecosystem as a harmonious system, with mutually restrained conflicts and constrained influences. An ecological design is a form of co-constrained construction, where the organisms, environment and designs are co-implicative, co-defining, and co-constructing. They all engage in a process of self-assembly, where the whole is the whole system.

To participate means to take part with others in some activity. Designers should participate in a complete design process, guiding involvement and commitment to the art of living together as a community. Adaptation is a process of making fit by adjusting to circumstances, environmental or cultural. Here it means fitting into a global system, within established large-scale cycles and functions. It is adaptation that improves the chances of survival for communities of species.

2.6.4.5. Loyalty Stability & Constancy

People often judge the health or wholeness of ecosystems, or the goodness of designs, by how they look. Traditional design has emphasized visual results above all else. Ecological design has achieved the same results by paying attention to the structure and function of the ecosystem. Global design has to be concerned with making balancing domesticate landscapes with wild ones. This may mean preserving, then restoring and creating, the conditions for wild ecosystems so that they are stable or constant within the processes shaping and affecting them. It may mean limiting or reconverting areas of domestic lands.

2.6.4.6. Harmony Productivity & Health

Most products of an ecosystem are produced and consumed and recycled within the ecosystem. Humans need to minimize the external inputs in the form of energy and exotic substances. The community must be restored to health. This means balancing human needs with bird or fish needs in an ecologically stable pattern.

Ecological design is the creative modification of ecosystems to repair or enhance their ability at self-organization and maintenance of their complexity and diversity.

Health is the overall ability of a system to maintain itself under a normal range of environmental conditions. Ecosystem health is one of the goals of design. The goal, of course, is not an end point that can be reached once, but is rather a continual striving.

Responsibility is the condition of being accountable for one's actions or obligations. Here it means undertaking all aspects of a design, regardless of the expected level of success. The only example possible is the planet, with its global characteristics and systems. The very purpose of global design is to respect limits (that means not trying to always exceed them and to be cautious) and especially to keep the capital intact, to allow cycles to operate in a constant building up and tearing down. These six characteristics have to be related to other levels, especially by global design (Table 264-1).

Table 264-1. Properties at Different Levels.

— Nature —		— Culture —		— Design —	
Field	*Ecosystems*	*Place*	*Culture*	*Good Places*	*Good Society*
Motion Process	Course	Dynamic Change	Conduct	Action	Method
Autopoesis	Identity	Self-making	Wholeness	Individuality	Self-extension
Differentiation	Diversity	Uniqueness	Flexibility	Richness	Variety
Integration		Integration Investment	Adaptation	Conviviality	Cooperation
Constancy	Stability	Regularity	Endurance	Consistency	Loyalty
Development	Productivity	Renewal	Vitality	Health	Harmony

For instance productivity of the ecosystem level promotes vitality in cultures and harmony in good designs in good societies, which are embedded in human places, which are embedded in ecosystems. And, diversity at the ecosystem level promotes flexibility at the cultural level and variety in good design in good societies.

2.6.5. *Principles of Global Ecological Design*
Principles of global ecological design reflect the differences between local and global systems. Principles, as fundamental rules, cannot be reduced to other principles, because they emerge from earlier ones and they expand into new forms. This expansion is why we cannot fit them into the old can of old principles. These design principles are based on basic principles of nature, such as the principle of change. Nature is in flux, culture is in flux, everything is in flux. The climate will change, the shorelines will change. Human understanding and behavior is changing. But, nature is also self-regulating. This is the principle of self-regulation. Nature has evolved to maintain its stability in the face of many kinds of disturbances from planetesimal impacts to changes in atmospheric composition. Nature will continue to regulate itself even as human beings make dramatic changes. The danger is not so much that nature

will collapse, as that humanity will lose those things that it values the most, from cool air to wilderness. Ecosystem ideas can be formulated into principles that can be used for designing and building cities. These principles of centering or timing. Connectedness to global cycles.

2.6.5.1. Global Ecological Principles

Principles of global ecology, such as stability, have to contribute to the design of landscapes because those landscapes have derivative principles, or maybe emergent principles. Principles must be flexible to mirror the flexibility of open systems; flexibility is provided by diversity in fact. Many global principles are similar to ecological principles, such as: The Principles of Being Flow, Separation, Growth, Self-regulation, Error, and Pattern. Other principles either emerge at the planetary level or become more important at that level, such as the Principle of Wholeness, which states that a whole emerges from the interactions of parts and that whole is more than the total number of parts.

Wholeness is a characteristic of the field of nature and of the planet. The whole is the environment of all parts. The parts cannot survive without the whole. Although the whole can survive without many parts, it is reduced in terms of completeness or resolution with the loss of any part. A whole is an organized system of subwholes. Each subwhole behaves as a whole to its components, as a self-contained whole, and as a dependent part in context. These subwholes, or holons as Arthur Koestler calls them, are stable in a hierarchy that displays rule-governed behavior and structural Gestalt constancy.

These principles, whether we accept or deny their applicability, influence our interactions with global patterns, cycles, and biomes. They influence our objectives, our standards, and our operations. The interplay of these principles with examples and exceptions will refine our approach to and understanding of global patterns. This is part of the process of living with and understanding global patterns. These principles are not a final presentation of a limited number. They are meant to be questioned, discarded, or expanded. You can help with that. From a limited set of principles, we can distill a definite set of standards and behaviors, for example: (1) Work within the scale and duration of the planet. Processes like long-term succession can be assisted, slowed, or speeded up, but not skipped or ignored. (2) Understand the patterns and connections. Make sure they are not broken.

2.6.5.2. Global Ecological Design Principles

Global ecological design principles emerge from knowledge of ecological principles. They must combine the constraints of ecology with the ascendant sensitivity of design.

2.6.5.2.1. The Principle of Spirit of the Planet: Act as if the Planet is Alive

The spirit of each place is unique. Place is not just location; it is the total sum of objects in the landscape combined into a unique whole. Every place has certain characteristics that enforce the spirit of place, for instance, a strong definition of place or indicators of great. A sense of wildness and water also contribute greatly to the spirit of place. Each place expresses a unique combination of elements, including contrasts, dramatic features, and the presence of water. Design can work to be consistent with the recognized spirit of place. If the design recognizes this aspect of the landscape, it may be stimulated by the spirit of it and may further enhance it.

The identity of place often leads to human identity, thus people call themselves by their place names. The more unique a place the stronger the emotional attachment of the inhabitants. One of the goals of global ecological designs is to relink people with regional and global cycles that support specific places. Identity with the planet allows people to save all the planet, not just the beautiful places and the charismatic species, the whole planet, with hurricanes and parasites and viruses.

2.6.5.2.2. The Principle of Sensory Force: Use Every Sense to Appreciate Life

Visual force is a psychological interpretation of perceived power in a landscape. As a principle, it is embodied in psychology, art, graphic design, and architecture. The human mind responds to visual force in predictable and dynamic ways, for instance, where lines in landscapes draw the eye down convex slopes and up concave ones—the strength depending on the scale and irregularity of the landform.

The effect of any landscape is not completely visual, however. Smell, sound, touch, and even taste play a large part of our appreciation of them. Crawling, which is highly recommended by Gary Snyder, climbing, listening, and tasting things, such as soil, bark, or lichen, can expand our perception of other aspects of the forest.

2.6.5.2.3. Principle of Valuation: Everything has Value to Itself or Someone

All beings have value in themselves (what is called intrinsic value in the Platform of Deep Ecology). These values are independent of the usefulness of the nonhuman world for human purposes. Life includes individuals, species, populations, habitats, and all human and nonhuman cultures. Like Deep Ecology, global ecological design expresses deep concern and respect for cultures, as well as for the well-being and flourishing of all life.

There are different ways to maximize value, according to John B. Cobb, Jr. One can act to maximize value for yourself at all times (this is selfishness); one can act to maximize value for yourself in the future (also called prudence); or, you can act to maximize value for all humans for the indefinite future (the greatest good for the greatest number). Cobb points out that these three ways are unstable and unacceptable. That leaves the final principle, which he suggests is to act to maximize value in general (the process view). This implies that the value of ecosystems and global cycles would also be maximized. But, should value be maximized? Would not that contradict a principle of limited good (that desired things only ever exist in limited quantities)? After all, nature contains good and bad (both human judgments). Perhaps we should aim for satisficing value, rather than a maximum or an optimum.

2.6.5.2.4. Principle of Preservation: Do not take everything for human use

Focus on what to leave, not on what to take. Ecologically responsible design leaves fully functioning systems at all spatial scales through time. In other words, ecologically responsible design identifies the parts of systems, local, regional and global, that must be protected to maintain short- and long-term system functioning, and then determines what can be safely removed for human uses.

These systems must include large wild systems, or wild areas, undominated by human processes, required to keep bioregional and global cycles operating, as well as for homes for ultrahuman species. This preservation of natural processes must transcend any economic, spiritual, or aesthetic needs. Preservation is essential for survival and self-preservation. Principles of preservation and protection suggest important standards, such as saving an optimum amount of habitat for the largest indicator species, limiting the use of species to a percentage of regeneration, or reducing the use of materials to natural rates of production.

2.6.5.2.5. Precautionary Principle: When harm is possible, take appropriate action
We should adopt a precautionary principle, which asserts that, if harm is threatened, and if there is uncertainty about the seriousness of the harm, then precautionary actions must be taken. Since the 1970s, in fact, this principle has been incorporated into Swedish and German environmental laws. This principle means that not doing something, "benign neglect," becomes a valid management option. Carl Walters suggests that inaction is an inappropriate alternative to gambling as a result of confusion, but inaction is a very appropriate alternative in the face of confusion—when in doubt about any large complex global system, we should not interfere. As appropriate alternatives, both inaction and action can be suggested by theory.

2.6.5.3. *Actions from Principles*
Principles must be flexible to mirror the flexibility of open systems; flexibility is provided by diversity and redundancy. These principles also result in a further qualification of standards and behaviors.

2.6.5.3.1. Protect Diversity
Ensure that all plans and activities protect, maintain, and, where necessary, restore biological, ecological, and planetary diversity (from genetic, species, and community diversity to large cycle and pattern diversity). Maintenance and, where necessary, restoration of all types of biological diversity is necessary to sustain life in wild ecosystems. Maintaining genetic diversity means ensuring that viable natural gene pools are restored to the site following human use. Maintaining species diversity means that viable natural populations of plants, animals, and microorganisms are maintained or restored, in previously degraded areas, throughout the various successional phases for each ecosystem type within a wild landscape. Maintaining community diversity means maintaining or restoring, in previously degraded areas, the variety of ecosystem types that result from natural disturbances at a variety of scales through short and long time frames in a landscape. Biological diversity must not be viewed as a frill or luxury. Ecologically responsible designers understand that maintaining natural biological diversity is an absolute requirement to ensure maintaining fully functioning ecosystems.

Protect, maintain, and, where necessary, restore ecosystem connectivity. Connectivity in ecosystems is maintained, in large part, by ensuring the protection of water movement patterns. This includes microscopic water movement patterns in the soil and in riparian ecosystems, from ephemeral streams and small wetlands to large river systems and wetland complexes. Connectivity is also maintained in ecosystems by protecting and, where necessary, restoring the full range of composition and structures from the large landscape level to the smallest patch.

2.6.5.3.2. Restore landscapes
Global ecological design involves solving problems with finesse and ingenuity, rather than with force. Rather than use an approach that label some parts as valuable and other parts as

worthless or harmful, global ecological design uses a softer approach that keeps all the parts in balance.

Restore whole watersheds and large landscapes. For example, degraded stream channels and fish habitat are often the results of timber management activities that emphasize clearcuts, throughout a watershed. A stream channel cannot be restored simply by replacing missing structures in the stream channel. Instead, all parts of the landscape that contribute to the loss of structures in a stream channel must be restored along with the stream channel. Pursue the satisfactory (*satisficing* in H. Simon's words) instead of perfect maximum.

Mimic natural ecological processes, particularly in time. People need to recognize that natural processes of ecosystem change are the way that wild ecosystem functioning is maintained, as well as the way that ecosystems rebound following disturbances. Wild design takes patience, since many ecosystems required hundreds and thousands of years to develop, and restoration will require hundreds and thousands of years to be effective.

Recognize the complexity of interrelationships. Not only is there a wide variety of plant and animal interrelationships, but also there are soil, water, atmospheric, and geological interrelationships. Each directly influences the others. Unique regional climates are related on a global scale. Any change produces a chain of effects, many of which are unpredictable or unwanted. Small changes can be more easily corrected.

Plan and carry out restoration activities in local contexts with local inhabitants. Act from the bottom and restrain from the top (act locally in parallel, within the constraints from the top and design). Sometimes destructive land use activities have often been designed by specialists and accomplished by powerful tools, using unlimited amounts of fossil-fuel energy. Respect traditional practices in place-based cultures. Effective restoration requires all kinds of people with different kinds of skills. People with shovels and ideas are as important as people with machines and plans. People who live and work in local communities are most likely to have both the commitment and the patience for effective forest restoration. Cultivate feedback from the participants. Work with errors, learn from, and allow them. Attempt to set up a transgenerational land tenure system to accommodate wild ecosystem time, which spans human generations.

2.6.6. *Applied Global Ecological Design As Lessons from Nature*

Design can imitate nature on many levels, from structure and process to landscapes. We can imitate the structure of mature forests by planting on every level of the forest hierarchy, from canopy to below ground. We can use native species. We can imitate the process of forests by allowing birds, bats, and other animals the opportunity to distribute seeds and energy to other areas or prey on "pests." We can create microclimates within the landscape that may shift the landscape in new directions. Planting trees, for instance, allows new species to become established under their protection. By doing these things within a global design, we can effect the global patterns that emerge from ecosystems. Ecosystem health is one of the goals of design.

The landscape provides its own metaphor for design. The landscape is a unique individual, a community, a dynamic system of interacting patterns—the human pattern is a part of it now and should be preserved as part of the whole pattern, but not necessarily as the only pattern or a completely dominant one. Most products of an ecosystem are produced and consumed and recycled within the ecosystem. Humans need to minimize the external inputs in the form of energy and exotic substances. The community must be restored to health. This means balancing human needs with birds or fish needs in a stable pattern (it is important to repeat this). Each element in a pattern relates to others and to the whole. The global system provides its own metaphor for design, also. The global system, however, has too many untraced connections to permit direct control through a design.

Natural patterns can suggest a number of regularities that could help us to understand the design of regional or global ecosystems, according to Michael Soule. For instance, well-distributed species are less at risk than concentrated ones. Large blocks of habitat are safer from species extinctions. Blocks close together are more effective than those far apart. Contiguous blocks are better than fragmented blocks. Interconnected blocks are better than isolated blocks. Corridors can make functionally larger blocks functionally. Roadless blocks are better. Human disturbances similar to natural ones are more likely to be resisted or accommodated in wild systems.

These rules work well with natural processes that operate in ecosystems, such as metalysis (building up and breaking down), animal movement, interelement flows, human interactions, and shifting mosaics. For instance, forest fragmentation can be reduced through the design of forested areas, taking into account the genetic diversity of the trees, catastrophic conditions, minimum viable populations, corridors, and edge effects. The survival of organisms usually depends on one of two factors in the web of relations. These factors can be modified by design.

Wild landscapes are affected by climate, soils, interactions, and disturbances. Domestic landscape is affected by land use as well. The greatest changes have been brought about by the destruction and creation of forests. With the predominance of artificial forests, it is important to consider the qualities of naturalness in the landscape. Forests are expected to meet the needs of society by producing timber, creating wildlife habitats, and providing recreational opportunities for people. But, forests are also expected to look natural.

The values of the land and forest must be most carefully assessed. The characteristic qualities must be identified and measured for uniqueness. Comprehensive landscape plans should be required when planting or extensive felling is planned on a large scale. The patterns established at these times may persist for many years or centuries. Good global design maybe able to resolve conflicts between characteristic qualities of the landscape and the changes from use. Of course, the design should last as long as possible and should be self-sustaining.

2.6.6.1. *Levels of Global Ecological Design*

As a design process includes planning of ecosystems, this means a sixth level of design— notice that after number four, all are properties of normal applications of design: (6) Planet as a whole, (5) Regions, landscapes, and watersheds (as well as airsheds), (4) The community, for instance forest, city, or corporation, (3) systems, such as ecosystems, traffic, or industrial areas, (2) products, as habitats, houses, roads, or plant sites, and (1) components, such as trees, fungus, bats, tools, rooms, cars, or land. Many problems occur at level three—more are to be expected at the new levels four and five.

All levels of design need to be addressed, from the conceptual to the political, and are involved in all stages of the process. This involves new challenges for ecological design, which has to:

6. Integrate a project in the global context (sixth level of design). This is the level of ocean and atmosphere, where all cycles participate in global movement and stability (as homeorhesis).

5. Relate a project to its regional context (fifth level of design). This is the level of regional cycles relating to precipitation and waste recycling.

4. Relate a project to its ecosystem context (fourth level of design); be concerned as much with cultural survival, justice, and wilderness preservation as with efficiency and aesthetics.

3. Consider the whole perspective (ecocentric, perhaps); the proper vision is of the whole community in which we dwell. Apply ecological concepts, such as networks and carrying capacity.

2. Make designs that are anticipatory, flexible, pluralistic, polyvalent, and polytechnic. Make open guidelines for long-term decisions.

1. Essentially, work backwards from values and goals, and from the bottom up and inside out, drawing designs from the genius of place.

0. Participate in place, care for all inhabitants, and assume responsibility for the designs.

One of the shortcomings of design has been the lack of consideration of the higher levels, which can result in specific problems with safety or recycling.

2.6.6.2. *Stages of Global Ecological Design*

Ecological designs, at the global, regional or local level, can be applied in eight stages. These stages should be typical of all designs:

- First, decide to design, then create a design matrix/management matrix, then review the situation, evaluate the ecological history of the site, observing patterns of movement, interconnectedness, population change, land use, building and development, boundaries, limits, and life. Conduct ecological and functional analyses.

- Then Inventory, record all of the resources, from physical resources to cultural resources. Survey the area and create base maps, from geological to zoological maps.

- Next, evaluate the interactions in terms of impacts, needs, goals, and limits. Assess the whole system. Delineate patterns and balance components.

- Create a series of plans, from the site plans to value plans. Integrate with biosystem services. Kinds of integration include: Horizontal, vertical, and temporal. Consider aesthetic-perception factors.

- Start to design, which is a community process requiring the participation of all people, including the elderly, handicapped, and poor, as well those ultrahuman beings who cannot voice their concerns. Synthesize simulations and models(conceptual, capability, and suitability). Make another series of plans, from landscape plans to policy plans, within a master design.

- Implement the design together. Use appropriate measures and techniques, emphasizing native species over an adequate time period to ensure the stable processes of transformation. Create new connections. Reduce unwanted effects, such as road effects or heat island effects. Integrate into context. Optimize passive mode; integrate water, biomass, and materials flows. Conserve water and energy.

- Maintain and manage design. Provide services for continuity and management.

Monitor and improve constantly. Monitor, Reassess.
• Prepare undesign and disintegration

Ecological design must work within the components, structure, and function of ecosystems. Unless it does, it will not be long-lasting or satisfactory. Because design has to work with systems, whose aspects are often ambiguous, fuzzy, changing, and general, design has to be able to work with these aspects. Furthermore, the design has to work within the constraints of the larger system, sometimes the global system.

Global ecological design takes far more time than graphic or automobile design, due to the complexity, size and longevity of its subject. A number of factors have to be carefully assessed before design work starts. Wild systems require a lot of observation before activities can take place. Wild systems can be highly reactive to change. Some people value different characteristics of wild systems than others.

2.6.7. *Applying Global Ecological Designs to the Planet*

All design elements are related psychologically by designers, as focus or frame, as contrast or uniformity, as dominant or recessive, or in a number of other pairs. Good global ecological design means not violating any of the aforementioned principles and ideas.

Geology, climate, disturbance, and stability all produce diversity. Landscape diversity is linked to ecological diversity, which depends on diversity of the substrate. Different ecosystems introduce diversity into a landscape, but different ecosystems often can look similar. The sum of regional and local systems yields a global system with a few large patterns composed of many smaller ones. Excessive diversity on the global level could lead to confusion in a global landscape design. Increased diversity also has the effect of reducing scale, so adding diversity can be used to do reduce the scale. A high level of diversity is acceptable if one element is clearly dominant or if the differences cannot be recognized from a distance.

Process applied to components yields pattern. Nature is composed of patterns. Organisms have characteristic patterns, such as the branching of trees or the cloud forms of tree crowns. Lichens have lobes, wood grain under stress has spirals. The cracks in tree barks form nets. Patterns are not still. A circular pattern through time can be recognized as a spiral (the earth's orbit for example). The pattern should allow for surprises and discontinuities; it can do this if it is flexible. The design of continental forests, for example, is vulnerable to surprises because nature is chaotic (unpredictable) and science itself is uncertain (by definition) about patterns of change in forests.

2.6.7.1. *Global Ecological Design Management*

Global Ecological design is not finished with the design. The system has to be managed in some way as a result of the design. Noninterference matrix management is proposed as a technique for design management (see discussion under ecological design). An ecosystem exists as part of matrix that many interacting elements. Any activity in the matrix can have some effect on these elements. The whole matrix needs to be managed with the local ecosystems in mind.

Understanding of the principles of ecology can lead to better management. One critical message of ecology is that if we diminish variety in the natural world, we debase its—and our own—stability and wholeness. Many ecosystems have been simplified and degraded.

Perhaps we do not have sufficient knowledge to manage a complex landscape because it is too complex to understand scientifically. But we can understand the pattern and drive it in a healthy direction with minimal intervention. We must do all that we can to restore its richness and the natural processes that created the richness.

2.6.7.2. Principles of Global Ecological Design Management

At the global level, management principles will be broader than the ecological or traditional. Two sample principles are listed below.

- Work at the Appropriate Scale of Space and Time. Scale management to the size of the pattern. At the global level, a manager will have to coordinate regional and local managers. This will involve putting limits on some regional or local actions. The design has to be a partnership at all levels. This will involve putting human wastes and human bodies back into the system. Work with global time; do not try to do everything in the two-year (or five-year) industrial schedule. Be as slow as necessary to introduce system changes. Attempt to set up a transgenerational land tenure system to accommodate forest time, which spans human generations.

- Accept Management Limits. Management is limited, by costs mostly, to regulating animal and plant populations in a local system, rather than climate, geology, or water, soil, mineral cycles—that is, global things. Manipulating plants and animals to the environment is more likely to succeed than the opposite. Management limits are determined by the ecological characteristics of the resources and the scales, before any economic, technological, or political limits are applied. The use must be limited by the ecology before monetary considerations.

Principles, combined with common sense and good judgment, are necessary as guides in the absence of definite knowledge. And, definite knowledge is lacking in systems, such as large natural cycles, characterized by ambiguity, uncertainty and chaos. Principles give us a broad predictive ability. For each principle, we have to ask, how will it affect our objectives for that pattern, and how will standards vary?

2.6.7.3. Actions from Global Ecological Design Management

Recognizing global level problems is easier than deciding on appropriate global level actions. Due to the complexity of interactions, many actions will have the opposite effects of the intended ones (similar to the operation of enantiodromia in tragic plays).

- Work backwards from known constraints. That will make sure that actions do not produce situations dominated by runaway positive feedback. At the global level, most action involves some form of restraint on technological interference or landscape conversion.

- If in doubt, do nothing. It is better to not do a right thing than to decisively do a wrong thing. Natural processes have billions of years of experience recovering from small and large-scale disturbances. So, the worst possibility of doing nothing is that we will not be able to correct a situation within a human time scale, which means that we may suffer the consequences.

Cautious actions can nudge natural processes in directions that may benefit human needs. Other cautious actions can restore broken connections or empty niches. If the management actions are slow and small-scale, in terms of actions rather than ideas and designs, then the system can do most of the work and take most of the control itself.

2.6.8. *Summary of Global Ecological Design*

Geology, climate, disturbance, and history all produce diversity. Landscape diversity is linked to ecological diversity, which depends on diversity of the substrates and historical developments. The earth itself is a complex total system.

Gyorgy Kepes compares a total system to the growth pattern of the human nervous system. The development of the brain increases the range and scope of perception, which leads to the need for a greater control within the brain to coordinate more information. The capital of perceptual knowledge offers a richer resonance to further perceptions. However, our social perceptions, with our tools of knowledge and power, are growing without any essential control. Our environmental crises occur because of the way things are made and used according to our simple, finite, isolated, noncontextual designs. Victor Papanek states that it is a design problem. Psychologists have recognized the need for diversity for people's quality of life and emotional well-being. An increase in ecological diversity would lead to an increase in diversity of the landscape. It would also tend to reduce the scale, but this would not be a problem in large landscapes.

Global ecological design is a five-dimensional, six-level, eight-stage discipline in which we describe particular actions. This level of design can help us adapt to place; keep track of everything (using ecological accounting); design with nature; and make the designs visible as art. Global ecological design has to exhibit diversity.

Global ecological design has to particular attention to emergent global phenomena, such as atmospheric gases (especially methane and CO_2), oceanic currents, and continental drift. It also has to pay attention to global human effects, such as ship, plane and auto transportation lanes, as well as communications technology, from wires and spectra to transmission lines and satellites. And, it has to pay attention to large-scale conversion processes, such as agriculture and urbanization, which change entire landscapes.

Global design can be a bottoms-up process within the limits of global constraints. To use an analogy, architects can build buildings, but the most successful ones are built within the limits of gravity and entropy, as well as within human psychological and social limits. Design has to have the courage or influence to limit production, reduce use, or even reduce human populations and use rates if necessary. It has to integrate systems in a global pattern.

In order to accomplish a global ecological design, we have to include the spheres of human cultures and global processes. Without a modest redesign of cultures, any actions to balance global designs may fail as cultures focus on immediate local needs, without regard for atmospheric and oceanic cycles, at the least. The totality of human experience and its institutional systems needs to be tweaked. Cultures need to become aware of specific purposes and meanings to become viable as participants in global actions. Cultures have been a form of unconscious self-organization, a shared experiential system communicated between generations. But, they have to incorporate large forms of design, now.

We may not want to think of design as the conscious creation of the planet, which is self-organizing, although design may contribute to the creation of a more encompassing image of the planet (literally a world as human-image). Perhaps that image would be of a 'garden,' as Bertrand de Jouvenel suggested. And, certainly human measure is important—it may be the only way we can measure—but, the system is being built on a wild planet, and it supports all human activities with aesthetics and services.

In designing the planet, it is important to design a frame that allows natural processes that are only integrated with artificial processes through larger natural cycles. In this sense we are not concerned with the design of a product or a structure, but a living constraint on natural and artificial systems. In this sense the design is to restore broken linkages, that is, to recreate what was interfered with, to restore the system with native parts or with equivalent niche-makers, and to reduce the interference of human activities.

2.7. Fitting Local to Regional & Global Design

Local designs are not going to work unless there are hooks at the regional and global scale to allow or encourage the designs to work. If a nation or international organization tries to enforce a top-down decision model for everything, then local decisions will be much more difficult. The challenge then is to start with local decisions and try to constrain and coordinate them based on regional or global requirements that are necessary for the continued health of local systems.

2.7.1. Local Ecosystems Emerge from Conditions in Place in the Planet

The planet originated as an aggregation of small pieces, within gravitational and electrodynamic constraints. Any local system seemed to resemble any other local system. At a point in its growth, the scale of the aggregation changed its interior and exterior conditions. The interior heated sufficiently to change the state of iron and nickel. Regions divided into separate spheres characterized by specific temperatures and pressures.

The sphere that was the molten core of the planet kept the heavier, hotter matter in the core. That part of the planet was not free to interact with the surface. The types of regions constrain their local systems so that they no longer interact outside the system.

Regions emerge from and constrain local systems. Other principles either emerge at the planetary level or become more important at that level, such as the Principle of Wholeness, which states that a whole emerges from the interactions of parts and that whole is more than the total number of parts. Regional patterns emerge from local ecosystems, from their properties and constraints.

2.7.2. Local systems Push Regional & Global Systems

The emergence of water from rock and from volcanic action created a global pool of water. The emergence of gases from the surface created a global envelope of gases. These pools then acted as constraints on regional systems in the planet. The suboceanic realm permitted anaerobic bacteria to thrive, as well as plants that could exploit shallow zones for light and animals that could use dissolved oxygen. The atmosphere limited most life at water, at first. As waste products, such as oxygen and methane, joined the atmosphere and built up, the scale of that change pushed the regional and global systems to change. Oxygen for instance, interacted with ultraviolet light, creating ozone molecules under those circumstances; this reduced the ultraviolet radiation reaching the surface of land and water, and allowed living forms to exploit the surfaces, especially terrestrial ones.

Regions are constrained by the global system. Global phenomena, such as atmospheric gases, can impose restraints on local systems, especially regarding carbon dioxide or fires.

2.7.3. *Are Local Designs Ever Independent of Regional or Global Designs?*

Because of the dynamics of the planet at least, local systems are not independent of the regional systems in which they are contained, as regional systems themselves are pushed by global flow patterns into distinct states with various degrees of temperature and moisture.

Local designs are never independent of regional designs. Scale-linking occurs with processes at different levels, such as evaporation. In a sense, no local system can ever be completely independent of the surrounding system, region, planet, or solar system.

2.7.4. *Leaving Hooks & Flexibility through Design*

No local life form has had to consider the regional or global effects of its accumulated exploitations and wastes. Changes in interactions or scale happened or did not happen. Some local forms, such as bacteria, became ubiquitous within the global system.

Humanity, however, has slowly, as individuals and small groups, become aware of the regional and global consequences of its wide-spread habitation, transformations and waste generation. As human designs, especially of transportation networks, waste disposal and habitations, have grown larger, people have recognized that the consequences to all systems have become negative and damaging.

A dynamic system can often reach a threshold or tipping point, beyond which the system enters a new stable state or an unstable state. Threshold: entrance, beginning point, when perceived. Tipping point turning point. The action that precipitates change is often called a trigger. A trigger is a mechanism that activates a release of information or energy (or an immediate change in scale). A trigger is also a small impulse that can release the stored energy of a larger impulse; it is a form of energy amplification. The changes triggered can be much larger than the expected results. We can suggest past triggers. The extinctions of woolly mammoths, subarctic horses and other megaherbivores, as a result of factors from climate change to overhunting, triggered the shift to less palatable plant communities after cropping stopped renewing vegetation. Fire triggered land conversion. Domestication triggered differences in animal behaviors, shapes and requirements. Fossil fuel use triggered waste, speed and atmospheric changes. We have been transforming the planet for 12,000 years. We choose what plants grow where. We grab minerals, redirect water, and spread wastes.

External disturbances such as asteroid impacts or basalt eruptions could have triggered significant transitions between different states. However, most transitions appear to have been generated internally with evolutionary innovation playing a role. There are many further considerations: To what extent is the Earth system self-regulating? What is the contribution of life to maintaining habitable conditions? In what sense can the Earth system itself be said to evolve? Are there reasons to explain why regulatory feedback should predominate at the global scale?

There are many kinds of physical, biological, psychological or cultural triggers. A crisis can be triggered by a change in scale. A trigger can have connections to a global system. A trigger often creates positive feedback, causing dramatic amplification (See Section 17326 for a further discussion). The novelty and complexity of interactions of innovations can lead to thresholds that are tipping points, triggering punctuations in the functioning of the system with consequences that may surprise humans.

Is there a way that abrupt changes in the operation of the system can be anticipated and predicted? Can we identify the changes are most susceptible to triggering by human activity? Many aspects of system dynamics are believed to exhibit multiple equilibrium states, and therefore may display abrupt transitions between equilibria. There is evidence for a transition from a green to an arid Sahara in the mid-Holocene, 5500 years ago.

We admit that design and its effects can have local, regional and global consequences from slums, to the destruction of regional forests to atmospheric change. That means that all designs are going to have to be studied for fitness with other scales. No global design to change atmospheric temperature should be tried without understanding the regional consequences of shifts in wind and moisture patterns that could destroy rainforests or agricultural designs. No design of pest control should be considered without understanding its consequences in poisons spreading through regional food chains. No design to restore the regional Sahara to forests should be considered without understanding the local consequences to endemic vegetation and local adapted cultures, as well as to global wind patterns.

Design has to leave sufficient hooks, in order to respond to triggers and changes. A hook catches something due to its design, or connect the parts. Regional designs can leave hooks for local design to respond to changes. All the constraints of a system can influence or limit design. Harmony, of course, is a constraint on the whole system. The design imposes constraints on the system, but if it is flexible, it can be altered if the system is affected negatively. Some constraints, for instance, on technological interference, have to be top-down. Flexibility, defined as the unused capacity for change, can be designed into the system. The parts of a system have to maintain the potential for all possible behaviors that could flow from any other part of the system. Although there is quite a bit of flexibility for local orders, at the global scale it would be better to keep flows in dynamic equilibria. Diversity at the ecosystem level promotes flexibility at higher landscape or regional levels.

Globalization as a general trend is triggering a profound shift in human consciousness. First by forcing us to realize we cannot do everything that we want, and next that cultural differences are less important than what is held in common. Human societies, like climate, are open systems that have chaotic and complex dynamics. Recognizing that our industrial civilization is a dynamic social system that can evolve or devolve may be the key to managing changes. At the same time our civilization has become unstable, we are acquiring the ability to design living systems. Buckminster Fuller said that to change something you have to build a new model that makes the existing model obsolete—not bother fighting the existing reality.

Joseph Tainter notices that while converging stresses can result from disparate developments, such as a harvest failure at the same time as the invasion, they are often caused by cascading crises, for instance, when a harvest failure causes famine which then triggers a rebellion or the invasion of some other society. This whole idea of cascading stresses is very important in many collapses such as the Mayan or Ik. Thus, there may be multiple triggers or cascading triggers. Autocatalytic loops keep self-organizing structures going. A trigger of one energy form sets off flow in another, which can trigger the release of another flow in the first, and possibly a chain of trigger/flow interactions.

Triggers can also be hot or cold, depending on the lag time between the trigger and the effect. Culturally, B.J. Fogg says that Facebook puts hot triggers in our path. The opportunity triggers our behaviors. Some behavior requires a trigger, although it can part of the path also. Cold triggers require later action, after a lag time. The goal with Facebook is connecting to people. Connection is important. Facebook and other interactive systems provide motivators, such as fear, hope or the desire to belong, for behavioral change. Change, however, has to be a satisficing amount to be immediate and effective, not maximal or optimal, which even if possible would require longer lag times.

Ecological design needs to make hot triggers, that is, the actions have to be immediate: Saving kilowatts, planting grasses or food, or lowering the thermostat. We claim the ability now to geoengineer the planet with large-scale ideas and projects. Even so, there are alternate ways of intentionally reforming the planet: An easy engineering way or a more difficult way of ecological design. If we experiment with engineering ways, we might want to localize the effects, by keeping the change to as few systems as possible, maybe in the southern or northern hemisphere only. For instance, gassing with SO_2 might have drawbacks: Acid rain, and shifting plant, animal and bird life. It might trigger uneven shifts. We rely on technology to 'change the game,' but we can change the game without involving more technology, by choosing ecological designs that incorporate conservation and frugality.

Ecological design requires understanding properties and principles that emerge from local and regional levels; some regional properties cannot be understood without knowing global history. Global ecological design also requires a sixth level of design, which relates everything in a global context, the level of global spheres, such as the atmosphere.

2.8. *Insurmountable Problems at the Local Regional & Global Levels*

Civilizations have experienced problems that seem insurmountable—because those civilizations subsequently collapsed and disappeared. Many of the problems are structural, logistical, or spiritual. Economic decline can lead to stagnation, disease and collapse. Many have to do with political power distribution, especially if related to the desire to conquer, control, and reform other cultures through war (and now the threat of nuclear war with nuclear winter). Over-administration has its own seemingly limitless costs that can lead to collapse. Imbalance as a general condition can lead to collapse. Some of these problems could have been solved or have been solved for a time.

Urban problems, and some national problems, seem insurmountable. Some problems, the largest ones such as earthquakes and tsunamis, seem insurmountable. The dedication of the legal system exclusively to human activities seems insurmountable. The problems listed below, having to do with water, heat, energy, crime, globalization, and structural maintenance, seem to be insurmountable for any civilization, industrial or ecological.

2.8.1. *Water & Drought*

Water has been a problem in many civilizations, from Mesopotamian and Indian to Chinese and American. The changes in wind and rainfall patterns or in river beds have resulted in drought. In Mesopotamia, for instance, cities compensated for declining rainfall by irrigating wheat and barley with canals. But that lead to salt retention in the soils. By 3500 BCE wheat and barley crops were equal. Wheat can tolerate salt at only 0.5% in the soil, but barley can take twice that. By 2500 BCE wheat had fallen to 15% of the crop, although overall crop yields were still high, then to less than 2% by 2100 BCE. By 1700 BCE no wheat at all was grown. Overall yields fell 42% between 2400 and 2100 BCE, and by 65% by 1700 BCE. It was written that the earth was white with salt. After much intensification, the land collapsed. Many of these things are long-term problems and do not become evident for several generations. They are also very difficult to reverse. For a society that needs surpluses to continue, with growing dependents and growing people, there is little flexibility to change. The only way to avoid the problems was to let the land be fallow for long periods until the water table fell. This alternative was impossible due to food demands.

When cities started to fail, as a result of attacks or droughts, people were able to emigrate. Many returned to herding, or when possible, hunting and gathering. For thousands of years, starting possibly 11,000 years before the present, people participated in a cycle of emigration to rural areas when time were bad and immigration back into cities when the ecosystems recovered and could be made productive again.

In complex, self-regulating systems very small changes have large consequences. In some cases, where conditions like drought, are cyclic, in the Sahel region of Africa, humans expand during the good times, only to perish when the drought returns. In other cases, human activities, such as deforestation or overgrazing of herds, can cause weather changes. The scale and rate of changes allows people to view the situation as natural, but once these catastrophes pass a threshold, the people and their cultures have been trapped by their demands, and only severe reduction or collapse can allow the system to regenerate.

Drought has been a major urban or rural problem. Even cities that have been located on rivers have been destroyed by long droughts (usually over ten years). Fresh water has become an intractable problem in the past 50 years, as more aquifers have been drained and more water sources are used for industrial purposes, such as cooling or washing away wastes. Areas of the Americas and Asia still rely on irrigation, and people steadfastly ignore the warning signs of drought and collapse.

2.8.2. *Energy & Desire*

Energy imposes different kinds of limits on civilization than ecological ones, which are concerned with productivity, exploitation and pattern forcing. Before the utility of coal was discovered, energy was limited by the number of trees or amount of dung that could be burned (although whole forests and soils were destroyed). Before oil was discovered, energy was limited by the difficulty of acquiring coal, especially in places with a high water table (although steam engines were invented to solve that problem) or by the amount of whale oil available for burning. Oil, however, being plentiful and rich, has addicted us to, and spoiled us with, a high-energy, concentrated-carbon fuel source. With oil, we have to face thermodynamic limits of energy generation and ecosystem introduction limits; and with peak oil in the past, we have to face high expectations of use and artificially high needs. As the energy for an industrial civilization, oil has been used to meet the perceived need for always on-demand peaks at the largest scale. Oil has allowed us to 'borrow' from the planet's savings account to meet any industrial need. And, the scale has expanded as well on demand. Trying to predict future energy sources and uses, with or without oil, requires making assumptions explicit.

2.8.2.1. Assumptions

The assumptions are based on physical science and human history. Climate change is a geological scale phenomenon. Even if we should control carbon emissions and limit them below 350 (or 240) parts per million in the atmosphere, we might not be able to avoid the next warm cycle or glacial cycle. Carbon dioxide is a natural product in a global carbon cycle. It is not toxic. It is necessary for the development of the biosphere.

We can imagine that we can control things, especially since we have seemed to have contributed to climate change with industrial emissions, as well as to an extinction spasm from ecosystem conversions, but we may not be able to. The system may be too large and complex. The laws of physics, chemistry and ecology have to operate in any system built by industry, but embedded in and supported by wild processes. Energy demand will continue to grow as long as the myth of growth holds sway; and this myth requires growth for prosperity and equalization, that is, for our overall economic success as a nation or civilization. As long as well-being is associated with growth, the difference between development and raw growth will be ignored, as it has been in many recent books on economics, such as *The Economics of*

Enough. These assumptions lead directly to problems.

2.8.2.2. Problems

Energy problems always include generation. How can we generate energy cheaply, after the oil is exhausted? And, how can we always supply energy for high demands? It has to be held or stored somehow. Oil or gas can be burned immediately, but the availability of solar would be limited to sunlight or back up systems (batteries or reservoirs).

For air transport, the source has to be carried on a lightweight, aerodynamic vehicle or beamed to it somehow. For individual vehicles, we may be limited to combustion engines or batteries, both of which have obvious disadvantages and costs. By contrast, buses, trains and ships would likely be more efficient with greater choices of power.

Our psychological desires are going to influence or force our decisions. With economic and social factors considered, people are going to demand low-priced, on-demand energy. And, it is likely that that demand is going to force acceptance of the least expensive option, coal, which is also the most abundant, dirtiest and carbon-intensive. We know and understand the health and economic issues of coal. But, is cheap cost a fair trade-off with climate chaos from carbon emissions?

2.8.2.3. Possibilities

Increased waterpower is limited by environmental circumstances, and it tends to destroy the local environment, specially the lakes, rivers and streams. Increased solar or wind power has higher startup costs, and has major impacts on wildlife and the environmental, also.

Nuclear power, fission or fusion, has high start-up costs, as well as high initial carbon releases (from concrete construction). There are other problems with nuclear reactions: The waste products require long-term storage or disposal, because they are relatively dangerous. The threats to human health are serious. The technology allows for tremendously powerful explosions; this is related to incompetence, accidents and terrorism. Natural nuclear material are limited also, like coal. Unless they were extended by breeder reactors, they would be exhausted in less than 200 years (maybe 100 years or sooner). Robert B. Laughlin (in *Powering the Future*) points out that the availability of cheap rods would make the economics of home-made rods too expensive. The rods could be traced, and they could be returned to the manufacturer whole. This could result in fewer problems with lost or stolen materials.

Storing compressed air undersea in tanks is possible, in technology and scale. A year's supply of electricity could be stored in 300,000 tanks in a group 50 km by 50 km. But, total energy would require two million tanks in an area about the size of Israel. The potential land area under the sea is could easily accommodate tank farms for energy storage. The storage would be relatively safe, despite its scale and potential explosive power.

Burning trash would not generate enough energy for civilization, even though most of it is paper. Burning plastic would not add enough energy (less than 5% of oil is used for plastic). Synthesis gas would not add enough energy for a planetary scale.

Laughlin suggests that hydropumping might be a viable source of large-scale on-demand energy. It has advantages over batteries or compressed air (in ocean storages). The advantages include 80 percent efficiency and reversibility, that is, it can absorb energy. It has a large surge capacity. The turbine technology is well-known. Dams and reservoirs already exist and could be used. Some of the pumping could be done with solar energy. Hydropump storage for the electric needs of our entire civilization would take about 60 Lake Mead volumes or half of 1 percent of the area of Canada or the Sahara.

2.8.2.4. Discussion

Laughlin predicts that cheap coal will work against green technologies, which have higher start-up costs and probably operating costs. There is a possibility that green technologies will be employed, even with higher costs, if they are regarded as environmentally necessary or 'sexy.' If the past is any indication, coal will replace oil when it gets low, and then nuclear or solar power (with wind and tide) will eventually replace coal. Nuclear power may be preferred for high intensity, large-scale power, while solar could work for decentralized systems.

Garrett Hardin's law states that political decisions that go against people's self-interest do not stick very well. And, people are interested first in cheap energy. Reducing energy costs seems more important than reducing carbon emissions. Carbon can be removed from the atmospheric system in other ways (and Laughlin notes that burying all trash, with no recycling or burning, would exceed the entire amount of carbon dumped into the air). Of course, if people were educated to understand their long-term self-interests related to climate, they might consider paying higher costs, but probably only if the self-sacrifices were spread out more evenly, so that the poorest people did not have to shoulder the burden of costs. Cheap energy is the goal and the desire, although it might not have a long-term benefit for us if our desire rationalizes the choice.

2.8.2.5. Recognition & Control

Life can be described in terms of heat differences, as well as of order and disorder. The process of ordering is like a river that flows uphill, but it creates a more massive downhill flow. Heat is necessary for indoor comfort in cold climates, but it usually comes from burning fuel. When the entire environment enters a higher heat regime, then people either adapt or emigrate. Our current form of adaptation is cooling the air in buildings by burning fossil fuels.

Individuals and groups desire to control the environment or specific events in it. Personal control is not only integral to the concept of health, but it is crucial to health. People in dependent situations, such as the elderly in rest homes, live longer if they feel they have some control. However, when it comes to large-scale temperature changes, we do not seem to have control. Living in hot climates has always limited the number of plants and animals, as well as humans. Controlling climate may not be possible with spiritual or technological techniques. Control of the environment on any but the very smallest scales may not be possible.

Heuristics, habits and shared illusions become shortcuts for understanding data and making judgments, but they can lead to systematic errors and biases, according to L. Mlodinow. Biases play an important role in decision-making, especially when related to kin or groups. When we have an illusion or bias, then we regularly interpret data to support it, rather than disprove it, even actively seeking evidence to support it, for instance, whether it is for or against global warming.

We may be efficient at recognizing patterns, but we neglect our ability to assess them critically. We need to act critically on uncertain and incomplete patterns. We can improve our decision-making by understanding our biases and illusions, and by understanding that we cannot control every circumstance or avoid every unpredictable event. But, we can recognize negative patterns and plan for them, and have a framework for making decisions about them. We can recognize irrationality and unpredictability in our behavior about energy.

2.8.3. *Liberty & Crime*

People like being free to do whatever they want, but they also like being secure or protected.

Usually the two conditions do not occur together in one political system. Even in a favored political system, crime and civic unrest do not disappear. Dangerous weapons, from automatic guns to tanks, and dangerous products, including nuclear reactors and biocides, are available to anyone to misuse. They are not strictly regulated. People choose to act badly sometimes. If a form of government is bad or ineffective, people can alter it. They can learn from mistakes or unintended effects. If the scale is small, then the catastrophe may be small. There will always be some injustice, inadequacy, and unpredictability. One law that Adolphe Quetelet described in his social physics was that vast inequities in wealth were responsible for social unrest and crime. Large institutions have made inequities worse currently.

A United Nations report has identified the world's rapidly growing herds of cattle as the greatest threat to the climate, forests and wildlife. And they are blamed for a host of other environmental crimes, from acid rain to the introduction of alien species, from producing deserts to creating dead zones in the oceans, from poisoning rivers and drinking water to destroying coral reefs. Cattle herds are the result of ignorance of the limits of local or global commons. Corporate cattle herds are worse. Corporations are especially bad at reacting to negative feedback from overshoot or drawdown—the lag time to the actual consequences is too great, as is the lack of a distinct trail leading back to individual or group poor decisions.

Extreme political views are divisive in a society not bound by minimum standards of behavior, especially regarding honesty or responsibility. Too little liberty results in as many problems as too much liberty, which is often the absence of agreed-upon cultural limitations. Too much division in a society, especially as regards material goods or luxuries, can result in crime as the perceived way of leveling differences. There are social controls on crime, such as shame, which requires a smaller community where all people are known and there is a threat of punishment through laws; but, shame is scale dependent, and only effective in small scale communities. Laws may be more effective if they are enforced at a large scale.

2.8.4. *Globalization & Localization*

What is globalization and what does it do to the local? People have been trading items for over 40,000 years, starting with ochre, stone, copper, furs, and continuing with gold, tea, or people. Trade involved travel by foot, camel caravans, or sailing vessels, then with steam-driven ships taking ores, grains and meats to many countries. Later coal and oil were traded, then shirts and manufacturing, electronics and computers. At first, there was a movement of people and luxuries. Later, there was a movement of materials and energy sources, followed by a virtual movement of money and capital. There was a specific movement of information.

Over the past 500 years, smaller economies have become incorporated in larger economies. The most recent phase of this development is now referred to as globalization. Globalization is a supranational trend and a gigatrend, as a result of exploration and trading. Globalization offers some advantages to many people and groups. Increasing globalization leads to more efficient production and to the recognition of the uniqueness of cultural creations. Because of the advantages of specialization, many nations produce and trade signature items. This kind of trade creates larger webs of interdependence, which is a crucial characteristic of globalization. This is not the same, however, as trade between self-sufficient nations.

Global trade allows anyone to have access to any product or luxury. This access allows many to assemble comforts and technological marvels to enrich their lives. A global network increases the size of the frame of reference, so trade and social relations have a larger context, and shrinks apparent distance. Globalization seems to break down barriers and enhance co-operation. Despite its limits, the North American Free Trade Agreement (NAFTA) in 1993 did integrate issues of environmental quality and protection with trade.

Globalization increases the levels and scale of economic order. It creates more direct connections between local levels, as well as with community, national and regional interests. Connections become tighter, more rigid and less flexible. Cities are unavoidably entangled in global networks, now. Cities used to be limited by the local carrying capacity, but can exceed that limit with advantageous trade. With global networks, self-correcting feedback has a significant lag time.

However, a countertrend of localization is growing and gaining momentum as a reaction to the threat of globalization to destroy local identities and homogenize unique local lifestyles. Localization responds to the increased abstraction of nature, commodification of places, and the modification natural processes. Localization emphasizes political governance of factors that impact the local systems. Localization can be typified by a number of trends: A slowdown of overall economic activity, population loss, urban shrinkage, an increase in the number of urban areas, and natural ecosystem rejuvenation. Localization makes feedback visible and more immediate, which could shape more self-reliant cities.

2.8.4.1. What are the negative impacts of globalization?

There are many arguments in favor of globalization. One argument contrasts it to the erroneous belief that the past was always nasty and short, teaming with suffering, disease, and tyranny from others. Thus, globalization acquires the patina of myth, that it has a destiny to save civilization. One myth of globalization is that it is saving and managing the planet. Another is that nature is a machine that can be simplified and streamlined. The metaphor of the machine is only a metaphor. Nature is not a machine and cannot be easily simplified to produce only positive and desired effects.

Globalization has always had negative impacts on people, places and economic exchange. As cities and communities traded over longer distances, for example, people were exposed to new diseases from other previously isolated populations. This exposure led to sudden epidemics and catastrophic losses of lives and cultures, especially in the American continents. This first wave of global trade destroyed many of the traditional societies of the Americas and produced incredible suffering. Globalization was destructive on other levels, not only from diseases, but also by conquests and exotic exchanges. American crops, for instance, grown under harder conditions, grew where native Euro-Asian crops grew less well. Exotic crops went everywhere, even manioc into Africa. This allowed expansion of populations into marginal areas. In general, the coming together of zones was destructive to native peoples, not just from trade and competition, but also from livestock, exotics, and diseases.

A later wave of globalization destroyed traditional economic systems in Africa, Asia and the Pacific. European exports, especially in textiles, undermined regional livelihood. These processes enriched the Atlantic shores, but also widened the inequities between and within nations. As a result of differential and interference trade (basically unfair), gaps in wealth increased and social inequality increased. Lester Thurow states that governments did not decide to start global trade and marketing. But, that is what happened through the establishment of East India, Hudson's Bay and other companies. There were also changes to local cultures, with clothing, foods, and songs being dominated by a few. These waves of globalization resulted in long-term changes to human immune systems, and to the composition of ecosystems. As national languages came to dominate local ones, many languages were lost, impoverishing knowledge and culture.

John B. Cobb Jr notes that after economic globalization was pressed under US Presidents Reagan and Clinton, the consequences for the developing world were terrible. Recognizing this, the World Alliance of Reformed Churches issued the Accra Declaration, which noted the convergence of suffering of people with damage to the natural world, locating the cause in an unjust economic system.

Donella Meadows has expressed reservations about the global trading system. It is a system where the rules are designed by corporations, and run by corporations for the benefit of corporations. The corporations have freed themselves from many local and national controls, to the advantage for profit and detriment to the environment.

This system limits feedback from anyone else. Meetings are closed to reduce feedback. The lack of local feedback weakens communities. Aspects of globalization tighten inter-system linkages and hierarchies. This tends to distance resource users from the immediacy of their dependence on systems. This also weakens feedback loops that are essential for responding to adjusting services. Increased interdependence can make the network more fragile. The lack of global feedback is a severe problem. We do not observe ecosystem responses yet to consuming the capital of local ecosystems or the global ecosystem, since we can always trade for more resources from somewhere else. The biosphere that supplies resources and services, as well as absorb wastes (materials and emissions), is limited by amounts and rates. That ecological limit is a limit to consumption, which is related to technology, population, and wilderness. Lack of ecological feedback is what allows limited resources, such as forests or fisheries, to continue to decrease in price as they are destroyed. The stock is not increasing, but technology is more efficient and economic rules are bound by their two-year horizon—and new global rules allow better access to remaining stock. Living on the capital of the entire planet, we feel we can transgress ecological limits indefinitely, or at least until the capital is exhausted, then we can go on to the rest of the solar system.

This system forces nations to race to the bottom, so to speak, weakening social and environmental safeguards in order to compete in the global market. Sturm and others have noted that 44 countries consume approximately one-third more ecological services than their capacity can provide. Thus, the global economy is poorly positioned for competition in the future. Significantly for their study, 16 of 20 eco-efficiency leaders are competitive, so overconsumption offers some short-term advantages. Some nations, such as Canada, are positioned well, due to their ecological remainder potential. Other nations, such as Japan, are competitive because of their efficiency, and because they save their own resources.

Many of the modern trade agreements are causing suffering and violence. The gross domestic product of the world increases, but what is produced? Who produces it? Eighty percent of the lowest produce only twenty percent of the global output. Globalization is a set of factors that offer advantages to traders or companies, but can weaken individual states (or possibly deterritorialize them).

Globalization disconnects identity from location. Globalization can de-territorialize culture, which has already happened to some cultures, through a disguised form of colonialization. Globalization presents the myth that cultures, through reterritorialization, can take root in new locations and develop new foods and new dances, without being bound by place-based foods, songs, or views.

The speed of globalization is a problem, especially for traditional cultures with traditional economics. When the speed is notably faster than traditional transactions, the culture can become less stable. With the globalization of trade and the domination of national languages, smaller cultures are being overwhelmed and absorbed into larger ones. Another concern with globalization is homogenization, as cultures shift to global materials and styles. Globalization could still occur without homogenization, but that would require respect and limits. Diversity and globalization are not totally incompatible.

Considering the clueless independent actions of corporations, the speed and novelty of the shifts of scale to a global system, with its social damages to communities and wealth, and the economic destruction of ecosystems for profits, the likelihood of some kind of collapse is growing. And, recovering from any collapse is going to be more difficult with globalized cultures in a completely occupied planet. We need—obviously—to rethink how we can redirect or slow these accelerating trends.

2.8.4.2. How can we Think of Globalization and Localization?
Globalization allows connections to other cultures and forcefully extends relations with them. But, we do not seem to have the moral development to live mutually and fairly with these other cultures, often constrained or limited by luck, geography, resources, or history. So, we take unfair advantage of them.

It is difficult for many people to respond to improve their lives or protect their places and ways of living. As local people become embedded in regional, national and international levels of relations, the number of connections and the scale of networks changes. Although some new information is shared between groups, much of the local information pool remains local. People became less knowledgeable about the distant center. Centers grow larger and dominate their peripheries, pulling them into a larger economy with less knowledge and power to act effectively.

Economic globalization allows design information to flow faster between nodes on a global scale. Consumption patterns can be recognized at a global scale. Globalism sucks values and people from their local context into the global context. Or perhaps destroys much of the local context, releasing people to the global. Capital becomes globalized. The economic nodes host exchanges that can bypass the control of sovereign nations.

Globalism allows most people to stay in place, but they still experience the displacement that globalism brings to them. Most human social existence continues to be local, of course, due to physical embodiment and attachment to home. Neighborhoods are based on proximity to start with.

Circumstances are more difficult than ever. Globalization stamps its undifferentiated image on the world. Traditional town based industries have largely disappeared as technology increasingly frees us from ties of place. The individual freedoms of the private car have had a high cost to the quality of the places where we live.

There are still many questions about globalization. Is globalization a universalization of the potential of one international society? Is international society just an aggregation of national societies? Are those just an aggregation of communities and individuals? Is aggregation the right word? No. Nations and communities are not thrown together. They develop in place through mutual concerns and shared values. Does globalization screw up the categories that we use for human behavior: Economic, political, social, environmental, or cultural? Global capitalism undermines traditional cultures by offering consumerism in the place of guides for behavior. Social roles seem irrelevant by comparison, if the good life can be bought without effort. This leads to a new kind of proximity. Inequities are made more visible. Threats to a shared environment become more visible. Dangerous conflicts are brought closer, which might set the stage for a violent clash of cultures. How does globalization react to Islamic militants? How does it react to sea level changes? How would it react to a possible collapse of China, Europe or the US?

An unquestioned, one-world, planned economy could become a great threat, if it continues to be based on unlimited industrial production, unlimited commodity consumption, increased exploitation of nature, and the free flow of resources and labor across cultural borders. This kind of planning requires the abandonment of local controls on development, trade, or lifestyles. All countries are expected to open their markets to outside investment, eliminate tariff barriers, reduce government spending (especially to the poor), convert small-scale, self-sufficient farming to agribusiness, and open all land to resource gathering. Planning is thus characterized by a utilitarian globalism that denies value to the systems that support it.

Localization is problematic, also, with its share of pluses and problems. Even the distinction of local can be fuzzy as centers at the local level are not recognized as local, unless each locale is the largest center of specialization, such as carriages or cheese-making; this seems to be a problem of focus and scale. Eventually, as a result of globalization, changes are made in behaviors and in public health strategies, from cleanliness and sanitation to isolation and inoculations, as hosts and agents both adjust. People in a strong culture may resist, which may revitalize the local. Local designs can communicate the identity of the place on a global scale. Terrorism, organized crime, and other emergencies will occur as often as other conflicts, but can be minimized in smaller nations.

The processes of globalization and localization have similarities with other binary or complementary processes, such as ektropy and entropy. Entropy is the running down of all systems. Ektropy can build order in a system but it produces and increases entropy in the surrounding systems. The processes cannot be separated. Globalization cannot proceed without the process of localization. And, it is likely that localization cannot exist now without some global exchanges. Both are evolutionary steps in human history. There seem to be tipping points now, especially related to runaway economic forces, that are favoring globalization. It is not hard to imagine tipping points that would favor increased and stronger localization; these points might rest in climate change and the scarcity of local resources.

Diversities are essential for human stability, capacity and resilience. Localization can suppress diversity; at first, globalization increases diversity, but then it works to suppresses many diversities. Human societies, like climate cycles and ecosystems, are open systems that have chaotic and complex dynamics. Localization often works to close a society. Globalization can make them more open for a while, but then the process disrupts society.

2.8.4.3. How can we Balance Localization & Globalization?
During the collapse of some regional centers, such as Rome, monasteries provided an opportunity for a different way of life. They provided isolation and opportunity for innovation and the security in uncertain times. Centralization was no longer practical. Localization became much more manageable and useful for people dispersing from a center. The recovery of isolation and self-reliance in small communities helped local cultures to continue.

We need to challenge our fascination with the techno-optimistic 'gigantisms' and question the need for a single set of global trading rules. Increasing localization counters the threats of globalization to destroy local identities and homogenize cultural lifestyles. It can lead to knowledgeable governance and can respond faster to environmental destruction. Local institutions should anticipate global problems and react accordingly. They can assess sustainability at the local level, as well as increase awareness of restraints and impacts of the global level on local levels. Within a more reasonable global framework of economic governance, with new rules based on better principles, local institutions could have the power to choose or refuse global connections (perhaps a form of deglobalization). Local institutions can demonstrate that diversity and high-value yield more profit in the long run. As globalization has accelerated the speed of transactions, local communities have worked to reduce speeds. The citizens of Bra Italy promote slow food. Italian municipalities have created a slow cities cause. France favors slow economics, which boasts fewer working hours, longer vacations, and job protection. European Union wants work councils to protect local products.

This is one good local answer to globalization, but it seems based on community anarchism more than a national movement to limit globalization to those things that it would be good at, and only to those things.

Institutions at the global and local level could monitor changes on different scales of STEM that might affect local sustainability. Global institutions could set standards for education and health; Global institutions could monitor all the major interconnections between systems at the level of the planet and coordinate the reporting at all levels of ecological accounting.

2.8.5. *Tourism & Virtual Reality*

People, even scientists put strains on animals and plants that hurt their ability to be healthy or reproduce. Capture causes stress and sometimes weight loss. Observing alters behavior in many. For instance, penguins delay returning to their nests after foraging, if tourists are in sight. The fledglings tend to have a lower weight and possibly survival ability. The heart rate of the penguins may start to double; they become much more vigilant with nervous behaviors like shivering, swallowing and beak dipping.

Bears, dolphins and dingoes can become disturbed by tourists and display odd behaviors—over vigilance may affect reproductive success. Owls and fox that are treated in wildlife hospitals seem to have shorter life expectancies afterwards. Bats, birds and wolves are often the object of study. The metal rings on bats, for instance, tend to warn moths that a bat is near; this reduced feeding efficiency and effectiveness. Metal bands on penguins can affect their ability to swim and catch fish. Banded penguins tend to arrive later to the breeding colonies. Animals that have been tranquilized and caught, wild sheep for instance, suffer reduced fighting skills and may lose to previously subordinate rivals. Many turtles fail to nest on a beach with tourists, returning to the water. The Florida scrub jay gets spoiled by peanuts, which are okay for adults, but the chicks can be malnourished and stunted unless the parents feed them grubs, also.

Tourism sometimes results in more interactions and pollution. Many moose and deer are killed by cars. Some animals are more susceptible to pollution. Gulls that ingest lead have trouble balancing and may fall over. Atrazine makes deermice more aggressive. PCBs retard the learning abilities of some monkeys.

Tourism also influences native cultures. It may distort the importance of some behaviors and ceremonies while ignoring others. It may artificially increase the demand for some kinds of art or souvenirs, which may interfere with traditional economic transactions—for instance, because the Masai do not sell shields or ceremonial clothing, the neighboring Bantu offer 'authentic' Masai items dor sale. The sheer numbers of tourists sometimes put stresses on water and other local resources.

Other forms of tourism might be acceptable and allow the recovery and isolation of wild systems. For instance, Remote Tourism is popular at some parks in South Africa and other places. Tourists can monitor cameras around water holes and at gathering sites, such as trees, to watch the animals feed, drink or interact. Virtual Tourism, where tourists play realistic games based on real sites, could easily be expanded with new software and massive computing storage. Although it might seem artificial in some ways, the interactions can be speeded up and made more interesting than real-world ones, which often require a lot of patience and disappointment. Of course, the attractions of virtual reality may distract attention from real ecological and cultural problems, especially the extinction of older languages and endangered species.

2.8.6. *The Maintenance of Civilization*

What is the relationship between amounts of complexity and amounts of design? Less is more, but so is more more.

In an ecosystem, the energy required to maintain the ecosystem is inversely related to its complexity; succession decreases the flow of energy per unit biomass until the system reaches maturity (Margalef's concept of maturity). In a mature forest, for example, almost 100 percent of the energy is required to maintain the state of the forest. Any system formed by reproducing and interacting organisms must develop an assemblage in which production of entropy per unit of information is minimized, that is, waste is minimized in cycles.

The same relation seems to apply to cities or civilizations. The simplest economic transactions were between individuals who gathered food or made tools and then traded. The number of artifacts seems to increase as the populations increase. The number of artifacts seems to increase as the complexity of a culture increases. Materials can be used to express power. With the increase in specialization and complexity, came individual traders, then guilds, and finally corporations. A complex (or mature?) civilization tends to use 100 percent of its energy to maintain itself.

In every case where civilization has become more complex, according to Joseph Tainter, the cost of maintaining that structure has required an ever-higher percentage of income be set aside for maintenance; that income is no longer available to increase the standards of living. Modern civilizations have avoided collapse only by overusing fossil fuels to hide or override their true costs.

The ecological design of civilization has to address the true costs of maintaining levels of control and luxury. Ecological design has to be the creative modification of ecosystems to repair or enhance their ability at self-organization and the maintenance of their complexity and diversity. But, most of all, it has to become capable of reducing some forms of complexity and reducing improper uses of energy. Design has to fit human exploitation within the limits of a changing planet.

2.9. *What is Beyond Design?*
(Being edited)

3.0. Naming Nature for Design

Nature is a name we assign to a complex of things that occur in the dynamic planetary system. The system can be described as a set of interacting spheres—the geosphere, the atmosphere, the hydrosphere, and the biosphere, which interpenetrate and interact. The system has specific characteristics that have developed over billions of years: The atmosphere, for instance, has a unique composition, which is kept that way by the cycling processes driven by the sun and equally by a diversity of living forms living in complex, changing ecosystems. As a result of these activities and the composition, the planet has a relatively cool temperature.

The environment of the planet, that is the solar system and extrasolar space, is relatively isolated from local high gravitational influences as well as from intense radiation. However, it is also dynamic and provides many kinds of surprises and challenges, from the shifting of the planetary orbit to interstellar gas lanes and colliding space objects.

The biodiversity of species and living environments has to adapt to the planet and its larger system to survive. The very slow changes from plate tectonics or climate shifts are also challenges, as are medium and large collisions with planetesimals. We tend to call these events catastrophes if a large number of species or families of species are driven to extinction. We think of them as anastrophes (positive turnings) if the environment is stable for long periods of time. Thus, life suffers mass extinctions, but responds with larger equilibrium levels of diversity.

Energy from the sun, combined with material cycles and living ecosystems, keep the global system in a mature high-energy state. Can these cycles and ecosystems be designed? Can we design forested systems that keep the cycles going? Can we do that with animal patterns or wilderness patterns? We have to try, because we cannot invent wilderness from rock, nor can we drive global material cycles with a technology of nanobots and computers. The designs may have to be limited to anticipating large events and disrupting them, to limiting human use and restoring ecosystems, to shaping human impacts and managing in partnership with nature (with benign neglect as often as possible).

3.1. *Design Factors: Elements & Materials*

Natural materials have been created and broken down for millions of years. Exotic materials, from human industrial processes, are different. In most cases, there are no predators or recyclers that can eat or break down the materials.

3.1.1. *Elements*
An element is a unique form of matter that cannot be broken down into other substances; oxygen and nitrogen are elements. To analyze a rock, it is necessary to break it apart and identify specific elements. There is an external supply of materials for the planet: Elements, such as carbon or water, are obtained beyond the boundary of the system in question. Elements are stable holons, from particles and molecules to organic molecules, DNA, and societies. Elements can exist in mixtures, solutions and suspensions, as well as in compounds, where two or more elements are chemically combined in definite proportions. Water and ammonia are compounds.

We need to understand that our local neighborhoods are linked in many ways by atmospheric and hydrological cycles, as well as by elemental cycles. Each of the continents is unique and takes a part of the cycling of elements. Africa receives 12% of the water from global hydrological cycle, whereas the United States alone in North America gets 33 percent.

A material ecosystem recycles important elements in circular paths within the system. Elements frequently associated with life—calcium, phosphorus, iron, and sulfur—are preferentially concentrated and deposited by living organisms. Chemical elements, especially those of life, circulate in the biosphere in characteristic paths known as biogeochemical cycles. The elements circulate in the biosphere in regular, more or less circular paths called biogeochemical cycles (bio=life, geo=earth). The movements of the chemical elements necessary for life are referred to as nutrient cycles. When one of these cycles varies, the other cycles are effected. Nitrogen and sulfur cycles, for instance, are effected by air pollution. In a mature forest very little material actually leaves the forest. It is held in cycles.

Chemical elements, especially those used by life, circulate in the biosphere in characteristic paths known as biogeochemical cycles. Very little is actually lost to space, but often the elements concentrate in sinks, where they may be unavailable for ecological or geological periods of time. Forests, for instance, act as sinks for carbon; the ocean bottom acts as a sink for phosphorus. The rapid release of sinks can affect other atmospheric or terrestrial cycles.

All chemistry is a matter of the electrical behavior of atoms. The planet started as a mixture of chemical elements (over a hundred, including hydrogen, helium, and nitrogen) circulating over an active geology, driven by solar energy. At the risk of making life seem too simple, complex molecules formed from the action of light on primary gases, such as methane (CH_4) and carbon dioxide (CO_2). Materials on the earth include elements, compounds, and combinations. The atmosphere is composed of elements and compounds, such as nitrogen and carbon dioxide. The ocean is composed of hydrogen hydroxide (water, of course), with many dissolved elements, such as carbon dioxide, gold, and sodium chloride (salt, of course). Materials can be extracted from places in the landscape: Metal ore from rock, wood from trees, oil from tar sands, or stone from formations.

Elements are molecules or atoms that make up the spheres of the planet, often cycling throughout the biosphere, atmosphere and ocean, changing the factors and conditions of the environment. Heterogenous elements are aggregated and linked self-consistently in an ecosystem, such that they participate in the functions of plants and animals. Organisms and communities are limited by elements and physical factors, such as gravity, light, water, gas, or salt. Living organisms produce food from sunlight, using various elements and compounds. Other organisms, bacteria and fungi, consume organisms.

Complexity emerges from the interpenetration of processes of differentiation and integration, processes running simultaneously from top and bottom and shaping the hierarchy from both sides. Microevolution generates macroscopic conditions for continuity and macroevolution generates microscopic elements for processes. Elements can be so complex that Jeremy Bernstein suggests that we cannot understand some of them, such as plutonium, which has six crystal phases between room temperature and its melting point.

3.1.1.1. Pure Elements

An atom is the smallest particle of an element that can exist in a compound. Atoms rarely exist alone. They can also combine to form molecules. Atoms can store energy in their electron orbits. Atoms are made up of smaller particles, such as neutrons and protons, which are made up of other particles. Hydrogen is the smallest atom, with only one proton. Possibly, there are 4×10^{79} atoms of hydrogen in the universe, based on the number of stars and galaxies, but not black holes or dark matter or dark energy. Ninety-two atoms exist, but in millions of quantum-mechanical energy states. Another hundred unstable atoms may be temporarily formed during stellar processes or in human experiments.

Hydrogen is the most abundant element in the universe. It is also a resource for human civilization. Any resource is abundant, renewable, slowly accruable (Basically nonrenewable), or Slowly dispersed (really nonrenewable).

Calcium is a poison. Calcium can also be formed by the weathering of rock. Living organisms take it up, putting it in shells or bones. In an undisturbed forest the calcium cycle is relatively efficient, with little leaving the system by flowing downstream (Bormann and Likens, 1967).

Sulfur is also present in gaseous form, in dimethyl sulfide and carbon disulfide. Sulfur exists in sulfur bearing rocks, which are weathered into the oceans. Marine organisms use the sulfates to build strong bodies, which then emit dimethyl sulfide (the aroma of fresh fish) into the air. Sulfuric acid droplets form the nuclei needed for condensation back to the land, where it is available to plants for growing. The large sulfur cycle is beneficial to marine and land organisms, important to the general pattern of production and composition.

Phosphorus is a relatively rare element. Asimov calls it the bottleneck of life, a good example of Liebig's law of the minimum, which states essentially that something always has to be in least supply. Phosphorus is a necessary constituent of protoplasm, but it is sometimes the element in least supply. Most of it is locked up in phosphates in rock, which is weathered and it "escapes" to the oceans. Sea birds play a large part in returning it to land. In fact, we mine guano deposits for fertilizers for our fields. Harvesting fish for food returns some to land, but we make little effort to recycle it.

3.1.1.2. Pure Metals & Machines

The availability of natural resources, such as minerals, is a common concern. We have had to search over the earth for them. In fact, the word metal is from the Greek meaning 'to search for.' Metals can be molded by fire and technology. Because of the exponential growth of technological exploitation, more primary metals have been consumed since 1940 than in all human history before then. When a resource, especially a metal, is in generous supply, it is used for many nonessential and trivial purposes. Thorstein Veblen described the use of waste as a sign of social success as "conspicuous consumption." When the metal becomes scarce, then it becomes more valuable and its uses are curtailed; it is then worth recovering from dumps. The Club of Rome assumes that metals are consumed like vegetables. Although the mines become exhausted, the metals can be recirculated. Some urban refuse is often richer in metal content than natural ores currently mined.

The Spanish extraction of silver from South America had considerable ecological consequences. The Romans had traded their silver to India for spices, jewels and textiles. Silver mining in Europe degraded the environment with air and water pollution, and deforestation and erosion. That boom started to decline by 1530. When the Spanish moved into Bolivia to exploit richer deposits and cheaper (slave) labor, they displaced the pollution there, along with destruction of wildlife and land. This kind of exploitative economy only 'worked' as long as there was silver, after which the frontier moved to Mexico.

Silver mining was one of the first commodities on the global frontier, and it also gave capitalism a boost, by turning wealth into capital. It also set the tone for the rapacious expansion of capitalist sequential overexploitation, and it heralded the new attitude towards nature, as something that could be transformed. Although the population increased, the purchasing power of the population increased more significantly, aided by the influx of gold and silver from the Americas.

Lead was often mined with silver. It is a malleable metal used for its properties of killing bacteria and fungi. The Romans used it in wine casks. Thomas Midgley Jr found that when injected into internal combustion engines it stopped knocking. It helped paints to be spread and lead pipes not to leak. As a result of the scale of use, lead has become a common air pollutant. It escapes from smelting and manufacturing processes. Most intake is with water. Lead poisoning causes brain and nervous system damage, and impairs mental abilities.

Early oceans were toxic, with concentrations of chromium, copper, iron, lead, and zinc. Metals are catalysts that can speed chemical reactions. Zinc is used by organisms for growing cells. They get zinc by ingesting wheat, whole grains, peas, fish, meats, and poultry. Without zinc, growth is retarded, and there are problems with anemia and healing wounds. In the North Slope of Alaska There is some copper, molybdenum, lead, zinc, and silver. Iron, at 10 million metric tons, and zinc, at 24 million metric tons, are the most common.

3.1.1.3. Relational Elements

We know that land-use and land cover changes are major elements in affecting the global carbon dioxide cycle, both the source and sink. Renewal occurs through a dissolution and recombination of the elements of a system. The system maintains its identity, despite the replacement of elements, molecules or organs. Humans are a geological force in the history of elements. There is movement of minerals, but also disruption of mineral cycles. There is the addition of novel elements into the atmosphere, but there is also a massive release of carbon.

3.1.2. *Molecules & Materials*

Novel elements can be created by humans and introduced into the biosphere. The introduction of novel elements into the biosphere—elements that have not been added slowly over time as the result of natural processes—can affect the operation of natural cycles.

3.1.2.1. Inorganic Materials

Inorganic molecules do not contain carbon. Inorganic molecules usually come from mineral sources; this includes metals and inert gases. Inorganic molecules, such as silicon, iron, and titanium, are often used by human technologies for special purposes.

Carbon dioxide is the second raw material necessary for photosynthesis. Forests not only assimilate CO_2, but CO_2 concentrations in forests may be three times higher than "normal" air. The higher CO_2 content is thought to contribute to undergrowth and seedlings growing in low intensity light under the canopy.

Impurity refers to a state of mixture. Impurities are often taken up by living organisms, who then have to separate the nutrients from the undesired elements. Volcanic action can produce sulfuric acid that can injure trees—smelters, cities, and fires also produce toxic substances.

3.1.2.2. Organic Materials

Bacteria photosynthesized away their time for a billion years, producing oxygen and organic matter from carbon dioxide and methane. Forests built soil and then protected it. Trees assisted soil formation with the addition of organic matter and the activity of roots. The original vegetation cover of the planet was extremely diverse. The potential natural productivity of wild vegetation may be 120 billion tons of dry organic matter per year. But, not all of that is available for human exploitation. The proportion of NPP above ground could only be 80 billion tons. These are all rough approximations. This is the theoretical amount that would be produced by wild vegetation with no major environmental changes.

Organic matter is a result of: Living tissues, such as roots or fungi; soil biomass, such as microbes and earthworms; leachates and exudates, such as sugars or enzymes; litter, such as twigs and leaves; coarse woody debris, such as logs; and, organic compounds, as a result of litter decay—this humus can be over 90% of the organic matter. Organic matter has the effect of gluing soil particles together. Wind and water erosion are both usually reduced. Organic matter in the soil acts as a reservoir for nutrients and possibly a sink for carbon.

Matter also participates in cycles—not gaseous, but sedimentary, since the reservoir is the earth's crust and not the ocean or atmosphere. Oil, coal, peat, and some woods are functionally nonrenewable. Geological time periods are required to produce them.

3.1.3. *Elements As Resources*

Virtually every material cycles through ecosystems, such as a forest: Nitrogen, carbon, phosphorus, potassium, calcium, sulfur, magnesium, and water. Nutrient cycling involves many of these materials. Nutrient cycles change with the succession of a forest. Table 313-1 shows the fluxes and accumulation rates of several nutrients needed to maintain a growth rate of 20 cubic meters per hectare per year in Corsican pine in Scotland. Note that half the nutrients for new growth are drawn from senescing needles and twigs.

The minerals and elements have been concentrated by plants so that those necessary for life are more abundant: carbon, hydrogen, oxygen, phosphorus, potassium, nitrogen, sulfur, calcium, iron, and magnesium. Some of these elements, such as phosphorus and nitrogen, are not abundant in rock and may be available only in soluble forms (which requires water); they are available for growth only as they are cycled.

Table 313-1. Nutrient cycling in Corsican pine (after G. T. Miller)

Elements (kg/ha/yr)	N	P	K	Ca	Mg
Overview					
uptake	69	6.0	28	34	10.9
accumulation in trees	18	2.1	11	12	3.6
litter fall	51	3.9	11	21	4.3
crown leaching	trace	trace	6	1	3.0
accumulation in humus	12	1.5	1	5	1.6
Sinks					
in new needles	92	9.4	45		
to replace lost structures	20	1.7	4		
net structural increase	26	3.0	13		
Total requirement	138	14.1	62		
Sources					
uptake from soil	69	6.0	28		
reabsorption from needles	61	7.2	32		
reabsorption from tissues	8	0.9	2		
Total	138	14.1	62		

Materials, often in the form of ions, are released from litter and soil depending on their chemical characteristics. These are then absorbed at different levels as they percolate through the soil. Each species of tree has a unique propensity for taking up different soil solutes.

These materials become involved in cycles of use, release, and reuse. Some cycles, such as the leaching of cations (negatively charged parts of molecules) from leaves, then being taken up by roots and returned to leaves, are completed quickly (under a month); other cycles, such as the deposition of carbon in coal, are measured in geological time.

3.1.3.1. Nutrients

Removing even single trees from a forest ecosystem can diminish nutrients in the system and affect or disrupt cycles. Other aspects of logging operations, such as skidding and clearcutting, can have an effect on nutrients, because of resultant exposure, compaction, erosion, and added chemicals.

The interactions among nutrient elements influences the nutrient cycle. Perry notes that divalent cations, such as calcium or magnesium, enhance phosphorus uptake by mycorrhizal and nonmycorrhizal roots. The concentration of soluble phosphorus in the roots determines how fast and how much is taken up. Phosphorus combined with calcium (for instance) forms insoluble granules in cells, so more can be taken up and stored (the sheath of ectomycorrhizae increases the number of cells that can be used for storage.

Phosphorus, for instance, is crucial to life—Isaac Asimov suggests that phosphorus is the ultimate bottleneck for life (remember Liebig's law of the minimum). Every living cell requires phosphorus. Why? Energy. Energy is required in all life processes, and all biological energy is converted from solar energy by green plants into chemical energy. Phosphorus is a key substance in the conversion; three molecules of phosphoric acid combine with one molecule of the base adenine and one molecule of the sugar ribose to form a nucleotide, ATP (Adenosine TriPhosphate).

3.1.3.2. Bound Energy

Energy can be bound in electron orbits, as well as in fats and lipids. Therefore, molecules, as well as plant and animals species, can be treated as resources. Organic energy can be released by consuming or burning. Fossil fuels are extremely concentrated organic forms. Some nuclear processes can release energy through fusion or fission. Almost all energy on the planet comes from the sun or from the original energy of the formation of the planet; energy cannot be recycled, but it can be trapped for long periods of time in various forms.

3.1.4. *Duration & Flow*

Stars can exist generally from one million to 10 billion years. Chemicals in interstellar space are linked to star formation. There are simple compounds like water, but also polycyclic aromatic hydrocarbons. Some molecules tumble in the form as gas or dust. These molecules can last weeks or years. The lifetime of a hydrogen atom on copper is much less than one nanosecond. Oxygen has a metastable state as a singlet. Because hydrogen atoms shuffle from one water molecule to the next, each pairing only lasts about a millisecond. The lifetime of a water molecule in DNA seems to be about one nanosecond. The lifetime of an excited system is usually shorter than a relaxed one. An mRNA molecule in a cell may last 30 minutes to 50 hours.

Water vapor in the atmosphere may only endure for 3 to 7 days. Carbon dioxide molecules can last from 100 to 500 years in the atmosphere. Proteins are long chains of amino acids. Proteins are polymers where the same molecular components are used again. The functions of proteins are connected to their practically infinite three-dimensional shapes. Sugar molecules have long chains, with branches and networks. Lipids are organic molecules like fats and oils; they are hydrophobic and will not dissolve in water. Cells have shapes held together by proteins. Singles cells can move or be part of an organ. Nucleic acids, such as DNA or RNA, are polymers with long base chains that carry information. DNA is the code from which organisms make new organisms. RNA moves the code from storage to execution at various times.

Species can last over a million years. Ecosystems can last millennia or millions of years. The Amazonian rainforest may be 11 million years old as a biome. Tree longevity can be up to 4000 years. Root longevity is often less than 1 year. The Great Barrier Reef ranges in age from 2 to 18 million years, although in its present form, it is only 8000 years old. Reef ecosystems may only last hundreds of years. Although the Sahara became a savanna about 10,500 years ago, it has been mostly dry desert for at least 22,000 years (and there is evidence of deserts in the area for 7 million years). Long-lived animals may rebound more slowly than short-lived. Some species can rebound in 5-7 years, while others may take much longer.

Ecosystems can rebound in 50-100 years, depending on the magnitude of destruction or time since collapse. Since the natural dynamics of ecosystems includes extinctions and invasions. Ecological patterns can be almost 4 billion years old. Natural processes—building up/breaking down, development, disturbance]—animal movement, inter-element flows, human interaction, shifting mosaics, operate in biomes, ecosystems and communities. Ecosystems and organisms are always interacting and changing. Conditions in an ecosystem are limiting factors that determine speciation or extinction. James Lovelock hypothesizes that every element in the system is related in a feedback network to every other element. For example, the biosphere can control the temperature of the surface and the composition of the atmosphere. On the other hand, soil types and the weather can limit vegetation; invasions of vegetation change soil types.

When you think about it, there are many keystone species in a forest: Trees, squirrels, millipedes, and mycorrhizal fungi. In a way every species is a key to the whole. Some species, such as mountain lions, can be removed from the forest with little gross change, but there are changes. Any species is an expression of variety, a niche maker—that enriches the ecosystem and expands the habitat for others—and locus of feeling. Any heterotroph, according to Eugene Odum, that consumes autotrophs and excretes matter (with inorganic ions) contributes to the circulation of nutrients and minerals. Furthermore, any being contributes to the energy flow of the system and is a link in the web. Although an ecosystem can survive without a species, it is reduced accordingly.

The kinds of resources and the possibilities of using them in production are considered in the scope of economics, as are flows and stocks in homes and businesses, the role of the government, business cycles, monetary details and policy, stabilization and growth, international trade, consumer behavior, production costs, pricing, and resource markets.

3.2. *Global Design Factors: Events Processes Flows & Connections*

Processes are responsible for distribution of elements. This results in renewal. Processes range from wind and fire to erosion and succession. Most all processes require energy acting on materials. The sun is the source of energy for most processes, but we have to recognize that the human use of energy has consequences to the natural systems, especially when too much energy is dumped into a system in terms of waste heat.

3.2.1. *Processes*

During the history of the planet, solar radiation 'worked' on the gases and compounds of the early atmosphere of the planet, which was composed of carbon dioxide, nitrogen, sulfur, methane, and other elements. Gradually, complex molecules built up from chemical processes driven by solar energy, the radioactivity of the earth and the geological processes set in motion by the formation of the planet.

A. N. Whitehead suggested that process is primary; living beings are self-organizing systems like vortices in an energy flow. Continual change is fundamental in renewing ecosystems. Nature is both an ordering phenomenon and a series of entropic processes, which result in hierarchies, where the system adapts to its medium and the medium itself becomes ordered. One thing in nature truly developing is the largest structure, the total system, the region or planet, which imposes constraints on entities and subsystems, such as ecosystems and communities. Natural processes are dynamically balanced; some take thousands of years, but all effect one another synergistically. The character of an ecosystem is maintained by a large number of processes. Natural processes, such as fire, wind, or species explosions, would be allowed to operate freely, even if they altered the functioning of the system.

Processes like homeorhesis occur on an ecosystem level (living systems do not display homeostasis—constant value—so much as a particular course of change in time—homeorhesis. The course is stabilized, not the constancy, according to Conrad Waddington. Changes to a system are symbolized by trajectories in a multidimensional phase space or landscape).

Communities and ecosystems are part of a process that is unending and imperfect, without a final state, and furthermore, that the attempt to perfect it results in disharmony. Ecoforestry accepts a constructive conflict in scale with the ecosystem. There are chaotic events, plagues and random frenzies in every system. Odum describes the ecosystem as a unit of organization undergoing an orderly process of development that is reasonably directional. Ecosystems, the essential unit of ecology, must be seen in dynamic and historical terms.

There are processes in organisms, also, physiological processes and metabolic ones. Transpiration of trees (a vital process) results in the air in the upper canopy having a higher humidity than in the open—although absolute humidity rises marginally, the relative humidity increases the most depending on the air layer and time of day (highest humidity is late night, when temperature is lowest).

Succession appears to be a process of self-organization occurring in every cybernetic system with the properties of an ecosystem. Process is the same as acquiring information. The ecosystem learns the changes, i.e., seasons, of the environment. Any system formed by reproducing and interacting organisms must develop an assemblage in which the production of entropy per unit of information is minimized. It is a general property of some systems that acquired information is used to close the door to further inflow. Climax needs less information, since it concerns only preservation. Limit of climaxes allows maximum variability between systems with slight external differences, like temperature. Ecosystems consist of different prefabricated pieces: species. The supply of species is limited, so succession becomes asymptotic. Climax is an information ladder; at the top, there is no more information. Margalef suggests replacing word climax with high maturity. But climax conveys as much if not more information than just maturity. However, climax is just a temporary state, dependent on a temporal combination of conditions, such as soil or climate, which are developing all the time.

Successions are stopped by fluctuations, by volcanoes or storms, Margalef calls the process exploitation, adding that its effect is rejuvenating. Whatever accelerates change and energy flow in ecosystem reduces potential maturity. In an exploited system, diversity drops and the ratio of primary production to biomass increases. Mature systems can regress to earlier forms when exploited; then new species can form. Ecosystems are constantly evolving under the influence of physicochemical processes poorly understood and so far more powerful than those that result from human activities. The reality is more complex than just systematic succession and climax. Plant diseases, plagues, floods, volcanoes cannot occur too often. Nature can purify air and water, reseed devastated areas.

Varela states that succession is the result of an indefinite iterative process with a stationary end state to which variables tend. This cannot be true, any more than a shark is an end state in the ocean. Succession in ecology resembles evolution in general biology.

The process by which species reorganize their structures to adapt to the environment is evolution, an integrated, partly open process that selects whole individuals in whole environments. Evolution flows upwards and outwards as well as inwards and downwards, from the simple to the complex, but also back again. Evolution uses different levels: the chemical has generally been replaced by electronic. Plants are generally chemical; electronics becomes important with motion for processing information. But electrical systems outran their design (hunting) in humans.

Evolution runs in a direction of enhanced individuality (genetic autonomy), the dynamic process of generations. Mayr held the view that genes mutate, organisms are selected, and species evolve. Therefore genes are selected only indirectly, if the organism survives to reproduce. But, according to theorists like Gould or Stanley, special groups (populations or species) can be selected.

Evolution is an integrated process; partly open-ended, involving choices, selection of whole individuals in whole environments. The cost of evolution by selection is so heavy that most of the time most populations are not perfectly adapted to changing environment. Evolution is never total adaptation; it requires destabilization; a risk accompanying all innovation, a self-presentation offering new symbiotic relations. At all levels evolution includes freedom of action as well as interdependence. The ultimate principle of evolution does not seem to be adaptation, but transformation and diversification of evolution. The death of individuals and species furthers evolution.

Evolving systems play between adaptation and nonadaptation; completeness of either would result in death of system. Portmann pointed out numerous functionless features of evolution. There is always extravagance and beauty of features. Margalef describes the baroque of the natural world, meaning that there are many more species in ecosystem than would be necessary if biological efficiency alone were an organizing principle. This is the misconception of the fullness of nature. Nature is not full. Plenitude does not obtain in nature because species do not evolve effectively and because of predation, which increases diversity and creates gaps.

3.2.2. *Events*

Western science is basically concerned with regular, reversible events. This kind of science has never been good with unique, irreversible, long-term, complex, catastrophic, or rare events, the kind that occur in historically-unique ecosystems, although some new approaches, such as chaos theory, are promising. There are chaotic events, plagues and random frenzies in every system. Ecosystems evolved through natural events, then as an effect of human activities.

Variations in the behavior of matter and energy enable rare events to occur. In a long evolutionary process, rare events are very significant. Events can be divided into the regular kind, like heartbeats or solar revolutions that occur in cyclic succession, and irregular ones, like earthquakes or mutations. The rarity of irregular events can be measured by a period of probability. Over long enough periods of time, many improbable events are certain to occur. Life is a rare, but very certain event. Given definite conditions and processes, it seems certain to arise. The enormous information capacity of biological molecules makes production by chance improbable. A random process could not have produced even a single organic molecule. But once one rare event occurred, another became possible, and so on, until organic molecules are inevitable. Humans have difficulty observing events on this time scale. Rare events are not really studied scientifically. Rare events are also revolutionary and may be very important. Ecological history has to address rare events. Evolution increases the levels of complexity through the operation of natural events.

The environment is constituted by a large set of events that are objectively definable by their outcomes. We tend to think that forests should exhibit regularity, but many forests, such as boreal forests, are arrhythmic, that is they are punctuated by surprise events (as Holling suggests). Events happen simultaneously. Disturbances in a forest are regular but unpredictable events. Many of them kill trees. Mortality is a normal part of the life cycle. Mortality in forests usually occurs from a combination of factors. A typical percentage of death is a normal condition, necessary for the renewal of the forest. The rate of death per year in an old forest is remarkably consistent at about 1-2 percent, even with wind storms, fires, disease outbreaks, and animal damage.

In the Navajo image of the world, events are primary, not actors (although the world is also personal and orderly). A sacred event may give sanction to a rock or tree. When we try to represent things that are not directly observable, because they are too small (quarks), too large (the universe), or too complex (forests), we use models (from the Latin modulus, meaning small measure). Models can be conceptual or physical; they can be plans, analogies, maps, diagrams, graphics, mathematical equations, or even sentences. Holling uses a graphic model (see Figure 322-1) to show how the release cycle in ecosystems is related to other cycles.

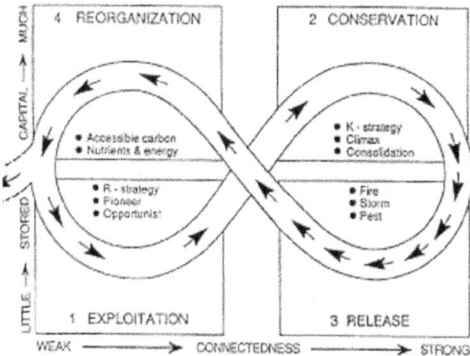

Figure 322-1. A model (after Holling) showing four ecosystem functions and the flow of events between them. The distance between arrows show differences in

the speed of flow. The cycle reflects changes in two attributes: the amount of accumulated capital (nutrients and carbon—the Y axis); the degree of connectedness between variables (the X axis). The exit from the cycle indicates a flip to a different stage, that is, a catastrophe or epigenetic change—see following explanation.

These trends are partly the result of our unconsciousness of large-scale, long-term events, partly the result of out cultural amnesia about things that make us unhappy, and partly the result of our cultivated indifference (doubtless from our remoteness from wild nature).

3.2.3. *Flows*
Ecological and economic processes and values are similar in ways. Both ecology and economics attempt to understand and predict the behavior of complex, interconnected systems where individual behavior and flows of energy and material are important. There are many other common or similar processes: Resource allocation, optimal behavior, adaptation, trophic flows, element flows, and energy flow.

3.2.3.1. Energy
Energy on (and in) the earth comes from two basic sources: From the radioactive decay of heavy atoms near the center of the earth as a result of the original formation of the planet, and from the sun. The energy from the sun is the result of nuclear fusion, where atomic nuclei are gradually fused from hydrogen to helium, oxygen, and iron under high temperatures and high pressures in the center of the sun. The constant energy from fusion pushes its way out of the sun as light, sometimes taking thousands of years to travel the 400,000 miles from the center to the surface—although once past the photosphere it only takes 8 or 9 minutes to get to earth.

Radiation is produced along a spectrum, with peaks depending on the kind of fusion process. Our sun, which is a medium-size main-sequence star, produces much of its light around 5300 angstrom units (yellow)—because life on the planet developed under light from this star, living organisms are very sensitive to this wavelength and in fact our mammalian sight organs have the features just the right size to collect this light. Having developed with other limitations, fish and insects, however, see things differently.

The amount of light hitting atmosphere in the short-term is 2 gram-calories per square centimeter per minute—this is the solar constant (a gram-calorie is the amount of heat required to raise the temperature of one gram of water one degree Centigrade at 15 degrees C; this amount is relatively small, so we use the Kilocalories as a more convenient measure), which varies minutely. This radiation is reflected or attenuated as it hits the atmosphere—on a clear day at noon, perhaps two-thirds of it reaches the surface.

The ecosphere acts as one system in which energy from the sun is cycled. The functioning biomass is integrated by feedback responses to extract the maximum of energy and still maintain a balance. Most of the solar energy is used for maintenance by the biosphere, which is an indicator of the biosphere's high degree of ecological maturity. Chemical equilibrium is a global regulator. The steady effect of light, the availability of oxygen, the thickness of soil, and the area and depth of the oceans is almost a steady state, as is the mean temperature of the planet.

3.2.3.1.1. Laws of Energy
The laws of energy, how energy acts and how much there is, apply to ecosystems. Energy is simply defined as the ability to do work. The first studies of energy had to do with heat and

motive force (thermodynamics). The laws of thermodynamics are as follows: (1) Energy can be transformed but not created or destroyed; energy is not created in the sun, just changed from its state in matter. "You cannot win," Harold Johnson used to say. (2) Energy transformation cannot occur unless some of it is degraded into a dispersed form where it cannot be used again in the same system. 'You must lose.' (3) The entropy of a system in equilibrium is zero at the absolute zero of temperature. 'You end up at zero.'

The first law is important because it means that when we use energy, we are changing its state from free (or useful) energy to bound (or useless) energy; we are not creating or destroying the energy. Einstein's famous equation, $e=mc^2$, related energy to matter (sometimes considered a primal state of energy). Nuclear fission releases energy by breaking apart nuclei; nuclear fusion is a result of a cycle that produces helium from hydrogen.

The second law is important because it tells us the cost of transformation; because of the cost, no transformations are ever perfectly reversible. The third law has some importance for very low temperature research, but is generally ignored by most physicists or biologists.

The laws have been called 'impotence principles' by the biologist E. T. Whittacker. They cannot be proved true as laws, but they limit what we can do, i.e., we cannot use all the energy in a system, and we cannot return the system to a previous state. The second law of thermodynamics states that the entropy of a system increases continuously and irrevocably. Entropy, as R. Clausius defined it, is the constant transformation of motion into heat. As a measure of heat loss, it is valid only under fairly strict conditions, using unlikely or highly artificial assumptions; for instance, the temperature can only be defined for a system in thermodynamic equilibrium, but no system in nature is in perfect equilibrium. It is held to be produced within an isolated system, but no system is completely isolated. Heat, for instance, is lost to a surrounding system, e.g., sent into the atmosphere from the forest.

3.2.3.1.2. Energy & Self-ordering

Life is part of a general order in the universe; stars and planets are increasing in order. Life can be taken to be a property of a thermodynamic system. Life is an open, exchanging system, composed of low energy atoms, that is part of overall entropy reduction of the earth. Life requires entropy as well as order. Life could only survive in a world where entropy increases; it could not survive either extreme—the bombardment of free energy or the complete lack of it. Order is a requisite for survival; it must be inbred through evolution. Entropy is not bad; dissolution is involved in life. Earth evolved from the entropy of the sun and other stars. The drift of forms toward entropy is the generation of a creative disequilibrium vital for change, transformation and evolution. Life requires death to release it from form. Since energy from a static form of life, that is, an immortal, would be unavailable for work, it is equivalent to entropy.

The concept of entropy will continue to be applied to biological systems as well as physical ones. The most important thing to remember about the second law is that it applies to a closed system. Fortunately the earth is not a closed system—energy is flowing through the system all the time. Furthermore, the entropy of one system may be useful to another system. The sun for example, produces light that causes pressure that balances the pressure of gravity. Once the light is diffused from the photosphere, a percentage of it is intercepted by the earth, where it is cycled in various ways. In a way, energy does not cycle because it is lost in equal quantities from the system; however, energy may be trapped by cycles on the earth for millions of years—in coal or oil, for instance. If we were to follow a unit of energy from the sun, it would contribute to the heating of the air, be bound in a tree by the action of photosynthesis, continue on in newsprint, be released in a home fire, be dissipated to the air again, adding a little more heat on its way to interplanetary space.

Energy is bound into organic material, measurable as productivity. Animals and humans are dependent on this productivity. Plant productivity has impacts not only on food and energy, but on the atmosphere, as well.

3.2.3.2. Wind Energy

Wind influences the environment. Wind can also cause catastrophic change. Vegetation holds soil in place, reduces wind speed at the soil surface, and improves water absorption and transport in the soil. Erosion, especially from wind or water, destroys soil and makes it difficult for plants to be reestablished. Recovery, if it occurs, may take decades. Erosion is an ecological catastrophe on a planetary scale, causing thousands of higher plant and animal species, and countless lower species to be lost forever.

Wind is a factor and variable influencing ecosystems. This is true also because the canopy trees change other factors, such as soil temperature and moisture, wind, and air composition. Wind is a climatic factor which can influence forests. Wind influences the shape of trees, the height of trees, the size of leaves, crown shape, and bole growth. Wind can cause continuous stress, branch death, and windfall—which depends on soil type, and age and composition of stands. Exposure determines the amount of sunlight or wind; the solar rays are more oblique on north-facing slopes (in temperate zones), while heat and evaporation are less. North slopes have other advantages relating to wind, drought and thaw-freeze, and dry-out protection.

Forests are dynamic systems that influence wind. A forest exerts direct and indirect influences on many physical and biological factors, from water to air temperature, soil temperature/moisture/nutrients, humidity, precipitation, snow conservation and movement, air currents (wind), and plant and animal life (including humans). Forests moderate the direction and force of air currents, protecting the leeside environment from high, cold, or dry winds. In general all conditions are ameliorated (made better) by trees acting as a windbreak, including reduction of evaporation and transpiration.

3.2.3.3. Water Energy

Here on the water planet (Arthur Clarke suggested calling it Ocean), water is a dominant feature, especially seen from space. For over 4 billion years, water has ebbed and flowed over the planet. One hypothesis is that water derived from the interior of the planet, separating out as the continents started floating up. Another is that it preceded the continents. A third is that it was carried to the planet during early planetesimals bombardment during the formation. In any case, as the continents have drifted around the globe for at least 3.5 billion years, the total volume of water has been relatively constant, although the ice caps have expanded and contracted, causing the sea level to rise or fall.

The oceans are part of the global energy 'engine' driven by the sun. The oceans, along with the atmosphere, by its motion, distributes energy around the planet. The oceans act as a reservoir of dissolved gases, which helps to regulate the composition of the atmosphere.

The action of water has shaped landforms, the shapes of embryos and the organs of plants (for a discussion on the power and spirituality of water, read *Sensitive Chaos*, by Theodor Schwenk). Water also participates in local and global cycles. It comes to the earth in rainfall and leaves through evaporation; it is thrown into the air by volcanic action; it flows into the ocean. By far the largest amount of water is contained in the lithosphere (roughly 20 times as much as in the oceans). The ocean, of course, has a lot of water. The smaller pools, in the atmosphere and circulating ground waters, are the most active and the most vulnerable to disruption and pollution. On the oceans, more water evaporates than returns as rainfall; just the opposite happens on land, causing a cycle of water flowing from land to sea, evaporating and falling back on the land.

3.2.3.4. Elements as Flows

The planet started as a mixture of chemical elements (over a hundred, including hydrogen, helium, and nitrogen) circulating over an active geology, driven by solar energy. At the risk of making life seem too simple, complex molecules formed from the action of light on primary gases, such as methane (CH_4) and carbon dioxide (CO_2). Elements frequently associated with life—calcium, phosphorus, iron, sulfur—are preferentially concentrated and deposited by living organisms. Note that molecular and atomic forms flow everywhere, after being broken down through physical or chemical action. Of course, the ocean carries larger items, such as seeds, wood, and plastic containers. Chemical elements circulate in the biosphere in biogeochemical cycles.

Many gases are made up of elements. Oxygen is a chemical element with 8 protons in its nucleus. In fact, oxygen was poisonous to early forms of life, the methanogenic bacteria. These bacteria photosynthesized away their time for a billion years, producing oxygen and organic matter from carbon dioxide and methane. Oxygen in the air increased dramatically, as the amount of methane fell dramatically and carbon dioxide decreased. Many species of bacteria died off from oxygen poisoning, and others learned to consume it. James Lovelock (the Gaia Hypothesis will be mentioned briefly then in more detail in Section 8.6) concludes that oxygen is poisonous, mutagenic (causes mutations), and carcinogenic (causes cancers), but that it has advantages to ecosystems of organisms: It enables organic matter to be recycled better; and it permits more rapid weathering of rocks, making more nutrients available. After another billion years or so, oxygen topped out at about 21 percent of the atmosphere, possibly because of the increase of forest fires (at 25 percent combustion is instantaneous and unstoppable—all forests and plants would be consumed). Lovelock suggests that the interaction between hardwood forests and softwood forests may act as a global regulator of oxygen for the planet. Fires in conifer forests burn cleanly, leaving little charcoal and decreasing the amount of oxygen (it is combined with carbon); as conifers become more successful, oxygen and forest fires decrease, allowing hardwood forests to increase (and oxygen increases again). For organisms, oxygen allows higher energy use, although oxygen is hard on bodies.

In addition to oxygen, the atmosphere is also composed of nitrogen, hydrogen, and argon. All organisms use hydrogen, oxygen, and nitrogen (as well as the solids like carbon, iron, sulfur, and magnesium) as structural elements, so these elements from through bodies from regional and global biogeochemical cycles.

3.2.4. *Conclusion*

Natural processes, such as fire, wind, or species explosions, would be allowed to operate freely, even if they altered the functioning of the system. After long periods without disturbance, a catastrophic disturbance is more likely—where wind and fire are absent, the probability of insect and disease outbreaks increases. Yet, even catastrophic disturbances like hurricanes rarely damage more than 5 percent of a forest. More than being agents of mortality, insects, diseases, and animals are native components of complex food webs in ecosystems that contribute to the selection of certain kinds (including healthy) and ages of trees (that determines the composition of the forest, which changes over time).

3.3. *Design Factors: Organisms Communities & Ecosystems*

Why should we be considering these factors—ecosystems, communities, populations, and organisms? An ecosystem is a discrete unit consisting of interacting living and material components. An ecosystem is the unit of survival for living beings and communities. Organic structures, are building blocks for organisms for eating or nesting. They also nourish the soil. Ecology deals with the highest levels of biological integration, from organisms to the ecosphere. Autecology, for example, is the study of individual organisms in an environment. A group of individual organisms of the same species in a particular place is studied as a population. The assemblage of populations of different species in a habitat is studied as a community. The community in its biotic environment is studied as an ecosystem. Ecosystems comprise the ecosphere. There are emergent properties at each level of organization, such as the diversity of species and the structure of a food web. Furthermore, each level of integration has distinct attributes.

A living system is produced by living beings. A system is a set of things—such as people, cells, or molecules—according to Donella Meadows, interconnected in such a way that they produce their own pattern of behavior over time. The system may be buffeted, constricted, triggered, or driven by outside forces, but the response is characteristic of the system. A system is not just a collection. It is an interconnected set of elements that is coherently organized to achieve something. The system maintains its identity, despite the replacement of elements, molecules or organs.

Human perception of nature seems to be hierarchical, regardless of whether nature is. Several theorists distinguish levels of hierarchy in nature. G.T. Miller includes particles, atoms, molecules, organisms, societies, and nations. Mario Bunge extends the list to include processes and knowledge: Elementary particles (atoms, bodies); physical systems (organisms, ecosystems); physical processes (chemical, biological, social); material production (ritual, culture); and knowledges (physics, history). Ervin Laszlo makes a distinction between a macrohierarchy, comprised of a space-time field, particles, stars, galaxies, and various aggregations; and microhierarchies like the earth, composed of molecules, crystals, cells, organisms, ecosystems, and Gaia. Hierarchies can be regarded as vertically arborizing structures whose branches interlock with those of other hierarchies at a multiplicity of levels and form horizontal networks; arborization and reticulation are complementary principles in the architecture of organisms and communities.

3.3.1. *Organisms & Populations*

Even if individuals can be described in terms of vortices, as the poet Pound did before the physicist Prigogine, they do exist materially, and they participate in the field because they exist. Organisms are composed of atoms and molecules that energy forms under certain conditions of temperature and pressure, and their emergent behavior is more complex than just "energy vortices" or "patterns of matter." Organic molecules process quickly out of the pattern of the organisms. Although the molecules themselves may outlast an organism, the organism needs a turnover of the molecules. The molecules cycle within the ecosystem and to a lesser extent regional and global cycles.

Atoms that came from stars and rocks make up molecules of seeds, flowers, defecation, and rotting leaves, which are cycled through our bodies. Bodies are open systems exchanging materials with the whole environment. It used to be thought that organic molecules could only be assembled by organic beings. Various experiments have shown that organic molecules, such as urea, can be created by the operation of energy, such as lightning, on mixed elements, such as carbon and hydrogen. Matter organizations are stable holons that are used and reused, from elements through organic molecules (and DNA) to societies.

Each participant creates an image of nature from what is meaningful to it. J. Von Uexkull suggests representing these unfamiliar worlds with a bubble model. The life image, or umwelt, of an animal is what has perceptual and operational meaning for the animal. All animals are fitted to their unique worlds with equal completeness—simple animals to simple worlds, complex ones to well-articulated worlds. Each is optimally fitted to a habitat.

Furthermore, the organism must adapt to the environment, which implies having a memory and being capable of learning, and must reproduce, that is, duplicate its pattern in a separate being. Organisms are goal-seeking, and often stability is sought above change or complexity. The individual is a subject centered in a milieu. Because of this implied point of reference, Rodman concludes that ecology is teleological.

The life-form communities and physical elements are related in a definite pattern, which is a real but untouchable property (structure). In general, this structure becomes more complex as time passes, as long as the environment is stable or predictable. The structure acquires a historical character. Maturation, as a function of historical processes, increases the levels of complexity of an ecosystem. Principles of organisms include:

• The individual is the unit of experience and reproduction.
• An organism is inseparably related to its habitat, the place where it lives.
• The niche of an organism depends on what it does in its place.
• Many organisms identify with place.
• The size of an organism is related to metabolism.

Living sets are self-organized by the actions of their members. Sets consist of matter organized at various levels. Sets of molecules make up genes; sets of genes, genomes; sets of populations, species. Each locality supports a segment of the total species population in a unique context, with a particular set of predators, competition, food, physical habitat.

Principles that apply to populations include:

• The population is the unit that evolves in nature (according Krebs).
• A species population has unique properties, such as density, mortality, natality, potential, dispersion, age distribution, growth form, and structure (isolation, territoriality).
• Populations interact in neutral, positive, or negative ways.
• Competition limits the number of species in a niche (the competitive exclusion principle). Garrett Hardin (1960) states the competitive exclusion principle as: complete competitors cannot coexist. Niches must be different for species. Krebs states that the fundamental niche of a species has an "infinite number of dimensions," making a complete determination impossible. Another difficulty in definition is the assumption that environmental variables can be ordered linearly and measured. Furthermore, competition is dynamic, whereas models freeze single instants.

A group of individual organisms of the same species in a particular place is studied as a population. The assemblage of populations of different species in a habitat is studied as a community. The community in its biotic environment is studied as an ecosystem. Ecosystems comprise the ecosphere. There are emergent properties at each level of organization, such as the diversity of species and the structure of a food web. Furthermore, each level of integration has distinct attributes. A population has a property "density," the number of individuals per unit area, which is not applicable to individuals; a community has "species diversity," which is meaningless at the population level; processes like homeostasis or homeorhesis, which involve a relationship with the environment, occur on an ecosystem level.

3.3.2. *Communities*

Henry Thoreau extended the idea of human community to animals and plants. In a romantic paroxysm, Thoreau proposed that nature was a vast community of equals. The word community took hold and was used to describe associations in nature. Botanists noticed that plants tended to live together in communities.

Plants and animals are always figuring out ways to live. If two plants happen to be living close to each other, they may compete for the same energy and nutrients, or they may start shifting their requirements. Over time, a long time usually, plants and animals that have similar general requirements, in terms of solar energy, heat, water, and nutrients, tend to live together in associations. The plants and animals that form associations in a forest do not usually have the same requirements. Thus, the ninebark beneath the ponderosa pine does not compete for the same nutrients or energy. Often, the plants and animals benefit from the presence of their neighbors, as alder benefits from nitrogen-fixing fungi.

The trees and other plants and animals evolve into a community of thousands of different species. The "checks and balances" of a complex number of predators, prey, and decomposers tends to dampen any one species from getting out of control (and becoming a pest). This is not to say that everyone lives in a disneyesque fantasy of good will. Organisms survive by defending themselves or attacking others. But, the defensive and attack strategies "coevolve" (Ehrlich and Raven's term) over time. Organisms specialize to avoid competing. Relationships become more intimate, as organisms cooperate for survival advantage.

Although botanists recognized that change was inescapable as a principle of the new science, Frederic Clements insisted that change was not an aimless wandering, but a steady flow towards a stable state that he referred to as a final climax. The climax community was thought of as the final state after a series of developmental stages. For example, certain places, with variables of wind and rain and temperature always produced forests; others deserts or grasslands.

Community is one level of a pattern. A community can be described through a number of properties and principles. Properties of a community include productivity and development. Principles of community include:

- Community is the level of survival
- Diverse species live in a stratified order
- Communities replace one another as a result of orderly processes (e.g., succession)
- Communities are named by structural features such as dominant species.
- Communities are stratified.
- Communities have a diversity of species.
- Communities are characterized by rhythmic changes in the activities of organisms, which produce regular recurring changes in the community (periodicity may be daily, lunar, seasonal, genetic, or climactic).
- Communities replace one another in a given area in sequence by an orderly process of change called succession (Succession appears to be a process of self-organization in a cybernetic system at the ecosystem level. It is primary for Odum).
- The final community in a successional series is self-perpetuating and homeorhetic, that is, in equilibrium with the physical habitat, that is, the energy/material budget is balanced in a mature community.

This concept of maturity, as an attribute of a community, is related to structural complexity and organization. Maturity increases with time in an undisturbed community. The species diversity, that is, the information content, of a community also increases with maturity, leading to a more complex spatial structure. Diversity incorporates species richness (how many different kinds are present) as well as a measure of abundance—how many of each, as individuals or biomass. Other aspects of diversity, such as life cycles, are less often considered. The energy in a mature system goes to the maintenance of order and less for the production of new materials. In general, diversity is higher, and life cycles are more complex; symbiosis between species increases, and nutrients are conserved. Complexity and diversity offer advantages for living forms. Complexity allows increases in size, which allows the colonization of harsh environments. Diversity allows more effective behavior through specialization; for example, a specialized organelle may digest less common molecules.

But, Odum points out, as some communities age, Wisconsin forests for example, there is a decrease in diversity (in the understory anyway). Also, diversity can decline with productivity, as in the eutrophication of lakes, for instance. While it is meaningful to speak of an optimum diversity, as the result of limits and the interaction of many factors, a maximum diversity may never be reached.

Conventional wisdom, starting with Charles Elton, holds that increased complexity in a community leads to increased stability. But in the 1970s, work with mathematical models tended to support the reverse, that complexity leads to instability. Robert May constructed simple mathematical models concerned with local stability, in which an increase in complexity lead to a decrease in stability. His connection, however, may have been a mathematical artifact, since his food webs were randomly assembled and sometimes unreasonable. May admits that his arguments are only true of mathematical models and that things "may be different in the real world." Ecosystems are the result of historical processes that are mathematically atypical. Furthermore, real communities are not randomly structured. A system drives to a nonequilibrium state as a mature ecosystem. The adaptively reorganized system is not necessarily more stable, but it is optimally resistant to the outside conditions that elicited the self-organization, a natural normalization process. The ecosystem learns the changes, periods, or seasons of the environment.

Every forest community is composed of a variety of kinds, numbers, sizes, and ages of plants, from unicellular organisms to trees. "Community" is a general term of convenience, like ecosystem, to designate complex units. The largest kind of community is a biome; in North America, according to Packham et al., biomes include tundra, montane coniferous forest, steppe/grassland, and temperate rain forest (for example, Vermont is in the cold-deciduous broadleaved forest with conifers; New Orleans sits in the subtropical summergreen coniferous swamp forest; and much of Arizona is thornbush/succulent). Biomes are then subdivided into regions, zones, and associations. Associations describe plants that grow in the same habitat. The deciduous forest biome, for example, is composed of distinctive associations, including beech-maple, oak-hickory, and aspen-birch. Associations may be subdivided further into layers (or unions).

A community is forced to accept an upper limit, beyond which it cannot grow any further. Further growth results in destruction or disruption of itself and its environment. This is the law of the maximum. Production could be stabilized in a steady state, where processes and cycles are constant.

Human communities are embedded in natural communities. Furthermore, land continuously occupied by humans may form analogs of natural communities, guided by trial and error, by unconscious values, and by random changes. The American tall-grass prairie is a case of the creation and maintenance of an artificial but desirable ecosystem. Unfortunately, it was dependent on a multiplicity of unintended accidents. Ecosystems evolved through natural events, then as an effect of human activities, now through deliberate social choice in some places. All members of an ecological community contribute to the integrity of the whole, which is vital to maintaining what we humans consider important: Visible animals, pharmacological sources, a moderate climate, and clean air and water (in a sense, nature has meta-economic values).

3.3.3. *Ecosystems*

Some ecologists, like Arthur Tansley felt that the word community was too anthropomorphic, that plants did not voluntarily live together. So Tansley coined the word ecosystem to represent the interrelatedness of nature and to emphasize the flow of energy through a system, such as a forest or pond. The systems concept, however, is concerned with more than just a number of elements or their kinds (species); it is concerned with relations between them. These relations are not explainable from the elements, hence, 'the whole is more than the sum of the parts.' The relations are emergent, that is, they are new 'things.' The theory of systems offers unique properties and principles, such as wholeness, growth and openness.

A thing that grows, repairs, holds itself together, oscillates, and capture solar energy is an ecosystem, a system resulting from the integration of all the living and nonliving factors of the environment. An ecosystem is a community of organisms interacting with one another and their environment. Odum describes the ecosystem as a unit of organization undergoing an orderly process of development that is reasonably directional. Ecosystems, the essential unit of ecology, must be seen in dynamic and historical terms.

Ecosystems are ambiguous in a sense, since anything from a log to a watershed can be an ecosystem. Furthermore, ecosystems overlap and interact. It is possible to define close but not exact subsystems. The vast number of interrelationships between systems keeps them open. For example, grassland is affected by climates, soil conditions, fires, surrounding communities, and human agents. Ecotones between systems are usually shifting. Each ecosystem is unique and original. Each patch (locality) supports a segment of the total species population in a unique context, with a particular set of competition, food or physical habitat.

Ecosystems build up information. There are three different channels of information in an ecosystem: a genetic (in replicable individuals); an ecological based on interaction between cohabiting species (expressed in changes in their numbers); and the ethological or cultural, transmitted through individual learning based on experience.

Feedback within the interaction of a species is expensive memory with little storage capacity. Whenever succession starts again, after a volcano eruption for instance, old information in form of interactions has not been saved. Genetic memory has much larger capacity and is long-term. Cultural memory enlarged with higher vertebrates.

Ecosystems are described with a variety of terms: energy, matter, entropy and ektropy, productivity, cycles, diversity, complexity, stability, and trophic structure. Many terms in ecology, such as biomass, stability and diversity are inexact. It is almost impossible to estimate the amount of degraded energy in an ecosystem (that is, entropy from transpiration or mixing of water).

One problem with describing ecosystem attributes is that both quantitative (measurable) and qualitative (conceptual) factors are included. Resistance to external factors, for example, is a qualitative attribute. This may be more the result of succession than a factor producing an ecosystem. All ecosystem properties are not equally important. There is no whole system without an interconnection of its parts, and there is no whole system without an environment. Behavior at any level is explained in terms of the level below, but its significance is found in the level above. Ecosystem behavior does not emerge from a set of organismic equations.

3.3.3.1. Properties of Ecosystems

Properties of an ecosystem include: Wildness, Productivity, Diversity, Complexity, Stability, Change and extinction, Historicity, and Wholeness and renewability.

3.3.3.1.1. *Course as Property of Ecosystem.* The interplay of material cycles and energy flows generates a self-renewing "homeostasis" in Odum's words. Processes like homeostasis, which involve a relationship with the environment, occur on an ecosystem level. Yet, living systems do not display homeostasis—constant value—so much as a particular course of change in time—homeorhesis (from the Greek *homeo* and *rheo*, meaning 'same flow'), according to Conrad Waddington. The course is stabilized, not the constancy. Changes to a system are symbolized by trajectories in a multidimensional phase space or landscape. Homeorhetic mechanisms protect the system from many disruptions. Negative feedback counteracts the effects of change to maintain the system in a steady state or homeorhetic state. A mature community is self-perpetuating and homeorhetic, with a dynamic balanced energy-matter budget.

Homeorhesis is a significant phenomenon in evolution. Waddington applies it to the tendency of a process to continue in its original pattern, even if disturbed. Homeostasis is tendency of spatial structures to remain the same. Like embryos, ecosystems have many properties and are affected by many environmental conditions. Their changes are symbolized by trajectories in multidimensional phase space; orderliness can be described in terms of constraints on trajectory courses, and these constraints are visualized as attractor surfaces. If the system starts from any condition, represented by a point in multidimensional phase space, the trajectory will move to nearest attractor surface and then move along it.

3.3.3.1.2. *Self-Making Identity as a Property of an Ecosystem.* Bertalanffy called life a system of self-organization, a developmental unfolding at progressively higher levels of differentiation and organized complexity. Living systems are autopoetic units in physical space. The system is autopoetic, that is, self-creating. It renews itself as its contents change, as disturbances change the parameters. As a mature system, it continues to move to a point of higher maturity, recovering from disturbance to its original trajectory, where productivity declines and stability increases.

The organism and environment are co-implicative, co-defining, and co-constructing, in a process of self-assembly, where the self is the organism/environment system. In an autopoetic framework, every being is embedded in a world and observed by an embedded observer. The material components of life move through physiological processes. Autotrophs (bacteria, algae, green plants) convert energy into organic compounds; heterotrophs reconvert into heterotroph flesh.

If there were no identity, there would be no differences and so no relationships. Without relationships, there would be no things, no events, and no universe. Objects precipitate out of relationships and are defined by them.

Ecosystems are part of an unending, imperfect process, without any final state. Furthermore, the human attempt at perfectibility through self-improvement causes disharmony, which is part of the same imperfect process. Each system is a practical application to place. Unknown factors determine a large part of the operation of any system. Furthermore, there is chaos in every system; there are plagues and random frenzies.

3.3.3.1.3. *Openness as a Property of an Ecosystem.* The ecosystem has to be open to flows of energy and materials. Closure, e.g., well-defined boundaries, with steady input output flow rate, is not a contradiction. Boundaries are produced by the system. The boundaries are permeable, allowing exchanges of energy and matter. Too much openness would allow the system to be overwhelmed by the environment. Too little openness would cut the system off from the environment and it would run down, in terms of energy. The system is a local system, with specific structures and functions. The vast number of interrelationships between systems keeps them open. For example, grassland is affected by climates, soil conditions, fires, surrounding communities, and human agents. Ecotones between systems are usually shifting.

3.3.3.1.4. *Stability as a Property of an Ecosystem.* Stability is the ability to maintain the identity of a system under the flow of external forces and disturbances. Stability can be refined through the specifics of constancy, resistance, resilience, and accommodation. Stability can be related to ideas of compartmentalization, communications, richness of interactions, and connections.

Stability is a complex topic. Ulanowicz suggests that stability might be explained by diversity flow topologies, where flow topology is a descriptor of how ecosystems develop. The homeostasis of the ecosystems, that is, stability, as originally proposed by Eugene Odum, becomes the result of regular flows of energy and materials. Growth and development are characterized by a qualitative formalism of increasing ascendancy, which explains the drive towards coherence, efficiency, specialization, and self-containment.

At least four kinds of stability can be identified. *Persistence* means continuity or the state of being in a continuous flow or coherent whole for a duration. The system is constant if it stays relatively unchanged in form or composition. This means a lack of change in a system parameter, like number of species. The system is persistent in time if it is self-maintaining, mature, and hysteretic (historical). The system changes and develops. It loses and gains species over time, but is still recognizable as a short-grass prairie, for instance. The integrity of its functions has not been greatly affected by changes in species. The system can have degrees of difference, malleability, and still be the named system. The system may have different states that oscillate. It may also have a trajectory to a mature state.

Resistance is the ability to withstand disturbance to a system and to continue. This is Robert McArthur's meaning of stability, from his analysis of food web structure; it is similar to Holling's concept of 'resilience.' McArthur's stability is less predictable, and it is more expensive to maintain.

Resilience is the ability to recover from stress. The ecosystem can be thought of as elastic. The elasticity refers to the time required to recover its initial structure and function after some kind of damage; it is the speed that a system returns to former state following a disturbance (as a measure in the community matrix).

Finally, *conciliance* (or reconciliance) is the ability to absorb change and still maintain the system identity. The system adapts to disturbance. It accommodates disturbances. It incorporates new things, such as new organisms. It tolerates new levels of things, such as increased heat. It fits the changes within the structure and function of the system. Amplitude is a measure of the maximum amount of damage a system can sustain and still recover.

These preceding terms are not comparable. Constancy and persistence are descriptive, implying nothing of underlying dynamics. Resistance to stress is a useful notion if one is interested in the maximum extent of the deviation between stressed and unstressed systems. Resilience is relevant to those who are concerned with the rate at which a systems returns to prestress conditions. And asymptotic stability is concerned with whether or not a system will eventually return to prestressed state.

Cyclic and trajectory stability have measures of inertia, elasticity and amplitude associated with them. Elasticity and amplitude seem to mean centripetal and centrifugal. Oscillation, or cyclicity around a point, or cyclical stability, is the property of a system to oscillate around central point; predator/prey systems, for example, have the property of a stable limit cycle. Malleability is the degree of difference from original state.

3.3.3.1.5. *Productivity as a Property of an Ecosystem.* Productivity is the ability to convert energy into living forms and the ability to incorporate materials into living forms. Productivity, in general, depends on the vigor, strength or vitality, of the system—that is, health. Health is the overall ability of a system to maintain itself under a normal range of environmental conditions (which may include hurricanes, volcanic eruptions, or fires). Obviously, a pioneer community may change the conditions to favor a new level of the system with new components. Health is a dynamic measure of ecosystem organization, vigor, and resilience. Organization is described by diversity and connectivity; vigor is related to the amount and speed of productivity; and resilience is a measure of reaction to stress. Too much stress, for example, leads to unsustainable patterns of behavior; continuous stress leads to a breakdown of processes that becomes irreversible—the system dies.

To relate health to growth and productivity, we could say that the capital of an ecosystem would be its physical environment and its gross primary productivity; its interest would be the net ecosystem productivity. The production percentage would be the amount necessary to keep the ecosystem healthy. Our measurements of productivity, however, are not adequate. We are measuring over a year or two only to establish a growth rate or productivity. We should be measuring over centuries. A forest is a long-term, dynamically-changing being. We cannot use a short-term industrial approach to measure a few parameters and then pretend we know enough about a forest to cut a large percentage of it. Forests are created by slow processes that take millennia to develop.

3.3.3.1.6. *Diversity as a Property of an Ecosystem.* The environment has been constant enough for organic evolution, but variable enough for natural selection to be challenged. Variability challenges organisms to adjust and thrive. Variability, even in small ways, leads to diversity. Diversity, as a measure of genetic variability in ecosystem, enlarges information. A mature system needs less information, since it works toward preservation. The limit of maturity allows maximum variability between systems with slight external differences, like temperature.

Ecosystems develop knowledge bases that reflect information on themselves and their environment. Species richness may stabilize some ecosystem properties, such as Net Primary Productivity (NPP). Frank and McNaughton (1991) showed that plant community structure (and productivity and biomass) is more stable when a greater diversity of species is present. Also, diversity can decline with productivity, as in the eutrophication of lakes, for instance. While it is meaningful to speak of an optimum diversity, as the interaction of limits and other factors, a maximum diversity may never be reached.

Most healthy ecosystems have high degrees of flexibility. As Gregory Bateson interprets Ross Ashby, any biological system can be describable in terms of interlinked variables, each of which has an upper and lower threshold of tolerance, beyond which the system acts pathologically. Within the limits the variables can be moved for the system to be adapted to the environment. Under stress, some variables move to maximum values near the upper or lower limits—the system loses flexibility, that is the "uncommitted potential for change" (Bateson's definition), and can be destroyed by further stress. The danger in each case is working near the maximum of the system. Adaptive ecosystems are not static orders; they are flexible, as well as historical and irreversible. The strengths of local systems lay in the diversity of values and in their fitness to particular places.

3.3.3.2. Principles Standards & Practices of Ecosystems

Remember, principles are fundamental rules that we can use to create images or models, while standards are repeatable models of quality, and practices are the actual work in place (see Section 3.8). Ecosystems can be described as a series of principles:

- The ecosystem is the level of integration and the unit of organization undergoing a directional development (after Odum).
- Energy is bound into organic material, measurable as productivity.
- Energy/matter is transferred through individuals as a food chain.
- The interaction of individuals in a food chain results in trophic structures (pyramids).
- Chemical elements circulate in the biosphere in biogeochemical cycles.
- Energy and material no longer used by a system is the waste of that system (entropy)
- Energy required to maintain an ecosystem is related inversely to maturity (Margalef's concept of maturity).
- An ecosystem has a minimum size.
- An ecosystem has a distinct pattern, related to health.
- An ecosystem as a level of integration and organization undergoes directional development
- That quantity of energy and material no longer of use to the system is wasted.
- Chemical elements, especially those of life, circulate in the biosphere in characteristic paths known as biogeochemical cycles.
- Life is limited by elements and physical factors (light, water, gas, salt); too little of an element limits (Liebig's law); too much limits (Shelford's law of tolerance).
- The transfer of energy and materials through organisms is referred to as the food chain.
- The interaction of individuals in a food chain results in the trophic structure of communities (ecological pyramids).
- The energy required to maintain an ecosystem is inversely related to complexity; succession decreases the flow of energy per unit biomass (Margalef's concept of maturity).

To maintain ecosystems, we can create standards based on those properties and principles. Sample standards (many of these based on Conservation Biology) include:

- Preserve the spirit of a forest place
- Preserve minimum species and habitat for wholeness
- Preserve diversity at all levels
- Retain appropriate shapes and corridors for pattern unfolding
- Retain hierarchy of all levels
- Allow for limits, disturbances, processes, adaptation, evolution
- Leave snags and fallen trees
- Leave 2x buffers around riparian zones
- Restrict roads/trails to appropriate terrain
- Use appropriate equipment, e.g. cable yarding, in steep areas
- Take/leave trees of all ages
- Avoid damaging activities, e.g., slash burning

Based on these standards, we can describe practices of exploitation, which would allow limited exploitation of an ecosystem, to keep it wild and healthy. Samples of practices include:

- Maintain size and completeness of forest
- Strengthen shapes and margins

- Keep density and openness in a balance
- Preserve the interior/protect riparian zones
- Plan paths to avoid sensitive areas and emphasize pleasing perspectives
- Bundle paths (power lines, utilities) to minimize intrusion
- Preserve the character and all aspects of the forest

Ecosystems comprise the ecosphere. There are emergent properties at each level of organization, such as the diversity of species and the structure of a food web. Furthermore, each level of integration has distinct attributes. Ecosystems are the result of historical processes that are mathematically atypical. Furthermore, they are not randomly structured. A system drives to a nonequilibrium state as a mature ecosystem. The adaptively reorganized system is not necessarily more stable, but it is optimally resistant to the outside conditions that elicited the self-organization, a natural normalization process. The ecosystem learns the changes, periods, or seasons of the environment.

3.4. *Design Factors: Landscapes*

A landscape is a heterogenous area composed of a mosaic of interacting ecosystems of various sizes. How are these local design factors like cycles and spheres? They recycle elements regionally. Landscapes contribute to cycles and flows as a result of their form and size. Transforming an entire landscape alters flows.

Landscape is a convenient idea that serves as a unit of analysis. Landscapes have a range of scales rather than an intrinsic spatial scale. As ecology undergoes a scalar shift, microecology, with ties and cellular biology and dominating the field, yields some attention to macroecology, the study of wildlife and landscape ecology. Macroecology has fewer practitioners. Russian scientists pioneered the concept and only recently have other scientists perceived the need for regional and global scale ecology. Ecology can learn much from geology by working at multiple scales.

The patterns of ecosystems are addressed best at a landscape level. The word landscape was used by the geographer Alexander von Humboldt as a scientific term in the 19th century. A German biogeographer Carl Troll used the phrase "landscape ecology" about 1939 to describe land—and not just living organisms—as an integrated holistic entity to be studied in its totality by geographers and ecologists, geography providing a horizontal approach to the vertical one of ecology. F. E. Egler in 1942 emphasized the active role of humanity on the holistic nature of vegetation in the landscape. Later, Raymond Dansereau noted that landscape was the highest integrative level of environmental processes. Like G. P. Marsh, Dansereau described landscape modification by humans throughout history.

The modern science of landscape ecology was developed in Europe in the 1970s and formalized as a discipline during that time. Landscape planning is interwoven with the interdisciplinary aspects of urban and local planning. Naveh and Lieberman list three factors that form a new science of landscape ecology: (1) Landscapes are recognized as natural and cultural entities whose health and integrity are vital for human survival; (2) As an approach to the study of landscapes, the conventional reductionistic scientific paradigms are replaced by integrative and holistic methods based on a systems view; and (3) Technological advances in remote sensing and satellite images, combined with the capabilities of processing large quantities of data, support the possibility of dealing with landscapes holistically.

Landscape ecology is a goal-directed science for studying the complexity of landscapes, as well as for preserving their integrity and health and natural and cultural diversity. In order to manage complex landscapes, it has to develop better methods for study and management. The conceptual framework for landscape ecology is derived from three connected theories:

1. General systems theory, a holistic description of the hierarchical order of nature as a complex of open systems of increasing complexity. J. C. Smuts described holism as the "whole is more than the sum of the parts." Ludwig von Bertalanffy, Arthur Koestler, Ervin Laszlo developed the foundations of systems philosophy.

2. Cybernetics, a theory of self-regulation of those systems through deviation counteracting (negative) and deviation-amplifying (positive) feedback cycles. Norbert Wiener generalized cybernetics, feedback, and information beyond computer technology. Shannon and Weaver's information theory and John von Neumann's game theory are also fundamental contributions.

3. Ecosystem ecology, a set of transdisciplinary concepts with the ecosystem as the level of integration. Eugene Odum promoted ecosystem theory and emphasized differences in ecosystem patterns. W. Haber extended the foundations to landscape ecology.

Landscape ecology addresses the overall patterns of large-scale ecosystems, considering the biogeochemical, atmospheric, and hydrological cycles in relation to the shape and extent of individual landscapes. Landscape ecology can identify: Candidate ecosystems for restoration; candidate ecosystems for preservation, conservation, or reservation; and, patterns of forestry to preserve larger functional islands. Landscape ecology, with its holistic, cybernetic ecosystems approach, can address the emergent features of large systems and large cycles.

Landscape is the level lower than the biome and larger than the site or watershed. Landscapes include stable communities and fragile habitats. A landscape has to be large enough to offer diversity of resources for cycles, and small enough for exchange between organisms. It as to have sufficient time for development, that is, changes cannot be too fast. It has to incorporate a number and diversity of species, trends and patterns.

Before satellite data analyzed by GISs, ecologists and foresters did not have tools that could address the scale of landscapes. Regional and global data was hard and expensive to collect. The International Geosphere Biosphere Program (IGBP) coordinated by the U.N. uses GISs for global and regional issues, such as deforestation or desertification. The large images from satellites are exceptional for identifying landscape patterns, especially those related to the scale of species behavior, e.g., home range or breeding dispersion. Pattern can also be measured at the level of patch size and spatial relationships (that is, inter-patch distance), which is critical for relating the size of a habitat to the species in it—to apply the theory of island biogeography. The data derived from satellite imagery and from field studies can be used to model the landscape at various levels. Some of these tools are used for specific problems in the classification of forest ecosystems and harvesting schedules.

3.4.1. *Morphology History & Productivity*

The morphology of a landscape has effects on nutrients and soil. For example, slope stability determines the potential migration of material into watercourses, the effects on landscape structure, changes in overland flow of materials and nutrients, and, changes in habitat conditions suitable to resident species. Soil erosion causes changes in physical properties of soil, such as structure, texture, bulk density, infiltration rate, depth for favorable root development, and available water-holding capacity. These flows determine the productivity of a landscape.

As humans began to settle permanently to exploit their surroundings continuously through agriculture and animal productivity, landscapes became more disturbance-dependent and became less resilient to climatic events. Human cultures attempted to cope with risks or to exploit opportunities, which required more management of the environment, although different parts of the environment operated at a range of scales, most of the natural dynamics and landscape occur slowly by comparison with human dynamics. As a result humans adapted themselves to a dynamics of the environment at the beginning, but over time cultures served their own needs by modifying the environmental dynamics. Human cultures thus become dependent on colonized systems, which required certain social institutions, especially those involved in organized production and storage.

Descriptions of historic landscape disturbance regimes, e.g., fire magnitude and frequency, and the ecosystem component patterns they maintained, e.g., vegetation composition, provide an initial template for descriptions of ecosystem health. Some regimes result from economic needs, such as large timbers, as well as from cultural influences and human values, especially fads. In a way, the internal disturbances have as much effect on the dynamics of forest communities and forest landscapes as fire and blowdowns. Castello, Leopold, and Smallidge suggest that pathogens, by eliminating less vigorous or genetically unfit (filtering them out of the stream of life), control the direction and rate of succession. Tree mortality from pathogens occurs on various scales: gap phases (small scale), forest development (large scale), and landscape patterns (immense scale). Pathogens are one of the determinants of growth and development. Pathogens determine tree mortality, which drives forest and landscape patterns.

3.4.2. *Human Uses & Impacts on Landscapes*

Human use of fire, as mentioned, can shape landscapes. Geologists have noticed an increase in fossil ashes from the Pleistocene (1 million years BP—before present) that may indicate that early hominids used fire. Peking man (250,000-350,000 years) definitely had fire. Deliberate fires for clearing land and accidental fires from lightning strikes and drought have all destroyed forests. Before settlers arrived in the western America there were few large fires—although large fires periodically occur in certain kinds of forests, e.g., in lodgepole pine forests every 300 years or so. Since then, large fires have become a regular feature.

Environmental factors have shaped the course of human history to a greater extent than had been realized. The decline of Rome is a study in forest ecology. There were previous catastrophes in the Tigris and Euphrates valley, Greece, Khmer, Maya, Midwest United States, and the Australian outback. Many peoples could not solve their problems; many immigrated to new lands.

Only in the 19th century, beginning with G.P. Marsh, did people start to realize that humanity has done as much to change the environment as the environment has done to mold human history. Marsh, the first American ambassador to Italy, was one of the first to study the role of humans in changing the face of the earth. When he visited the near east in the middle of the 19th century, he was shocked to find deserted cities, silted harbors and wastelands instead of flourishing civilizations. He concluded that ecological errors had led to the deterioration of agriculture in Mediterranean countries. To avoid future deterioration, he advocated agricultural conservation practices.

Land continuously occupied by humans may form analogs of natural communities, guided by trial and error, by unconscious values, and by random changes. The American tallgrass prairie is a case of the creation and maintenance of an artificial but desirable ecosystem. Unfortunately, it was dependent on a multiplicity of unintended accidents. Ecosystems evolved through natural events, then as an effect of human activities, now through deliberate social choice in some places.

The archaic universe was regarded as the creation of order out of chaos, with humans contributing; nature and man are harmonious. Now, there is arrogance from Plato to Dubos, that all human action improves the spontaneous course of nature. Dubos was disturbed that farms are overgrown and does not like to see vegetation revert. Dubos dreaded forest regrowth in New York as barren and uninteresting, but Leopold felt in Wisconsin it was a welcome prophecy of nature's second coming. This is the same Leopold who had once wanted to raise deer in a wolfless world, but later realized the necessary function of predators in a balanced community. Dubos and others have claimed that "nature knows best is wrong," that nature is inefficient and wasteful. But, ecosystems that seem inefficient and wasteful are many times extremely redundant, and therefore stable and flexible. Natural processes that seem destructive are cyclic and preservative also.

3.4.3. *Landscape Management*

Once a landscape has been exploited or modified, it has to be managed for as long as human use is considered. Of course, landscapes can be allowed to become wild again, but the process is lengthy. In the late 1970s, C. S. Holling described Adaptive Management as a response to natural disturbances at the landscape level. Matrix management (Noninterference) can be used to deal with designed landscapes, especially forested landscapes exploited with ecological techniques (Wittbecker, 1992). Shortly afterwards Alan Savory applied his ideas of Holistic Resource Management to preserve the quality of the landscape. The 1980s also saw a comprehensive Integrated Resource Management (Mitchell), Permaculture as the expression of a permanent responsibility for the landscape (Bill Mollison), and the Wholistic Timber management (Herb Hammond), which was dedicated to maintaining the structure and functions of diverse forests.

We must pay attention to the processes that make up the habitat, for example, the role of herbivores on trimming vegetation (and diversifying it by predation). The design of the forest and its management must ensure that the processes operate to maintain a dynamic state. Furthermore, the context must be conserved. The forest, however, cannot be considered outside of the context of the entire landscape, including human images and institutions.

3.4.4. *Conclusion: Importance of Healthy Landscapes*

Healthy landscapes reflect a balance of processes, including extinction, colonization, and connectence. Colonization is necessary to equalize local extinctions. The loss of colonization, for any reason, can allow species in a local system to be depleted. Island ecosystems, surrounded by poor areas, may lose species without colonization. If the matrix is rich enough, there should be successful colonization.

Wild landscapes are affected by climate, soils, interactions, and disturbances. Domestic landscape is affected by land use as well. The greatest changes have been brought about by the destruction and creation of forests. With the predominance of artificial or managed forests, it is important to consider the qualities of naturalness in the landscape. Forests are expected to meet the needs of society by producing timber, creating wildlife habitats, and providing recreational opportunities for people.

How we treat large-scale communities is an important question. Deep ecology is one form for asking questions about landscapes, especially forested ones. For example: How can we design forests for neutral elements in interrelated processes in a landscape? What are we restoring when we restore an ecosystem without all the parts or good knowledge about the ones we have? These questions highlight the uncertainty we face in dealing with large, wild, complex, long-lived entities at the scale of landscapes. Managers have to live with uncertainty; this means that management decisions are essentially gambles. Gambling is a profession that acknowledges the operation of chance and makes conclusions in the absence of facts— few people are successful at it. This is an important admission, that we do not have facts to base our actions on, that nature is a stochastic process, and that landscapes and biomes always changing. Furthermore, we do not know for sure what effects our actions will have on landscapes, which live so long, in such diversity, in many places. Successful gambling suggests that the proper attitudes for gambling with nature are awareness, humility and courage, not arrogance, fear and maximum use.

Perhaps we do not have sufficient knowledge to manage a complex landscape because it is too complex to understand scientifically. But we can understand the pattern and drive it in a healthy direction with minimal intervention. Perhaps we lack sufficient courage or willpower to manage large-scale forms. According to Garrett Hardin, many of the ideas necessary to fit humanity into the pattern of nature are known but not yet popular. For instance, exponential growth cannot be maintained very long. Human communities cannot grow 4 percent per year without disastrous consequences to the infrastructure and the quality of life. Growth cannot be continued because the landscape is limited, in terms of productivity, energy, and resilience. Thus, we need to fit our population into the limits of a landscape or biome, although some limits can be expanded by technology or by lowered expectations. Redundancy at this scale, with local or regional self-sufficiency, keeps global limits less critical. The carrying capacity of the area is not only a function of the limits of the community, but it is equal to the number of people multiplied by the level of comfort (quality of life style). Changing the scale of concern does not solve any problems relating to overconsumption or drawdown; it just displays the regional limits sooner.

3.5. Design Factors: Local & Planetary Cycles

In a mature ecosystem, most nutrients and materials are held in cycles; very little leaves the system. A cycle is defined as a pathway along which an element moves through biotic and abiotic compartments. All the chemicals, nutrients, or elements—such as carbon, nitrogen, oxygen, phosphorus—used in ecosystems by living organisms stay in a relatively closed system, which means that these chemicals are recycled instead of being deposited or lost. Virtually every material cycles through the forest: phosphorus, potassium, calcium, sulfur, magnesium, and water. Nutrient cycling involves many of these materials. Nutrient cycles change with the succession of a forest.

Materials cycle above and below ground, between the atmosphere and trees, between trees and insects, and squirrels and fungus. Chris Maser is fond of saying that most of the cycling is invisible because it is underground or in the air. Many cycles are investigated through ecosystem analysis, where energy and materials are traced through transfers through compartments in an ecosystem. For example in the nitrogen cycle, the compartments are the atmosphere, vegetation, forest floor, and mineral soil, while the transfers are precipitation, throughfall, leaching, litter fall, mineralization, fixation, and denitrification. Each compartment keeps the nitrogen for a certain time, termed the residence time; for a hardwood forest floor, for example, residence time is about 17 years. If the compartment keeps the element for a long time in large quantities, it is called a reservoir, like the ocean; if the residence is short the compartment is called an exchange pool, like a cloud.

Cycles require the movement of elements. Some reservoirs, such as the atmosphere, allow rapid movement. For example, almost any gas or fine particulate matter released into the atmosphere can spread across the planet in days. And, because of the movement the atmosphere stays in a pattern of dynamic equilibrium.

Cycles are nonlinear systems, with limits and thresholds of which we are relatively ignorant. They allow unstable and improbable reactions to keep taking place. These cycles are not only interesting, but necessary. Consider the oxygen cycle, for instance: Without cycling, all reactions would take place, and the atmosphere would reach an equilibrium (as on Mars).

Cycles are connected; for instance, the water cycle contributes to the operation of the carbon cycle; in fact, both meet in photosynthetic activity of plants. Carbon dioxide is part of the carbon cycle, which is tied to the heat cycle of the atmosphere. Carbon dioxide is the source of carbon for plants and it plays a role in the weathering of rock. All chemical elements occurring in organisms are part of biogeochemical cycles. In addition to being a part of living organisms, these chemical elements also cycle through abiotic factors of ecosystems such as water (hydrosphere), land (geosphere), and the air (atmosphere); the living factors of the planet can be referred to collectively as the biosphere. These cycles form the metabolism.

3.5.1. Major Individual Cycles

Some cycles move at higher speeds than others. Some speed up seasonally or daily. Microbial activity speeds up an iron cycle. Some cycles grow or decline. Tectonic activity can renew some cycles, such as the phosphorus cycle. Some material cycles between other ecosystems or in larger patterns around the planet. Some of the cycles are daily; some are seasonal or annual; others are years or decades; a few, like the carbon cycle, are centuries or millennia long. Human beings do not pay much attention to very long cycles or to those we perceive as not affecting us. Sometimes we deliberately or inadvertently interfere with cycles. An excessive concentration of carbon dioxide in the atmosphere is an example of a disrupted cycle. The nitrogen cycle is being disrupted by runoff of fertilizers. Nitrogen fixation has doubled on the global scale, and this favors plants with higher nitrogen needs. Sulfur emissions to the atmosphere have doubled, also. Human interference in forest cycles can collapse the residence time; for example, clearcutting alder in Washington State causes very high nitrogen losses. Human interference in one cycle can affect other cycles.

3.5.1.1. The Water Cycle

Water, in its different forms, cycles continuously through the lithosphere, hydrosphere, atmosphere, and biosphere. The energy for the water cycle is supplied by the sun, which drives evaporation. Evaporation is the process in which liquid water becomes gaseous—precipitation is the reverse of this. Water evaporates into the atmosphere from the land and the sea. Plants and animals use and reuse water and release water vapor into the air. Once in the air, water vapor circulates and can condense to form clouds and precipitation, which fall back to ocean and earth. At one time or another, all of the water molecules on earth have been in an ocean, a river, a plant, an animal, a cloud, a raindrop, a snowflake, or a glacier, before being moved by living or geological processes.

Figure 3511-1. The Water Cycle

The sun provides the energy that drives the climate system, with its weather systems, which move the water vapor from one place to another, and from ocean to land. Once water condenses, gravity pulls water to the mass of the earth. Gravity continues to operate, as water flows across the surface or underground. It can be temporarily trapped in lakes or oceans. The main path through which water leaves oceans is through evaporation, which leaves behind salts and minerals. When water precipitates, it can pick up pollutants and acids and deposit them; over land, water can pick up minerals and pollutants—thus, evaporated water is relatively clean, but other water takes whatever is dissolved into traps or sinks. When water freezes, it can remain in a solid state for long periods.

Organisms participate in the water cycle, and most organisms contain a significant amount of water, although it tends to evaporate or transpire quickly. Animals and plants lose water through evaporation from the body surfaces, such as skin or leaves. Evaporation from leaves is responsible for significant amounts of water entering the atmosphere.

3.5.1.2. The Carbon Cycle

The carbon cycle tracks the water cycle; water is the vehicle that carries the complementary biological reactions of respiration and photosynthesis. Respiration combines carbohydrates and oxygen to produce energy, carbon dioxide, and water. Photosynthesis produces carbohydrates and oxygen from carbon dioxide and water. The outputs of one are the inputs of the other and vice versa. The reactions are also complementary regarding energy. Photosynthesis stores solar energy in the carbon-carbon bonds of carbohydrates; respiration releases that energy. Only plants (and a few other producers are capable of photosynthesis, but respiration is part of the activity of every living being.

Neither oxygen nor nitrogen can absorb either infrared or visible radiation. Bigger molecules, like carbon dioxide (or methane or water vapor), vibrate at the same frequencies as infrared radiation, absorbing heat and reradiating it, warming up the atmosphere. The early atmosphere had a high percentage of carbon dioxide, estimated from 3 to 10 percent, much of it from volcanoes. The gas also reacted with rock to form limestone. Over billions of years, microorganisms and trees pumped carbon dioxide from the air and into the soil (it is only a 300th as common now—air spaces in the soil have 10-40 times the CO_2 as the air above). As trees die, much of the mass is oxidized by decomposers to carbon dioxide. As carbon dioxide dropped, plants responded to lower levels. Grasses, for instance, can photosynthesize with lower CO_2 levels. (As the solar output increases over the next billion years, less CO_2 will be needed to maintain higher temperatures—but, what kind of plants will be living then?) The carbon dioxide cycle seems to be a very slow cycle as the carbon ends up in limestone and chalk at the bottom of oceans.

The carbon dioxide produced by plants accumulates in a variety of reservoirs such as the ocean or rock formations. Carbon dioxide dissolves readily in water, and can precipitate naturally as calcium carbonate (limestone). Corals and algae build up limestone reefs with the process.

Table 3512-1. Reservoirs of Carbon (in billions of metric tons)

Atmosphere	720
Ocean	39,000
Carbonates	100,000,000
Fossil fuels	4,000
Land plants	560
Soils	1500

The carbon in plants now has three possible paths. Some of the carbon is released to the atmosphere by the plant through respiration, where it can stay in the atmosphere, be taken up by other plants, or dissolve in the interfaces of water bodies; some is consumed by animals and can be kept in flesh or released through respiration; and some is in flesh when the plant or animal dies, and can be recycled by decomposers or buried. All carbon in biological systems ultimately comes from plants, usually from predation. In the animal, the carbon also has the same 3 possible fates. Carbon from plants or animals that is released to the atmosphere through respiration will either be taken up by a plant in photosynthesis or dissolved in the oceans.

Buried carbon, treated by tectonic processes, ultimately forms coal, oil, or natural gas (fossil fuels). The fossil fuels are recovered and burned by traditional or industrial human processes, releasing carbon dioxide to the atmosphere. Carbon in limestone or other sediments can only be released to the atmosphere after they are brought to the surface, and weathered or released by volcanoes. Increased carbon dioxide in the atmosphere can cause atmospheric heating by trapping solar energy; it also pushes more into the oceans, making them more acidic.

Table 3512-2. Residence time of Carbon

Atmosphere	3-5-50-300 years
Leaves	2-3 years (Amazonian forests)
Fine roots	3-10 years
Microbes	6-10 years
Living trees	200-260 years
Soil	25 years
Shallow ocean	350 years
Deep ocean	1000 years
Rock	8-10,000 years
Fossil fuel	Millions of years
Carbonates	150 million years

Because reservoirs are interconnected, the hyphenated short residence times are underestimates. The reservoirs with longer residence times, like the ocean, release carbon back to the atmosphere, increasing atmospheric residence time is over 100 years. The anthropogenic flux of carbon, from fossil fuel burning and deforestation, to the atmosphere is 8 billion metric tons per year, but the atmospheric increase is only 4 bmt/yr. Where does the other 4 bmt/yr go? There are two possibilities: Land Plant Uptake (Excess Photosynthesis) or Ocean uptake. Because the residence times are so different however, it is important to know which. Storage on land is much shorter-lived than storage in the deep ocean. These fluxes are important to understand.

Table 3512-3. Fluxes of Carbon (in billions of metric tons/year)

Land Plants	-
Photosynthesis	120
Plant respiration	60
Soil respiration	60
Plants to soils	60
Fossil fuel formation	0.0001
Fossil fuel burning	6
Deforestation	2
Ocean Dissolving from atmosphere	107
Exsolving to atmosphere	105
Carbonate formation	0.3
Weathering	0.6
Volcanoes	0.1

The flux from volcanoes is relatively low, but the clouds from volcanoes can contribute to global dimming, with their cooling potential. The amounts of carbon dioxide in and out of oceans are close to balance due to the biological processes, such as photosynthesis and respiration. Anthropogenic processes have a much greater effect than weathering or volcanic action. Fossil fuel formation is a very slow, constant process that can create huge reservoirs of carbonates over millions of years. Fossil fuel burning, however, is a very fast process and a new source of atmospheric carbon. It is a relatively small flux compared to plant respiration, only one-tenth, but it is an additional source, and carbon dioxide is a potent greenhouse gas that can trigger temperature changes in relatively small amounts.

3.5.1.3. The Oxygen Cycle

The oxygen cycle parallels the carbon cycle, since the atoms are often combined, as in carbohydrates and carbon dioxide. Oxygen of course is an important component of water. The oxygen cycle is driven by living beings. Oxygen is released from water to the atmosphere by autotrophs during photosynthesis and taken up by both autotrophs and heterotrophs during respiration. After two billion years of biological activity by autotrophs, mostly cyanobacteria, the oxygen content of the atmosphere increased significantly, allowing multicellular plants and animals to take advantage of the increased energetic reactions that oxygen allows.

3.5.1.4. The Nitrogen Cycle

The nitrogen cycle is more complex; there are many important forms of nitrogen, all created by the interconversions of organisms. Nitrogen is critical in forming the amino portions of the amino acids that form the proteins, which make up skin, muscle, and other important structures of organisms. All enzymes are proteins, and enzymes carry out many of the chemical reactions in the organism.

Nitrogen is a chemical element with 7 protons in its nucleus. Because of its electron structure nitrogen bonds strongly with itself (N_2); the action of lightning can break this bond and allow nitrogen to combine with oxygen. The resulting nitric acid falls in rain and would eventually be locked up in the ocean as nitrates—except that nitrogen is fixed by bacteria in proteins and DNA (deoxyribonucleic acid); some bacteria break up detritus and return the nitrogen to the atmosphere. In fact, almost all nitrogen resides in the atmosphere, where it sustains atmospheric pressure and dilutes other gases.

The movement of nitrogen into and out of the atmosphere is a nitrogen cycle. The main reservoir of nitrogen is the atmosphere, which is composed of about 78% nitrogen. Nitrogen gas in the atmosphere is in the form of two nitrogen atoms bound to each other. It is a relatively non-reactive gas, that is, it takes a lot of energy to break it up, so it can be combined with other elements, such as carbon or oxygen. Atmospheric nitrogen gas can be fixed in two ways: Lightning provides enough energy to burn the nitrogen with oxygen and fix it in the form of nitrate. The other form of nitrogen fixation is accomplished by nitrogen-fixing bacteria, using special enzymes to fix nitrogen. These nitrogen-fixing bacteria come in three forms: some are free-living in the soil; some form symbiotic, mutualistic associations with the roots of bean plants and other legumes (rhizobial bacteria); and the third form of nitrogen-fixing bacteria are the photosynthetic cyanobacteria (blue-green algae) which are found most commonly in water. All of these fix nitrogen, either in the form of nitrate or in the form of ammonia (nitrogen with 3 hydrogens attached). Most plants can take up nitrate and convert it to amino acids. Animals acquire all of their amino acids when they eat plants (or other animals). When plants or animals die (or release waste) the nitrogen is returned to the soil. The usual form of nitrogen returned to the soil in animal wastes or in the output of the decomposers, is ammonia. Ammonia is rather toxic, but, fortunately there are nitrite bacteria in the soil and in the water which take up ammonia and convert it to nitrite, which is nitrogen with two oxygens. Nitrite is also somewhat toxic, but another type of bacteria, nitrate bacteria, take nitrite and convert it to nitrate, which can be taken up by plants to continue the cycle. We now have a cycle set up in the soil (or water), but what returns nitrogen to the air? It turns out that there are denitrifying bacteria which take the nitrate and combine the nitrogen back into nitrogen gas.

3.5.1.5. The Phosphorus Cycle

The phosphorus cycle is a more simple cycle, because the heavy phosphorus molecule can only be carried by water or living organisms. Combined with energetic oxygen atoms, it has a basic form, phosphate in rock; otherwise, it is found dissolved in water or as part of an organism. When phosphate is exposed to water, it weathers out of rock and goes into solution in watercourses or soil, where it can be taken up by organisms. Phosphorus is an important constituent of cell membranes, DNA, RNA, and ATP, the cell's chemical battery.

Animals obtain phosphorus by consuming plants. Fungi are efficient at taking up phosphorus and also form mutualistic relationships with plant roots, which gets more phosphorus and nitrogen into the plants. When plants die, the elements are returned to the soil. Phosphorus is a component of bones, teeth and shells. Marine birds compose an epicycle of phosphorus, eating fish, which contain phosphorus, in the ocean and defecating on land, transferring that element in their guano. When animals die or defecate, phosphorus is returned to the soil or water by decomposers. Some of the phosphorus stays in the water and ends up in the deepest parts of the ocean, where it becomes part of the sedimentary rocks forming there. Without living organisms keeping the cycle rapid, all phosphorus would end up there and only be released as the weather operated on the surface rock. Human beings now accelerate the phosphorus cycle by mining guano and phosphates to use for fertilizers, which can cause local abundances that allow algae overgrowth and eutrophication. With eutrophication, algae use up oxygen so fast that other aquatic organisms, e.g., fish, cannot live.

Phosphorus is a limiting nutrient. Because it has no gaseous form, it has limited availability as a nutrient—Isaac Asimov suggested it was the bottleneck for life—and is a good example of Liebig's law of the minimum, which states essentially that something always has to be in least supply. Its geochemical cycle is slower than the others. It is also vulnerable to being lost through erosion to the ocean. Phosphorus is a necessary constituent of protoplasm, but it is sometimes the element in least supply. Most of it is locked up in phosphates in rock, which is weathered and it "escapes" to the oceans. Sea birds play a large part in returning it to land. In fact, we mine guano deposits for fertilizers for our fields. Harvesting fish for food returns some to land, but we make little effort to recycle it.

Nitrogen is also a limiting nutrient in ecosystems. There are also cycles of hydrogen and calcium, as well as of trace elements, such as molybdenum, sulfur, magnesium, and iron. Other organic nutrients, such as sugars, cycle through ecosystems. There are even cycles of certain artificial exotic molecules, such as from plastics, which are broken down and end up in ocean gyres or spread throughout water columns, where they interfere with the food chain.

3.5.2. *Interactions of Cycles with Other Cycles & Ecosystems*

The water cycle is linked to other cycles, especially the carbon and oxygen cycles. When the water cycle changes, as a result of naturogenic or anthropogenic actions, the other cycles change also. The interaction of cycles means that if one cycle is disturbed or interfered with by human activities, other cycles will also change.

Although the phosphorus cycle is pushed by the water cycle to the ocean, the activities of animals, especially fungi, fish and birds, bring this element back to the land or into ecosystems. Croplands may sequester carbon for instance, although current models may overestimate the extent. Old growth forests also sequester carbon, although current models may underestimate that amount. The uptake of carbon by forests, however, is limited by the availability of nitrogen, as well as water and nutrients. The sulfur cycle regulates the phosphorus cycle by converting it from an insoluble to a soluble form usable to organisms.

The interactions of biogeochemical cycles with ecosystems and organisms means that disruption or interference of cycles can affect ecosystems and organisms; conversely, disruption or interference of ecosystems and communities of organisms, at a sufficient scale, can affect the global cycles.

3.6. *Design Factors: Planetary Spheres*

As a whole planet with global processes, the planet can be reduced to various spheres for analytical purposes. The planet is composed of interacting spheres that interpenetrate and overlap. We tend to think sometimes of the planet as a rock decorated with a few fluids and gases, but the ocean and atmosphere emerged early from the denser elements as a result of the formation of the planet. The energy from the sun and the formation produced complex chemicals that learned to replicate their patterns. As these patterns reproduced on the scale of regions and the planet, their activities and wastes influenced and drove cycles that promoted more complex forms, which further modified global states and cycles. Although we identify major spheres on the planet—the geosphere, hydrosphere, atmosphere, and biosphere—they are very much interactive and their boundaries are open and shifting. These spheres intersect and interpenetrate. The biosphere produces certain kinds of rock, and the oceans produce atmospheric effects, so that we cannot say that each sphere does only one thing. Regions function as subsets of the spheres, with unique patterns and rates of movement and exchange.

3.6.1. *Geosphere*
The solar system formed through the gravitational collapse of a dense molecular cloud. Planets aggregated in the plane of rotation of the solar nebula, as a result of collisions of large and small planetesimals. The growth of the earth was impacted by several giant impacts, one of which most likely resulted in the formation of the moon. The planet continued to grow as a result of planetesimal impacts and cold dust accretion.

Impacts made the earth very hot. Metals formed a protocore. A solid inner iron-dominated core formed with a fluid overcore. As a result of these processes, a silicate mantle and then a light, floating crust formed. The mantle is a thick, flowing solid on which the oceanic and continental plates float.

3.6.1.1. Lithosphere
Heat is conducted from the core to the mantle and then through the lithosphere. Radioactive isotopes produce heat in all the levels (possibly exceeding the heat from the mantle). Of course, the surface of the earth receives heat from solar radiation, which powers most of the pyramid of life.

The planet produces three general kinds of rock. Igneous is produced by volcanic action or intense heat; gabbro is formed deep in the crust, while basalt and pumice are formed at the surface. Under water, granite and obsidian are formed. As these basic rock weathers, they become sedimentary, such as sandstone or shale or gypsum. Mixed with the content of living organisms, they become limestone, chert, or coal. When these rocks are pulled under by tectonic action, they become metamorphic, like slate, schist, gneiss, quartzite, or marble.

Minerals are limited from the formation, but many of them are used as nutrients for life.

The crust of the planet is dynamic. Mountains are limited in height by the thickness of the crust (which is not a problem with Hawaii which rests below the crust). The planet started as a mixture of chemical elements (over a hundred, including hydrogen, helium, and nitrogen) circulating over an active geology, driven by solar energy.

The action of the ocean and atmosphere tends to level the surface through erosion, but crust movements tend to build mountains. The division of earth into water and land and air determines directions of evolution, or rather limits the possibilities. Plate tectonics keep a significant portion of the planet surface above water, permitting the evolution of land and land-air based life. The changing locations and sizes of continents, from the supercontinents of Rodinia, Gondwana, and Pangea, to smaller continents like Australia and Antarctica, produce different patterns of oceanic and atmospheric circulation, and thus changing environments, that push the emergence and diversification of terrestrial plants and animals.

Intensive plate tectonic processes continue today and recycle the crust, along with organic material that has not been recycled by living organisms. This organic material is transformed into petrochemicals containing high-quality energy.

3.6.1.2. Pedosphere

Radiant energy from the sun flows around the surface of the planet, contributing to the heating and cooling, and eventual breakdown, the disintegration and alteration called weathering, of rock. Water also flows over rock, getting absorbed, trapped, stored and released. Plant roots find purchase in the rock and absorb water, send it to the leaves, and let it transpire back into the atmosphere. Plants also emit oxygen, which is used by other organisms. Dead plant matter forms a layer of the rock, a loose assemblage of animate and inanimate particles, where other organisms recycle residues from plants and animals. The remaining water follows the gravitational gradient to streams, aquifers and the oceans.

The soil acts as a buffer to keep water in the system longer for the use of plants and animals. This slows down erosion and flooding, and sustains the gradual constant release of water into streams. Many nutrients in the water also move downstream. Water increases the mix of mineral particles, gases and nutrients that make up the soil substrate. The soil is a medium where recycling of nutrients takes place. The soil is a living filter that mixes nutrients, dilutes pathogens and toxins, and purifies water. It is very porous and has an active surface thousands of times larger where biophysicochemical processes continuously take place.

A small amount of soil contains the living roots of plants, as well as a large community of organisms of various sizes, from microscopic to macroscopic—in fact, billions of organisms, from viruses fragments and bacteria to worms and insects. Although the soil forms a relatively thin layer, possibly less than one meter averaged across the planet, it is exceptionally active. It is strongly interconnected to the ocean and atmosphere through water. These interactions allow soil processes, such as sequestration of carbon, to contribute to the composition and properties of the atmosphere.

Soil formation is a slow process in human terms. It is gradual and progressive, as living forms collect and are decomposed. That material is churned by earthworms, insects and small animals. Eventually, several horizons are differentiated. It may take thousands of years to form a millimeter of soil. Humanity inherited incalculable soil resources that allowed the entire enterprise of agriculture. Unfortunately, agriculture tends to simplify the soil ecosystems and allow increased erosion. Although not necessarily an unnatural process, it is amplified and accelerated by the scale of agriculture, which converts entire grasslands and forests to 'smiling fields' by removing vegetative cover for long periods, altering the biotic community, and changing the topography and microclimates. Regions may be characterized by a suite of soil types that result from regional climate and processes.

3.6.2. *Hydrosphere* (on Planet Ocean)

The cooling of the earth's surface allowed oceans to emerge within several million years. Vaclav Smil states that major oceans formed as early as the early Hadean era, in a hot reducing environment, and they would have been strongly stratified, with a deep anoxic bottom. Subsequent collisions with large objects may have vaporized all or part of the ocean, eventually resulting in a new ocean after condensation for a thousand years.

Water is formed under conditions where molecules of hydrogen and molecules of oxygen react (that is, their elemental bonds are broken in a chemical reaction and recombine as water, releasing energy in the process). Water has some extraordinary properties: A high boiling point, high specific heat, and high heat capacity (due to its high heat of vaporization and low viscosity). Low viscosity makes it a good medium for swimming. The heat of vaporization allows evaporation to carry large volumes of water with a large thermal energy.

There is a vertical heat exchange in the ocean, as cold dense water sinks and is replaced by warmer water flowing poleward. This happens basically in two partially independent cells in the Atlantic and Pacific. Minor changes in ocean surface temperature can affect the expansion of surface waters (such as with El Nino).

The ocean stores over 96 percent of water on earth; it is the source of most evaporation (86%) and receives most of the precipitation (78%—the difference is that much precipitation is caused by mountain ranges and falls to earth, and there are regions such as the north Pacific that have surplus precipitation). Every other reservoir of water is limited to below two percent of the total storage, including ground water (1.688%), atmospheric (0.0009%) and biota (0.0001%). Evaporation is influenced by mean sea level as it changes annually as well as during ice ages or supercontinent formation. The importance of these small reservoirs is important. For instance, water vapor in clouds covers 60% of the planet in clouds at all times, which reduces solar radiation reaching the ground. And, living organisms pump water through their leaves, which evapotranspire it to the air. Much of the fresh water (65%) on earth is in underground reservoirs and aquifers.

Core heat flow and the radiogenic decay of elements perpetually creates new oceanic lithosphere as hot magma ascends and causes crustal spreading. This heat flux, especially from ocean vents, also supports bacteria and chemotropic archaea. Life either began near these vents or near the surface.

Water was the basis or medium of life. Water is necessary for life. It is essential for metabolism. It makes up most of living biomass by weight. Water makes up most of life, as life found a way to incorporate water in its cells. Water donates its hydrogen in photosynthesis.

Life would be confined to pools of water, except that water is evaporated and then precipitates back to pools. Water cycles between oceans and pools to the atmosphere and through the land. This cycling allows photosynthesis to occur in land and air as well as in pools. The sun drives the cycle and the ocean dominates it. A small amount is stored in living beings, although evapotranspiration pushes more water vapor into the atmosphere (~10%).

Water is a wonderful solvent. The basic elements of life, especially carbon, oxygen, nitrogen, and sulfur, can be transported in water in solution. The cycling of these and other nutrients keeps the biosphere functioning. These cycles are carried to some extent by the water cycle, although they are driven by plant and microbial metabolisms rather than by the sun. Mineral cycles participate in the water cycle, although they are pushed by a tectonic cycle that binds or liberates them.

Although the overall productivity of the oceans is about the same levels as deserts, some areas of the ocean are as diverse and productive as tropical forests or temperate wetlands. Life in the ocean is limited by physical factors, from the abundance of nutrients to light, gas, and salt. The ocean has effects on the erosion of land as well as on the amount of water in air and the circulation of the atmosphere.

3.6.3. *Atmosphere*

The atmosphere is composed of gases, mostly nitrogen and oxygen, with argon and traces of others gases and particles. The first atmosphere was thought to be Argon and other inert elements. Continued development added CO_2, H_2O and N_2, which would create a weak reducing atmosphere. After life formed, it added more ammonia and methane (from the anaerobic decay of biomass), which would have contributed to a strongly reducing atmosphere. CO_2 would have been reduced from 1000 times current to 100 times current levels by 2.5 billion years ago. At the same time, there was burial of organic carbon as a result of tectonic movements. After that the oxygen content rose from the activities of phytoplankton. The rise in atmospheric oxygen to 5-18% of current level may have triggered the evolution of animals.

The atmosphere is the place of gaseous cycles, such as water vapor, oxygen, and nitrogen. Changes in the atmosphere is driven by solar energy. This is one reason why the atmosphere is so dynamic; The distribution of heat drives very rapid circulation patterns, much faster than oceanic or terrestrial changes. The composition of the atmosphere creates a greenhouse effect that traps solar energy.

3.6.3.1. Anatomy of the Atmosphere

The atmosphere stratifies at different densities. The temperature profile is not monotonic; in fact the temperature decreases into the tropopause, but then starts increasing to the stratopause, where it starts decreasing again until after the mesopause, it increases to over 100 degrees Centigrade.

Nitrogen comprises the bulk of the atmosphere (approximately 78%). Nitrogen is the one element found almost entirely in the atmosphere—there's very little on land or in the sea. Nitrogen cycles slowly through the earth's system. A molecule of nitrogen gas is made up of two atoms very tightly bound together. It takes tremendous amounts of energy, such as produced by lightning or fires, to break the bond.

Nitrogen is a key element in proteins and nucleic acids and so is essential to life. But, if nitrogen is not available in a usable form, living beings and ecosystems would fail to thrive. There are bacterial species that specialize in taking nitrogen from the air and converting nitrogen into different usable forms. This conversion is difficult due to the high stability of gaseous nitrogen, which has two nitrogen atoms triple bonded together.

These bacteria also release nitrogen from organic material back into the atmosphere. This process, denitrification, returns nitrogen to the atmosphere where it is once again available for fixation. Some bacteria carry out denitrification under anaerobic conditions.

Oxygen is found in the atmosphere at a stable concentration (approximately 21%). Because it is a very reactive element, it can quickly combine with other elements and disappear from the atmosphere. Yet it persists, and in high concentration. It is the cycling of oxygen through photosynthesis and respiration that accounts for its presence and stability. A world without cycles, without life, would retain little if any oxygen in its atmosphere. Oxygen is a direct product of the metabolism of living forms.

Some atmospheric oxygen is bumped high into the upper reaches of the atmosphere, the stratosphere, where, it is converted into a new compound, ozone, in a series of reactions powered by solar radiation. Ozone absorbs ultraviolet (UV) radiation from the sun, which has the effect of protecting many life forms from damage. Ozone molecules unform and reform, as they are carried around the upper atmosphere. The formation and destruction of ozone by various reactions have been in equilibrium for hundreds of millions of years.

3.6.3.2. Physiology of the Atmosphere

Microorganisms can be lofted high into the atmosphere (beyond the troposphere). They can exist on their own or in aggregations. Viruses, bacteria, fungi, spores, and pollens may be carried far in the atmosphere, before returning to lower and more comfortable levels to reproduce. Many do not survive the temperatures or the UV radiation. Some forms are resistant to UV-rays, X-rays, and gamma rays. Birds and insects frequent the atmosphere for movement and hunting. Most travel and hunting occurs at low altitudes. Migratory birds (and a few insects) hit higher altitudes, which allow higher flight speeds.

Living ocean processes contribute to massive landforms as well as to the composition of the atmosphere. For instance, sedimentation of a small fraction of the biomass removed carbon and released oxygen to the atmosphere. This biomass formed large deposits of hydrocarbons that were cooked and distilled to forms of gas, coal, and oil. The process of biomineralization put carbon, calcium and phosphorus into great volumes of sediments that were elevated by tectonics to form large mountain ranges. The atmosphere, through reflection, can limit the amount of solar radiation received. Living organisms can change the content of the atmosphere, especially in terms of water vapor, nitrogen, and oxygen.

3.6.4. *Biosphere*

Vladimir Vernadsky developed the idea of the biosphere, expanding and refining the term from Eduard Suess. Vernadsky pointed out that none of the works in geology treated the biosphere as a whole. The biosphere is the network of living organisms as they live in groups and patterns. Vaclav Smil notes that the atmosphere, hydrosphere and planetary crust are defined by obvious physical discontinuities, but that living organisms have invaded all three. He also notes that the boundaries are moveable and interpenetrating. In the atmosphere, microorganisms can be found throughout the troposphere and in the stratosphere (50 km in vertical distance). Birds can also fly to impressive heights (5-9 km). In the oceans, the entire water column (to over 10 km) has eukaryotic organisms, such as fish and whales, as well as viruses and prokaryotes. Hydrothermal vents host ecosystems on the ocean floor, where some organisms can tolerate high water temperatures. Some microorganisms and insects can be active in temperatures below freezing. In the land surface, soil-forming organisms work in the first 20-30 meters; Smil suggests that the extent might be as far in as 7 kilometers. Viruses extend the reach of the biosphere. Cryobiotic forms also extend it. And, autotrophic ecosystems near thermic vents, without sunlight or chlorophyll, energized by the oxidization of sulfides.

In general, without knowing an exact path, life seemed to have risen out of self-defining processes. Christian de Duve suggests that life began from a 'protometabolism,' after short polypeptides were formed from thioesters of amino acids in more acidic water (in volcanic lakes or near hydrothermal vents). A high-energy thioester bond may have cycled to form cells or RNA. Perhaps the origin was cold, however, and life formed on a metallic substrate under different conditions, based on inorganic nutrition and progressing to photosynthesis and organic matter. Smil notes that we know that the earliest organisms were prokaryotic, anaerobic, and tolerant of ultraviolet fluxes and planetesimal collisions. Many early living forms were cyanobacteria, which survived in extremes of salinity, acidity, extreme temperatures, desiccation, radiation, and low oxygen. Other Archean microorganisms were methanogens and sulfate-reducing bacteria.

Bacteria dominated the planet up to 1.8 billion years ago, when the first eukaryotic micro-algae formed. Bacteria still dominates in terms of biomass. Eukaryotes became short-lived specialists, rather than slowly-evolving generalists. Life converts solar energy into foods. Vernadsky concluded that life forms in indivisible and indissoluble whole, where the parts are interconnected with living beings as well as the environment. The total of life is the biomass, which grows and shrinks, depending on the global environment, on seasonal cycles, and on long-term cycles and changes. This biomass has energy flows.

Living organisms drive global cycles of nutrients and elements. These cycles provides elements and nutrients to living organisms, including humans. Furthermore atmospheric gases are biogenically formed, especially oxygen. Life produces sulfides and bind calcium. The elements circulate in the biosphere in regular, more or less circular paths called biogeochemical cycles. The movements of the chemical elements necessary for life are nutrient cycles.

Human beings have started to evaluate these 'free' services' of cycles and biomass in monetary terms. Regarding ecological services, the biosphere can supply resources, as well as absorb wastes (materials and emissions), however, it is limited by amounts and rates. As George Woodwell (1976) put it, humans live as "one species in a biosphere whose essential qualities are determined by other species." Humans metamorphosed with a technosphere. Perhaps a biosphere reproduces at the end of a chain of metamorphoses.

3.7. *Design Factors: Intricate Patterns*

What are the big patterns in local systems? At various times, ice has covered a medium to high percentage of the land and water area. The continents move over deep time and will continue to do so until the energy of the planet core is exhausted. At some time during the development of the earth, natural processes created a pattern that guaranteed the maintenance and reproduction of a system of processes. These processes were considered living. The forests move across continents following the patterns of water. Mid-range patterns include vegetation, from oceanic algae to terrestrial forests. Small patterns include food webs and worm tunnels.

3.7.1. *Distribution & Density*

A species population has unique properties, such as density, mortality, natality, potential, dispersion, age distribution, growth form, and structure (isolation, territoriality). Studies in secondary production are difficult; the species must be measured for population density, age distribution, food consumption and utilization, growth, and reproduction; bacterial, fungal and parasitic populations must also be considered. Trees penetrate the soil, which can be penetrated much easier than the parent rock, to get water and nutrients to grow. The nature of soil determines the kind of vegetation on a site and the distribution of vegetation between sites. Thus, soil limits the occurrence of a species within the range allowed by climate. Thoreau also spent a lot of time correlating land use and history with forest composition. He described differences between wilderness and forests disturbed by insects and fire. He studied the role of the soil in the distribution of trees. Soil moisture, even in a homogenous stand of trees, has a mosaic pattern, due to species and individual differences in water distribution. Thus, a diversity of microsites can govern the pattern of future tree establishment.

The distribution of forests is influenced by climactic, edaphic (soil), and physiological factors. The first two are also effected by the third, which creates microclimates and soil particularities. Physiological factors include configuration (such as mountains and valleys), altitude, slope, and exposure. The amount of water influences the occurrence, growth, development, and the distribution of forests, because water is vital for life.

Wildlife that inhabit the soil environment are sensitive to soil contamination. Air emission can cause reductions in soil organisms and shifts in trophic structures, such as insectivorous bird species. A reduction or change in decomposers can result in a decrease in litter decomposition and nutrient cycling. The distribution and abundance of salamanders may be influenced by soil pH. In the United States, approximately 50 percent of the species of frogs and toads and 30 percent of the species of salamanders use ephemeral forest ponds for reproduction. These small pools and ponds can be acidic because they receive snowmelt and spring rains that have little contact with the soil buffering system.

In their studies of biogeography (simply the distribution of life over an area of the earth), MacArthur and Wilson hypothesized a general theory of island biogeography based on the observation that the number of species (species richness) on true ocean islands is lower than it is on the mainland. The theory states that, assuming all other factors are equal, large islands have more species than small islands (according to a species/area equation). The number of species on an island reaches an equilibrium between the extinction of existing species and the immigration of new species from other areas. The rates, however, are a function of island size and remoteness. MacArthur and Wilson suggest that small islands have a reduced habitat variety, reduced immigration of species, and higher extinction rates of species; furthermore, the climatic variability is altered. These variables affect the stability and longevity of populations.

It would be difficult to develop strategies to prevent the depletion of genetic information, especially with intense public and industry demands on forests, without knowledge of the diversity and distribution of genes in the tree population. Although study of gene flow and genetic recombination is needed, a lot of this knowledge can come from careful observation of the forest, the mating systems of trees—e.g., pine species require pollination as a mechanism for outcrossing—and the shape and size of a forest area. Smaller organisms with limited distributions tend to become extinct faster.

Diverse shapes for the preserves would minimize the dangers from physical and climactic changes. The greenhouse effect could drastically alter the species distributions in preserves, with the loss of many species. Placing the preserves on heterogeneous soil types and topographies increases the chances that the temperature and moisture requirements of species would be met. A range of elevations across areas would minimize the effects of climactic change—and the possibility of extreme change is rarely considered in wilderness design.

The biosphere varies considerably in its density, from deserts and tropical seas to marshlands and tropical forests.

3.7.2. Surprise & Joy at Diversity

There are emergent properties at each level of organization, such as the diversity of species and the structure of a food web. Furthermore, each level of integration has distinct attributes. A community has 'species diversity,' which is meaningless at the population level.

Ecosystems are what they are as the result of the diversity of life—all forms of life contribute some value to the system as a whole. The whole system is already self-ordering and self-renewing. Any ecosystem not subjected to outside disturbance changes in an orderly and directional way: The complexity of structure increases and the energy flow per unit biomass decreases. The physical environment limits the type of change. Homeostatic (or homeorhetic) mechanisms protect the system from many disruptions. Thus, maturity is self-preserving. When conditions change, the system may go to a state of lower or higher diversity.

This concept of maturity, as an attribute of a community, is related to structural complexity and organization. Maturity increases with time in an undisturbed community. The species diversity, that is, the information content, of a community also increases with maturity, leading to a more complex spatial structure. Diversity incorporates species richness (how many different kinds are present) as well as a measure of abundance—how many of each, as individuals or biomass. Other aspects of diversity, such as life cycles, are less often considered. The energy in a mature system goes to the maintenance of order and less for the production of new materials. In general, diversity is higher, and life cycles are more complex; symbiosis between species increases, and nutrients are conserved. Complexity and diversity offer advantages for living forms. Complexity allows increases in size, which allows the colonization of harsh environments. Diversity allows more effective behavior through specialization; for example, a specialized organelle may digest less common molecules.

The number of species in an area is related to the idea of diversity, which is a description of the variety of life forms (or the condition of being different in quality—the greater the number of differences, the greater the diversity). Diversity can also be related to the increase or decrease in ecosystem measures, such as productivity, biomass, and stability. Diversity is related to genetic information as well as to relationships of species, such as predation or mutualism. In fact species number may increase depending on interactions, resources, niche overlap, redundancy, or exploitation efficiency in a community (as in old-growth). For example, deciduous forests in Europe typically have fewer species than deciduous forests in Asia or North America (from ice-age effects).

Diversity, as a measure of genetic variability in ecosystem, is decreased by domestication and agriculture. Diversity is basically the bag of tricks for organisms facing environmental perturbations. Coevolution with humans reduces or destroys integration with other species; then, stability depends on human control. And human control is not always certain. Natural selection is the process for strengthening biodiversity, while the unnatural selection prompted by people artificially robs the Earth of its most important genetic resources.

The idea that diversity promotes stability has been defended and attacked elsewhere. Concepts of stability often include consideration of diversity. Using information theory, Margalef claims that higher information content and increased interactions promote stability. Margalef was the first to propose that diverse systems were more stable than less diverse. Robert May and L. Ashby argued that the opposite may be true. However, by many definitions of stability the former is true—there is still a problem with the term in ecology. The structure of food webs may enhance stability. May suggests that communities are more stable if they are compartmentalized (as holons, perhaps), that is, where subunits within a unit have stronger interactions than those between units.

Biomass seems to increase with diversity. The native animals in Kenya, for instance, are diverse and adapted and support a greater quantity of biomass than the domestic animals that are replacing them. Diversity is an expression of the dynamic properties of a complex system. It seems common that a complex of circumstances allowing a high diversity also permits high stability or constancy in taxonomic composition: "nature tends to become baroque in situations permitting high maturity, with little energy left for large changes," according to Ramon Margalef. Stability and diversity are not a matter of definition for Margalef, but reflect crude impressions of the behavior of physical systems.

Margalef states that biomass and primary production increase during succession; but the ratio of productivity to total biomass drops. According to E.C. Pielou species diversity decreases and pattern diversity increases during succession. There is also an increase in the proportion of inert matter, and an increase in structures like paths and burrows.

We are actors in a tremendous presentation. This metaphor formed part of the basis of a worldview where nature was a theater of violent competition. The frame where a metaphor originates carries the conceptual baggage of the time. So, over time the play supported the idea of superiority of "favoured" races in the struggle for existence and emphasized the role of competition in biological and cultural situations, at the expense of other interactions.

Alas, the metaphor needs to be expanded. All species play a temporary role in the local stage in the ecological theater. Herb Hammond notes that all the actors and acts are essential. We are foolish to think some species are more important than others—that is ignorance or wishful thinking. All species and things contribute to the functioning of the whole, including rocks and gaseous elements. But, some are invisible to us because of their size or longevity. Some play their roles in a clump of soil, others in continental landscapes. Some acts last less than a second; others take millions of years. Even if an actor seems to leave the stage or the act is over, they continue to influence the play with their corpses and elements.

There are many stages playing simultaneously in the theater, and it is not a one-act play. The human play has converted some of the stages, subverted others. The human actors are ridiculously egotistical and ignorant, pretending that the stages and theater is for them only, and others are support characters. They pretend that the less important people and freaks are in the audience, but the only audience is other stages with partial perspectives.

And, some plays are embedded in others. One play is the evolutionary play, where autopoetic beings drift through filters in the morphogenetic landscape. Another play is the human conversion of ecosystems and the urbanization of human communities (with some companions, familiars and pests).

We have trouble understanding the theater or the plays because of our physical, temporal and psychological and cultural limits. We see species and ecosystems as individuals, when most of them are in fact communities. We see walls and barriers, and not permeable filters. We see philosophical constructs of classes in isolated locales. We have problems dealing with motion, indeterminacy, ambiguity, and vagueness. We are seduced by logic and fallacies to believe that we can understand and control the play.

Although the scales involved in a planetary theater are linked by processes, in fact the scale of the planet and its evolving life forms is significantly larger and longer. As important as human changes are to us, and to the ecosystems from which we emerged and on which we depend, for the planet those changes may only bump the global system to another stable state, well within the range of those from the past 2 billion years. We are the stewards only of ourselves and companion forms, not of the planet.

On the other hand, Francis Bacon noted that we faced four major obstacles in understanding nature: The idols of the tribe, cave, marketplace, and theater. The perspective of the tribe is our inherent tendency to interpret and measure nature through our human senses, with their limits and scale. The idols of the cave refer to our personal peculiarities. Those of the marketplace refer to errors from language and culture. Most of the errors of the theater have to do with the fallacies of our philosophies, even those based on good metaphors.

Where does joy come in? We feel joy that things are always new and different, that there is enough to continue to provide and to inspire. We are actors in a process, but we are genealogical actors in ecological roles in the evolutionary play in the ecological theater. As long as we realize we are not everything or the center of existence, we can continue to feel joy.

3.7.3. *The Reality of Patterns*

Process applied to components yields pattern. Nature is composed of patterns. Organisms have characteristic patterns, such as the branching of trees or the cloud forms of tree crowns. Lichens have lobes, wood grain under stress has spirals. The cracks in tree barks form nets.

Regularities in systems are patterns. Patterns can be seen in things or even cultures. For instance, laws of genetics are natural patterns; human customs are artificial patterns. Where the natural ratios of females to males are altered by female infanticide or other action, the pattern is semi-natural.

Bunge distinguishes four kinds of real patterns: Laws, trends, correlations, and rules. A law is a stable pattern inherent in things; it is discovered; laws like gravity are boundless; biological laws are bounded, he says. After the 'Big Bang,' gravity was bounded. There are examples of social or cultural laws: "The inertia of a social system is directly proportional to the number of components and inversely proportional to its cohesiveness." Or: "Higher culture does not emerge in society until the basic needs of some of its members have been satisfied."

A trend is a temporary pattern, such as the globalization of capital or fertility. Trends can be reversed. A correlation, usually statistical, is a covariation of two properties, e.g., a correlation of sickness and education—but this correlation is problematic, due to fact that educated are wealthier and report their sickness more than the poor; the real correlation could be reversed. A better correlation is: Single-species forest stands can be correlated with standing armies in the northern hemisphere; standing armies were thought to be characteristic of people in tougher climates, which encourage large stands of trees. A rule (or norm) is a social convention set up by people, in force in a social system. The analysis of patterns is the strength of systems analysis.

Paul Shepard describes living natural 'objects' in terms of events that constitute a 'field pattern.' Relations are not prior to objects; they arise together. The wasp and the yucca coevolve; they are not co-linked by prior relations. Furthermore, a specimen is more than the sum of its species' relationships to an environment; it is an intentional being that, with other members of the species, can create niches, as well as adapt to them. Because the STEM field produces life, the qualities of life cannot be separated from its physical qualities. While it is true that living subjects are at a different level of description than events in field patterns, they should not be treated as ontologically subordinate. All of the aspects of the field have equal status. The ecosystem model, as a reaction to 'superorganismic' metaphors of early ecologists, attempted to be a field theory, but has been limited by its parentage, thermodynamics, and has been rejected by new practitioners.

Patterns are not still. A circular pattern through time can be recognized as a spiral (the earth's orbit for example). These patterns can be analyzed into events. Everything that exists has its place in the order of nature. This does not mean that reality is an organism or that everything is reduced to biological terms. It does mean that every thing resembles a living organism since its essence depends on the pattern in which it occurs, and not on its components. In some ways, patterns are prior to things, in helixes, light, fields, and ecology. Paul Shepard and others have written that relationships are as real as the objects that result from them. The science of ecology attends the overall pattern of relationships, beyond the details. The pattern should allow for surprises and discontinuities; it can do this if it is flexible.

3.7.3.1. Natural Patterns

Organisms are material patterns in space as well as energy moments. Even if energy is considered primary metaphysically, organisms are still composed of the atoms and molecules that energy forms under certain conditions of temperature and pressure, and they act differently than just 'energy vortices' or 'patterns of energy.' Furthermore, the organism must adapt to the environment, which implies having a memory and being capable of learning, and must reproduce, that is, duplicate its pattern in a separate being. Organisms are goal-seeking, and often stability is sought above change or complexity. Nor is the gene a permanent entity, although Dawkins argues that it is. The pattern lives longer than the molecule, and even the pattern changes with mating.

Nature consists of moving patterns whose movement is essential to their being. As a rope makes the knot visible, so the body is a pattern made visible. The body is a movement that maintains a topologically stable pattern; it is a vortex but not the water. The thing, the pattern, is a cross section cut through the movement. The mind is an invisible knot that is capable of recognizing both visible and invisible patterns, that is to say, a rope is not always necessary for the demonstration of a knot. Culture is also this kind of pattern. Culture can be analyzed into smaller blocks; the pattern of the whole organization is reflected at every division in differences of organization on either side of boundary. The wholeness of the character of a culture is reflected at every level. Patterning relates symbolic meanings in the context of a cultural system as a whole. The patterns form another level of meaning that has to be addressed in understanding a culture. Although the physical environment imposes limits and sometimes determines patterns and rates of change, the community controls the development of the system.

An ecosystem has a distinct pattern, related to health. According to E.C. Pielou species diversity decreases and pattern diversity increases during succession. Very small changes in complex, self-regulating systems, such as forests, that have large and important consequences, as when rainfall patterns shift after a forest is removed. Weather patterns on North America, for instance, have created 'nation states' (see Paul Colinvaux) of trees that surprised the first European naturalists because they were so different from the European forests. The large sulfur cycle is beneficial to marine and land organisms, important to the general pattern of production and composition. Some material cycles between other ecosystems or in larger patterns around the planet. Some of the cycles are daily; some are seasonal or annual; others are years or decades; a few, like the carbon cycle, are century or millennia long. Ramon Margalef proposes maturity as a quantitative measure of the pattern in which the components of an ecosystem are arranged. The life-form communities and physical elements are related in a definite pattern, which is a real but untouchable property (structure). In general, this structure becomes more complex as time passes, as long as the environment is stable or predictable. The structure acquires a historical character.

A Principle of Uniqueness can be recognized in living systems. History creates unique patterns, especially in forests. Each forest is unique in its parts and structure, in its matter, energy, forms, information, and in its dynamics and history. Some patterns in forests are scale-dependent; for instance, hemlock trees may dominate small clusters, but be scattered all across the entire forested landscape. Pathogens determine tree mortality, which drives forest and landscape patterns. That is to say, the pattern changes with the scale. This is true of processes in forests as well. A typical forest is composed of many patterns, including vertical layering (stratification), horizontal segregation (zonation), activity patterns (periodicity), food web patterns, social patterns (including reproductive), interactional patterns (from competition or mutualism), and stochastic patterns (from random events).

There are long-term trends in ecosystems, that is, patterns of: Primary productivity, organic matter accumulation, inorganic inputs and movements through soils and water, disturbances, and populations in a trophic structure. The actual substance of which the forest environment is made consists of patterns rather than things or individual species. The forest environment is generated by a patterning of ecological ebb and flow of energy, substances, individuals and species across a suitable landscape.

The forest ecosystem is a large-scale pattern of millions of minute events. The environment requires an enormous amount of minuscule local adaptations between the earth and its users. The landscape is a unique individual, a community, a dynamic system of interacting patterns—the human pattern is a part of it now and should be preserved as part of the whole pattern, but not necessarily as the only pattern or a completely dominant one.

The patterns of ecosystems are addressed best at a landscape level. Landscape ecology addresses the overall patterns of large-scale ecosystems (biota), thus considering the biogeochemical, atmospheric, and hydrological cycles in relation to the shape and extent of individual landscapes. The large images from satellites are exceptional for identifying landscape patterns, especially those related to the scale of species behavior, e.g., home range or breeding dispersion. Pattern can also be measured at the level of patch size and spatial relationships (that is, inter-patch distance), which is critical for relating the size of a habitat to the species in it—to apply the theory of island biogeography.

As Richard Hart mentions, patterns are the key to understanding the nature of ecosystems. Nature, for Alfred North Whitehead—this is one philosophical foundation for considering patterns—consists of patterns whose movement is essential to their being. These patterns are analyzed into events. Everything that exists has its place in the order of nature. This does not mean that reality is an organism or that everything is reduced to biological terms. It does mean that every thing resembles a living organism since its essence depends on the pattern in which it occurs, and not on its components. In some ways, patterns are prior to things, in helixes, light, fields, and ecology. Paul Shepard and others have written that relationships are as real as the objects that result from them. Ecology attends the overall pattern of relationships, beyond the details.

3.7.3.2. Cultural Patterns

Human beings create patterns, for food, places, events, and histories. Cultures create different patterns of living, from eating to building. Of language, relating, and changing. Many of the patterns are adaptive; some are not. Many patterns modify natural patterns. Humans have modified animal and plant associations in a different way, simplifying patterns of energy and chemical exchange, solidifying themselves at the end of many food chains as a dominant species. Patterns of eating have influenced the constitution of species and the very contours of the earth. Throughout their history, humans have used animals and plants for food and clothing. Animals were followed, herded, corralled, tamed, and finally bred. Plants were domesticated later. As technologies developed, human relationships with animals and plants changed. Hunting, grazing, and agriculture provoked large ecological disturbances.

The general patterns of living landscapes— Patches, Corridors, Matrix, Connectivity, Extension, and Geometrization—have been duplicated in some extent by humans in agricultural fields and cities. Places are patterns of things and webs of relations that can be understood by observing and participating. Human cultures have other big cultural patterns as well, such as: Overshoot, reproductive success, ecosystem conversion, stupidity and violence, stagnation, and the asymmetry of sex and handedness.

The problems of cultures, of natural ecosystems, and of modern, industrial, corporate, urban civilization, have been documented quite thoroughly. We have identified most of the problems in the problematique, from erosion, pests, and fertility loss, to population migration and diseases, and we have addressed them separately, using technological innovations or political adjustments. But, we have not dealt with them in a whole pattern. We have not understood them as complex large dynamic systems.

The intent of describing large-scale trends or patterns is to have human patterns fit with observed patterns in nature; patterns have a form, sometimes repetition, and sometimes regularity, but each of these is caused by some limiting factor. Fitting the pattern can lead to both continuity and predictability, and both of these are needed to adapt human activities to natural limits.

Figure 4-1. Natural Farming at Altazor Forest
(orchard with mixed crops, grasses, herbs, and flowers, after M. Fukuoka)

4.0. Designing Nature

On the assumption that nature might be more amenable to large ecological designs, we can start by addressing the most dramatic of recent changes, the atmosphere. Up until recently, most of our concern has been on local problems, such as pollution, which we have been able to correct or avoid by improving industrial processes. But, the atmosphere seems to be a global problem, with no simple or even complex solution.

It is equally important to consider water problems and water flows. Our planet is a water planet. Life is water; water is life. Life is water dependent. Living beings average 75-90% water by weight. Trees are standing water. Water is a molecule, a perfect solvent, a medium for dispersal, a resource, a commodity, and a habitat. Water defines a sense of place; its state and form can measure seasons and time. It can be a solid, a liquid, or a gas. Water as a critical element of life; it is a means of renewal and regeneration. Water is a link to other ecological issues. Water is a useful metaphor for designing.

4.1. *Global Problems: Atmospheric Change*

Due to its rotation around the sun, and to imperfections in mass distribution, the earth tilts on its axis. The moon stabilizes this tilt. This obliquity of the ecliptic creates seasonal variations, to which most plants and animals have adapted. Any changes, even relatively small ones, could be catastrophic for climate. Furthermore, plate tectonics can contribute to changes in global climate change, ocean circulation, and ecosystem disturbance. Climate change is controlled primarily by cyclical eccentricities in Earth's rotation and orbit, as well as variations in the sun's energy output. 'Greenhouse gases' the atmosphere also influence Earth's temperature, but in a much smaller way.

There are several major causes of global temperature shifts. Astronomical causes include the three Milankovitch cycles:
- A 21,000-22,000 year cycle of the Earth's combined tilt and elliptical orbit around the Sun (precession of the equinoxes)
- A 41,000-42,000 year cycle of the +/- 1.5° wobble in Earth's orbit (tilt). 2004 is in the middle of one.
- And, a 100,000 year cycle of variations in shape of Earth's elliptical orbit (cycle of eccentricity). In 2004 there was only a 6% difference between January and July.

The sun has two cycles of variability of sunspot activity, 11 year and 206 year cycles. And the solar system follows paths that cut through intersolar gas clouds and lanes every billion years.

The atmospheric effects on temperature changes include heat retention due to atmospheric gases, mostly gaseous water vapor (not droplets), and also carbon dioxide, methane, and a few other gases—referred to as the "greenhouse effect." The atmosphere also controls solar reflectivity due to white clouds, volcanic dust, and white polar ice caps.

Tectonic changes can also cause temperature shifts. For example, continental drift can cause changes in the circulatory patterns of ocean currents. This cycle can initiate an ice age only when continental drift puts land near one of the poles, north or south, apparently regardless of the other cycles. 'Sea floor spreading' at central ridges, associated with continental drift, can cause variations in ocean displacement.

Global climate and temperature cycles are the result of a complex interplay between a variety of causes. These cycles and events overlap, sometimes enhancing one another, and sometimes interfering with one another. Therefore it is difficult to identify a statistically significant trend in climate or temperature patterns from just a few years, decades or centuries of data. At the extreme of cycles, the variation in sunlight received is less than a tenth of percent, but that difference can change atmospheric temperature by 9 degrees F (12.7 C).

Using various techniques and evidence to reconstruct a history of past climate, climatologists have found that during most of the Earth's history global temperatures were probably 8 to 15 degrees Celsius warmer than our current era. In the last billion years of climatic history, warmer conditions were broken by glacial periods starting at 925, 800, 680, 450, 330, and 2 million YBP.

Around 540 million YBP, organisms began building skeletons of carbonate, as they absorbed CO_2 from seawater. This factor alone may have decreased the number of ice ages. Andy Ridgwell and his colleagues hypothesize that shell-forming plankton 300 million YBP stabilized the planetary thermostat. Planktonic calcifiers changed the positive freezing loop because they floated on the ocean and were not tied to continental shelves. This prevented too much CO_2 absorption and further cooling. About that time forests were covering the land, and this also tied the carbon cycle down somewhat. Finally, 55 million YBP, modern coral reefs drew volumes of CO_2 from the atmosphere.

During the period from 2,000,000 to 14,000 YBP, the Pleistocene Ice Age, large glacial ice sheets covered much of the Northern Hemisphere for extended periods. The extent of the glacier ice during the Pleistocene was not static, however. There were periods when the glacier retreated (interglacial) because of warmer temperatures and advanced because of colder temperatures (glacial). During the coldest periods of the Ice Age, average global temperatures were probably 4-5 degrees Celsius colder than now, in the Holocene epoch.

The current glacial retreat began about 14,000 years ago. CO_2 in the atmosphere increased steadily since then. The warming was interrupted by a sudden cooling, known as the Younger-Dryas, at about 10,000-8500 BCE. This event may have been caused by the collapse of the ice dam holding back Lake Agassiz. The release of massive quantities of fresh water into the North Atlantic Ocean may have altered vertical currents in the ocean, especially the Gulf Stream, which exchange heat energy with the atmosphere. Warming resumed by 8500 BCE, and by 3000 BCE average global temperatures reached their maximum level during the Holocene—1 to 2 degrees Celsius warmer than now. The atmospheric temperature and CO_2 levels reached levels similar to a previous interglacial cycle of 120,000-140,000 BCE. From beginning to end this cycle, the Emian Interglacial Period, lasted about 20,000 years, before the next ice age returned.

A shift in the orbit of the planet between 10,000 and 4000 BCE brought 7-8 percent more sunlight to the northern hemisphere. Rainfall in Mesopotamia increased 25-30% and available moisture increased by about 700 percent. After 3800 BCE the orbit reverted and the rainfall pattern shifted north. This event forced changes to tilling fields, double cropping, and irrigation canals by 3100 BCE. In this "Climatic Optimum," many ancient civilizations began and flourished. The city became a key adaptation to drier climates.

The global climate and the biosphere have been in constant flux, dominated by ice ages and glaciers for the past several million years. Approximately every 100,000 years Earth's climate warms up temporarily. These warm periods, called interglacial periods, appear to last approximately 15,000 to 20,000 years before regressing back to a cold ice age climate.

Statistically, we are nearing the end of this interglacial. Global atmospheric heating during the current interglacial warm period has greatly altered the environment and the distribution and diversity of all life. Approximately 15,000 YBP the earth had warmed sufficiently to halt the advance of glaciers, and sea levels worldwide began to rise. By 8,000 YBP the land bridge across the Bering Strait was drowned, cutting off the migration of men and animals to North America. Since the end of the Ice Age, the temperature has risen approximately 16 degrees F and sea levels have risen 300 feet. Forests have moved north, reclaiming land areas free of ice.

During the last 100 years there have been two general cycles of warming and cooling recorded in the U.S. We are currently in the second warming cycle. Overall, U.S. temperatures show no significant warming trend over the last 100 years. Ground- based recording stations, show an upward distortion of increases in ground temperature over time, due to heat island and other effects. Satellite measurements are not limited in this way, and are accurate to within 0.1° C. They are widely recognized by scientists as the most accurate data available. Significantly, global temperature readings from orbiting satellites show no significant warming in the 18 years they have been continuously recording data.

4.1.1. *Anthropogenic Change & Global Climatic Chaos*

Is the climate entering a chaotic stage? Global warming is an unfortunate term, on the order of words as global toasting or tanning. It implies a comfortable, warm future where plants grow more and people need less heat from coal. Global burning or suffocation might raise more alarms. How we use language influences how people react to catastrophes.

Human additions to total greenhouse gases play a relatively small role, contributing about 0.2% - 0.3% to planetary greenhouse effect. Seemingly small but effective in causing change. Total human contributions to greenhouse gases account for only about 0.28% of the greenhouse effect. Anthropogenic CO2 comprises about 0.117% of this total, and man-made sources of other gases (methane, nitrous oxide (NO_x), other misc. gases) contributes another 0.163%. Of course, CO_2 and methane are triggers, and do not need to be present in massive amounts. CO_2 is a trigger for a more powerful greenhouse gas, water vapor, as heating allows the atmosphere to take up more water. CO_2 is very long-lived, over 100 years. 56% of it is still aloft. CO_2 levels may be as high as in the Eocene, 50 million years ago.

Do we need icecaps? Do we need permafrost? Low sea levels? Should we worry about the health of the planet? Is it just us that is going to get discomforted? Climate change could slow human economic and political progress, especially when combined with scientific uncertainty. More international cooperation on things like global climatic chaos is needed, in the form of the Kyoto Protocol (climate) or the Montreal Protocol (ozone). Better organization is needed. Global designs are needed to help restore forests and grasslands and develop alternative energy sources. Should we try a high-tech solution, with massive programs, or common-sense conservation?

The optimum growth for plants is 25 degrees C. For humans it is about 14C or 41F. Optimum CO2 is higher for plants than for animals; the plants would make more tissues but they would be tougher and less digestible. Plants might pump down CO_2 but that might increase O_2. That might cause a runaway positive feedback. This interglacial is warmer now, so it is more fragile in terms of interference.

What are we doing? We are transforming ecosystems, such as forests and grasslands. We are inventing and using novel chemicals, such as CFCs and plastics. We are producing ozone and methane. And, we are expanding our impacts, by using more energy and making more people. The impacts of human cultures on ecosystems (and the planet) are tremendous. To the atmosphere, we add carbon dioxide, methane, trace gases, carbon monoxide, nitrogen oxides, chlorofluorocarbons, acid rain, and trace elements. To the soil, we change the salinity and acidification; we let it erode. We pollute water, then changes its courses and levels. We change the shape of the land, excavating, dumping, and stabilizing and destabilizing landforms. We change vegetation cover by cutting or converting, by grazing or burning, by extinguishing species or adding exotic invasive species, creating deserts and wastelands. We domesticate some animals and extinguish others, then we move around many others. All of these impacts and interactions have consequences, such as warming, altering cycles, making irreversible changes, simplifying, and starting extinction spasms.

4.1.2. *Greenhouse Gases*

There are several possible causes of global climatic chaos or planetary heating. The first is increases in Greenhouse gases, such as carbon dioxide, from burning fossil fuels, volcanoes, deforestation, and respiration. What could be safer than carbon dioxide; after all, plants need it? Methane, from increase in agricultural field, rice paddies, cow pastures, termites, mining, and gas leaks. Methane is the simplest hydrocarbon, CH_4 which could be increased by the switch to a hydrogen economy. Gasoline is made of hydrocarbons, like C_8H_{18} (octane).

Water vapor increases from evaporation, and changes in water vapor are triggered by changes in CO_2 and other gases. Nitrous oxide increases from fertilizers, deforestation and fossil fuel burning. Sulfur dioxides increase from industrial and utility emissions. Ozone increases from engines and petrochemical processes, photochemical processes, and solvents.

Chloroflourocarbons (CFCs), Bromines & HFCs increase from propellants, coolants, and solvents. The HFCs and CFCs are 10,000 times better at capturing heat energy and last for many centuries. CFCs are wonder gases invented by Dupont in 1930s; their chemical properties made them good refrigerants and propellants in aerosols, and they were also used in solvents, plastics and foams. There is now about half a pound for every living person on earth. But, it had a side-effect, it destroyed ozone, by mimicking the ozone molecule and leaving only oxygen, not ozone. And many last decades or centuries, such as dicholodiflouromethane (Ccl_2F_2). Dupont and many countries, especially European, are banning halocarbons and replacing them with less hazardous compounds, such as $C_2H_3C_{12}F$, which are more easily destroyed by low surface oxidation. These noncorrosive, nonflammable, low-toxic chemicals have fewer risks, except for scale effects and "side-effects."

Deforestation can be related to changes in the atmosphere, as rainforests, tropical forests, northern, and pacific forests are heavily cut. The evaporation from forests is a form of local and planetary cooling. Amazonia as a cooling system is worth 150 trillion dollars according to Lovelock. That means each hamburger, taking 5 square meters of land, really costs $65 USD (assuming air conditioning costs of $1300 per hectare and perfect efficiency).

Ozone loss is a contributor to heating. Ozone absorbs ultraviolet light, which splits O_2 into O which can combine to make ozone (O_3). Ozone is lost from photochemical reactions with pollutants and electrical discharges. Pollutants also absorb infrared radiation.

Urbanization and overpopulation are contributing factors. Human-modified environments change the surface area and increases the change. Energy comes from burning organic compounds and fossil fuels is 90 percent of that.

What do we know about the heating? Yes, temperatures are rising, about 1 degree F since 1600 and maybe 1 since 1900. Maybe 1 degree more by 2030 or 9 total by 2100. And temperatures are linked with carbon dioxide change. Is it an abnormal heat increase? Have humans interfered? Human production of carbon gases has affects on the atmosphere. Carbon dioxide is recognized as a greenhouse gas, and increasing levels in the atmosphere are linked to heating and climate change.

Greenhouse gases are increasing. This is certain. The average temperature is increasing. This is certain. Extinctions are a certainty. The extinction of species is certain. There is a high degree of certainty. Temperatures of mountain habitats are easily measured, as are conditions that mountain species can tolerate. As climate warms they have nowhere to go but up, and we know the height of the mountains. We know the planet will heat by 2 F this century, or 5 F if business continues as usual.

What is uncertain is the rate, magnitude, and patterns of global climate change, as well as the long-term impact on the biosphere and the anthrosphere. Things are complex. By doubling methane and adding CFCs it is the same as an additional 40 PPM of CO_2. The interglacial only added 100 PPM of CO_2. Because the system is so large-scale, there is a lag time of hundreds of years. The length of the cycles varies. The entire CO_2 cycle may be 1 million years long.

4.2. *Local Regional & Global Problems: Water Flow*

Our planet is a water planet. Life is water; water is life. Life is water dependent. Living beings average 75-90% water by weight. Trees are standing water. Water is a molecule, a perfect solvent, a medium for dispersal, a resource, a commodity, and a habitat. Water defines a sense of place; its state and form can measure seasons and time. It can be a solid, a liquid, or a gas. Water as a critical element of life, a means of renewal and regeneration. Water is a link to other ecological issues. Water is a metaphor.

Water defines and shapes habitats. The quality, quantity, form and fluctuation of water significantly influences the species and population of a watershed. The interaction of air, water and energy shape the weather and storm patterns of the planet. Water is critical for survival. Water is a crucial, although little-understood, resource. Human economic survival depends on water and its interconnections of place. Water ways and water sources have historically defined the locations of cites, communities and cultures—and been a limit of their prosperity. Water furnishes recreation, enjoyment and awe.

There is an uneven distribution of water on the planet. Water occurs in lakes, pond, streams, rivers, rapids, pools, currents, bayous, bays, oceans, estuaries, glaciers, snow, ice, frost, clouds, mist, fog, humidity, tides, floods, storms, springs, aquifers and ground water. It impacts the physical and ecological environments. It defines and connects ecological systems. It is collected and ingested by every living being.

Humans require water. In A. Maslow's needs pyramid, water is a basic physiological need, although it can contribute to safety and esteem needs in various ways. Water has to be in the connections at all levels for people and systems. That is, people have to be connected to the source of their water, as well as to other people to provide social comfort.

Human beings may have developed as a semi-aquatic species, according to E. Morgan. We may have used it for birthing as well as protection. We have used water for transportation, from swimming to boats and ships. We have used it for drinking and washing food. Most environmental issues are water issues. Like land, water can be owned in a legal system; rights can be assigned, fair use can be allowed. There are issues with salt water intrusion, storm water runoff, yard and agricultural run off, non point source pollution issues, waste water treatment, freshwater and marine habitat loss, aquifer recharge, ground water protection, ground water contamination, desalinization, water diversion projects (dams and conduits), causes and impacts of changing water table heights, world wide water supply issues and projections, water demand issues, changing water use patterns, alternative water treatment options, alternative water catchment options, alternative irrigation techniques, watercourse restoration, aquatic habitats and watershed management. All of these issues and problems have economic and political dimensions.

4.2.1. *The Water Cycle Restated*

The water cycle is mostly driven by climate, meteorology, and geology—evaporation and movement down hills. The sun drives climate and meteorology. The sun provides the energy that drives the weather systems that move the water vapor (clouds) from one place to another—otherwise, it would only rain over the ocean or plant communities. Precipitation occurs when water condenses from a gaseous state in the atmosphere and falls to earth. Evaporation is the reverse process in which liquid water becomes gaseous. Once water condenses, gravity takes over and the water is pulled to the ground. Gravity continues to operate, either pulling the water underground (groundwater) or across the surface (runoff). In either event, gravity continues to pull water lower and lower until it reaches the ocean or the lowest land basins. Geology and topography influence the cycle, by slowing or accelerating it.

Frozen water may be trapped in cooler regions of the Earth (the poles, glaciers on mountaintops, etc.) as snow or ice, and may remain as such for very long periods of time. Lakes, ponds, and wetlands form where water is temporarily trapped. The oceans are salty because any weathering of minerals that occurs as the water runs to the ocean will add to the mineral content of the water, but water cannot leave the oceans except by evaporation, and evaporation leaves the minerals behind. Thus, rainfall and snowfall are comprised of relatively clean water, with the exception of pollutants (such as acids) picked up as the waster falls through the atmosphere.

Organisms play an important role in the water cycle. Water moves out of organisms quickly in most cases. Animals and plants lose water through evaporation from the body surfaces, and through evaporation from the gas exchange structures, such as lungs. In plants, water is drawn in at the roots and moves to the gas exchange organs, the leaves, where it evaporates (and is called transpiration) to the atmosphere. In both plants and animals, the breakdown of carbohydrates (sugars) to produce energy (respiration) produces both carbon dioxide and water as waste products. Photosynthesis reverses this reaction, and water and carbon dioxide are combined to form carbohydrates.

The amount of water on the planet is relatively constant. Juvenile water, which originates deep in crust, does no add much to the biospheric water cycle. Little water is lost to the reduction the water molecules.

4.2.2. *Interactions of Cycles*

Certain substances move endlessly throughout the earth's biosphere, hydrosphere, atmosphere, and geosphere, existing in different forms and being used by different organisms at different times, but always moving, always circulating. In addition to water, many other substances such as nitrogen, oxygen, and carbon cycle through the earth and atmosphere. These cycles are important to individual animals and plants and even to entire ecosystems. But we're less familiar with the notion that these cycles fundamentally influence the planet as a whole, dramatically and unmistakably altering the earth's atmosphere. When you think about it, this influence only makes sense. The atmosphere is the greatest, fastest, and most reliable global transport system we have.

The carbon cycle and heat cycle of atmosphere are limits, as are heat and water evaporation. The ability of carbon sequestration by plants is limited by water and nutrients. The health of a water system is more important than access or use. Water and phosphorus intersect in the water cycle at various times.

4.2.3. *Problems with Disrupted Cycles*

The biogeochemical cycles of oxygen, carbon, nitrogen, phosphorus, sulfur, water have stayed within certain limits, and the environment has been constant enough for organic evolution, but variable enough for natural selection to be challenged. The key term is limits. The carrying capacity of the ecosystem can sustain only a given amount of life, and there are other limiting factors that may not have a short time scale.

Humans are modifying the cycles. We are changing the rates of nitrogen fixation, which skews the plants, especially due to leakages, applications and losses. We are increasing phosphorus 12-fold in systems. Waterborne phosphorus and nitrates cause eutrophication of streams and lakes, and even some estuaries and seas. We are changing albedo of planet, draining aquifers, diverting rivers (with dams and canals), changing evapotranspiration (from deforestation). Water withdrawals, globally, have increased 15-fold, although the population only increased by six fold (the same kind of increase holds true with energy use, also). These actions affect the integrity of the biosphere and global cycles. And, they do so rapidly. They may alter the course of biosphere development, away from the human comfort zones. How can we appraise the possibility or probability of change?

There are problems with water scarcity and water stress, not just in Africa but in Europe, Spain, Italy, and Germany. Some of these scarcities are hidden under national reasons. The demand for water increases as population grows. Cities may become more vulnerable to drought and heat waves. Water reservoirs are being drained. Water reservoirs are a significant feature of the planet. Water reservoirs take up an area of 0.5 million km^2 (less than 1% of ice-free land area). Stocks of water are part of systems. They usually change slowly. They can act as drags, delays, lags, buffers, ballasts and "sources of momentum in a system." Changes in stocks can set the pace of the dynamics of a system. The stock is a source that has in inflow and an outflow. A reservoir is a stock that lets farmers get water in dry times. Systems can have more than one stock.

The temperature or shape of water is important. Could the Gulf Stream stop? Vaclav Smil states that the Gulf Stream is not driven by thermohaline circulation. Like the Kuroshio and the Agulhas, it is a wind-driven (due to solar radiation) and torque-exerted (from the rotation of the planet) flow. He further says climate does not require a dynamic ocean.

However, local weather does. The average may be the same but the details determine habitability and crops.

4.2.4. *Valuing Water*

Donella Meadows notes that any growing system runs into some kind of constraint. In the outflow or inflow or in the stocks. The higher and faster a population or system grows, the farther and faster it can fall. With exponential growth of a nonrenewable resource, even a doubling or quadrupling of finds of the resource, such as water, offers little added time for developing alternatives. Also the more the growth loops evade the control loops, the worse the fall after production.

In any commons system, there is a resource commonly shared. If it is eroded, then it can become a tragedy. The tragedy of the commons arises from missing, or long-delayed, feedback from the resource to the users of the resource. Some systems are thus allowed to get worse; this is a drift to low performance. Rivers get dirtier, the tragedy slows, and we adapt to it. All of the ways to avoid the tragedy are cultural ways, whether through education or privatizing to regulation. Rich countries transfer technology and capital to poor, never questioning that capital or technology might not be the limiting factors. It may be clean water and ecosystems.

We need to learn to value water. There are three use values for water. Direct use value refers to goods that are consumed or enjoyed, such as water, food, or views. Indirect use value refers to environmental services, from ecosystems rather than individual animals or resources, such as climate or river flow. Option value has to do with the possibility or potential of getting benefits later. Nonuse values—there are three also—derive from the benefits of ecosystem services without using them in any way. Bequest value means passing on ecosystems to future generations of humans. Existence value is what people derive from knowing that something exists regardless of potential use. A final category of value, self-value is the value of nonhuman beings for themselves and their needs.

Simple conservation makes the most sense, as we just reduce the flow and have less to bury or deep-place in the ocean, or avoid with solar screens or aerosols. Efficiency alone promotes more consumption. Conservation may include additional rules on the size and efficiency of vehicles, no different than other restrictions on speed or driving laws.

We can identify nontrivial risks with water. We can identify areas of ignorance and fill a few in. But, we have to be willing to live within certain ecosystemic and biospheric limits. We also have to accept precedence of the health of the whole system. We have to accept multigenerational equality and commitment to general dynamic goals for health and fitness of the whole system. And, of course, have to deal with current inequities between groups and nations, as part of the program to fit in the planet.

Cheating is a trap; it is avoiding the rules. The solution is to redesign rules in direction of achieving purpose of rules. The trap is seeking the wrong goal. The way out is identifying goals that reflect the real wealth. Focus on result not effort. Leverage points. Where and when to intervene. Meadows cites Forrester's leverage point of growth, economic and population.

Systems goals are parameters, which can be leverage points, and can make big differences. Buffers are leverage points. Buffers are large stabilizing stocks. Delay length is a high leverage point. Balancing feedback loops is a leveraging point. Governments and corporations are drawn to a price leverage point, but sometimes push in the wrong direction.

Meadows suggests that strengthening market signals, such as full-cost accounting, does not go far due to the weakening of another set of balancing feedback loops—those of democracy. The self-correcting feedback has been disrupted by leaders who control information. Designing government here might help. Taking away the power to influence elected representatives with money might be a start.

Figure 424-1. Restored Noname Creek at Altazor Forest

4.3. *Design Factor: Animals Species & Diversity*

Animals, including zooplankton and insects, are mobile organisms that survive by preying on plants or other animals. Each organism is inseparably related to a place; breaking the bond with place may mean death for the organism. Organisms and places shape each other. Even if individuals can be described in terms of vortices, as the poet Pound did before the physicist Prigogine, they do exist materially, and they participate in the field in which they exist. The organism must adapt to the environment, which implies having a memory and being capable of learning; the organisms must also reproduce, that is, duplicate its pattern in a separate being. Organisms are goal seeking, and often stability is sought above change or complexity. The individual is a subject centered in a milieu. Because of this implied point of reference, J. Rodman concludes that ecology is teleological. Often, organisms strive for well being beyond just survival. Their goal is to come into the fullness of being; A. N. Whitehead considered that all organisms have three urges: To live, to live well, and to live better. Living better is being more attuned, stimulated, receptive, flexible, and integrated into a milieu.

4.3.1. *Wild Animals: Individuals, Species & Ecosystems*
Each participant in a field creates an image of nature—or world— from what is meaningful to it. Jakob Von Uexkull suggests representing these unfamiliar worlds with a bubble model. The life-image, or *umwelt*, of an animal is what has perceptual and operational meaning for the animal. All animals are fitted to their unique worlds with equal completeness—simple animals to simple worlds and complex ones to well-articulated worlds. Each is optimally (or suboptimally) fitted to its habitat.

Animal populations vary according to cycles or unknown causes; they move and concentrate or disperse. Because animals fluctuate in such a cyclic manner on the tundra, for instance, management by the maximum sustained yield concept is even more difficult and questionable. Furthermore, the ranges of caribou and birds cover more than one ecosystem; wilderness designation for areas outside any one focus must be coordinated. Animals move between ecosystems, carrying their biomass and experience. Wolves, elk and other species use different systems for feeding, mating and sleeping. Other species, such as birds, can connect ecosystems separated by thousands of kilometers.

Microbes are vital to the environment because they participate in the Earth's element cycles like the carbon and nitrogen cycles. Microorganisms are involved in the production of oxygen, biomass control and 'cleaning' the Earth of remnants of dead organisms.

4.3.1.1. Evolution & Domestication

The process by which species reorganize their structures to adapt to their environment, as well as modify the environment for their benefit, is called evolution, an integrated, partly open process that selects whole individuals in whole environments. Evolution can be said to flow upward and outward, as well as inward and downward, from the simple to the complex, but also back again. For moths that mimic bark patterns, there are others in the same area that are conspicuous; for herbivores with complicated stomachs, such as deer, there are those with simple stomachs, such as elephants or horses. It cannot be shown that a particular evolution took place by necessity, only that an adaptation had value and the species survived. The consideration of observations can lead to the conclusion that evolution is converging to a single end. Darwin himself did not want anyone to consider evolution a purposive movement towards a goal. Rather he regarded evolution as a bush, growing where it can. Evolution can be considered as a building up of complexity, or an unfolding of patterns. Evolution seems less a hierarchical ladder than a history of forms adapting to changing environments.

The adaptation of beings to a changing milieu cannot be perfect. Over-specialization reduces flexibility and the ability to change, but underspecialization reduces efficiency. Beings that survive tend to have a satisfactory level of specialization. Beings that are not optimally (or satisfactorily) adapted are eliminated through competition and stress. Evolution can increase the levels of complexity through the operation of natural events.

Species are defined by their position in the environment and thus are in internal relations within the environment, but it is also true that they define the environment through positive and negative feedback. While some species adapt to a niche, others create new niches. The species is much more than a passive group—it is intentional and flexible, sometimes stress-seeking and maladaptive. Species in their milieu are in dynamic relationships. While relationships are as real as the organisms, the relationships are not necessarily or logically prior. The whole part, or holon, creates the whole as the whole creates the part. The organism creates the ecosystem as the ecosystem creates the organism. The multiplicity of beings and relationships create and are created by the field.

Humanity created new niches for wild animals when people started to domesticate animals. However, candidate animals were distributed around the planet very unevenly. For instance, with mammals over 100 pounds (45 kilograms), Eurasia had pigs, five species of cattle (Aurochs, water buffalo, yak, guar, banteng), horse, and mouflon sheep. Africa had 3 wild pigs, buffalo, and zebras. And, the Americas had bighorn sheep, bison, and boars. In Eurasia, almost all the candidate animals were domesticated. In Africa and the Americas, only one in each region was domesticated.

Table 4311-1. Approximate dates for Domestication (after Diamond 1999)

Species	Date	Place
Dog	60,000-10,000 BCE	SW Asia, China, N. America, Australia
Sheep	8000	SW Asia
Goat	8000	SW Asia
Cat	8000	Egypt
Pig	8000	China, SW Asia
Cow	6000	SW Asia, India
Horse	4000	Ukraine
Donkey	4000	Egypt
Water buffalo	4000	China?
Llama, alpaca	3500	Andes
Bactrian camel	2500	Central Asia
Arabian camel	2500	Arabia
Guinea fowl	2500	Central Africa
Reindeer	1500 BCE	Russia, Scandinavia?
Rabbit	900 AD	Europe?

4.3.12. Wild Centers

The disappearance of wildlife and wild places to live is a problem for the whole planet. Many people do not miss wolves, for instance, but they are very important to keeping the deer populations adjusted and healthy. Pets and domestic animals also need to be considered as competitors for wildlife. But, they also need care, even when it means sterilizing them to keep them from breeding freely. The solutions to this are to save some habitat—do not convert it to tree farms, farms, or roads—using an ecological plan. Then, limit the takes of all wild species, not just wolves but also deer.

Except under rare conditions (such as heavy snow), wolves do not determine deer populations, whose numbers are limited by food supply. Instead wolves cull weak and diseased individuals that lag behind their herds. Wolves help strengthen herds by killing such animals. Wolves are also efficient scavengers of domestic animals that are sick or have died. Old or unhealthy animals can be a burden on a herd, for example, by eating browse that a healthy animal might need or by infecting young deer. By eliminating such animals, wolves perform an important natural function in wild ecosystems. Wolves are keystone species and an indicator of the quality of wildlife habitat. Their actions are crucial for maintaining the long-term viability of ecosystems. Other predatory species are also supported by wolf kills, including ravens, foxes, wolverines, vultures, bear, and eagles. Wolves contribute to an ecological 'balance' and prevent overpopulation in deer and other grazers. Wolves are evidence of the diversity that we value so much, from genetic diversity to the full spectrum of an ecosystem. Their persecution could reduce the complete functioning of ecosystems in immeasurable ways.

All of nature is not human nature, however. There are many other sentient species. All animals, 'two-legged and four-legged,' are equals in the view of Black Elk. Science is only beginning to support this idea in terms of ecosystem functioning. Adolf Portmann shows that every form of life appears as a gestalt, developing in a specific place. All living forms create an image of their environment. Genetics provides the proper image choices for some—frogs, for instance, focus most closely on objects that have the same size and trajectory as flies. Others must learn what is valuable. Von Uexkull implies that the human world is only one of the many possible. Animals are not suboptimal beings relegated by evolution to second-rate habitats. They are optimally fitted to places that humans are not.

The theory of life-images is a basis for a new, genuinely nonanthropocentric metaphysics. Natural processes take on an expression of significance of their own without reference to humanity. All things have an ultrahuman value of their own. There are other life-images that are measuring parts of habitats. There are other centers, which are equals. Humanity is not the center of all. The planet cannot be controlled for only human benefit.

Humanity is exploiting nature recklessly, without attention to the minimal health of ecosystems. Yet various societies are working to preserve animals, species, and habitats. Their efforts are described, according to three levels: individuals, species, and habitats. An ecological philosophy is outlined as a basis for the united effort of these societies. Ecology supports the uniqueness of individuals in their life-worlds and the interrelatedness of species in communities. Psychological and geographical studies support the importance of healthy places for human beings. The concept of earth as home is proposed as a metaphor for the development of appropriate attitudes and participation in appropriate ways of living.

4.3.2. *Interactions in Nature: Disturbance, Exploitation, and Interference*

This section examines the parallels between the interactions of processes, of animals and of humans in ecological systems. It concentrates on disturbance and exploitation behavior and contrasts them with the interference behavior that characterizes the nonecological activities of the dominant human, industrial culture. Examples of each will be taken from wild ecosystems, forestry, animal cultures, archaic human cultures, and industrial culture. The word interactions is used, instead of words like 'events' or 'catastrophes,' to describe the feedback and cyclic nature of actions.

Humanity is exploiting nature recklessly, without attention to the minimal health of ecosystems. Many ecologists, such as Eugene Odum, have observed that complex communities have existed for thousands of years in relatively stable environments, even though these environments are characterized by regular disturbance and constant exploitation. These environments are now vulnerable to human interference, which is a different thing from disturbance or exploitation. Disturbance, by definition, is an event that can be caused by climate, biological entities, or other actors. Exploitation is the normal use of a resource or of a species by another species, including the human species (this ecological definition differs from a sociological definition, which means 'selfish or unethical use,' although it may suffer from negative connotations due to the latter); in fact, ecological exploitation has a rejuvenating effect on populations. Exploitation is contrasted with interference, an activity that can degrade, destabilize, or destroy entire ecosystems. Interference is not a form of disturbance, exploitation, or competition; it is destruction without gain to any species; sometimes it is caused by planetary events, but in the case of human interference, it is the destruction of the structures and processes of evolution for large-scale, one-species, short-term gain.

The interactions of living beings in ecological contexts may have positive and negative effects on themselves and other species, as well as constructive and destructive effects on ecosystems and the operation of biogeochemical cycles. Human interactions are also considered. The pandominance of ecosystems by humanity is related to the biological and cultural characteristics of the species. Ignorance and indifference are identified as major reasons for continued interference.

4.3.2.1. Interactions

Living organisms in a given area interact with the physical environment so that an energy flow leads to the defined trophic structures and material cycles that comprise an ecosystem, according to Eugene Odum. An ecosystem can be analyzed into parts, including organisms, energy circuits, food chains, diverse patterns, nutrient cycles, development and evolution, and control. No organism can exist by itself or without participating in an ecosystem.

Organisms interact in a number of ways. Interactions can be positive, negative or neutral in effect. In 'neutralism,' for instance, neither population is affected; in 'competition' each group adversely affects the other for resource use; in 'parasitism' one benefits and the other is adversely affected; and, in 'mutualism' both benefit in a necessary relationship (See Section 4.4.1.1.). These kinds of interactions are basically forms of exploitation. Disturbance or interference are also possible.

4.3.2.1.1. Exploitation

Animals and plants, algae, bacteria, fungi, live together in ecosystems. Living together involves many kinds of interactions, from competition and conflict to cooperation and mutualism. Interactions may be reciprocal or complementary. They may dominate or control. Interactions are multidimensional. A wolf, for instance, may howl to communicate, or to restore proximity with a mate, or for simple pleasure. Many animals, such as wolves and caribou, develop together over time, adapting to each other's strategies. Paul Ehrlich and Peter Raven refer to this mutual adaptation as coevolution. Coevolving systems never completely adapt.

Every species uses some part of other species or of the environment. This use is termed exploitation. Insects, diseases, and animals, more than being simply agents of mortality, are native components of complex food webs in ecosystems, and they contribute to the selection of species. In a Ponderosa pine forest in the Pacific Northwest, insects exploit trees; they pollinate some trees and overwhelm others, but rarely more than 1 percent of a forest. Diseases exploit trees; they remove stressed trees—also probably a low percentage on the order of 1 percent. Their effect on the long-term health of a forest, however, is positive.

Birds and mammals eat foliage and seeds; they also disseminate seeds. Mammals, the best regulated of more recent species according to Frank Golley, change their habitats to suit themselves by chewing, digging, and burrowing. Rodents can dislodge earth at a tremendous rate. In many cases, these activities improve the conditions for the growth of vegetation. Mammalian grazing promotes vegetative regrowth and the movement of seeds. Bison and prairie dogs were responsible for much of the character of the American plains. Rodent caches may account for a good percentage of pine seedlings; possibly 15 percent of a Ponderosa pine forest rises from such seed caches. Beavers and other rodents create their own microsystems. Wide-ranging caribou and wolves transfer energy between systems.

Predation increases the survivability of a prey species. Predation also increases the diversity of species, according to Steven Stanley, by limiting the most populous prey species. Rarely do predators kill all of the prey. Rarely do animals interfere with the operation of the biogeochemical cycles in the environment. The exploitation of the system by plants, insects, and animals contributes to the health and continuity of the ecosystem. Exploitation is not chaotic; there are limits and rules.

4.3.2.1.1.1. *Rules.* Animals obey rules of behavior. Many animal communities have codes of behavior that regulate interactions. In birds and less complex mammals, these rules may be very rigid and predictable. With increasing brain complexity, however, learning takes a larger role. For example, young white-tailed deer in Idaho have to learn to cross highways in Idaho and Washington They appear to use rules of thumb (not the best phrase), finding a proper balance between safety and reasonable progress, between no traffic in sight and bumper to bumper congestion.

Animals like wolves have behavioral inhibitions against killing too many prey or killing their own kind, against coupling with a mated or disinterested female, or against attacking nonprey species. Animals that break such inhibitions are usually sometimes attacked or ostracized. In general, food is shared by all members of a wolf pack. Adults will regurgitate part of their food for adults who stay behind with juveniles. The members of a wolf pack cooperate bringing down an elk, but then compete for the choicest parts of the prey. Rules are not always strict; wolf mates raising pups may consciously deceive one another to get a break from the responsibilities, according to Michael Fox.

The social structure of a wolf pack is most important. Breeding, playing, hunting, feeding and territoriality are tied to social structure. Wolf pups are taught how to behave and how to hunt. Much of the behavior of wolves is directed to keeping the animal's status in the pack or to raising it. Quarrels take place often and the entire pack seems to take an active part. Actual battles, however, are rare. Ritualized squabbles result in few physical injuries. Wolves do kill each other, however. Wolves that behave strangely, such as epileptic pups or adults crippled in the chase, are sometimes killed by the pack. Disputes over the alpha position may end in death. Foreign wolves may be killed if they do not flee. Prey may be killed in excess during times of denning. Rules of encounter are complex and the outcome depends on numerous circumstances, such as the abundance of prey, the size and health of the pack, and stress; that is, the rules often depend on limits.

4.3.2.1.1.2. *Limits.* Mammalian behavior is controlled and population regulated through the use of space in general. Most mammalian populations, wolves for instance, regulate their density well below the limits of the food supply, often by as much as 50-70 percent. Territoriality limits populations, but populations can also be limited by specificity of prey or plant source, size of prey or plant populations, predators, natural events, or even individual tastes.

Wolves in the Arctic disperse with the migration of the caribou. According to David Klein, they prefer Caribou to other often more easily obtained species, such as mice. This preference reduces their hunting efficiency, however.

The goals of an organism are limited by the life-images of its species. Each animal is a participant in a field of existence. Using its senses, each participant creates an image of nature, or world (*umwelt*, life-world, is the term used by von Uexkull), from the sensations that are meaningful to it. Each animal fits itself to its unique world as completely as it can—simple animals to simple worlds, complex ones to well-articulated worlds. Each fits its place as well as it can. Konrad Lorenz, Michael W. Fox, and others have elaborated this kind of fitness in more detail.

Each organism is inseparably related to a place; breaking the bond may result in death.

Organisms and places shape each other. This is true of archaic human cultures as well.

4.3.2.1.1.3. *Traditional Ways of Archaic Societies.* Most human cultures are located in a particular territory. This is especially true of the Campa, who live in a tropical forest in Peru, and for the Ituri pygmy, who live in a tropical rainforest in Africa. The features of their cultures are unique to their place. They literally could not live with images of desert or ocean, like the Taureg of the Sahara or the Samoans of the Pacific. The very circumstance that makes each culture unique—being in a unique place—ensures that it can fit a place. This fitness ensures a limitation of exploitation.

Particularly in agricultural societies, cultures are gauged closely to seasons. The culture makes the world manageable by limiting it. A local culture is also tuned to the limits of the local ecology, within the knowledge of interactions—the long-range ecological consequences of drainage, irrigation, or overexploitation can contribute to the success or failure of a culture. Many, but not all, archaic cultures are a form of fitness and limitation. Most archaic groups try for adaptation before domination. For instance, according to Gerardo Reichel-Dolmatoff, the goal of the Desana Indians is cultural continuity in the rainforest.

The Chipewyan people in Northern Canada occupy the same territory as wolves and compete for the caribou, although their niches are not identical. Both social systems are adjusted to the hunting of barren ground caribou. Chipewyan hunters depend on animals other than caribou, which migrate out their Indian grounds. The cultural decision to hunt caribou as the primary item of subsistence, however, has produced many similarities between the two species in their utilization of land and in the formation and distribution of social groups. The cultural decision to hunt caribou results also in a population density lower than what would result through other decisions regarding the utilization of resources.

For hunting, Chipewyan use dog teams, snow shoes, and boats to increase their mobility and rifles and bows to supplement the traditional spears. The strategy of the Chipewyan, according to Henry Sharp, is to kill caribou at any opportunity. They increase their opportunities by walking aimlessly, watching, and driving the caribou, although the Chipewyan expend less energy by watching and ambushing. Wolves follow the more active strategy because of their increased and superior mobility. Both species adopt a pattern of dispersing and concentrating with the caribou. The basic choices regarding subsistence patterns, social organization. demography, terrain usage, and yearly cycles are made on the basis of the internal logic and structural characteristics of the two cultures (wolves do have culture, in the general anthropological sense, according to Sharp and Fox).

Although the two species do not compete directly, both Chipewyan and wolves are predators that put pressure on caribou populations. Sharp suggests that the commitment of both to caribou hunting is ecologically inefficient, since both species could spend more energy on secondary sources of food. For the Chipewyan, a deliberate "underutilization" of moose, rabbit, grouse, birds, and fish, is the result of their cultural values, including their willingness to live below the carrying capacity of the local environment—a characteristic of most hunting/gathering societies—the complex practice of drying caribou meat, and the reciprocity of their kinship system, i.e., caribou is a better basis for future relationships. The cultural decision to hunt caribou as the primary item of subsistence has produced a unique pattern in the utilization of land and in the formation and distribution of social groups. Wolves also underutilize their resources.

Regardless of cultural order or cooperative interactions, part of the process of life is uncoordinated, unfitting, disorderly, unbalanced, and destructive. Therefore, suffering often occurs. Suffering is an unavoidable part of disturbance or exploitation. We cannot intervene in every case, nor can we eliminate the possibility of suffering. We cannot maximize the self-realization of every being, and we cannot make evolution into a perfectly functioning machine; the functionless features of evolution are part of the process, but, we can protect the process. The mode of operation of nature consists of a rhythm of dissolution and reformation. The extravagance and beauty of the natural world features many more species in an ecosystem than would be necessary if exploitation alone were its organizing principle.

4.3.2.1.2. Disturbance

Disturbances in nature are regular but unpredictable events. Disturbances are caused by geological events, climatic events, physical processes, and biological agents. Hurricanes, for example cause disturbances, as do volcanic eruptions, windstorms, tidal waves, disease outbreaks, and acid rain.

Disturbance is one thing that causes change in an ecosystem. On a small scale, a single tree falling over is a disturbance. Although an individual dies, species continue. Mortality is a normal part of the life cycle of the forest. The disturbance may be necessary for the ecosystem to continue to mature; for example, according to David Perry, without windthrown spruce that expose mineral soil seedbeds, the northern forest ecosystem would shift to bogs.

Disturbances, if sufficiently regular, become a 'known feature' of the ecosystem. In Florida, some species such as Cypress, need the complete inundation provided by hurricanes to remain healthy. Yet, even catastrophic disturbances like hurricanes rarely damage more than 5 percent of a forest; for instance, the 1938 hurricane in the eastern US blew down less than 4 percent of the trees.

As the frequency of a disturbance increases, the forest becomes adapted to the disturbance; even pine plantations in the southeastern United States that are managed with controlled burns are less damaged by lightning-caused wildfires. Many disturbances in forests, such as insect explosions or fires, kill low percentages of trees.

After long periods without a major disturbance, however, a catastrophic disturbance becomes more likely. Where wind and fire are absent, for example, the probability of insect and disease outbreaks increases. By trying to prevent one kind of mortality, ecosystem managers often establish conditions for another kind.

Fire is regarded as catastrophic. In some ecosystems, for instance tall grass prairies in Illinois, fire is required to suppress competition from trees. In some forests, such as lodgepole pine in Washington state, fire is required for the cones to open and the trees to regenerate. Forest fires rarely damage more than 10 percent of the whole forest. Even the Yellowstone fire of 1988, still regarded by some as a tremendous disaster for "Smoky the Bear" policies of prevention (resulting in dead material forming fire ladders), caused limited damage as it leapfrogged along, leaving healthy untouched stands that became the source for the regeneration that is now being observed.

A typical percentage of death is the normal condition of an ecosystem, necessary for its renewal. The rate of death per year in an old forest is remarkably consistent at about 1-2 percent, even with wind storms, fires, disease outbreaks, and animal damage. In some cases, a larger percentage of the forest is affected. For instance, high elevation balsam fir forests are subject to bands of dieback that progress up the slopes parallel to the contours of slopes.

These "fir waves" seem to be triggered by cold winds striking exposed forest margins. A new stand regenerates where the trees have been killed.

Disturbance may change the direction of maturity in an ecosystem, but it also is stimulating for those species adapted to it. Disturbance may continue succession or it may deflect it, according to Bormann and Likens. Because of the range of scales and intensities with disturbance, it is a complex concept.

Disturbances that are not part of the history of an ecosystem may cause irreversible changes to the system, because the system has not evolved a defense or response mechanism to such a rogue disturbance. A meteor strike would be such a disturbance, especially if the landform was altered by a crater. Human disturbances, in the form of acid rain or clearcutting, are both novel and threatening. If they are small enough or rare enough, however, the ecosystem may rebound.

Very large scale disturbances, such as volcanic eruptions or meteor impacts, can destroy entire ecosystems or disrupt global biogeochemical cycles. However, such very large-scale disturbances are rare, and the ecosystems often have thousands or millions of years to become reestablished, although changed.

4.3.2.1.3. Interference

Although rare large-scale or novel disturbances can interfere with ecosystem processes, the term 'interference' is reserved for constant large-scale or novel effects. The destruction of ecosystem processes in nature by the action of one or more species is rare; any species that did so would become extirpated or extinct, unless it was not dependent on a single ecosystem, as is the case with wolves. Many commentators have accused mammals, wolves for instance, of overkilling their prey. It is fairly well established now, by David Mech and others, that wolves will take prey in excess of their immediate needs. This behavior has been interpreted as useful in maintaining not only the wolf but also secondary predatory and scavenging populations, for example, foxes and ravens. Indian informants are aware of this aspect of the wolf's excess kill, but they attribute to the wolf sufficient foresight to kill an excess of caribou near the den site in order to have an adequate food supply when the caribou are absent. Regardless of the wolves' intent, excess kills of caribou by wolves seem to be linked to the pup-rearing part of the pack that follows behind, as well as providing some food for the reverse seasonal migration—wolves can eat the remains of kills that are up to a year old.

Like wolves, human beings, as part of the process, interact with the individuals of other species or with entire species. Human beings are mammals who, as George Woodwell put it, live in a biosphere whose essential qualities are determined by other species. Mammals are bound by biological requirements that must be met for a population to survive.

Like other mammals, humans change their habitats to suit themselves. Humans have modified animal and plant associations in a different way from other mammals, simplifying patterns of energy and chemical exchange and solidifying themselves at the end of many food chains as a dominant species (a dominant is a species with greater influence than others in its biotic community, changing the lives of other species and the character of the habitat.)

Human populations have increased exponentially, with billions in giant urban ecosystems. Agriculture has produced monumental yields, but only at the cost of tremendous erosion and great subsidies of fertilizers and pesticides. Dams have been built all along rivers, and riverine forests have been cut, altering rivers and fishing grounds. Changes have been made without regard to the long-term impact on the ecosystem or on its human population. We dominate entire ecosystems.

4.3.2.1.3.1. *Human Pandominance.* By its influence of all ecosystems, humanity has become a pandominant species. As such, humanity reclaims, overgrazes, clears, depletes, and wastes at a scale that interferes with the stability, processes, and existence of many systems. One of the ecological consequences of human activity is the degradation of wild habitats for human developments and the introduction of novel elements into the biosphere—elements that have not been added slowly over time as the result of natural processes.

The biomass of the human species probably exceeds the biomass of any nondomestic mammalian species, and that biomass is supplemented by the tremendous biomass of domestic animals, which is far greater than the human biomass and consumes much of the same food as humans, including milk, fish, and grain. The domination of humanity is related to other characteristics as well, its large annual population increase (over 2 percent), high structural organization (of information and matter), and high energy use (globally 13 times the total of all other mammal equivalents).

This dominance has major effects on ecosystems: Transient perturbations in energy relations (from oil spills and burning); chronic changes/shifts of systems (from dams, irrigation, and chemical wastes); species manipulation (from the import and export of exotics); and interference competition with wild species. None of these effects are exclusive to humans as a species, but they are excessive, rapid, compounded, and very large-scale. Humanity has upset the balance of nature in favor of its own needs. Animals, plants, and habitats are being destroyed because of short-sighted, short-term economic interests.

Human beings have contributed to the extinction of species and to the destruction of ecosystems. Human hunters are hypothesized to have wiped out the most of the large mammalian species of the Pleistocene through overhunting—not for future food, but rather from the style of hunting—by driving herds over a cliff. There are other instances. In the 1880s, soldiers and cowboys slaughtered buffalo as a political strategy to reduce the resources of native peoples. Farmers and loggers destroyed the dense forests of Ohio and other states. Settlers and industrialists in the Amazon are destroying vast tracts of rainforest, as part of a political strategy to move peasants out of cities. Industrial forestry in the Northwest is content to take a high percentage (well over 90 percent) of a forest for wood and pulp, destroying the basis for the continuity of the forest, as well as all beings that depend on the old-growth, fungi, and physical properties of the forest to live.

Human exploitation at the tremendous physical scale that occurs in industrial states is different from exploitation by other species, because it results in the destruction of the entire system, the very basis for renewal of a system that human beings (as well as other species) need for life. Human actions are damaging global biogeochemical cycles, such as the carbon or nitrogen cycle. For instance, deforestation, burning, wetland loss, and industrial processes are releasing massive quantities of carbon dioxide into the atmosphere, which disrupts the carbon cycle. Although the destruction of large species, from whales to frogs, has a dramatic effect on ecosystems, the destruction of microbes, which generate oxygen and recycle nutrients, has a critical impact on the entire food web. These actions are global, like a large volcanic eruption, but, unlike a volcanic eruption, they are constant and hourly. These human activities are best referred to as interference.

4.3.2.1.3.2. *Human Industrial Ways.* The cosmologies of archaic cultures have been limited to historical places and by human perception, tradition, and technology. Modern technological cosmology, beyond being another kind of order, more linear and abstract, is wrongly considered the evolutionary successor to traditional cosmologies, and is displacing them rapidly—although we cannot afford to suppress the diversity of thought necessary for adaptation to the diversity of environments or to eliminate ecosystems and the societies adapted to them.

Our modern problems reflect an unbalanced and immature image of the earth, the earth as a machine, for instance. People sometimes constructed their worlds from preconceived notions, and many of these worlds did not survive, because they could not adapt to the environment. Our modern cultures are defective for this reason. The modern attitude toward nature as a resource has resulted in pollution and depletion of resources. It has allowed humans to overpopulate many of their habitats.

Even worse, decisions regarding resources are still made exclusively on short-term economic rationalizations and lead to material shortages and environmental degradation. The crises of environmental degradations are crises of cultures. Monocultures of the industrial kind lead to 'dedifferentiation,' that is, the decomposition and destabilization of complex structures. A species or culture that destabilizes its ecosystem through misbehavior risks its own extinction. Human beings make changes that endanger themselves.

Humanity is calculated by Norman Myers and others to be using over 40% of the ecosystem productivity for the entire earth (56% by 2004, according to Stuart Pimm and others). Humanity influences virtually every ecosystem to some extent, destroying some, interfering directly with many, and exposing the rest to exotic chemicals and materials. Species normally use a percentage of system productivity without disrupting the processes of production. The human species interferes with the processes.

Based on limited scientific and cultural perspectives, humanity fails to value those beings and communities for which no use is known. But, as Aldo Leopold (1949) notes, the majority of the beings in nature have no human uses. Even ecologists cannot think of uses for many large birds and mammals. The real danger is genetic loss, which is frequently grossly underestimated. As wild areas grow smaller, even wild species interbreed. As species are lost, the ecosystems become simpler or start to collapse.

4.3.2.2. How Should We Live? A Discussion of Rules Limits & Actions

The problems of ignorance and inappropriate images are multicultural, ecological, and cosmological, and must be solved on those levels—the entire activity of culture is guided by metaphors. Metaphors emphasize likenesses between living things and languages (or human constructs). A metaphor furnishes a label and emphasizes similarities. It not only defines and extends new meanings, but redescribes domains seen already through one metaphoric frame.

New metaphors already exist for interactions in an ecosystem. These include a process view (A.N. Whitehead, from the 1920s) in which organisms are dynamic structures that are immanent and simultaneous with the process, rather than consequences of natural selection of past random mutations; a field concept (C.H. Waddington, in the 1960s) for development, emphasizing dynamic transformation (form as organized spatiotemporal domain), in contrast with the particulate concept of an organism—understood in terms of group dynamics rather than selective advantage or cost/benefit; self-organization (Franceso Varela, in the 1970s) or autopoesis, which refers to the dynamic self-producing and self-maintaining activities of living beings, which incorporate materials through physiological processes; and reciprocally constrained construction (R. D. Gray, in the 1980s), according to which the organism and environment are co-implicative, co-defining, and co-constructing in a process of self-assembly, where the self is the organism/environment system.

Combining metaphors, we can see that organisms put together (enfold) structures based on historical patterns, and move (unfold) through a filter of limits, like minnows through a fish net, rather than behaving as interchangeable units competing for resources. These metaphors could form the basis of a new image for humanity, where we are an integral part of food chains and part of an organic cycle of birth and death. We humans need to recognize that we automatically participate in everything, and that we cannot unparticipate by choice. Participation starts at the quantum level and extends through the ecological and cultural. Human nature does not find meaning in an absurd world, but discovers its structure through interaction with the surrounding order. Human identity exists partly in relation to nature; the destruction of ecosystems may lead to the destruction of human identities mediated through cultures.

A culture that fits a local ecology is adapted and more likely to survive. Fitness is a way of reducing negative effects to make cultures more flexible and longer lived. An understanding of ecology, with an emphasis on limits, can lengthen the life of a culture, but ecology is not enough. Good metaphors are necessary, as are good rules of behavior.

4.3.2.2.1. *New Human Rules.* Human beings have no complete guidelines to interacting with other species in an ecological context. Cultural ethics are usually restricted to some members in a local ecosystem; such ethics are assembled inductively, from experiences from living in specific places. Philosophical ethics have been traditionally restricted to the human species and human situations. The areas of concern of ethics are not broad enough; their foundations are not deep enough. Philosophical ethics is the ethics of the human species living alone, without wild animals or plants in modified ecosystems. An ecological ethics has been developing that describes rules based on expanding knowledge and expanding ideas of the self.

An ecological ethics is a set of rules for living together with other beings. It is based on ecological knowledge, grounded "in the breadth of being," in Hans Jonas' words, founded on principles discovered in existence. An ethics based on ecological knowledge places human behavior in vital social and biological communities in nature. The frame of reference of ethics is enlarged, as Albert Schweitzer predicted, leading to appropriate behaviors in a larger context, through reverence for life. Skolimowski and Callicott recommend a reverence for life ethic. We must develop specific rules to live with other species, more formal than isolated cultures like the Campa and more comprehensive than modern cultures like the French.

Ecological ethics is a series of rules for living together. Most sets of ethics make the rules easy to follow. They emphasize the differences (relativism) or similarities (absolutism) of human beings only; or of the individual or the group; or of good feeling, reason, or desire. But ethics has to confront the individual, embedded in a community, located in a bioregion, on earth. And, the rules really are not as easy as human systems have presented. Schweitzer made them too difficult, with a constant valuing, but neither are they that difficult. An ecological ethics can be detailed only on a local level—even when it uses a global strategy.

4.3.2.2.2. *Observing Ecological Limits.* Life involves a vast number of interacting structures. Living consists of complex behaviors whose limits are defined empirically. The earth is suitable for life because of three kinds of limits: (1) solar radiation has stayed within certain limits for 4 billion years; (2) the biogeochemical cycles of oxygen, carbon, nitrogen, phosphorus, sulfur, water have stayed within certain limits; (3) the environment has been constant enough for organic evolution, but variable enough for natural selection to operate.

Animals and plants stay within limits of an ecosystem; for instance, Klein concludes that caribou populations are limited by food supply. Wolves are sometimes limited by stress. Trees are limited by water supply.

Traditional cultures have often stayed within the limits of an ecosystem. A sense of place, with its beings and features, is necessary for information on how to live, get food, and stay dry. The ecological benefits of rootedness are that people will take care of a place if they realize they are going to be there for a long time. Having a place means that the inhabitants have stock in it and participate in its unfolding, through planting and caring. Detailed understanding of the plants in a locale allow gathering of food and medicine. People in place—meaning in a human scale in unique surroundings—acquire a sense of community, share a set of values and concerns, and reap physical and spiritual benefits.

4.3.2.2.3. Practicing Noninterference. Exploitation, in the ecological sense, is necessary and beneficial to biological populations. A machine metaphor approach, with its assumptions of interchangeability and quantity, apparently has difficulty distinguishing between exploitation and interference. An ecological metaphor, that is more receptive and reverential, may be more appropriate to understanding organisms and nature in general. Such an approach stresses noninterfering observation rather than controlling manipulation.

Applied to nature, human intelligence discovers the significance of natural rules of interaction and exploitation. The reverence for beings as they are results in the rule of noninterference (Wittbecker 1984). A rule of noninterference states that human beings ought to avoid behavior that disrupts essential ecological processes or destroys biotic communities. As Paul Taylor states his rule of noninterference, it requires a "hands-off policy" for whole ecosystems and biotic communities; the rule stated here is concerned with limited and sustainable exploitation of ecosystems already shaped to some extent by human activities. Many other ecosystems, perhaps covering 50 percent of the land area of the planet, would be reserved by law for predominately natural ecosystems or adapted first nations. Noninterference also means "letting be" (after Martin Heidegger), or "letting alone" in the words of E. O. Wilson. Noninterference is not indifference, which is diffuse. It is caring. Noninterference will not lead to chaos, poverty, or stagnation. It can permit rational exploitation of resources.

We need to practice the rule of noninterference so that all beings can enhance their lives and habitats. Noninterference can be derived from nonviolence (or taoistic nondoing, a metaphoric expression for the nonbeing of nature) or even from English Common Law, which is well-established in Western law; it includes a precept: "Use what belongs to you in such a way as not to interfere with the interests of others" (*Sic utere tuo ut alienum non laedas*). This rule could be defined by positive laws and by negative restraints on behavior. This attitude would entail using what is necessary, exploiting some ecosystems completely, changing a place to fit human aspirations, and killing plants and animals for sustenance. But, it would also mean limiting humanity and its technological effects, limiting human use to local impacts, and letting other beings live without interference. It is not necessary to dominate or terraform the earth completely. Humanity could contain itself to a small percentage of the planet's surface and ecosystems and only visit or ignore the remainder.

4.3.2.3. Conclusion: Dealing with Interference

Interference has been a rare phenomenon on earthly ecosystems; it has happened in the past as the result of global catastrophes, such as meteor impacts. Now, interference, as opposed to more limited and predictable disturbances or exploitations, is threatening the stability of all ecosystems. It is dangerous to interfere with the processes of ecosystems because it disrupts the communities on which other species, and ultimately human communities, depend. Furthermore, in the deepest sense, it violates the idea of living together with other species on the planet. The proper relationship of humanity with nature includes competition and exploitation and mutualism, but not interference.

We kill millions of animals in laboratories to insure our safety, we kill billions of plants and animals for food and clothing and products, while indulging in the sentimental preservation of some individuals of other species. Animals do not need to be saved from natural death, a great regulator of life, but from unnecessary suffering, experimentation, and premature extinction. The world would not be a better place without sharks, silverfish, rats, cockroaches, or hyenas. They need their own places. The places, entire ecosystems, need to be saved. If we diminish variety in nature, we debase its stability and wholeness. To save ourselves, we must preserve and promote the variety of nature.

How do we incorporate this variety of interactions into design? Should we ban interference activities entirely? What do we design for? Wild animals in place, good domestics? Many original ecosystems supported good numbers of mammals, from mice and coyotes to deer, antelope, and elk. To some extent they have been replaced by domestic species: Dairy and beef cattle, swine, sheep, and poultry. Livestock often outnumbers the human population. Over 60 percent of the cropland production is devoted to feeding them, and over 30 percent of raw materials to housing and transporting them (for example, it takes 5 to 20 calories of fuel to produce 1 calorie of meat). Should design consider the numbers of feed animals. Should it arrive at an optimum? Would the optimum be determined by dietary preferences or by the amount of land available for raising healthy free-range domestic animals?

Although many animals, such as cattle and sheep are raised on ranges, they often spend months in feedlots being fattened with grain for human consumption. About 95 percent of this food goes for respiration or ends up as manure. The 95 percent loss is acceptable when an animal is raised on rough ground or when native populations, antelope for example, are used for food.

Harvesting some wild animals may be a better alternative than agriculture; this would be cheaper than improving the pasture degraded from overuse. Wild species might be more appropriate on marginal soils. The husbandry system whose forms underlie the foundations of modern thought, excludes wild nature as chaotic and other. We commit biocide on the wrong (or not 'our') animals. The living world faces a massive failure of interspecies dynamics, with great destruction of life and devastating psychological effects.

Pets, as domestic animals and a few wild but caged birds, reptiles, and mammals, need to be considered for fitness into nature. Because of pets, it could be argued, we are less concerned with wild animals and allowing them to slip into extinction. Then, too, we maintain our pets out of context; they are fed food from cans, kept indoors or in cages, and taught to defecate neatly. Zoos, aquaria and aviaries keep animals alive out of context, out of the conditions that shaped them. Denied the possibility of biological meaning, many of the animals may go mad. We will never understand why they are the way they are if we only study them in our homes or their prisons.

Aldo Leopold's chief concern was the need to reestablish the personal coexisting relation with nature, rather than large-scale impersonal management of resources by a professional elite. What the preservation designer wants is to preserve natural cycles, not a frozen state of habitat like scenes from a Disney movie. As civilization staggers, we can find an ecology of the psychological environment that can be trusted to find balance and compensation; it is part of life-support system.

4.4. *Design Factor: Wilderness*

The scope of this section is to characterize, on local to global levels, to prevent, or reverse, increasingly significant annual losses of biodiversity, species and ecosystem services. For the moment wilderness is pragmatically defined as areas necessary for global maintenance that have minimal human industrial or technological impacts). In his previous studies (1970, 1972), Eugene Odum suggested that forty percent of ecosystems needed to be set aside to provide ecosystem services for human populations. In his lecture just before this one (at the IV International Congress of Ecology, 1986), Odum emphasized the importance of wilderness for the planet, as well as a general knowledge of ecology by the people living on the planet.

There are significant problems in dealing with wilderness, however. To begin, we are not sure how to define wilderness. Nor do we know exactly how much of it remains anywhere. Next, we are saving wilderness from the bottom up, as separate species or habitats, rather than from the top down, which would identify entire wild areas to represent each kind of ecosystem, especially those that do not have grand viewscapes or charismatic animals. Furthermore, we are losing wilderness faster than we can understand it, and the more endangered the wilderness ecosystem is, the faster it seems to be declining or disappearing. Finally, we are not even identifying wilderness fast enough, and we do not have plans to survey and monitor all wilderness ecosystems. We know that most wilderness is not protected, but we have not studied the missing pieces of wilderness, such that we can save the processes that create and maintain the great diversity of wilderness.

This discussion of wilderness will expand two points: Wilderness definition and the idea of a minimum—or an optimum if we prefer to be safe and thoughtful—area of wilderness for the entire planet. First, wilderness definitions have tended to be restricted to banning human activity in mountainous areas. This kind of definition needs to be expanded to include every kind of ecological system, including those shaped by traditional cultures, as well as ecosystem dynamics of all systems.

Second, the idea of a minimum needs to be refined. A static minimum is easier to calculate, but a dynamic minimum is necessary to consider. The idea gets fuzzy because of the variables. Therefore, it is better to try for an optimum, rather than risk missing variables and adhering to a figure that is relatively meaningless.

4.4.1. *What is Wilderness Again?*

Wildernesses are ecosystems that are self-creating and self-maintaining. An ecosystem itself can be defined several ways, as a community, as a place, or as a process. As a special kind of ecosystem, wilderness is a human classification of an area (including air, water, land, and species) that recognizes that the content is determined by natural physical and evolutionary processes. Of course there are many classifications of wilderness and other definitions. This article does not address the definitions of wilderness that refer to human inspiration, recreation, purity, separateness, fiction, or myth. We have placed so much baggage on the word that it may not be useful. Wilderness is not a pure ecosystem. There never has been, for the past ten thousand years or more, a pure, humanless nature. Nor should wilderness be considered a simple playground for human needs for entertainment or therapy. Here, wilderness is simply a designation of natural systems that are not dominated or significantly modified by human processes.

Wilderness is a metaphor, after all, an invention of city people to describe the kind of nature existing outside their walls. This definition anticipates that nature was being converted to human nature rapidly. The metaphor has a psychological dimension. For instance, the forests of Greece were wild. But, the contemporary degraded landscape is also considered wild. No definition can be completely objective, due to this psychological dimension. Wilderness is a state of mind combined within a state of nature.

Wilderness is designated by boundaries, usually formal ones where governments have set aside wild lands. But, the boundary does not have to be formal, as long as there are rules. It could be recognition of places left alone because they are cold or remote. The boundary can be a limitation on interference or conversion, and perhaps a limit on the number of people or on their use of technology.

At least eight kinds of wilderness can be distinguished: Sacred areas, Foundation areas, Reservation areas, Preservation sites, Restoration areas, Neopoetic communities, Conservation parks, and wild agriculture and forests.

4.4.1.1. *Sacred landscapes* are those areas set aside out of respect, for habitats for others, where humans do not belong, and for species that cannot coexist with humanity. Examples of sacred landscapes are mountains, fisher/marten habitat, and large areas of tundra. In many archaic cultures, significant parts of the landscape are regarded as sacred (and, although not used, the wilderness is considered a part or center of their territory). This may be a mountain at the center of the world, a burial ground, or the portion of a river that belongs to the fish gods. Sacred spaces are highly valued locations, with limited access and usage. Such spaces have great symbolic value to societies where they are associated with significant events (or divine manifestations). The assignation of sacredness may result from functional necessity, to prevent flooding, limit population, or to balance the economy. In sacred landscapes no direct human impact should be permitted; mapping would be possible from satellite.

4.4.1.2. *Foundation areas* are zones that contain active wild communities where the integrity of nonhuman life is not challenged by human interference. Foundation Areas house the depth of wilderness. The depth of wilderness is what allows recolonization after disruption or human influence. Foundation areas could include common lands, that is, one of each kind of ecosystem on the planet, regardless of aesthetic value. Some visitation could be allowed, but without permanent residence or machinery. Human impact would be limited to noninterventive science (perhaps based on Goethe's method of contemplative nonintervention). Foundation areas would include large areas of the Amazon.

4.4.1.3. *Reservation Areas* are zones for the support of nonindustrial native cultures expressing traditional ways, that is, natural systems in which humans play a limited role—limited by a percentage of net ecosystem productivity, for example, the Miskito in Central America. Many traditional or archaic human cultures play a large part in the design and maintenance of these areas, not only by using fire, but by their traditional hunting practices. Limited human use often increases the diversity of a system, although the diversity tends to flow from human preferences for some plants and animals. Removing humans from this kind of wilderness would alter it and perhaps diminish it, so human use is retained.

4.4.1.4. *Preservation Sites* are places where the system is maintained by human activities, such as burning or planting. Specific kinds of preservation areas could include: Boundary areas, ecotones or lands bordering agricultures, small patches influenced by human activities, or corridors. Human impact could be limited to traditional, experimental science. An example could be the central long-grass prairie in the United States. In these areas, human activity is consciously limited to maintaining the desired state of the system, where the system has lost keystone species or where the area is too small or oddly shaped for the large-scale processes of an ecosystem to work independently.

4.4.1.5. *Restoration Areas* are zones which are set in a former native pattern by human activities, but which may not need further intervention. This may include the experimental restoration of ecosystems disturbed by human activities or natural events, for example, Lake Shagawa or larger restored areas of Tall-grass Prairie. Once the system has been reestablished, it would be allowed to develop on its own. This zone requires large enough areas for the natural processes to continue.

4.4.1.6. *Neopoetic Communities* are areas that have been created through the introduction of exotic species over time, but have become naturalized. Most of these recent areas have been started by human activities, but saved as self-regenerating systems and not built up or interfered with, for example the California grasslands, which have been changed with Mediterranean plants, e.g., wild oats, wild mustard, wild radishes, and wild fennel, or the *Stara Planina* of Bulgaria, modified by grazing over thousands of years, so that grasses replaced the shrubs and trees.

In a sense, every ecosystem and community is neopoetic. North America once boasted a wide suite of large mammalian species, including camels, horses, jaguars, and mammoths—a combination of circumstances, that may have included climate change and human hunting, resulted in their extinction. It seems these niches were never quite filled completely by the immigrant species from Eurasian continent. Species compete or partner for space in ecological systems. They are always making and remaking the assemblages, causing new combinations or new extinctions. Creating new wilderness in North America by reintroducing similar megafauna from South America and Africa would be an interesting experiment in community building. The Wildlands Network has been promoting such an experiment.

4.4.1.7. *Conservation Parks* are areas set aside for multiple use of resources without interfering with the operation of the ecosystems. Research may be conducted to answer questions as to whether the park is big enough and shaped correctly to constitute a proper habitat for its inhabitants. Human recreation would be permitted in temporary camps and with some light machinery, for example, the Boundary Waters Canoe area in the United States. These areas have been adapted to human presence and exploitation of many species.

4.4.1.8. *Wild Lands (*forests & fields), as well as other wild systems managed for the harvest of wild species, such as antelope or fish, are a final category. Although forests can be lightly harvested for wood and other products, they are predominately wild, and would be left wild, rather than being clearcut and replanted with a single species. This is an unusual category of wilderness, since we do not always consider forests and some kinds of agriculture as wild. In fact, wild species in wild systems have been shown to be more efficient in their use of resources and more resistant to other wild species that we humans consider pests. In fact, as regards forests, wild areas are preferred to agricultural areas such as tree farms, for many reasons, from automatic regeneration to the quality of wood. In this category, we recognize that areas of wilderness can be used without being destroyed or domesticated.

By just wilderness, the first three areas are included indiscriminately. They all have similar values, as mirrors of existence, examples of natural, complex processes, expressions of love for nature, and wild, nonhuman places. What they have in common is that they continue natural cycles without having to be supported by massive human intervention. In this discussion, all of these refinements are included under the umbrella term wilderness. In a complete zonagraphy, these categories would be a critical part of all separate (limited) areas.

4.4.2. *Do We Humans Need Wilderness?*

Human needs have been placed by the psychologist Abraham Maslow into five groups: Physical (food, shelter, clothing), Safety (law, order, security, freedom), Psychological (belongingness, love), Esteem (strength, self-sufficiency, competence, attention, prestige), and Self-actualization (achievement, creativity). I reason that wilderness provides the basis for each group.

We humans need wilderness physically. It is the matrix that is producing and supporting us. Even if we understood all of the intricacies of the process, and thought we could duplicate it in artificial environments, on spaceships for instance, we would still need wilderness—as a source of basic ecosystem services (that keep the planet functioning) and as a source for ecosystem and human health.

Wilderness, as an ambihuman ecosystem, provides safety through its intrinsic order. In fact, wilderness has provided security for people because of its resources and because by definition it has not been under cultural control. Traditionally it has been a source of freedom from the many kinds of human political tyranny.

We depend completely on the natural environment psychologically. Psychologists, such as D.O. Hebb, have conducted experiments that show the effects of a limited environment. Cut off from external stimuli, the mind becomes strange. The external world is needed to keep us alive and sane. This world is composed of remote occurrences, on polar icecaps and distant stars, as well as immediate personal events. Wilderness provides a constantly presenting environment. The human body links this exterior to a mental interior. The person is inextricably woven with the world; the body is a mode of presence to others.

Living within many kinds of wilderness, as do many foraging and horticultural cultures, allows the development of strength and self-sufficiency. And, this results in feelings of competence; although, this may be true also where wilderness is used for recreation by industrial cultures. In Archaic cultures, wilderness, as sacred space, is reserved as sacred; it confers more prestige on the sacred chiefs, e.g., Tonga or Samoa (where prestige is the power to influence others, due to success or status). Wilderness can provide sources of prestige that are outside of inheritance or social rank.

We need wilderness mentally and spiritually. Wilderness, basically as sets of places where nonhuman beings could continue, is necessary to the human psyche, where "otherness" continues to exist. It is a source for inspiration from the nonhuman. It permits creative expression and spiritual development, outside of the human cultural limits. Wilderness is the source for all human creativity. It is the experimental ground where forms are recombined.

4.4.2.1. *How much Wilderness do we need?*

Wilderness is an ambiguous and indefinite thing. It is a need like the sun is a need. But, we do not need to live in it or even visit it. It supports and cycles our human ecosystems. We need it because it is not us, not human and not human-made or maintained; it is the source of all nonhuman being.

It is one of those things that we need as much as possible of. We can not know exactly what the minimum is, either for the planet or for human physical or psychological requirements. Even if we could determine a minimum, we should consider a more conservative number, or an optimum.

For ecosystem services, that is, for clean air, water, food, shelter, we could calculate a minimum (or optimum), based on a number of criteria. To ensure the continuity of many habitats and species, we need to restore many systems that have been degraded; and save every one that is left. The hot continental region has been transformed for agriculture and little natural vegetation remains. Nevertheless, many of these areas could be restored as common areas, and support a suite of species that is adapted to the system.

Perhaps, there is enough wilderness for inspiration, but to inspire 10 billion people may take more than is left on the planet. Humans are inspired by different things, so a token wilderness will not likely serve as inspiration for every one who is part of the current or projected population.

There may be enough wilderness at this time, especially if we employ better technologies and more discipline in their use. For growing populations, we will require more wilderness. Without knowing how much wilderness we need, it may be possible to calculate an amount that might serve our needs, after considering a few more questions.

Figure 442-1. Old growth western redcedar at Altazor Forest 2000

5.0. Facing Culture Through Design

Human beings are mammals who live in groups and are good at imitation. These talents have allowed humans to create cultures to adapt to stressful environments. Cultures became useful in many ways, from externalizing memory to training the young. Many adaptive patterns were developed over the past 12,000 years, from agriculture and cities to technology and industrialization. While these adaptive patterns allowed great and dramatic improvements, their application in every circumstance is resulting in tragedies. Perhaps one solution is a global culture or a conscious framework for traditional cultures, but this solution would depend on understanding cultures and human motives.

5.1. *What is Humanity? Is Humanity Local or Global? What are Human Limits?*

Human Beings are mammals—omnivorous, social, bipedal, featherless, symbol and tool-using, game-playing, neotonous, bilateral-hemispheric, culture-making generalists—who, as George Woodwell (1976) put it, live as "one species in a biosphere whose essential qualities are determined by other species." Mammals are bound by biological requirements that must be met if a population is to survive, although these functional requirements are rather minimal for humans, however, being only food, clothing, shelter, and reproduction. Like other mammals, humans change their habitats to suit themselves. Humans have modified animal and plant associations in a different way, simplifying patterns of energy and chemical exchange, solidifying themselves at the end of many food chains as a dominant species. Unfortunately, they often dominate their habitats. A dominant is a species with greater influence than any other in its biotic community, changing the lives of other species and the character of the habitat.

Humanity is a pandominant species. As such, humanity reclaims, overgrazes, clears, depletes, and wastes at a level that threatens the stability and existence of many systems. One of the ecological consequences of human activity is the degradation of wild habitats for human developments (food, housing, and recreation) and the introduction of novel elements into the biosphere—elements that have not been harmoniously worked in over time. The biomass, or demomass, of the human species probably far exceeds the biomass of any non-domestic species, and that biomass is supplemented by the tremendous biomass of domestic animals, which is four times greater than the human (Borgstrom, 1975). This biomass forms an equivalent population that consumes much of the same food, such as milk, fish, and grain. The domination of humanity is related to other characteristics as well, such as our large biomass (6×10^{14} Kcal), large annual increase (2 percent), high structural organization (information, matter), and our high energy use (globally, 13 times mammal equivalents).

This dominance has major effects on ecosystems: Transient perturbations in energy relations (from oil spills, burning); chronic changes/shifts of systems (from dams, irrigation, chemical wastes); species manipulation (from the import and export of exotics); and, interference competition with wild species, as opposed to exploitative competition, which can be stabilizing). None of these effects are exclusive to humans as a species, but they are excessive, rapid, compounded, and large-scale.

Probably no consequence of human development has had a greater impact on the natural landscape and ecological processes than the production of food. Patterns of eating have influenced the constitution of species and the very contours of the earth. Throughout their history, humans have used animals and plants for food and clothing. Animals were followed, herded, corralled, tamed, and finally bred. Plants were domesticated later. As technologies developed, human relationships with animals and plants changed. Hunting, grazing, and agriculture provoked large ecological disturbances. Early domestic animals were revered, but nondomestic animals were considered competitors or nuisances. Now, animals are treated as commodities processed in factories and wildlife is regarded as useless. Hunting persists, but mainly as recreation. A few plants provide the bulk of human diet; the rest are considered ornamentals or weeds. Humans are omnivores, although they have been represented as carnivores, vegetarians, or fructivores by different factions. In view of our control over animal and plant populations, a reexamination of our eating habits, and our use of animals and plants, is critical. The functional biological requirements are rather minimal for humans, however, being only food, clothing, shelter, and reproduction. Other requirements, such as respect, comfort, and self-fulfillment, depend on socio-cultural systems.

5.1.1. *Living Together*

People living together develop a common image of the place. That common image, along with traditions or formal rules, determines how people behave, not only towards each other, but towards the place. They work to fit a place by dividing up labor in the place, by arranging their relations in certain ways, and by moving throughout the place according to habits or resources. Forms of social behavior, for instance, may reflect the difficulties of acquiring food and resources. A place is a part of the environment claimed by feeling. Emotion binds together motion and perception. Emotion creates an 'in-place'. Emotion can transcend distance. Living together results shared ways of going together (which is the definition of ethics).

People put things together in complicated arrays because they are possible, before considering any kind of formal or ideal design. Living together involves all kinds of interactions, from competition and conflict to cooperation and mutualism. Kropotkin maintained that progress occurred by people working together, not through competition. In Fields, Factories and Workshops Tomorrow, he argued for self-sufficient local economies and an integrated urban and rural society with decentralized industries. Civilization is held together by practical and political activities. Building and planting, taking care of the self and others, is primarily a prosaic activity that varied with place.

5.1.2. *Adaptive Patterns of Living*

While some species adapt to a niche, for a way of eating, others create niches. The individual is much more than a summation of all of its historical responses—it is intentional and flexible, sometimes stress-seeking and maladaptive. This kind of behavior can result in new adaptive patterns.

Humans started out as scavengers then became hunters and gatherers, foraging in a territory identified with a band or tribe. As people started gathering wild plants, they may have settled in semi-permanent villages and have expanded their ways of eating and living.

5.1.2.1. Hunting & Gathering (& Fishing & Wild Gardening)

We have a long experience as hunters and gatherers, at least forty thousand years. Unfortunately, we know little of how hunters and gatherers built a satisfactory way of life. Hunter-gatherers possibly were completely conditioned by the features of their immediate surroundings. Through their senses and reason they acquired an empirical, holistic knowledge so well fitted to their environment that they could cope effectively with the wilderness in which they lived. Ortega conveys how humans can still learn to function as organic parts of a wild place, instead of just observing it.

Hunters perceive all species as having roles in a vast society, according to Levi-Strauss, whereas farmers perceive human groups as species. Change of climate and overkills of animals possibly contributed to the necessity of farming. In evaluating ecological situations, we cannot avoid making judgments and giving preference to human values.

Hunters eat at the natural location and at the densities of resources. Gathering seems to be more important in warm climates, and hunting in cold climates. Hunters tend to be nomadic. Mobility and food limits are incentives to remain in small groups, which tend to have low population densities (the world average is 1 person for 1 square mile). They tend to share resources equally, as a result of egalitarianism and reciprocity. They have "natural" leaders, who lead by the example of being a good hunter or a good shaman. Territory was shared with neighboring bands.

Size favors an egalitarian political system. How can you control mobile people with few possessions, no herds and no wealth? How can you dominate them militarily? You would have to be much larger, but there are penalties for size in those ecosystems. How can you economically dominate them? Or linguistically? They tend to be localized, decentralized, oral, and egalitarian. Therefore language groups would be small.

5.1.2.2. Herding & Domestication

After the ice age, animals were captured and gradually domesticated. Domestication has been used to imply one simple process, but it is various grades of relationship in which humans have degrees of control over reproduction and composition of animal groups. Factors leading to increased domestication include food, ritual, play, and transportation.

Different ranges of activities in hunting could lead to herding. Random hunting and controlled predation, followed by attentive herd following, could lead to loose herding, then eventually close herding. As hunters gradually encouraged some characteristics in wild animals, they eventually acquired greater physical and genetic control.

Herding domestic animals would have advantages, such as knowing how many animals were available for milk or meat. There were some disadvantages, such as an increase in the amount of responsibility for forage over the entire year, which doubtless led to following, then circulating the animals by season (transhumance).

Domestication profoundly influences the diversity and integration of a species. Diversity, as a measure of genetic variability in ecosystem, is decreased. Diversity is the bag of tricks for organisms facing environmental perturbations. Coevolution with humans reduces or destroys integration with other species; stability depends on human control. Human control is not certain.

Changes in fitness probably occurred following the later stages of closer man-animal relations, with adaptation to captive and restricted environments resulting in gene pool changes. Domestic animals would then have different characteristics compared to wild animals. Possibly, mutation and gene frequency would also contribute to domestication of animals. Sex and age ratios changed. People ate younger animals and kept females for breeding stock. Goats and sheep would have been bred to produce excess milk. Other changes lead to wool production.

At many stages of herding, hunting, gathering, and fishing, may still have been important. In the first stages of herding, fire may have been used to move animals or keep them together. As technologies developed, human relationships with animals and plants changed. Hunting, grazing, and agriculture provoked large ecological disturbances. Early domestic animals were revered, but nondomestic animals were considered competitors or nuisances. Now, animals are treated as commodities processed in factories and wildlife is regarded as useless.

Much of the earth is no longer occupied by wild vegetation, either pristine or degraded. Pastoralists have had a profound influence on native vegetation. Much of the savanna in the tropics would be more heavily wooded if not for fires and grazing practices.

5.1.2.3. Agriculture

Foragers understood the principles of gathering seeds and planting them. They knew when and where their favorite foods were. They engaged in broad-spectrum collecting. Young or elderly collected snails, turtles, clams, crabs, and seeds. Most meat came from ungulates, like gazelle or deer. In the western foothills of the Zagros mountains, the wild einkorn wheat grew so thick and tall that a family of six with a stone sickle could gather enough in three weeks to last for a year.

Many of the activities in the gathering of plants could lead to plant domestication and agriculture. Foraging and collecting wild plants resulted in knowledge of the plants and their requirements. Tending wild plants in burned or dry areas resulted in more knowledge. Throwing unused seeds into a kitchen midden might have resulted in some crops growing in richer soil. Living near the plants might encourage reliance on those plants. Selecting specific plants that required intervention might have made the plants dependent on human collection and planting. Selecting for various features, such as closeness of seeds or gigantism might have led to selecting those plants only, or monocropping.

Plants selected for domestication would have desirable characteristics, such as ease of establishment and tolerance of a wider range of conditions, such as drought or poorer alluvial soils. The plants should have good production, big seeds or leaves, good biomass. They should have the ability to get or trap nutrients, with shallow, wide or deep roots. They should be able to tolerate climatic changes, resist drought or salt. They should be able to resist pests and diseases. They should be compatible with other plants. They should provide a balanced nutrient content, with less toxicity. They should be easy to control and have detachable seeds, fruits or leaves.

After domestication and agricultural development, the plants would have increased size in seed heads, tubers, and roots. The normal chemical defenses, such as toxins, would be suppressed. The plants would be tried in a wide range of environments, which might result in an equally wide range in shape, size and taste. Selection for strong rachis or seed heads would reduce reproduction by wind and insects. Monocropping might reduce cross-pollination, such that humans would have to take over as agents of propagation.

In its early stages, agriculture is constrained to early stages of succession, that is, it uses nutrients and energy for quick growth and reproduction, rather than invest in efficiency, durability or maturity.

5.1.3. *Attachment to Place*

Ecosystems are embedded in places. The making of places by human beings is an ordering of a distinct structure and center. Attachment to place is a form of deep love, from which many other virtues for living well, such as frugality and humility, spring. The organization of perception, meaning, and thought is intimately related to specific places. Place is a focus of meaningful events and a platform for ordering a world. The individual image of a place is modified by memory, experience, emotion, imagination, and intention.

A place is a part of the environment claimed by feeling. Emotion binds together motion and perception. Emotion can transcend distance. Emotion creates an 'in-place'. A place must be found and made; it does not exist before a priori. Humans, like plants and animals, identify greatly with local environments. Maybe this is a function of the limbic system of the brain, a function we share with territorial mammals. Human emotion creates an 'in-place.'

Neil Evernden asked what it does to a person to see his place, context, destroyed? What does it do? It creates violence or apathy. Modern populations are rootless, moving about from city to city. We are suffering from a placelessness, which arises from our style of efficiency and proclamation of mass values. So far, no psychologists have studied what happens when a person sees her/his place, their very context, destroyed. These catastrophes may be the basis for diseases, depression, or cancers.

The word nostalgia was coined by Johannes Hofer, a Swiss medical student, in 1678 to describe an illness characterized by insomnia, palpitations, stupor, fever, and persistent thought of home. The disease could result in death. For the Northern Aranda in Australia, as well as for émigré Russians, it is not possible to stay away from home indefinitely and still live. Nostalgia can be a fatal disease. Thus far, the sense of place cannot be gleaned from an analysis of the nervous system. Yet a place shapes the nervous system, somehow.

Maturity is linked to the increase of identification with, and care for, others. Albert Schweitzer noticed the expanding circle of care from family to humanity to animals, although different cultures have different emphases.

Humans love place (topophilia), as well as life (biophilia) and home (ecophilia); perhaps this love is an instinct or a meme. Four elements in loving have been identified by Eric Fromm: (1) Care, the active concern for life and growth; (2) Responsibility, the desire to respond to others needs; (3) Respect (meaning to look at), the recognition of others' uniqueness; and (4) Knowledge, combination of objectivity with participation and intimate identification. These elements define a loving relationship. The inexhaustibility of a living being or of our relationships constitutes much of the nature of love. Human beings are compelled to seek other beings and love is the most rewarding approach.

5.1.3.1 Love of Home (Ecophilia)

Adolf Portmann observed that insects and animals displayed a powerful attachment to places; that it was best understood as home. What does it mean to be at-home? The fundamental ambiguity of existence is that humans have different capacities for feelings and awareness. Some feel strongly about a place or home; others never do. Several metaphors have been used to describe the human place on earth. The earth is a storehouse, property, or a spaceship. But the earth is not a spaceship or storehouse; it is home. Victor Ferkiss proclaimed that: "The world and humanity are one entity, one system in equilibrium. Earth is humanity's only home; humanity is one people in relationship to the earth."

Home living is simultaneous on different levels; the importance may shift from city to nation, or nation to state, or house to bioregion, or state to habitat. Each level is a metaphor for the next. There are parallels between nature and a house, as the basis for home. Solar space is like the landscaping; wilderness is the foundation; conservation areas form the shell and provide services; and each bioregion is a unique room. The analogy cannot be carried too far. But it shows that a house is not, as Le Corbusier said, "a machine to live in." It is a matrix for home. Home is not just a house, either; it is a complex of significant events centered in place. It is the foundation of our individual identity on one level, and our role in the community, on another. What makes home different from house? One thing is participation in making it, commitment to it. People invest parts of themselves in a place, to make a home.

The concept of home has a mixed reputation. Paul Shepard (1974) finds humanitarians obsessed with the 'homelessness' of stray pets and wild animals. He points out how the fixation on shelter is taken over by the advertising of wood industries, who describe their meager reseeding efforts as 'creating a home for wildlife.' He sees protective organizations swaying to the tunes of propagandist lullabies. Perhaps he is partially right. But that is a misunderstanding of home. A home is not the house, not an undifferentiated place. Animals accustomed to rich woodlands are not at home in a replanted clearcut. The use of the word is a cheap advertising device; yet, it shows the importance of the meaning. A home is living, a house is not.

In English the term for dwelling is to stay. This is the symbolic opposite of moving or changing. It means to withstand time. Dwelling resists and persists. Permanence is important element in idea of home. One ecological benefit of rootedness is that people will take care of a place if they realize they are going to be there for a thousand years. Having a place means that the inhabitant has stock in it and participates in its unfolding, through planting and caring. Detailed understanding of plants in a locale allow gathering of food and medicine. People cultivating a sense of place are people in place. Their work can be appropriate; appropriate growing, logging, mining, or building. The Royal Commission on Local Government in England and Wales found that people's attachment to home area' increased over time. Gaston Bachelard (1969) has written much about the significance of home: "For our home is our corner of the world. As has often been said, it is our first universe." The home is a springboard to understanding the universe.

5.1.3.2. Love of Place (Topophilia)

Paul Shepard (1967) suggests that for each individual the organization of thinking and meaning is intimately related to specific places. Experience focuses on a place, which acts as the background for specific events. The features of world are experienced meaningfully. The place is a matrix for ordering experience. The specificity of place is important. The earth extrudes itself into particular plants and animals; flexes mountains; and sweats weather. Places animate (from the Latin anima, meaning 'inspire') people. The inspiration of the sentiment of dependence is called impregnation. Animals and humans are imprinted early in life to particular places (philopatry). Each difference in the landscape has meaning, as when the aborigines of Northwest Australia perceive physical differences and even a symbolic landscape. In fact they structure space according to myth, where Europeans use buildings and roads. Every place has a unique identity, a persistent sameness as a result of combinations of factors. Topophilia, love of place, is the recognition that all human beings have affective ties with the material environment (the word was coined by Yi-Fu Tuan).

The attachment to a place is rootedness. Von Uexkull describes the importance of rootedness in his concept of life-world. Simone Weil (1955) regarded rootedness as "perhaps the most important and least recognized need of the human soul." Rootedness arises from participation in a place. It is the need for order, liberty, security, status, and responsibility. A deep relationship with a place is necessary. Without it, existence loses much of its significance. Caring for a place involves concern and responsibility. This attitude is similar to one described by Martin Heidegger (1960) as caring (Sorge). Care is the recognition that a human being is a participant in the world. It is tolerance for the essence of a place; absorption in a place; concern, the willingness to not change or exploit. Care is presented as the total structure of the person: in the projection of being, in being in a particular situation (with limits), and in absorption in the world by virtue of concern for it. Care is the recognition that a human being is a participant in the world. It is tolerance for the essence of a place; the willingness to not change or exploit. A home can only be realized through caring; to have a home is to dwell on the earth.

Care, is, according to Erich Fromm, one of the four elements in loving (along with responsibility, respect, and knowledge; it is the active concern for life and growth. The building of a place to dwell on earth is hard earned; civilization is held together by practical and political activities. Building and planting, taking care of the self and others, is primarily a prosaic human activity.

5.1.3.3. Life (Biophilia)

We mistakenly conclude that our skin is the boundary to ourselves. But intuition also senses the interdependence of nature. We extend the boundaries of personality to other things and people. The human skin is like a pond surface, according to Paul Shepard. The skin's interpenetration ennobles and extends the self—the beauty and complexity of nature are continuous with ourselves. We know subjectively that we are not separate from the earth, that wolves are capable of love and tenderness, that trees are beautiful. Edward Wilson (1984) argues that the essence of humanity is inextricable tied to life on the planet. He claims that biophilia is the natural affinity for life, and central to the evolution of the mind.

5.1.4. *Human Ecology & Political Limits*

Physical and ecological limits are important for design. Human psychological and cultural limits may be equally important, especially when we ignore them. Human beings have many kinds of limits, from obvious physical limits, such as height and bone shape, to psychological and emotional limits that keep people from listening to every single item of bad news. Some of these limits are relatively strict, while others are malleable. We are only finding these kinds of limits as we exceed them. However, when it comes to a local design or planning, we cannot afford to operate this way. We need to anticipate these limits and act accordingly.

5.1.4.1 What are Human Genetic Limits?

Our genetic make-up predisposes us to some things and pushes us in other directions. It does make limits on our plasticity (for possible limits, see Table 5141-1). It could promote behaviors that damage the environment, and hence our long-term interests. If behavior is limited by "stone-age" genes, then some pro-environmental behaviors may be ineffective, while others are effective. For example, we must have some contact with the natural environments where we evolved or suffer some psychological and physical harm; this is suggested by several hospital studies. In general, however, behavior is determined by immediate personal consequences, that is, short-term egoism, regardless of eventual consequences in modern world.

The limited genetic potential of a species limits success in general. Species limitation ensures the diversity and integrity of the whole. A species that was too successful, perhaps like humanity, might endanger the interactions of many other species.

As members of cultures and physical beings, humans have very real limits. Our physical limitations are well-known. We cannot run or jump as fast or far as we want, for instance. We cannot lift too great a weight. We seem to be allocated only a finite number of heartbeats, according to Isaac Asimov, in a single lifetime.

There seem to be mental limits, also. We are unable to remember everything that impacts us. We are unable to give attention to everything at once. The upper limit for measurable relations seems to be about seven, plus or minus two. George Miller found that seven was a magic number in human psychology; it represented the maximum number of items that a subject could reliably remember, as well as other variables. Possibly it applies to the number of subjects having an intelligent conversation as well. In general there may be limits of receiving or processing information.

Table 5141-1. Genetic Behavior Potentially Relevant to Environmental Problems (after Gardner and Stern, 1996)

Rene Dubos	Humans have a genetically based need for the stimuli from a natural environment, absence of which may bring harm
Paul Ehrlich	Humans have genetically-based urges for sex and reproduction that cannot be limited, causing overpopulation
B. F. Skinner	Genetically-based "short-term egoism" leads to the environmental Tragedy of the Commons
Edward Wilson	Egoism is tempered by a genetic tendency to live in groups and to behave altruistically towards kin (extended egoism)
Garrett Hardin	A genetically-based denial causes underestimation of probability and severity of environmental threats
Robert Ornstein	Our old mind does not perceive or respond to gradual environmental deterioration
Jay Forrester	The mind cannot comprehend the complexity of social systems, which act in counter-intuitive ways

Perhaps some of these limits have to do with the structure and history of our brain. The bilateral vertebrate, mammalian brain offers an important advantage; it allows two tasks to be addressed simultaneously. Recent brain research (MacNeilage and Peters, 2009) indicates that the left hemisphere specializes in top down control and self-motivated behavior, usually well-established patterns, while the left allows bottom-up control, which is environmentally motivated. The body-crossing nerves resulted in preferences for handedness. The divide allowed specialized behaviors, such as language and tool making, to evolve. Surprisingly, the left hemisphere responds to global patterns to detect and respond to unexpected stimuli. And, the right hemisphere integrates local details. Apparently, immediate global perceptions of change and threats do not translate into true global phenomena, that is, those affecting the planet with slow, invisible, large-scale changes. Perhaps the brain could be trained to respond to those, after recognizing and being affected by planetary change.

5.1.4.2. What are Psychological Limits?
As members of cultures and physical beings, humans have very real limits. Physical limitations are well-known. We cannot run or jump as fast or far as we want, for instance. We cannot lift too great a weight. We seem to be allocated only a finite number of heartbeats, according to Isaac Asimov, in a single lifetime.

There seem to be mental limits, also. We are unable to remember everything that impacts us. We are unable to give attention to everything at once. The upper limit for measurable relations seems to be about seven (plus or minus two). George Miller found that seven was a magic number in human psychology; it represented the maximum number of items that a subject could reliably remember, as well as other variables. Possibly it applies to the number of subjects having an intelligent conversation as well. In general there may be limits of receiving or processing information.

Some of our problems are the outward manifestations of inner causes, as Ervin Laszlo argues. Our perception and thought structures have limits. Science is logically limited (with a predicate logic), committed to the simple operation of cause and effect and the idea that things cause other things in a linear way. Analytic science has reached its limits. Data and information developed by hard studies have undercut the paradigms that guided their investigation. The compartmentalization of scientific fields has exposed the complex connections.

There seem to be limits to our personal space and levels of tolerance to human intensification, also. Urban intensification leads to the question: Is there a limit to human numbers? Perhaps there is a limit in terms of space, but is there a psychological limit? People in cities seem to do well with high-contact, high-proximity living. What happens when people are crowded or feel crowded? Physical complaints, emotional complaints, sexual dysfunction, or feelings of fear, seem to be expressed often. There may be limits of crowding. Are there social limits, in terms of the number of people one can tolerate? We may have a requirements for personal space, home space, and wild space.

Psychological limits may be the basis for some of the great failures of human life. The first is the "failure of perception." We cannot see slow change or anticipate it. No one really sees the incredible interdependence of humanity and nature, of diversity and success (or stability). We do not seem to be able to see others as feeling human beings. Then, there is the "failure of imagination," that limits our understanding and visions of future. We can explore planets and modify genes, but cannot seem to offer functional education or meaningful jobs, dignity in retirement, or goals for living. We insist on individual rights and freedoms, but neglect the whole framework for individual success. We cannot imagine the importance of difference or challenge. Our international system is going to produce a boring uniformity and a painful collapse.

Humans can even create virtual worlds by limiting what could be received; for instance, if a being could see in the x-ray part of the spectrum. Yet, human imagination is limited, as is human knowledge. Many organisms exist of which we know nothing. Their worlds have little meaning in a human world. E. Hall's space bubble or Lewin's personality field extrapolate the umwelt to humans. The "failure of feeling" keeps values and morality as local effects. We do not extend respect or love to distant others. Our personal values and beliefs do not let us. Finally, the "failure of nerve" (or will) dooms us to cleave to the familiar, to ignore other alternatives, to fear change or equality.

5.1.4.3. What are Cultural Limits to Human Cosmologies?

The very circumstance that makes each cosmology unique—being in a unique place—ensures that they have limits. The entire cosmology of a culture is concerned with some form of adaptation or modification. Particularly in agricultural societies, cosmologies are gauged closely to seasons. Cosmologies make the world manageable by limiting it. They can be tuned to the limits of the local ecology, within their knowledge of interactions. Some cultures ignore the long-range ecological consequences of drainage, irrigation or overexploitation, and these cultures may decline and die. But many archaic cosmologies are a form of fitness and limitation. Like the Tukano people, most try for adaptation before domination, according to Gerardo Reichel-Dolmatoff. Many of their rituals limit the hunting of fish and game. The entire cosmology of a culture is concerned with some form of adaptation. Particularly in agricultural societies, cosmologies are gauged closely to seasons.

Furthermore, Jeremy Rifkin states that cosmologies are a way of hiding the unimaginable: Extensive voids, confusing gaps, or sheer size. A cosmology can relieve apprehension. They make the world manageable by limiting it. They make the world comfortable and small. Rifkin claims that humanity inflates its daily activity into its image of the universe. In some ways we do.

Yet, human cosmologies are limited and contradictory. All cosmologies cause destruction and waste; all produce the opposite (evil or enantiodromia) of the good intended. Archaic and modern, occidental and oriental world views are complementary but not complete. Often, a cosmology is accepted as unquestioningly as a language, technology or place. Cosmologies can influence a culture to accept or ignore limits.

The adaptiveness of religious belief is related indirectly to ecology. Religion is sometimes responsive to ecological and economic conditions, for instance, when the Catholic Pope decreed fish to be eaten on Friday to help fishermen. Many archaic religions are even more a form of fitness and limitation. The ritual cycle of the Tsembaga can be interpreted as a regulating mechanism to maintain limits on fighting and using resources, according to Roy Rappaport. Their rituals also facilitated trade and distributed local food surpluses.

Once powerful cosmological ideas are adopted, they can influence many cultures over centuries. The principle of plenitude, restated in Christian terms, says that an intelligible creator gave an earth of unlimited bounty to humanity for their use. This principle seemed to be confirmed in the Renaissance with the discovery of the richness of heaven, of microscopic life, and of unexplored continents. Many modern political ideologies and economics have been shaped by the principle of endless wealth. Adam Smith once calculated that the real price of anything was just the toil acquiring it. Inequality in a world of abundance could only exist through human suppression and exploitation of other humans. The invalidity of this principle came with the recognition of limits.

The frontier of myth no longer exists. And we cannot understand new frontiers until we are divested of the old one. One new myth has to be stronger, the myth of participating in organic beauty, where development, not growth, is without limit. Once doubt is sown—let another recession do that—then the cognitive dissonance from poverty proclaiming that it is wealth will transform the old myth. The universe is a frontier, the mind is a frontier, but they are based on a whole and healthy planet, a nature of which humanity is a special part among special parts.

Our cosmologies influence how we respect or exceed such carrying capacities. If for instance a global culture had a cosmological image of the earth as a desert planet, the carrying capacity for humans would probably be reduced to only 100 million people. If a global culture saw itself as completely technological, it would consider that technology could extend the capacity to 10 billion through conversion and substitution. Many modern cosmologies, modified by advances in technology, pretend that the limits can be exceeded. The cosmology of biotechnology is still economic in a primitive sense. Only the myths have changed to include greater manipulation. It is still concerned with utility, growth and efficiency, as short-term goals. The problem with efficiency is that it is defined within such narrow limits. True efficiency means continuity over long periods of time, as with natural processes. Long-term exchanges in nature are not efficient in the industrial sense. The large sense of economics is the measurement of nature.

5.1.4.3.1. Are There Economic Limits?

The acknowledged fundamental problem of economics is the contradiction between scarce resources and unlimited human wants. The kinds of resources and the possibilities of using them in production are considered in the scope of economics, as are flows and stocks in homes and businesses, the role of the government, business cycles, monetary details and policy, stabilization and growth, international trade, consumer behavior, production costs, pricing, and resource markets.

The focus of economics, however, is rather narrow, in that the concept of resources is very limited and the unlimited wants of human beings are not much discussed. There is no psychology or ethics; there is no ecology or aesthetics (thoughts about beauty). There is no concern with the triviality of the free choice of a worthless 'doodad.' There is no concern with the welfare of the other beings that share the ecological community. Economics has to attempt to understand and predict the behavior of complex, interconnected systems where individual behavior and flows of energy and material are important.

Is there a limit to economic growth? Some theorists, like Paul Samuelson, have concluded that growth is necessary to rid the economy of disparities. The need to grow is intrinsic in this kind of economic system. A large literature has treated perpetual growth as the only conceivable state of affairs. Capitalism depends on growth for stability. There is some analogy with plants (to be elaborated on later). Some stability can be gotten from growth in early stages; later stability must result from limits and metabolism. Growth in plants can delay the onset of senility by ridding the plant of waste products in more diluted form. However, too much growth produces a strain on tissues and early decay. In fact, one herbicide promotes excess growth as a means to kill weeds.

Development instead of growth would equalize wealth more efficiently—after all, economies have been growing for at least 400 years and increasing the disparities. There is no necessary association between development and growth, as Daly and others have shown. Development means the introduction of an innovation. Economic development will require technology. Ecologically sound technologies will minimize stress to the environment. Economies could be modeled after climax vegetation and not successional vegetation, where diversity in scale is greater. A community is forced to accept an upper limit, beyond which it cannot grow any further. Further growth results in destruction or disruption of itself and nature. This is the law of the maximum. Production could be stabilized in a steady state economy, a mature economy, like a climax system, where processes and cycles are constant. A steady state economy must be based on natural laws and ethical principles. Natural laws include thermodynamics and ecological theories. Rules of economics, laws of nature, and ethical principles must be related.

There is another distinction between growth and development. The ecological social approach (or a redistributive environmental strategy) to development makes it irrelevant to discuss global limits to growth. Local limits are far more significant to majority of population. Regardless of how much food exists, people will starve unless they can get it. Redistribution of resources and improvement of environmental quality (home environment.) are more important than increased production by sophisticated technology. The natural capacity of local photosynthesis must be limiting factor in development, especially in tropical and subtropical areas.

There are two basic roads to wealth in a culture, according to traditional economics: (1) producing a bigger pie (supply) or (2) reducing each portion (demand). This assumes that wealth is defined as supply divided by demand.

Equation 2721-1. **Supply ÷ Demand = Wealth**

Demand, as Paul Ehrlich and others have pointed out is physically a result of the number of people and their expectations. If supply is limited, then wealth can still be increased other ways: (1) reduce the expectations of individuals, or (2) reduce the number of individuals. Supply may be mostly material things (but not status, for instance); demand has the more psychological dimension. Therefore, because of demand, wealth will always have a psychological dimension. Gregory Bateson thought that economics may be founded on a fallacy because of that dimension.

5.1.4.3.2. Are There Political Limits?

Politics is the art and science of human government. The first goal of politics is to ensure the survival of the human community. Then it has to maintain the affairs of that community. Politics is the interactive means of providing the basic food and necessities of a community. As survival is survival in nature, politics rests on an ecological foundation. The organization of a community must be in accord with natural laws. Political participation depends on information, much of which can be provided by ecologists.

The politics of community is small-scale and local. Moral consensus is applied to daily operations. Robert Bellah and associates make three distinctions of politics, of which politics of the community is the first. It is followed by politics of interest, where different interests are pursued, according to agreed-upon neutral rules; conflict tends to overwhelm consensus. Finally, the politics of the nation addresses the "higher" affairs of the nation. It is more concerned with leadership than citizenship. Like the politics of interest, it accepts the status quo of relations of power or distribution of inequality. Symbolism becomes more important (perhaps because now the citizen diminishes in importance and the symbolism unites and includes them minimally).

Government has become subservient to economic actors, according to John B. Cobb. Partly because the ideology of economics is so positive. It proclaims that continued growth will solve most of the problems of modern civilization, from poverty to conflict, although the promise has not been fulfilled in the past 400 to 800 years. The problems have increased: Food shortages, housing shortages, energy shortages, unemployment, inequality of opportunity or goods, environmental deterioration, increase in weapons, and insecurity.

These problems continue due to the limits of politics. For example the size of society is a real limit; if there are too many people, within a limited territory or limited system, politics cannot provide them with an identity or control them. Size limits the distribution of things, food, housing, jobs and wealth, also. The time-frame of the political society is also a limit; the short-range visions of national interest are often inimical to the long-range ecological requirements of the support system.

The participation of the members of a society is also a political limit. Communities have always had face limits, in terms of numbers of face-to-face encounters. In terms of distance, communities have limits, also. The politics of communities is small-scale and local, of regions, it may be national. Moral consensus is applied to daily operations.

Leadership is another limit; the pool of applicants is usually relatively narrow. The use of power is a limit. How do we describe the will to power? This does not seem limited to the community, as does participation. The will to power can be observed in human communities for at least 13,000 years. It seems to have evolved from the simple domination of other community members to all of nature.

Security is a problem for local communities and national governments. It becomes more difficult to protect against most any kind of weapons. Perhaps the solution is a global one, with the coordinated change in national policies and world-wide distribution of excess wealth. Some facts result in motivation or fear, fear of the future in general, or the reactions of others, in general. Our response is a defensiveness in face of an unpredictable future. Thus, psychological limits intersect with political limits.

Policy is an intuitive value judging process. Planning has to acknowledge uncertainty. Policy cannot be based on one analytic method. Ozone loss will create patterns of good luck and bad luck, as will species losses and ocean rises. If we do not act responsibly, we are gambling that the cost can be paid later. Even if we act responsibly we are gambling.

5.1.4.4. How do we Deal with Limits?

There are many ways to deal with limits. One obvious way is to ignore them. Ignoring limits usually results in cultural collapses, as happened to the Easter Islanders and others. Recognition of limits, however, does not mean that we are forced to live as bacteria. Limits allow a considerable amount of freedom within their numbers.

Another way to deal with limits is to try to expand them. Eric Jantsch states that limits to growth can be overcome by the evolution of dynamic structure; they appear as widened limits. Expansion has happened before as the result of ecological evolution. To some extent, technology can expand limits.

What is the solution for climate change? Restore forests and grasslands, use alternative energy. Should we try a high-tech solution? A massive program, like the atom bomb, only on alternative energy technologies, might work. Global warming could lead to reversal of ocean currents.

Or, we can respect limits and work within them. This is an ethical way. The limit of Buber's ethical dialogue, I-Thou, is that it is concerned only with the present; posterity is ignored. Human duration is too short. The systems and processes that we deal with are intergenerational. Economic rationality is even more limited. Future values are discounted at current rates. Planting a tree that takes 30-200 years to mature is calculated to be uneconomical (the giant sequoia takes almost 200 years to flower, for instance). Who benefits from it? If our ancestors had planted trees for our benefit, it would be easier to justify continuation. But identification through time is related to that with place, locational stability. People not living in place have no vested interests.

The land ethic Leopold described was a sense of ecological community between humanity and other species. When we see land as community to which we belong, we will use it with love and respect. Such an ethic would change the human role from master of earth to plain member of it. Predators are members of the community; no special interest has the right to exterminate them for the sake of benefit for itself. This attitude is important for habitat protection. Aldo Leopold describes the extension of ethics as "actually a process in ecological evolution. Its sequences may be described in ecological as well as in philosophical terms. An ethic, ecologically, is a limitation on freedom of action in the struggle for existence." An ethic, philosophically, is a differentiation of social from anti-social conduct. These are two different definitions of one thing. The thing has its origin in the tendency of interdependent individuals or groups to evolve modes of cooperation.

Humanity is part of nature, as valuable and unique as cranes or lousewarts, but not more valuable or unique. This attitude would entail using what is necessary, exploiting some ecosystems completely, changing a place to fit human aspirations, and killing plants and animals for sustenance. All animals and plants alter their environments to fit to some extent. But it would also mean limiting humanity and its technological effects, limiting human use to local impacts, and letting other beings live without interference. It is not necessary to dominate or terraform the earth completely. Humanity could contain itself to five percent of the earth's surface and ecosystems and only visit or ignore the remainder. This ideal requires change, even different ways of being.

Care and compliance are limited kinds of being compared to love. Maslow presents 'love-knowledge' as unlimited. Love creates an openness to experience, without judgment. Beings unfold. Contradictions abound in love's completeness. Love expands the awareness of the self and other beings. And its intensification of feeling pulls the frame through the focus, yet preserves the original distance. Love binds space and time in miniature. Its intimacy permits distance. Its duration reaches future generations of beings. Love personalizes the universe, but keeps it free.

Is there a limit to human ethical feeling? How many individuals can a person care for? 100? 50,000? A billion? Or, how many animals or plants? How many living beings? Are these fair questions? Obviously, there is some kind of limit—otherwise, everything alive would be cared for and protected. People seem to be limited to their extended families and groups, as well as some kind of larger groups, a race or nation.

New ethical principles may develop out of the convergence between ecology and theology. The principle of reciprocity in ecology is that no entity can exist by and for itself; everything is connected to everything else. In religious terms, this is the golden rule (see Aldous Huxley's Perennial Philosophy). Reciprocity is the recognition of mutual obligation. All things are bound in bonds of mutual dependency.

We could respect limits by being wise. Jonas Salk defined wisdom as the art of disciplined use of imagination in respect to alternatives, exercised at the right time and in the right measure. Judgment is required as to what is right, and judgment may be an innate art. It is a new kind of fitness, supplanting the biological kind of evolution. Humans have made radically different conditions that they must now accommodate. If the mind is exposed to economy of nature, as revealed through living systems, humans may recognize the necessity of balancing values. Total win-lose conflicts are unwise. Value systems concerned with dynamic equilibrium, aesthetics, complementarity, reciprocity, justice, interdependence, reconciliation, and intuition are the language that biology speaks.

Myths (with transformations) and metaphors (with structure of integrated differences) are modes for conveying ecological wisdom; they are less concerned with survival than the survival value of a good fit between dualisms of life. Equilibrium is needed between self-restraint and self-expression, between self-protection and self-restriction. Not self-expression or self-restraint, or exponential growth or plateau, but all in the finest fit. Fitness attunes us to limits. Wisdom cannot be dependent on perfect knowledge; it does not exist. Humans must act "as if" they were wise, circumspectly, with caution and respect.

5.1.4.5. Why are Limits Important or Necessary?
Limits to growth, even talk of limits, is regarded as a defeatism, as pessimistic, and a blow to human growth. Unseen limits have as real effects as seen limits. Denying limits does not make them go away. Furthermore, as William Catton has pointed out, there is a difference between raising the limits of carrying capacity and simply permitting greater overshoot of the limits, with the threat of a greater and more catastrophic collapse.

Limits define locality, local spaces, and local systems, from the global. Limits are not only important for life, but also are implicated in diversity and maturity. But, even these limits are based on the physical. Although physical and chemical limits are real limits, they are aggravated or increased by psychological limits, which are aggravated by socio-cultural limits, which are further increased by political limits. Each "higher" more complex level makes other levels worse. The attempt to exceed limits becomes less efficient. One of the most important psychological limits is our ability to process data and to draw conclusions from it. We are so ignorant of the complexities of ecosystems that it is suicidal to pretend to "maximize" their use for resources. A free market civilization has to be limited by conservative calculations of ecological balance. It is almost impossible to estimate the economic value of this natural balance. The discussion of limits is expanded in the following chapters, especially when the concept is applied to wilderness or human populations.

5.2. What is Culture? Is it Adaptable? Is There a Global Culture?

The word culture has been used to identify human groups, as well as describe their unique beliefs and artifacts. The word is also applied to emergent properties of other living groups from wolves and termites to bacteria and fungi.

5.2.1. *Animate Culture*

At first, humans were considered special because of their divine spark. Then, they were special because they were disconnected from nature by culture, a magical reification of the opposite of nature, a social structure that we could produce at will to overcome the constraints of nature. It has been argued that culture let us push out of the constraints that limit other hominids, like chimps and apes. Culture was once considered an attribute exclusively in the human domain.

If one defines, as Imanishi Kinji does, culture generally as a form of behavioral transmission that does not rely on genetics, then animals, birds, fish, and perhaps some insects can be said to have cultures. This definition is not limited by the kinds of mechanisms, such as copying or imitation, and it spreads across the continuum from guppies and apes to rich human culture, which need not be devalued by being part of the continuum with animals. Value systems and technical achievements are sometimes the results.

5.2.1.1. Copying & Animal Culture

Culture can be defined as the transmission of information through behavior; imitation is one behavior, teaching is another. Culture involves a mix of: (1) trial and error learning, (2) social learning through observation and imitation, and finally (3) teaching. Culture is one way of transmitting information. Other ways are genetic coding and individual learning, although individual learning is difficult and easily lost without a context.

Some birds learn through imitation. For the bower bird, the bower is an extension of its plumage and size, for courtship displays, but the extension changes faster than the biology. Evolution accelerates in this extension. When the environment supplies tools, new interactions and transactions occur. Culture is like an external bundle of secondary characteristics, according to Edward Hall quoting E. Thomas Gillard. Bowerbirds no longer pair off to mate nest and raise young. Males gather in clans and get ranked hierarchically. In spring, clans form around arenas for display. Each display court is called a lek. Females mate with the dominant males, those with the most baubles. How does female mate choice change?

5.2.1.2. Imitation & Human Culture

Human culture, and human behavior, need to be defined against a long evolutionary backdrop of deep history, the common ground we share with other animals. Perhaps human culture is transmitted through imitation.

What is required for culture? Many theorists require the following things for a group to have a culture: Language (for communication); dexterity for tool making and using; brain power for artifact design and making; social skills for home building; capability of governing or self regulation; and external memory (in customs or things). It can be argued that none of these things is exclusively human, but only humans easily qualify for having all of them. Wolves easily have language and social skills, but they have difficulty making and using tools (although they do sometimes use sticks or stones for training pups) or keeping external memories; some ethologists argue that wolves have culture, while others deny that they do.

5.2.1.3. Extension & Culture

For humans the extensions are more elaborate. A simple extension enhances the function of an organism, as knives are better than teeth for cutting. Extensions permit faster changes to answer environmental challenges without waiting for the body to change. Hall points out that externalization allows one to examine what was inside the head, to study it and change it as a thought experiment. Initially, extensions are low-context and this is what allows dramatic change. That means they are easy to learn and easy to change.

All organisms alter their environment and try to control it. Edward Hall describes two complementary processes of control: Externalizing and internalizing, which are ubiquitous normal, and continuous. The conscience is an example of internal control. In Northern Europe sexual control was internal and left to the woman. In southern Europe, the control was considered external, and doors and locks were used.

Culture is an extension system, but any extension leaves out some things. It can acquire an identity and a history. It can also allow sharing, not just of tools like knives, but tools like cultures, radios, poetry, and plays. Because extensions change so fast, they can seem lifelike and more important than the biological and ecological environment. They can also destroy the natural environment. Extensions, once they are outside the body and mind, can lead to detachment. This allows dissociation. They allow changes in perception, so that others are less alive or can be killed without moral problems. Theoretical systems are treated as real, and everyday life problems are dismissed or undervalued. Cultures can be ranked by use of extensions into high-context or low-context, as extensions decrease context and allow things out of context. Low context cultures are better able to use extensions without screwing up the integrity of the culture.

Culture provides a filter between humans and environments. So, culture can serve as another form of evolutionary filter. It designates what we attend or ignore. It has to screen less valuable information (to avoid overload, even in archaic times). Like a trigger, culture allows less information to activate the system. And this is the only way to increase information handling without making the system larger and more complex—of course, this is what stereotypes and metaphors do, also. Culture programs the individual or institution.

5.2.2. *What is Human Culture?*

What is the world like? How did the world get the way it is? And, what is the role of humanity in the world? All cultures ask and answer these questions. Some of the questions could not be answered from direct observation. And, many of the answers were not limited to observable events. Ideas concerning humanity and the nature of the universe tend to form a coherent system in which ideas are integrated or rejected over days or centuries. Culture includes all of the expectations, understandings, beliefs, and commitments that influence the behavior of a human group. Culture exists in minds, signs, and things, but most importantly in places. The word culture, from the Middle English, meant 'place tilled,' from the Latin *colere* meaning to 'till, care for, inhabit, worship.' For the Romans and English, to have a culture was to inhabit a place and cultivate it, to be responsible for it.

5.2.2.1. Definitions of Culture

The classical definition was put forward by Sir E.B. Tylor (1871): "Culture ... is that complex whole which includes knowledge, belief, art, morals, law, custom, and any other capabilities and habits acquired by man as a member of society." Although the definition seemed to limit culture to 'man' alone, it definitely restricted it to human behaviors and artifacts.

Others, including Claude Levi-Strauss and Leslie White argued that culture required imitation, teaching, and language, and therefore the concept could not apply to other species. In 1952, A. L. Kroeber and Clyde Kluckhohn identified 164 definitions of culture. Most of these definitions are refinements of the Tylor definition. Robert Boyd and Peter Richardson defined culture as 'information capable of affecting an individual's own traits; they acquire the information from other individuals through imitation.' Information is a broad enough concept, especially as information, to include behaviors and artifacts, and the vehicle is identified as imitation.

Peter Berger defines culture as the totality of human products. Culture is everything created as a group, tribe or nation, physical or ideal, in the past or present. This embraces cookware, arrows, steam engines; artworks, books, legal codes; symbols, values, social structures. A cultural system surrounds the network of human interactions with raw materials, forms of life and other humans. A general definition of culture identifies it as the ensemble of values, worldview, aspirations, customs, technologies, techniques and physical artifacts that characterize a people and distinguish them from others.

5.2.2.2. Metaphorical Definitions of Culture

The use of metaphors might contribute something to the understanding of culture. Culture is an 'organism' that grows and matures, that is, it is organic and whole, with a finite life span. It processes energy and matter to survive. But, it does not have a skin or genetic limits—nor does it have a genetic way of passing on its characteristics, although it has the memory of its members, and they can communicate.

Perhaps, culture is an 'ecosystem' that is a self-maintaining, stable and changing pattern. But, that neglects an ideational component, as did 'organism.' Like systems, cultures develop over time; they are capable of evolving. There are similarities in their evolution: Similar dynamics and machinery, and a direction towards complexification. Evolution, biological or cultural, depends on two things: The rate at which useful changes arise and the rate that they spread in the communities. But, this neglects internal states and forms of ideas.

Culture is a pair of 'glasses' that allow us to see some details or patterns; that is, it is an add-on to the human animal that focuses and shades perception to permit survival. This metaphor implies that many do not need culture or that culture can be removed or replaced at will. Feral children, beings without culture, have difficulty surviving; removing culture, even if possible, would reduce people to a feral level. Another culture can be grafted onto a first culture, as people immigrate into a second culture, although that can be a long, incomplete process of transformation.

Culture is a 'hologram.' It is a whole that arises when the reference beam of nature shines through the mental patterns of a human group. This metaphor could capture the interior and exterior aspects of culture. The place of the reference beam would color and shape the culture differently. The culture could develop, however, only with decay over time, not through a transformation. These metaphors, although interesting, have limitations. A broader attempt at definition might be more fruitful.

5.2.2.3. Synthetic Definition of Culture

Culture stretches vertically to include the physical, economic and political. It stretches horizontally to include society as a whole. In fact, culture is concerned with all things and beings. It is organic, like Aristotle considered a work of art, and whole. So everything in it is interrelated in some degree. Many relationships are encompassed by the holistic perspective of culture: The relation of people to themselves, to each other, to objects they create, and to their natural environment, and to their cultural environment. These bear on psychological well-being, social bonds, material legacy, and on the association with other forms of life.

These many definitions overlap and can be combined in to a synthetic definition, with common characteristics: Culture is a symbolic system of shared beliefs, values, customs, behaviors, and artifacts that emerges as a unique, coherent whole pattern that orders the experiences and meanings of its members and that members of society use to adapt to changing environments and to adjust to each other, and that is transmitted to succeeding generations through learning behavior and language, so that culture and the environment constrain and construct each other over time as a complex of ecological, social and historical processes. This definition might be refined later.

5.2.2.4. Properties of Culture

Culture is imbibed through a process of social interaction. People acquire culture unconsciously through social interaction, as well as consciously, through apprenticeship or formal learning. The young, or new members of a culture, observe others and imitate that behavior. Cultural models are internalized by individuals, so that part of culture resides in the mental sphere of people. Beliefs and knowledge are not shared equally by all individuals, thus culture is shared differentially.

Human beings use conceptual devices, such as symbols, to communicate abstract ideas about nature or society to one another. Through our linguistic capacity, we can use symbols as meaningful representations of reality. Public shared meanings provide a set of designs that allow an educated individual to survive nature and society. The understanding and practice of culture is shared in a culture. A culture has properties that are related to ecosystems and places (Table 5224-1).

Table 5224-1. Contrasted Properties of Different Levels of Patterns

— Nature —		— Culture —		— Design —	
Field	*Ecosystems*	*Place*	*Culture*	*Good Places*	*Good Society*
Process	Course	Dynamicism	*Conduct*	Action	Method
Autopoesis	Self-making	Identity	*Wholeness*	Individuality	Extension
Differentiation	Diversity	Uniqueness	*Flexibility*	Richness	Variety
Integration	Construction	Investment	*Adaptation*	Conviviality	Cooperation
Constancy	Stability	Regularity	*Endurance*	Consistency	Loyalty
Development	Productivity	Renewal	*Vitality*	Health	Harmony

5.2.2.4.1. Conduct as a Property of Culture

Conduct is the course of cultural behaviors through a behavioral landscape or field. New mathematical treatments of fields tend to be three-dimensional. Rene Thom's catastrophe theory and Conrad Waddington's epigenetic landscape are two such theories. The epigenetic landscape field can be used to explain why some people cannot kick a bad habit even though they know they should. The need path or chreod is deeper than the path of a cognitive chreod to better health. If the need chreod is too deep, one cannot explore alternatives with a cognitive chreod. The field also explains why necessity cannot be the mother of invention; the necessity path, starvation for instance, is too deep to allow exploration of a shallower path of designing. The broader landscape of leisure is needed. Conduct can be described as the stable path of culture through a landscape of possibilities. Once the course has been set, it is most likely to channel subsequent behavior, unless some event triggers a deeper course.

5.2.2.4.2. Wholeness as a Property of Culture

Culture provides an identity for its members. It tells them who they are, where they came from, and why they are special. Identity is basic to human existence. People are identified by their roles. A person is an incarnation of his and her group—even in industrial culture, one is identified as an astronomer or farmer. Identity is that persistent quality that serves nothing; it is. Identity can be described apart from its performance in interactions, but not isolated. The relationship between identity and wholeness is a rhythm, with unique patterns. In fact, culture is concerned with all things and beings in a whole. To paraphrase Arthur Koestler, culture is a holon, a stable subwhole in a hierarchy that displays structural gestalt constancy and rule-governed behavior.

5.2.2.4.3. Flexibility as a Property of Culture

The rules of a holon lend order and stability to the whole, as well as flexibility. Flexibility means not being over connected, or not being too rigid or efficient. Than means that culture is able to slough off people or community structures and to incorporate new people and structures. Culture is able to keep some options and unused connections open. Some of the flexibility comes from different ways of establishing connections in specific places. Although culture is bounded, it is open to flows of energy and materials. In fact, it requires a steady input and output exchange flow. It requires order, but not too much order. It is a loose-fitting patchwork of ideas, relationships and things. It is tolerant of discontinuities and contra- dictions, and this gives it flexibility. Humans can tolerate inconsistency and operate with contradictory beliefs: Soldiers fight for peace; ministers save the unborn for starvation. If the contradictions become too great and maladaptive, however, then the culture can collapse. Cultures that become too isolated often stagnate, then collapse, even if they do reconnect with other cultures.

5.2.2.4.4. Adaptation as a Property of Culture

The patchiness of culture is parallel to the co-constrained construction of a species and its environment. Culture has to balance between embracing change and resisting change. People show a desire for new things, but often fear and resist change. According to most theories of cultural adaptation (or integration or evolution), resistance to change is normal as a cultural process. Groups like the pygmies have specialized to fit the requirements of the environment, successfully. This makes it difficult to adopt other cultural arrangements. On the other hand, resistance to change itself is an adaptive mechanism. According to Betty Meggers, it works as a successful "cultural isolating mechanism." Isolation remember is what allows a culture to develop in the first place. But, then can it force a culture to become stagnant?

A primary culture is adaptive because it aids survival in the ambihuman world. Its ways of living are sophisticated survival mechanisms. Each way of life is a set of adaptations to the limits of the environment. Mutually constrained construction enforces coevolution, the emergence of a highly ordered complexity to full structuration. The cultural practice of polyandry in Tibet may be an adaptation to limited resources. With monogamy and every woman married, there would be more children to strain the resources available.

5.2.2.4.5. Endurance as a Property of Culture

The cultural system is stable and persists through time. It is a general property of some sys- tems that acquired information is used to close the door to further inflow. A mature culture needs less information, since it works toward preservation, and closes itself off to information that does not fit the shape of the culture. The effect of maturity is to allow a maximum variability between systems with slight external differences, such as place or initial conditions.

According to Harding (1960), when forced to act on changes, a culture will only accommodate those changes that preserve its fundamental character. Stability is the ability to maintain identity under the flow of external forces and disturbances. A culture has to be able to resist disturbances that are too disruptive. Culture also has to be resilient enough to recover from intermediate and small disturbances. It has to accommodate changes that contribute to its identity. Stability can be related to compartmentalization, communications, richness of interactions, and connections. With order and integration come stability and security, without which no one can survive. When human societies were small, the amount of control and security required was small. Security came from knowledge of place and environment. Although societies have grown, human security has not.

5.2.2.4.6. Vitality as a Property of Culture

To be constant and stable, a culture has to be vital; that is, it has to be productive, to be able to convert energy and materials into foods and structures for survival. The system is self-creating. It renews itself as its contents change, and as disturbances change the parameters of the system. The ideas and metaphors of a culture have to be vital, also, capable of adapting to a dynamic environment and to other cultures. What barriers are there to cultural renewal? How can they be overcome?

5.2.2.5. *Functions of Culture*

Culture has to deal with the facts of life, from change to death, so that people can survive in their communities in their places. Freeman Dyson distinguished three crucial biological inventions in the transition from unicelled to multicelled organisms: Death, which allowed differentiation of the future from the past; Sex, which enabled the characteristics to be shared and mixed; and Speciation, which increased diversity through separation and genetic barriers. Life can experiment with the diversity of forms and functions.

Dyson noted also that each biological invention has its analog as a cultural invention in human societies: For death it is tragedy; the fact of death is made a theme in ritual and drama. Great cultures have distilled great works from the fact of death. For sex, it is romance, where sex is turned into a thing of mystery and beauty, in dance and poetry. Speciation has been transformed into cultures and languages at the human level. The flexibility of social institutions grew out of their differences and places.

5.2.2.5.1. Culture Grounds

A culture orders a whole cosmos. A culture selects what is important from the previously undifferentiated phenomena of nature and presents it in myths and stories to be learned by people. People in a primary culture share a common image of their world—from the German word meaning 'man-image.' The image is a construct of human knowledge that reflects human awareness of a local environment. The image is constructed metaphorically, but treated 'as if' it were true. A traditional way of living evolves with people's experience and knowledge. The image guides their behavior.

Grounding is more than the expression of a local place through culture. It is also the expression of individual behaviors through culture. Culture can be observed through individual actions. Individual and collective entities have to be understood and balanced. Both are important parts of a whole system. Culture limits and constrains individual behavior with its conventions, but individual behavior can extend culture, by acting and communicating.

5.2.2.5.2. Culture Orders

A culture orders a whole cosmos (from the Greek word meaning 'to set in order') and applied order to the human face as well as to all of nature, and what was beautiful was also morally admirable. A culture selects what is important and what is important is often beautiful. Every culture strives for beauty that goes beyond utilitarian values. People structure their worlds with their own group at the center. With order and integration come stability and security, without which no one can survive. When human societies were small, the amount of control and security required was small. Primordial security came from knowledge. The industrial sense of security has been centered in the state, and now in the corporation, in retirement, insurance, housing, recreation, and even in credit cards. Primary peoples were responsible for their own food clothing and housing. The economic well-being of modern industrial families depends on wages alone. People are dependent on the goodwill of capital corporations for personal security. We have substituted insulation and insurance for knowledge.

5.2.2.5.3. Culture Explains

Culture explains the universe. It also explains the behavior of its adherents. By explaining reality, culture binds the human and ambihuman, and the past and the present, into a meaningful whole. Culture explains why traditions are necessary. It explains why things are as they are, and how they came to be that way. It can do this in stories, and sometimes stories need a special language to explain, such as Wataluma or Latin. In modern culture, a scientific theory is a statement that explains some aspect of the world and allows one to ask certain questions.

Cultures occupy a particular territory. This is especially true of the Campa, in a tropical forest in Peru, and the Ituri pygmy, in a tropical rainforest in Africa. The features of their cultures are unique to their place. They literally could not live with images of desert or ocean, like the Taureg of the Sahara or Samoans of the Pacific. Regardless of the features of a place, myths are created to give it special significance; giving mythical significance to a place strengthens a people's identification with it. People identify with their place and often equate their own characteristics with it; the Ituri consider themselves as bountiful as their forest, while the Mongols are as undeniable as the wind from their plateau. Knowledge allows survival in fragile habitats. The !Kung San of the Kalahari desert exist in small communities that enable them to continue traditional hunting and gathering without depleting their resources; they hunt eighty types of animals. The Hanunoo of the Philippines distinguish 1600 plant species, where scientists only know of 1200 in the same area.

5.2.2.5.4. Culture Integrates

Culture provides an identity for its members. It tells them who they are, where they came from, and why they are special. Identity is basic to human existence. People are identified by their roles. Culture integrates the roles and the individuals into one society. Jurgen Habermas argues that the task of culture is to understand the meanings attributed to objects and events by individuals under concrete circumstances. He identifies culture as a set of subjective meanings held by individuals about themselves and their world. A primary culture is adaptive because it aids survival in the ultrahuman world. Its ways of living are sophisticated survival mechanisms. Each way of life is a set of adaptations to the limits of the environment, which are integrated into the culture.

5.2.2.5.5. Culture Justifies

Another function of a culture is to justify human activities, in order to have those activities continue. Unless an activity satisfies a basic human need, it may not be repeated. The needs may be physical, psychological, or social, such as acquiring status. According to Abraham Maslow, the needs are hierarchical and higher needs cannot be addressed until basic needs, such as food, are satisfied. Jeremy Rifkin suggested that psychological needs can only be satisfied by the cosmology of a culture. In ancient China, people found justification of their world view in the matrix of nature; perceived cruelty in nature justified real cruelty by human beings.

A common culture provides an ideal framework for public and private decision making. The Sami in northern Scandinavia have institutionalized ways of avoiding conflict, for instance, by shaming those who would impose their will. People in a culture may enlist ideology or religion to justify transfer of wealth to the rich and the leader. The shared religion also makes strangers act more peacefully without kinship. It gives people a reason to sacrifice their lives for an institution, nongenetic.

5.2.2.5.6. Culture Controls

Some human behavior is controlled and regulated through the use of space. Most foraging populations, furthermore, regulate their density well below the limits of the food supply. Culture controls many behaviors, from hunting to birthing. Women control sex, birth, weaning, and often population size. For Australian aborigines, women are regarded as controlling social harmony, health and connections to land, while men, with more time and strength, tend to control creative activities and hunting for prestige foods.

Cultures have been self-sufficient for thousands of years. Although some of them fail, others last for quite a long time. The Desana, for example, have existed in the Amazon for over four thousand years by maintaining an equilibrium with the environment. Ecosystems are local, not global. Although we regard communities as being tied together globally, each community is alone. Each culture has to be responsible for its own welfare. Preserving the local environment is one requirement. A culture is a way of doing things by a unique people with a unique identity, with a unique history, in a unique environment, communicating in unique ways, and using materials in unique ways.

5.2.2.6. *Effectiveness of Culture*

There are similarities of concepts between a culture and an ecosystem. Culture is a sloppy concept, like an ecosystem. Culture is also scalable, like an ecosystem, and can be nested. Properties of a culture resemble those of any system, such as an ecosystem. They include: Identity, openness, productivity, co-constrained construction, and vitality.

Like an organism, a culture has to satisfy needs to continue. The functions of a culture reflect its needs, which when expressed, are also the strengths of the culture. Like an organism in an environment, a culture engages in activities and interrelations, such as change and development. It can compete with or take-over other cultures. It can cooperate and trade. There will always be some conflict or misunderstanding, from having different images and metaphors, limits and rules of behavior. The strengths of a culture may allow it to persist for a long time; its weaknesses, though, ensure that it will eventually fall apart.

5.2.2.6.1. Strengths of Culture

A culture has social and ecological functions. The image a people have of their world makes sense of the overwhelming confusion of nature; it gives people a unique identity and justifies their behavior. Culture ties them to a place, whose qualities are known and preserved. These things are the strengths of culture, when they work.

5.2.2.6.1.1. *Personalizing a Place*. Primary cultures think of nature as presence: Things are personalized first, before being categorized. A noise in the woods at night is the voice of a living being or person. The natural world is seen as having human qualities. A core of anthropomorphism is necessary for understanding the ultrahuman. This core actually sponsors diversity in individuals and species. We understand other beings by expanding ourselves, not by shrinking them. This leads to relational knowledge. Symbolic associations and transformations are made between diverse entities. The social life of humans and other beings is not separate. Anthropomorphic thought increases the dimensions of the human intellect. By rejecting this thought, the experience of others is restricted and the scope of self-knowledge is reduced. The Pygmies of the Ituri Forest of Zaire see themselves as a part of the forest, an abundant provider; they regard the forest as mother and father, conceive of themselves as children of the forest, and live in harmony with it. The Pueblo Indians consider the sun to be their father and earth to be their mother. Earth and sun create an endless series of cycles that govern life; people and animals and seasons are part of the cycle.

5.2.2.6.1.2. *Knowing a Place*. Knowledge allows survival in fragile habitats. The Kayapo in the Brazilian rainforest practice seed selection and crop rotation; they fertilize the soil with wastes. They gather 250 types of wild fruits, 650 medicinal plants, and 100s of tubers and leaves. The Karen of the Thailand practice shifting cultivation, which is only possible with low population density and sufficient land for a rotation of decades or generations. Oddly, industrial culture considers that primary cultures have a shortage of economically-relevant knowledge. But primary people know how to find or grow edible and medicinal plants; they know how to make appropriate houses and cooking utensils with a minimum of effort and materials. Traditional communities have a rich biological knowledge of animals and plants, as well as a rich mythical knowledge of animals and plants.

5.2.2.6.1.3. *Adapting to Place*. A primary culture is adaptive because it aids survival in the ultrahuman world. Its ways of living are sophisticated survival mechanisms. Each way of life is a set of adaptations to the limits of the environment. The myths of the Tukano, for instance, do not describe their place in nature in terms of mastery of a subordinate environment. Instead, the Tukano learn that they are part of a larger system that transcends individuals. Survival and maintenance of the quality of life are possible only if all other lives are allowed to evolve according to their specific needs, which are described in myths and traditions. Myth limits the impacts a people can have on the habitat. For example, the Tukano are taught that only a limited number of animals were created and that these were placed under the care of the spirits. Tukano fishing is limited because the fish own the riverbank, so humans have no rights over it. In fact, humans are held accountable by the shaman if the area is overfished. The punishment for taking too many is also told in myth—it is human extinction. The role of the shaman is not to seek more animals or exemption from overhunting, but to restrict human use. Myth presents rules for regulating the birth rate and for social behavior, as well as for the harvest rate.

Over many generations, people learn what kinds of food to eat and where and when to find it. Particularly in agricultural societies, cultures are gauged closely to seasons. They are also tuned to the limits of the local ecology. The pastoral economies of the Taureg of West Africa are as ecologically efficient as intensive ranches in industrial countries. The !Kung San of the northwest Kalahari are likewise efficient; individual adults need only devote 12-19 hours per week to support themselves.

5.2.2.6.1.4. *Preserving Place.* We know that cultures have been self sufficient for thousands of years. The Desana, for example, have existed in the Amazon for over 4,000 years by maintaining an equilibrium with the environment. An Acoma town in the southwestern

United States has been inhabited for almost 900 years. Industrial culture offers great control over many aspects of the environment. One form of this control has been the setting aside of parks and wilderness areas for future generations; functioning ecosystems, such as wetlands, are being preserved, sometimes haphazardly, as the basis for human activities. Each culture has to be responsible for its own welfare. Preserving the local environment is one requirement.

5.2.2.6.1.5. *Offering Security*. With order and integration come stability and security, without which no one can survive. When human societies were small, the amount of control and security required was small. Although societies have grown, human security has not. Primordial security comes from knowledge of the environment. Narrowness of experience is one source of insecurity. Although a culture can help people cope with fear, it cannot offer security against the biological fact of death. Michael W. Fox has pointed out that death awareness motivates much of human behavior. Many primary groups have worked acceptance of death into their myths. The Campa attitude toward death reflects acceptance: since the first Campas were made of earth, they return to earth after death; had they been made of stone, they would have been immortal. According to an Inuit myth, the first human beings lived on an island, Mitligjuaq, in the Hudson Strait; no one ever left and eventually there were so many people that the island began to sink under their weight. An old woman shouted: "Let it be so ordered that human beings can die, for there will no longer be room for us on earth." Her wish came true. Death was the solution to survival. Death is natural. But, the human longing for immortality, from Neanderthals to postmoderns, is also natural.

5.2.2.6.2. Weaknesses of Culture

Culture does not fit together into a perfectly integrated whole. A culture is a loose-fitting patchwork of ideas, relationships, and things. It is not complete. There are discontinuities and contradictions. Many ideas are arbitrary. A culture can be indifferent to its place or fate and overexploit natural systems. The balance of freedom and necessity ensures that no culture will ever fit its environment or its members perfectly. Fortunately, humans can tolerate inconsistency and operate with contradictory beliefs: Soldiers fight for peace or ministers save the unborn that they may starve (weaknesses are expanded in Section 5.8)

5.2.2.7. *Diversity & Universals of Culture*

These functional requirements of culture are rather minimal for humans, being only food, clothing, shelter, and reproduction. Throughout their history, humans have used animals and plants for food and clothing. They have been able to convert animal and vegetable resources into all their needs for food, shelter, and clothing. Other requirements, such as respect, comfort, and self-fulfillment, depend on socio-cultural systems. But, the differences in local environments and cultures ensures a diversity of those things.

The origin of clothing, for warmth or prestige produced ecological changes in local animals. According to Mark Stoneking, body lice may have evolved from head lice after a new ecological niche, clothing in this case, opened for them. Using a molecular clock, he calculates that clothing appeared about 75,000 years ago in Africa. Perhaps it appeared first for status, but clothing allowed the move to colder climates. Much of the diversity of human clothing has to do with the diversity of localities. And, much has to do with the aesthetic sense and human invention. Clothing is often the first thing noticed when cultures meet. In the eyes of another culture, clothing differences are often exaggerated. For instance, westerners seeing Africans or Polynesians for the first time focused on the absence of clothing or on body adornment. Naturally, Africans seeing westerners for the first time emphasized the elaborate clothing and facial hair.

Behavioral difference may be noticed after a longer acquaintance. Behaviors and symbols may take decades to understand. Many gross operations of a culture seem familiar and may be comforting in their familiarity. All cultures seem to have certain things in common. Of course, many of these are referred to as soft universals, since everyone needs shelter. Other differences may be the result of contacts between cultures, because cultures are mimetic, that is, they copy one another. This may be the secret behind agriculture, cities, and industrialization. Other universals range from age-grading and athletics to weaning and weather control.

Perhaps cultures seem to have universal characteristics because humans are of the same species, have similar requirements and similar bodies. All people live in groups, have some form of shelter, and have an incest taboo. Even though a process may be universal, the implementation and symbolization of it may be quite different and unique. Some scholars have argued that these universals could allow a global human culture. That may be true, although the relationship of a local culture to a global culture may be problematic.

5.2.3. *Dependence & Constraints*

Our minds are not only nature dependent, but culture dependent as well. The wind, trees, birds are sources of signals and symbols, and so are gestures and words. A community implies that the experiencers share ways of experiencing or the same experiences. This enables an individual to go beyond a finite view, to see the embedded culture as one of many ways of relating self to universe. Culture evolves from the interactions of humans with nature; both are in a constant state of flux. Cultures may be thought of as parallel to other species [memes (ideas)=genes]. The study of human adaptation to nature is cultural ecology.

Culture is a codification of reality, a symbolic system that transforms physical reality into experienced reality. Culture codifies reality through expressions, which can be preserved and transmitted through generations through language. Different languages program events differently, therefore no culture or belief system can be considered entirely apart from language, or language entirely apart from place.

Four relationships are encompassed by the holistic perspective of culture: The relation of people to themselves, to each other, to objects they create, and to natural environment. These bear on psychological well-being, social bonds, material legacy, and on the association with other forms of life.

Culture puts real constraints on our actions, as well as our imaginations. The structure of the earth also puts constraints. The structure of our bodies and minds adds more constraints. Culture may be referenced to the discrete events that brought it into being. But, the process is chaotic and arbitrary. We do not know all of the events or their progressive development. Even without a theoretical framework, the process of culture is abundantly inventive and beautiful.

5.2.4. *The Health of a Culture*

A healthy culture has many of the same properties as a healthy ecosystem. Unlike an ecosystem, however, cultures tend to grow beyond maturity to a point of collapse. At that point or sooner, another major difference manifests itself: Violence in the form of war. Big problems from the success of human culture include: Reproductive success, ecosystem conversion, then overshoot and stagnation, followed by stupidity and violence to try to rebalance the culture. Violence does sometimes work for a culture, usually by reducing the population and redistributing some of the wealth. In fact, every culture goes through a cyclic that may have from two to six phases. For example, George Modelski describes long-term cycles in cultures in terms (1) preference for order and (2) availability of order, each of which may be high or low.

Table 524-1. Modelski's Long cycles in politics

Cycle phase	Preference for order	Availability of order
1	High	Low
2	High	High
3	Low	High
4	Low	Low

The time structure of the operation of a complex system is a set of phases. Moving through a sequence of phases can create a whole cycle. Phase one could be global war; afterwards, phase two initiates a new world power; then, plenitude and security erode, and the system becomes delegitimized; finally, a low pressure for order and a preference for wealth, combined with low available order, deconcentrates the system. Modelski suggests that nation states are in the second part of the long cycle, a phase of decline. The model predicts increasing salience and rising nationalisms associated with increased conflicts. It should be leading towards a new global war.

In normal stochastic processes, from molecules to nations, things bunch together or move apart. We can identify cycles of this. For instance, great wars and dominant political powers can be also correlated to major innovations in discovery or technology. Cycles in politics are not just repetitions—they spiral outward, taking up the old ideas and processes—who says we do not learn unconsciously from history.

Based on this system, cultural historians William Strauss and Neil Howe suggest that modern history moves to the rhythm of life rather than just to institutional or economic cycles. They suggest an 80-year cycle, perhaps related to a human lifetime, composed of four phases: (1) The High, or first turning, is confident expansion as a new order replaces the old. Perhaps the most recent one began in 1945. (2) The Awakening is an inward turning away from the outer order, which is rebelled against. This phase may have started in 1964. (3) The Unraveling is characterized by waning public trust, fragmenting culture, changing values, and alienation. This phase may have started between 1983 and 2003. (4) The Crisis represents a major discontinuity, leading to a new order that replaces the old order. They suggest that this phase began around 2004; it might continue for twenty more years.

Interestingly, C.S. Holling and his group used a more ecological approach to identify four phases of a cultural cycle. These phases are based on high or low values of capital and connection.

Table 524-2. Two Models of Cultural Cycles

Cycle Phase	Model 1 (after Holling et al.)	Model 2 (after Strauss & Howe)
1	Low Capital	Expansion
2	High Capital	Inward turning
3	Low Connection	Unraveling
4	High Connection	Crisis of discontinuity

These kinds of cycles may explain many kinds of phenomena in cultures. Order is certainly a major factor in a culture, as are connections and capital, ecological or financial. One wonders if a cycle of the life of a culture or one cycle within its life, could be explained by combinations of population growth, security, stagnation, overshoot and environmental challenges, such as drought. Stability seems a characteristic of archaic cultures. We can even state it in Newtonian terms: A culture at rest tends to remain at rest. According to Harding (1960), when forced to act on changes, a culture will only accommodate those changes that preserve its fundamental character. Groups like the pygmies have specialized to fit the requirements of the environment, successfully. This makes it difficult to adopt other cultural arrangements. Since cultures have been traditionally self-reliant, resistance to imbalance is a positive act. Resistance to change or other cultures, is an adaptive mechanism that may encourage isolation. Yet, isolation is what allows a culture to develop in a unique way. Yet, isolation may lead to stagnation.

5.2.4.1. The Growth of Local Cultures

Some cultures grow, and in growing, decide that growth is a good thing to have without limits. Many lucky accidents, such as the rediscovery of the Americas, gave some cultures an impetus to keep growing. Other positive developments, such as science and industry, followed the conditions created by plagues and environmental restraints. The Spanish became a megaculture after benefiting from their exploration and exploitation of the Americas and western Pacific. The English became a megaculture after establishing an empire from North America and the Caribbean to Africa and Asia. North America became a megaculture in the Twentieth-Century. Each megaculture was able to dominate part of the planet with its influence and to see some of its products or rules become ubiquitous.

Global capitalism undermined many traditional cultures by offering consumerism in the place of traditional cultural guides for behavior. Social roles seemed irrelevant by comparison, if the good life could be bought without effort. Yet, it did not seem to work in Europe and the U.S. Instead of being free from economic want to develop their potential as creative human beings, people became trapped in a consumer cycle. Self-actualization was postponed for self-gratification. Democracy seemed to be good for balancing a middle class in some cultures, but it ignored other cultures and economies.

By virtue of its demomass and political system, China is becoming a megaculture. Chinese products are dominating the economies of other nations.

5.2.4.2. Emergence of Connected Local Cultures

Trade and exchanges by various local cultures extend to distant lands. Over time, regional systems start to develop, with newer connections and technologies, which draw the cultures and civilizations, of all ages, into a tighter pattern. Some global civilization may result, but what effect will it have on the earlier patterns? So much is lost, especially ways of being and acting in a more natural less technological environment. The world is acquiring a global economic structure.

A global political system is emerging, as there is increased vertical differentiation, evolving from nations and regions, that leads to differentiation into political groups and economic interests. The world does not function very well politically at a global level. It could be made better. It functions in the absence of a common culture or language. There is no world law, but there is a set of rules regulating international behavior; these are generally observed and understood. There is a small homogenous subculture, which belongs to the rich elites of every nation. This serves to integrate cultures to some extent. The modern world has become a very interactive system, especially with computer and communication technologies, which can lead to totally integrated mass communication and the extreme compression space and acceleration of time.

It is said that the emerging global system has no center. That is a good thing. But, it is expanding without limits and that is a bad thing. It could cause the disintegration of natural systems that are interlinked with our economic exploitation. Perhaps this emergence can be linked with a major cultural revolution, as identified by W.I. Thompson, Planetization, as there is absorption of a new consciousness surrounding the old. Other earlier revolutions resulted had a similar attendant process of miniaturization. The forest was miniaturized in clumps of trees; animals were miniaturized in artistic images; time on a lunar tally stick; plants in a garden; and, nature by culture in 1800 (under the glass roof of the Crystal Palace as Thompson notes). The question is whether consciousness can create a global culture.

5.2.5. *Is there a Global Culture?*

Human societies have tended to grow larger over time. With the change in scale of populations, other changes have occurred in the structure of settlements, fields, governments and religion. These changes may foreshadow some kind of global culture or government.

5.2.5.1. The Growth of Megacultures

Some cultures grow, and in growing, decide that growth is a good thing to have without limits. Many lucky accidents, such as the rediscovery of the Americas, gave some cultures an impetus to keep growing. Other developments, such as science and industry, followed the conditions created by plagues and environmental restraints. The Spanish became a megaculture after benefiting from their exploration and exploitation of the Americas and western Pacific. The English became a megaculture after establishing an empire from North America and the Caribbean to Africa and Asia. North America became a megaculture in the Twentieth-Century. Each megaculture was able to dominate part of the planet with its influence and to see some of its products or rules become ubiquitous.

Global capitalism undermined many traditional cultures by offering consumerism in the place of traditional cultural guides for behavior. Social roles seemed irrelevant by comparison, if the good life could be bought without effort. Yet, it did not seem to work in Europe and the U.S. Instead of being free from economic want to develop their potential as creative human beings, people became trapped in a consumer cycle. Self-actualization was postponed for self-gratification. Democracy seemed to be good for balancing a middle class in some cultures, but it ignored other cultures and economies.

By virtue of its demomass and political system, China is becoming a megaculture. Chinese products are dominating the economies of other nations.

5.2.5.2. Emergence of a Simple Global Culture

Trade and exchanges by various cultures extend to distant lands. Over time, a global system starts to develop, with newer connections and technologies, which draw the cultures and civilizations, of all ages, into a tighter pattern. Global technology draws the cultures and civilizations, of all ages, into a tighter pattern. Some global civilization may result, but what effect will it have on the earlier patterns? So much is lost, of ways of being and acting in a more natural less technological environment. The world has acquired a global economic structure.

A global political system is emerging, as there is increased vertical differentiation, evolving from nations and regions, that leads to differentiation into political groups and economic interests. The world does not function well politically at a global level. It could be made better. It functions in the absence of a common culture or language. There is no world law, but there is a set of rules regulating international behavior; these are generally observed and understood. There is a small homogenous subculture, which belongs to the rich elites of every nation. This serves to integrate to some extent.

The modern world has become a very interactive system, especially with computer and communication technologies, which can lead to totally integrated mass communication and the extreme compression space and acceleration of time. It is said that the global system has no center. That is a good thing. But, it is expanding without limits and that is a bad thing. It could cause the disintegration of natural systems that are interlinked with our economic exploitation.

Perhaps this emergence can be linked with a major cultural revolution, as identified by W.I. Thompson, Planetization, as there is absorption of a new consciousness surrounding the old. Other earlier revolutions resulted had a similar attendant process of miniaturization. The forest was miniaturized in clumps of trees; animals were miniaturized in artistic image; time on a lunar tally stick; plants in a garden; women in a household; and, nature by culture in 1800 (under the glass roof of the Crystal Palace). The question is whether consciousness can create a global culture.

5.2.5.3. Is Global Culture Possible?

Is there such a thing as a global culture? In his Notes Towards a Definition of Culture, published in 1948, T. S. Eliot suggested that a "world culture which was simply a uniform culture would be no culture at all" and that humanity would be de-humanized in a miserable nightmare. But, since we could not give up the idea of a world culture, knowing that we could not imagine it, and conceiving it only as a logic logical term of relations between cultures: "We must aspire to a common world culture, which will yet not diminish the particularity of the constituent parts."

The local and the particular are required parts of culture, so global culture has a contradiction. Eliot noted that difficulty in a national culture of Britain: "We have not given enough attention to the ecology of cultures. It is probable, I think, that complete uniformity of culture throughout these islands would bring about a lower degree of culture altogether." He concluded that a national culture should be a constellation of cultures that have to benefit each other and the whole to flourish.

The disappearance of peripheral cultures, in Britain and elsewhere, might be a calamity, but, modernity, in the predominant versions of liberalism and Marxism, sees the goal of history in a universal world culture. Alas, one wonders if a global culture would be worse than the national cultures that are so irreverent with their peripheral cultures. In destroying world peripheral cultures, there might not be any culture left to be global.

The 'McDonaldization' of the world is about economic dominance, the homogenization of trade networks and consumer rituals. This may undermine the identity of other cultures, but does not make a global culture, the idea of which has become a consumer object as much as burgers, fries and coke. Other local phenomena, such a Hollywood movies or Chinese slippers, that extend around the planet are local things that have been globalized. Cosmopolitan travel has the same flavor. Some cultures have more power because of these competitive advantages.

Although modern communications technology could work the same, it also may allow cultural differences and complexity to remain. The center seems to yield to the periphery, as other cultures all have hotel rooms, burgers, slippers and things. People can react by emphasizing the uniqueness of their ethnicity. No real globalism can exist until cultures and nations create a framework for the use of and protection of the capital of the planet, until cultures enter into dialogues about a global culture.

5.2.5.3.1. *Upside of Global Culture*. A global culture, that could emerge from the interactions of local cultures, would allow for more rapid exchange of ideas and things between individual cultures. It would provide the opportunity for trade in special goods, that have the potential to benefit every culture. It would provide paths for communication that might stimulate cultures. Cultures could learn from one another. A global culture could present a global morality, built on universal human tendencies, that would not be prescriptive of our private and public behaviors, but it would be proscriptive of damaging behaviors from murder to obsessive greed. Global functions and problems, such a atmospheric warming, would be easier to address. New global economic or political structures would also be easier to address, as would global problems, such as the overconnection of markets.

5.2.5.3.2. *Downside of Global Culture.* According to most theories of cultural adaptation (or integration or evolution), resistance to change is normal as a cultural process. Groups like the pygmies have specialized to fit the requirements of their environment, successfully. This makes it difficult for them to adopt other cultural arrangements.

Nature makes divisions and diversity. We need no more a unity of all people any more than of all wolves or fleas. Is it wise to have a single world? A single market in a single culture with a single control? It might create a suite of problems (see Table 5432-1).

Hyperadaptivity is a serious condition that allows humans to adapt to poverty, bad diets, crowding, stress, suffering and immense natural loss. Of course, we are unconscious of many of these problems. Capitalism does not seem to be adequate as the only global economic strategy. It breaks down useful limits and boundaries. The homogeneity of forms promoted by a global communications and advertising campaigns leads to loss of diversity. A severely limited number of ways leads to a lack of flexibility (stagnation). Then, we have the hyperpersistence of error, in the forms of stupidity and violence, which forms a positive feedback loop. At the same time, interactions are accelerating.

5.2.5.4. Considering a Global Frame for Small & Large Cultures

The desire to refine the focus has allowed the frame of reference to be neglected. An adaptive holoculture could place the human values of all cultures in a global framework. A new world order is a cultural problem more than an ecological or economic one. A global culture could incorporate the positive features of traditional civilizations: Personal security, respect for the individual, responsibility for actions (self-discipline), social integration, concern for others, and reverence for nature.

Such a global framework for culture would also consider important principles drawn from ecology. The proper attitude of an ecological global culture would be care (in Heidegger's word), a positive spontaneity, but also a "letting be," a reverence toward the wild alienness of nature, a willingness to comply with the limitations of natural systems, and a willingness to reduce the dominance of natural systems and to set aside wild areas. In addition, a global perspective might define: An authentic concept of humanity, rational economic development, a holistic education beyond that of a native culture, the responsibilities of societies to themselves, others, and the earth, and respect for all cultures.

The truth of human cultures and wild ecosystems is apprehended through myth. Cultures, through myth, need to fit our growing knowledge of the geology and biology of the earth into our hearts. Mythology is constructed as a poetic system. Campbell states that "Mythology—and therefore civilization—is a poetic, supernormal image." Mythologies and religions are great poems. When recognized as such, they point through things and events to the ubiquity of a presence that is whole in each.

5.3. Design Factors: Adaptive Patterns—Agriculture

Agriculture is a system of plant use that people developed towards the end of the ice age, when the weather and other conditions became much more variable. Changing climate after the last ice age forced many animals and plants to migrate or adjust. It is possible to look at agriculture as an adaptation to those conditions, which included drought, species shifts and extinctions. Changes in wind patterns brought more moisture to some areas and new populations of plants. Growing plants allowed people to survive longer droughts and the disappearance of large game animals. Of course, it is also possible that after 40,000 years of gradual but constant population growth, people had filled all the available open niches for hunting. In that case agriculture was an adaptation to the shrinking availability of land for hunting and gathering.

5.3.1. Historical Pattern of Agriculture

Agriculture is the use of plants in a place. Human groups came to understand details of plants as they collected them for use. With changes in the distributions of plants and animals after the end of the last ice age, some groups settled near lush concentrations of plants, such as wheat grasses. These groups gradually started selecting and replanting some varieties. As they traded surplus grains with other groups (for minerals, plants or animal products), the other groups planted their grains in areas where they did not normally grow, often dry areas with poor soils. The plants required care, in the form of water, fertilizer or protection. The large pattern of agriculture has been the selection of a few plants, an increase in the scale of planting and the assumption of control of all phases of the plant life.

5.3.1.1. Advantages of Agriculture

Agriculture offered many advantages. The source of food was more stable and more reliable. The plants could be isolated in fields or behind walls. Travel was reduced, since the plants would be brought to people, instead of people traveling to select plants when they were ripe. The select plants would have increased food value and decreased toxicity. There would be a surplus that would allow people to live in one place, where more people could be supported on a smaller area of land. The surplus would also allow intoxication to be domesticated and expanded. Humans learned to cultivate soil, plant seeds and wait until crops grew. A hunter needed 10 square kilometers per person for nourishment; intensive agriculture lowered this to 1 hectare ($1/100$ km^2). Agriculture could be 1000 times as efficient in terms of area.

5.3.1.2. Ecological Changes from Agriculture

Agriculture is an integral part of the environment in which it is practiced. Agricultural systems are distinctive types of human-modified ecosystems. An agroecosystem is a natural ecosystem that has been modified, or arrested at an early stage of succession, an immature stage. This is a form of disturbance by humans.

    In an ecosystem, humans are just one more group exploiting the system. But, in agroecosystems, humans are manipulating the entire system, that is, they are modifying the system and isolating from neighboring systems. They do this in many ways, by: Simplifying the system by removals; managing the system with domestics; and harvesting the bulk of productivity. Each of these has effects on the system. Simplifying the system, or lowering diversity, allows fewer species in the system. Not just the crop species, but pest species. Crops are analogous to early colonizing plants, that is weeds, which exploit ephemeral resources following disturbance and maximize their seed output. Plowing favors crops that grow rapidly.

5.3.1.2.1. Changes in Diversity from Agriculture

The diversity of use narrows with agriculture. There are roughly 250,000 described higher plants. Perhaps 30,000 are known to be edible. Of those, about 7000 have been used as food by foragers and horticulturists, and only 150 have been used as important crops—but only 15 are used as crops to supply over 90 percent of agricultural foods. Three cereal crops—wheat, rice and corn—provide over 50 percent of protein. The early agroecosystems were more diverse. In 1949, Chinese farmers had 10,000 varieties of wheat. In 2002, 300 were used. In Java, tropical gardens had between 100 and 350 species per hectare. Tree densities in Javanese gardens may be 2000 stems per hectare, with a biomass of over 120 tons per hectare. The Javanese used interspecific and intraspecific polyculture, that is, not only different species, but different cultivars of the same species, such as corn. Game theory shows that this strategy increases the chances of a good harvest.

Diversity is low because people try to plant monocrops. But, uniformity is also a problem. All individuals are the same kind, size, age, and use the same resources at the same time. It is a perfectly uniform predictable food source for insects and pathogens. So, there is no protection from diversity in genetics, age, and size. The selection of the cultivars for what we want to harvest has eliminated some defensive traits in wild plants. From a weed's perspective, it is a boringly predictable competitor for light and water. Weeds can compete because of the bare ground disturbance.

5.3.1.2.2. Changes in Disturbance from Agriculture

Disturbance can be created by ground disturbance, such as plowing. Energy is used to maintain the early stage. Soil carbon is also oxidized and lost with plowing, as are phosphorus and potassium, also. Plowing eliminates members of the soil community such as earthworms. Plowing also breaks down the soil types, so that erosion increases as the surface is exposed. Carbon, nitrogen, and phosphorus decrease from the upper horizons. Disruption inhibits mutualistic links such as fungi. With disturbance, energy and material subsidies are needed to be productive.

This limits efficient nutrient cycling, diversity, and resistance to disease. It ensures that the system will have low amounts of organic matter, small organisms with short life cycles and broad niches, open mineral cycles, undeveloped detritus communities and linear food webs. Nutrient flows are simple, one-way flows, from the industrial source of chemical fertilizers to outputs in harvested materials or mostly water runoff. Nitrogen for example, is applied in pulses and ends up being leached out. Applied fertilizer, usually anhydrous ammonia, which is oxidized to nitrates, can acidify the soil. Synthetic fertilizers inhibit the action of nitrogen-fixing organisms and detritivore, which work best when soil nitrogen is low. Erosion of the soil itself is a problem. Techniques expose the soil to wind and water at critical times.

5.3.1.2.3. Changes in Scale and Recycling from Agriculture

From an ecological perspective, there are local and regional consequences of monoculture specialization. Most large-scale agricultural systems exhibit a poorly structured assemblage of farm components, with almost no linkages or complementary relationships between crops and soils, or between crops and animals. Cycles of nutrients, energy, water and wastes have become more open, rather than closed as in a natural ecosystem. Despite the substantial amount of crop residues and manure produced in farms, it is becoming increasingly difficult to recycle nutrients, even within agricultural systems. Animal wastes cannot economically be returned to the land in a nutrient-recycling process because production systems are geographically remote from other systems, which would complete the cycle. In many areas, agricultural waste has become a liability rather than a resource.

Recycling of nutrients from urban areas back to fields is difficult. Part of the instability and susceptibility to pests of agroecosystems can be linked to the adoption of vast crop monocultures, which have concentrated resources for specialist crop herbivores and have increased the areas available for immigration of pests. This simplification has also reduced environmental opportunities for natural enemies. Consequently, pest outbreaks often occur when large numbers of immigrant pests, inhibited populations of beneficial insects, favorable weather, and vulnerable crop stages happen simultaneously. As specific crops are expanded beyond their "natural" ranges or favorable regions to areas of high pest potential, or with limited water, or low-fertility soils, intensive chemical controls have to overcome limiting factors.

5.3.1.2.4. Managing the System with Domestics

We try to control the system. We decide what species grow and which are denied. Our level of concern is usually plant-to-plant, rather than with the whole system. We try to control light, water, and minerals as well. But, there is no thought to the overall landscape. What does the ecosystem want to do? Or try to do? To develop, to grow, to grow old, and to be efficient?

5.3.1.2.5. Harvesting the Bulk of Productivity

We harvest as much as possible of all the productivity. Therefore, we use machines to get more, even if they destroy soil and relationships of living beings. Harvesting is limited to cultural choice. We choose not to harvest dandelions or sowbugs, ants or spiders. These new systems are based on ecological systems that have been rearranged for human purposes (food). However, they still depend on solar energy, photosynthesis, biogeochemical cycles, stability and movement of the atmosphere, and the services of nonhuman organisms. They are subject to the laws of ecology.

5.3.1.3. Cultural Consequences from Agriculture

How did agriculture affect culture in early Mesopotamia, for instance? Just as a result of irrigation? From changes in religion, from otherworldly gods to specialist gods? There are many general associated changes from shift to agriculture; they can be considered cascading effects from positive feedback loops. For instance, clearing larger areas of land and irrigating to transfer water to land lead to technological improvements in the channeling and manipulation of water. New tools extended efficiency. More crops required new ideas of storing foods. There were more innovations in general. Domestic animals were used to haul loads and then work in the fields for preparation or harvests. This required new kinds of digging sticks and new kinds of harnesses. The kilns for ceramics required more wood, which often came from newly cleared fields or transport from remote forests.

The creation of more kinds of material goods led to the accumulation of more material goods, that is, more things. Things had to be measured and tracked, and that required counting and sorting. Books had to be kept. A currency of exchange had to be invented to allow things to travel less and be more still, in place (see Table 5313-1).

There was an increase in the size of labor input into agriculture. There were specialized occupations related to tools, specialized occupations related to trade, and specialized occupations related to conflict, protection, control, and project management. There were changes in distribution of food substances and things. Not only things, but people concentrated into villages and towns. Luxury goods and wealth accumulated. Society may have stratified first by specialization, some of which had more prestige, then later by prestige itself. To protect things and places, new specialists in protection were developed. Control and conflict increased. At some point control of social activity lead to control of the poor, control of women, of animals, and control of wild areas and animals.

Table 5313-1. Agricultural Summary of Effects

Agricultural Change	Economic change	Social/political change
Seeds/resources	Trade	Standard value, writing
Manipulation	Increase size	Domestication dependence
Personal fields	Ownership, limits	Conflict, Protection / taxes
Working	Increase in hours	Changes in status
Water needs	Canals	Cooperation
Timing	Calendars	Schedules
Tools	Technology	Specialists
Specialization	Specialists	Pay, records
Distribution of crop	Pottery	Standard sizes, records
Surplus accumulation	Storage places	Protection, buildings
Sedentary Life Styles	Field exhaustion	Field expansion
Permanent fields	Settlement	Mobility decrease, Growth
Permanent settlement	Cities	Concentration, critical mass
Distribution of crop	Pottery	Standard sizes, records
Surplus accumulation	Storage places	Protection, buildings
Property accumulation	Larger homes	Protection

Commercial farmers have to cope with a constant parade of new crop varieties as varietal replacement due to biotic stresses and market changes has accelerated to unprecedented levels. A cultivar with improved disease or insect resistance makes a debut, performs well for a few years, typically five to nine years, and is then succeeded by another variety when yields begin to slip, productivity is threatened, or a more promising cultivar becomes available. A variety's trajectory is characterized by a take-off phase when it is adopted by farmers, a middle stage when the planted area stabilizes and finally a retraction of its acreage. Thus, stability in modern agriculture hinges on a continuous supply of new cultivars rather than a patchwork quilt of many different varieties planted on the same farm. The need to subsidize monocultures requires increases in the use of pesticides and fertilizers, but the efficiency of use of applied inputs is decreasing and crop yields in most key crops are leveling off. In some places, yields are actually in decline. There may be several underlying causes of this phenomenon. Yields may be leveling off because the maximum yield potential of current varieties is being approached, and therefore genetic engineering must be applied to the task of redesigning crop. On the other hand, the leveling off may be because of the steady erosion of the productive base of agriculture through unsustainable practices.

Cultural differences between ecology and agroecology include: Different perspectives, for instance, human-managed systems; different goals, such as productivity and efficiency; different principles, like subsidization. Crop ecology for agriculture is limited by intense labor. Special cultural practices are developed to engage the work and celebrate its successful completion. There are many personal and social consequences from the adaptation of agriculture (see Table 5313-2).

Table 5313-2. Other Consequences of Agriculture

Psychological consequences	Increase in compliance and obedience
	Stress due to uncertainty of yield
	Increase in hard work
	Reduced food choices, malnutrition
Sociological Consequences	Sedentarization of people
	Altered relations, status competition
	Change in world views
	Increase in movement of goods
	Rise in central authority, hierarchy
	Use of irrigation to increase yields

5.3.1.3.1. Agriculture & Population

Agriculture did produce one unexpected change. Initially, farming peoples suffered from increased sickness and mortality. But, they also saw greatly increased fertility. Populations in Europe increased one-hundred-fold. Of course, agriculture allowed people, by virtue of remaining in one place, to have more children than they could previously, in hunting bands, carry or travel with. So, the sedentism of agriculture allowed larger populations to exist in place. Furthermore, the nature of work in agriculture meant that more children were useful for labor.

Population increased. Perhaps the fertility of women increased when they stayed in place. Being in place may have removed those controls from prolonged breast-feeding or sexual abstinence. Agricultural populations increased and spread, bringing their crops with them, pushing hunters into the periphery where crops could not grow. Once the initial agricultural startup costs were paid, such as truncated diversity and nutrition, as well as decrease in human size and health, other advantages accrued, such as specialization, creating a different form of political control, which allowed armies. Agricultural society now had a massive social advantage over foraging societies.

Perhaps related to population growth and size, have come changes in cultural status, especially related to equity, distribution, and dominance. Status has to do with standing in society, and with appearance and ownership. Status may come from longevity, of the self or ancestors, as well as from the results of good decisions, from hunting for example, or from owning more than others, or just more things or more people. Status is a powerful human need, and may drive the growth of goods or populations. As distribution becomes more unequal and as conflict occur more between groups, dominance and slavery appear, and are both related to status.

5.3.1.3.2. Population Size after Agriculture

The maximum number of individuals that an environment can support is its carrying capacity, which fluctuates with various factors. Physical limits may be absolute or flexible. Population control can either be accomplished through crowding, by the food supply, or by natural resources. There tend to be as many people as there are utilized natural resources for, and unpolluted room and food for. The greatest number of people over time implies that the living population be limited to biological carrying capacity. Any permanent destruction of carrying capacity to allow a larger living population implies a future reduction in population indefinitely, or until ecosystem restoration (if possible).

5.3.1.3.3. Property Increase after Agriculture

Agriculture, by creating permanent places, allowed people to accumulate far more property than they could carry or put on a horse or buffalo—or move. This new property allowed people to have things that were luxuries rather than needs. It let them collect things that might normally be given away or reworked. Agriculture set in motion many kinds of physical and technical changes, that resulted in further social, economic and political changes. Many of these are intricately related and developed out of previous changes. Trade, for instance, expanded from a few necessities, such as seeds or ochre, to many forms of tools and luxuries.

5.3.1.4. Disadvantages of Agriculture

Disadvantages of agriculture include a dramatic increase in work hours, especially during harvest time. All the work that was previously covered by natural processes, from propagating to irrigating, was now accomplished with human labor. In addition to that were requirements for tools and storage devices. Despite this, there was a decrease in nutrition, as the overall diet was less diverse and less complete. There was an increase in fertility and a population increase, perhaps at a rate thirty times as fast within generations.

The form of agriculture, from simplification of the system to an increase in scale of the system, not to mention intensity of energy and use, resulted in the degradation of ecosystems, including erosion, siltation, salinization, and an explosion of pests. Modern agriculture replaces the biota, or transforms the ecosystem into an artificial system. Furthermore, with fertilizers robbing foods of nutritional values, more of each food is required to meet nutritional needs; for instance, the nutritional value of broccoli has declined almost 500% in 50 years, which means we have to eat 5 times as much as our grandparents.

Highly specialized systems are accelerating the change from 'palaeotechnic' to 'neotechnic' agriculture. This change minimizes costs and maximizes profits, but it also narrows the ecological basis of world food production and decreases human livelihood. Life on the planet cannot afford the continuous genetic narrowing generated by agriculture. The application of modern agricultural techniques in tropical areas disregards the local realities, and is socially and ecologically disastrous. Traditional ways have operated successfully for thousands of years, although not without serious problems of their own. But, these problems may be overcome, unlike industrial problems, which have fundamental flaws and cause rapid conversion.

5.3.2. *Farmland Designs*

The lands used for farming took on various typical characteristics depending on its history. Land with native grasses was shifted by selection, although as a landscape it looked much the same as it did with native grasses (the Palouse short-grass prairie, for instance). Many prairies were shifted from diverse tall grasses to monocrops grains, resulting in soil destruction and erosion. Temperate forest conversion in Europe and North America resulted in dramatic changes, as the forest was cut and burned. As tropical forests were converted to farmland on the European model, people encountered diseases without having evolved resistance to them. A large percentage of wetlands was filled in for agriculture, which changed entire hydrological cycles and made the areas vulnerable to changes in climate, also.

Because of its scale, industrial agriculture tends to produce monotonic landforms. The land is often unbroken by natural bunchings of trees, shrubs, and flowers. Almost no land is left for any kind of buffer between areas. Often, crops 'leak' or flow into other areas. The land is often broken by roads for access. The nature of this type of agriculture encourages erosion.

The modern mutant agriculture promotes the myth that megafarms are more efficient with output per unit area than small farms. By pushing mega-technology into a marriage with megafarms, the costs of agriculture are inflated beyond the reach of most groups or even small corporations. The scale of larger acreages requires more massive machinery and amounts of chemicals. The scale of the monoculture undermines the genetic integrity of the crop, so it is more susceptible to diseases, pests and changes in conditions, requiring more pesticides and water—a classic case of the law of diminishing returns. Yet, the myth survives because of seriously flawed methods of measuring farm productivity. In fact, small farms have greater output per unit, with smaller chemical and mechanical inputs. They decrease the potential for adverse environmental and health effects.

5.3.3. *Timberland Plantation Designs*

Natural forests were husbanded in some places for thousands of years. Starting in Germany in the 1800s, entire forests were replanted specifically with harvesting as a goal. The trees were placed at 'optimum' distances apart, in rows that could be accessed more easily and safely by loggers and machines. Other tree species, shrubs, flowers, and herbs were removed or kept out. This model of plantation spread through Europe and North America to virtually every other continent. By this time the movement to preserve natural forests was advanced enough to have millions of acres set aside as parks, reserves or wild areas.

Harvesting a forest on the scale of a landscape can be represented mathematically as a catastrophe surface, where changes are most often discontinuous and negative. Initiating a plantation on that scale would also create discontinuous change. Destruction of soils and erosion would effect climate and the growth of the plantation. In addition to the Catastrophe Theory of Rene Thom, the application of Island Biogeography Theory of McArthur and Wilson can be used to explain the behavior of isolated areas as the result of timber harvesting patterns. As cutting patterns and plantations cover larger areas of a landscape, old-growth or natural areas lose territory, interior area and species.

By the 1970s, some foresters were suggesting group cuts, with minimal plantings, as a strategy for harvesting many forests on more fragile soils or in harsher climates. Others were allowing plantations to assume the characteristics of mature wild forests, by not trying to control all the invading species. The advantages of wild forests are so much greater than for plantations—for instance the ability to resist wind storms, pest outbreaks, droughts, and many other events—that many groups advocate harvesting wild forests on an ecological time scale (50-500 years) rather than planting more plantations.

With the predominance of artificial forests, it is important to consider the qualities of naturalness in the landscape. Forests are expected to meet the needs of society by producing timber, creating wildlife habitats, and providing recreational opportunities for people. But, forests are also expected to look natural.

The use of forests has to change, also. Use of forests exclusively for timber reduces opportunities for nontimber uses. Foresters need to broaden the base of materials taken from the forest, and not concentrate on one or two commodities, such as wood or mushrooms. The scale of cutting has to change, so that cutting is below the renewal rate of the forests. Perhaps foresters could invoke the Principle of Least Effort (also a principle in economics and cybernetics), by encouraging natural regeneration; let the forest do as much work as possible—it has millions of years of practice in some cases. They could sponsor contests to make the fewest cuts, or to go the least distance.

5.4. *Global Design Factors: Adaptive Patterns—Technology*

What does technology do, basically? It creates extensions, as tools, to our bodies, letting us access new resources and create new things. Something as simple as a harness can replace human labor with animal labor. Something as simple as a knife can replace the muscular effort of chewing large pieces of food, which can be rendered more tender by another tool, fire. The Greek root words, from which the word 'technology' was taken, meant art and study. Technology can be broadly defined as the material entities created by the application of mental and physical effort to nature in order to achieve some value. In its most common use, technology refers to tools and machines that may be used to help solve problems. Technology is a technique that lets us use resources to produce products and solve problems; this technique feeds back into culture.

5.4.1. *Historical Pattern of Technology*
People make tools to get food. Then, they make clothing, homes and other things for making life easier. Art, luxuries and sculptures follow. Clothing and tool-making is a universal in human cultures. In later agricultural societies, permanent settlement lead to the growth and intensification of settlement. The engineering of an infrastructure lead to massive buildings, monuments, and fortifications.

As tools increase in complexity, from knives and levers to computers and space stations, so does the knowledge needed to support them. Complex modern tools require libraries of information that has to be continually increased and improved, then spread. Technology first simplifies life, then complicates it. Digging sticks led to plows and tractors. Lean-tos lead to pit houses and balloon-framed houses. Domestication led to horses and to horse wagons. Paths led to trail and highways.

5.4.1.1. Considerations of Technology
Technology is defined by tools, as well as by its processes for handling materials, energy and wastes. Technology is also an intentional practice.

5.4.1.1.1. Materials in Technology
To be constant and stable, a culture has to be productive, to be able to convert energy and materials into foods and structures for survival. As culture extends human ideas into place, humans use energy and materials to create different structures in place. Materials were first limited to plant and animals. Materials cycle above and below ground, between the atmosphere and trees, between trees and insects, and squirrels and fungus. The material aspects are the physical products or imprints of culture. These range from paths and plant patterns to tools and structures. Tools start simply as choppers, scrapers and cutters. Structures include pots, clothing, and houses. Found materials gave way to modified materials and finally constructed materials, that is materials rearranged from elements with the use of energy. In fact, the extraction of materials, and later energy, became a primary economic activity in many cultures. The processing of materials became secondary.

5.4.1.1.2. Energy in Technology
Technologies require energy, and many kinds are available: Solar, nuclear, fossil, fire, animal, and human. Each has dominated some historical epoch. The industrial revolution increased the quantity of energy, but decreased the variety of energy resources. Carbon products are the

largest source of energy. The earth is also a source of energy from tides and hot springs. The discovery of other forms of energy, such as fossil fuels, lead to technologies that could leverage them. Technologies require energy. The industrial revolution increased the quantity of energy, but decreased the variety of energy resources. Since phosphorus and fossil hydrocarbons may be in short supply, a good power source would be required, perhaps solar or nuclear. The three main uses of energy for industrial civilization are motor traffic, heat for buildings, and manufacturing.

5.4.1.1.3. Substitution in Technology

One technological process is especially important, substitution. Substitutability of resources is neglected, as well as new forms of power or price mechanisms. Plastics are used to substitute for wood, leather and metal in many industries. This substitution may temporarily ease pressures on biological systems, but it depends on the supplies of petroleum. Goeller and Weinberg, in an article the "Age of Substitutability," said that this age will be characterized by an asymptotic society that settles into a steady state of substitution and recycling, using renewable resources and practically infinite resources, such as iron or aluminum. They suggest that society could exist on infinite materials indefinitely. Sophisticated technology can create huge savings on resources. Fuller noted that a one quarter ton satellite outperforms the transoceanic capabilities of 175,000 tons of copper cable, although he may not have included all the ancillary costs, such as launch or recovery costs.

5.4.1.1.4. Waste in Technology

Nicholas Georgescu-Rogen described the economic process as entropy-producing, where entropy is visible as heat, waste and pollution. Every life process generates waste that is recycled in biological or geological cycles. Nature wastes countless lives and materials. Resources may lie buried for millions of years by landslides or volcanic eruptions. Unfortunately, some technological waste is different: many materials have no processes or organisms to break them down and recycle them—no solutions have evolved; sometimes the quantity, of a toxin for instance, is immeasurably greater than natural toxins. Even natural disasters, such as volcano eruptions that devastate whole areas, are recolonized from outside. The impact of human eruptions around the globe mean essentially that there no longer is an outside. Humans have the capability to fill the entire atmosphere with radioactivity or organophosphates.

There are no sinks on the earth where waste vanishes. Things are only moved around; eventually they return. Early technologies dealt with waste by moving and storing it. Some things are dissipated into the earth's environment, the cosmos, which is an energy source and sink. Even this large environment may not be inexhaustible, as many analysts such as Eric Jantsch claim. Later technologies tried to burn waste to break it down into component parts that could be sifted out of ashes; the burning, however, resulted in air pollution and some toxic wastes that had to be isolated. Pollution is a symptom of imbalance and improper resource utilization. A serious problem is our lack of understanding of the extensive, long-term effects of pollution on the atmosphere. More recent technologies have found ways to separate out valuable components of waste. Others have found ways to neutralize toxic wastes with chemical treatments.

5.4.1.2. Technology as Extension & Intensification

Technology can be considered as an extension of human body; knives extend teeth. Or, it can be an extension of animal labor, such as the harness that allows single and multiple animals to pull a heavier plow than a man. Technology can be considered an extension of fire: Fire herded animals and made grasslands, Wood fires cooked food and provided warmth; wood fires also shaped metals and provided steam for power; coal fires provided electricity, and oil fires provide electricity and motion. Steam donkeys and steam engines provided work, heat, and motion.

A tool is an extension of the individual into a new kind of specialized animal (perhaps because it has a new essential property it can be a new pseudospecies. Neither the ecosystem nor the animal knows what it ought to do, hence the explosion and chaos. For humans, this is what a new ethic and cosmology is needed for: to tell us what to do. We need direction. Technology must become an art again to make good places.

Technology intensifies. There has been an intensification of tool-making during the past 3000 years. Population increased and new areas were exploited. The agricultural revolution was a technological revolution. The industrial revolution, by definition, was a technological revolution that resulted ultimately in men being replaced by machines. Only the Luddites faced this limitation, unsuccessfully. Human labor is increasingly dispensable. Industrial culture confuses mechanical with personal power. According to W. I. Thompson, industrialization is really an intensification of civilization.

Table 5412-1. Parallel Events in the Recent History

Ideas	
	Mechanization (Clocks)
	Quantification (Grades)
Technology	
	Inventions of tools: Telescope, Microscope, Camera
	Communications tools: Printing Press, Telegraph
Management	
	Bureaucracy
	Social Science

Because of the exponential growth of technological exploitation, more primary metals have been consumed since 1940 than in all human history before then. Technology is bound intimately with the exploitation of a convenient source of energy, oil. Unless the trend is changed, reserves will disappear. There is a finite limit on actual numbers of molecules of a given resource in the earth's crust, but the practical limits to exploiting the potential are a function of technological activities. So we are importing materials from the past (oil) and the future (soil), for now. Humans are hunters and gatherers of materials whose renewal times are geological, not days or decades.

Recent industrial history reflects new tools, new livelihoods, changes in settlement and behavior patterns. All tools, from the simplest word to the most complex computer, are disturbers and rearrangers of primordial nature and reality; they are implements for working on something. They have addicted us to purpose. We look for purpose in everything; to seek an explanation of nature, and to justify the seeking. Humans are tied to their tools and machines. The basic difficulty with the quality of life resides in machines. Machines pollute air, water and earth. Machines of war threaten human lives. Machines displace forms of life; they take up space. They are a competitive species, whose members die, reproduce, evolve, and sometimes think.

5.4.1.3. Effects & Consequences of Technology

What are the effects of tools on human cultures? The complexity of tools leads to rules for their use, and for those who uses them, such as trade unions and clubs like masonry. Society is very adaptable to technological change, according to Kenneth Boulding. Perhaps too much so according to Rene Dubos, and the risk is shaping human behavior to tools.

The use of technology has a great many effects; these may be separated into intended effects and unintended effects. Unintended effects are usually also unanticipated, and often unknown before the arrival of a new technology. Nevertheless, they are often as important as the intended effect. The most subtle 'side effects' of technology are often sociological. They are subtle because the effects may go unnoticed unless carefully observed and studied over large areas and long periods of time. These may involve gradually occurring changes in the behavior of individuals, groups, institutions, and even entire societies. A nation needs to address the effects, all of the effects including unwanted ones, of technology on business, culture, management, and the environment.

New technology has been a primary force for change for decades; but some technologies, like computing or genetic engineering, may lead to enantiodromia. The clock itself, one of the basic parts of a computer, was developed by monks to regulate and limit the hours of prayer, to serve God better, before it was used for secular purposes, beginning in France, to control the trading hours of merchants. The use of new tools can be expected to have unexpected effects. For instance, mass-produced computers may lead to individual autonomy. New energy technologies could have the same effect. Technologies have the capability to minimize the use of resources.

The implementation of technology influences the values of a society by changing expectations and realities. The implementation of technology is also influenced by values. There are major, interrelated values that inform, and are informed by, technological innovations. These realities and expectations may alter the world image of a culture, especially as regards efficiency, bureaucracy, or progress. We should use new technologies, but as Peter Drucker says, make sure that we know that the technology matters less than the changes it triggers in substance and content. It should not replace an ecological perspective and critical thinking. And, we need to remember that knowledge itself can trigger change.

5.4.1.3.1. Control

Can technology be separated and kept under control? No likely. Technology has become autonomous. Can the effects of technology be controlled? Tools have effects on human culture, and on nature. What are the effects of tools on ecosystems? Disturbance of soils, and scale effects. What are the effects of tools on humans? The use of knives reduced the need for some teeth. Tools, simply by being intermediate between the hand and the object, may increase psychological distancing from things. Intimacy with the tool can replace intimacy with the thing. The increase in physical depositories, for memory, changes the kinds of memory capabilities. Tools may contribute to the loss of hand-eye coordination and perception of wild, yet increase hand-eye coordination of tools.

5.4.1.3.2. Risk

Energy demand has resulted in the use of high risk sources. With nuclear power, the burden of proof for safety is on the agencies themselves; its use in the absence of complete assurances of safety is a terrible threat to living beings. G. Hardin is even more adamant in his conclusion: Guilty until proven innocent. Even with inefficiency in traditional sources, there is no need to use high-risk energy generation. Buckminster Fuller claims that by using only proven energy resources, only proven technologies, and only at proven rates, within ten years all of humanity could enjoy energy income equivalent to United States in 1960, certainly adequate for a good level of luxury for all inhabitants. Nuclear and fossil fuel energy could be phased out during that time, as alternate sources are developed.

We ought to reestablish earlier energy patterns for regions, and use combined systems of wind, water, solar, organic, and fossil fuels for energy. Singly, these may be inadequate, but as a mosaic they could meet decentralized needs. The energy pattern should be pieced together organically from the potentialities of a region. The establishment of decentralized communities based on ecologically sound organically based agricultural practices, with local technology using local energy sources and recycled nonrenewable raw materials. Some of the energy crisis and risk could be avoided by using less consumptive settlement patterns and natural energy utilization. More risk would come from the scale of those operations also as they struggled to meet demand. They would also have more negative impacts on the environment at a larger scale. The modern industrial crisis is a crisis of too much energy use. Consumption and production of energy must balance, safely.

5.4.1.3.3. Several Consequences of Technology

The whole idea of technology, according to Evan Eisenberg, is not to eliminate work or to replace nature with synthetic artifacts; it is to restore the balance of work and leisure that hunters understood—it is to find the best way between the dirt and the mouth.

As technologies developed, human relationships with animals and plants changed. Hunting, grazing, and agriculture provoked large ecological disturbances. Early domestic animals were revered, but nondomestic animals were considered competitors or nuisances. Now, animals are treated as processed commodities, and wildlife is regarded as useless.

Machines are part of human ecology, like other tools. They too contribute to flows of elements, such as sulfur and nitrogen, although the machine contribution may exceed the total of all natural living and nonliving sources. Machines also increase flows of rare elements, such as lead or mercury to tens of times the natural flow. Machines also create new compounds that are suddenly introduced into a system that has no pathways to deal with them. They displace wild ecosystems, modify food webs, and create new energy flows within the system. Cars replace horses, and mowers replace deer and other grazing animals in simplified ecosystems. Video games may be the cockroaches, leaf-blowers may be the rats of the new world, taking more energy than they yield benefits. Television can be seen as a form of cancer that encourages the replacement of every living thing with new machines.

Buildings have impacts on ecosystems. They become kinds of ecosystems themselves, but not self-sufficient ones yet. We shape our culture, then culture shapes us, we build buildings, then they shape us, as Winston Churchill recognized of Parliament.

5.4.1.4. General Problems with Technology

Due to the increasingly widespread use of ever more complex technologies and the frequently unintended consequences, problems may arise in their use that are unrecognized or only partially addressed. Technological innovation, combined with accelerating population growth, lead to clearing of many forests for agriculture. Technology also promotes land degradation. Plowing causes erosion.

The combination of cheap goods and complex tasks can lead to sweatshop slavery and unsolved wastes. Some problems are solved, but new ones are created by unconsidered use—problems such as toxic waste or radioactive waste. Time and leisure are needed for technological innovations. People stressed or starving rarely invent their salvations.

5.4.1.4.1. Unintended Effects

Technology has reduced the globe to a single, closed system, which humans can share according to their financial resources. Our direct experience of the world has become shallow, in spite of faster travel. Travel used to broaden the mind, but now it narrows it. We travel in sealed corridors like boxed goods, comforted by homogenized foods and several 'world' primary languages. Technology has distanced human experience from the meaningful time and extent of experiences. Technology or social structure can mask the internal stress from fast economic growth. Technology has made the suffering of many domestic animals invisible to consumers of animal products.

5.4.1.4.2. Development & Dominance

We often arrange cultures in a taxonomic scheme, sometimes based on the technologies of cultures. Thus, we have the stone age being replaced by the bronze age, then the iron age, and steel age, the industrial revolution and the post industrial and computer revolution. Lewis Mumford suggested three basic divisions: Eotechnic, paleotechnic, and the Neotechnic. Ortega distinguished between the technology of chance, of the artisan, and of the technician. Walter Ong suggested the distinction between Oral cultures, Chirographic, Typographic, and Electronic. Marshall McLuhan had a similar division: Oral, Gutenberg, and Electronic communication. Finally Neil Postman defined three different divisions: Tool-using culture, Technocracy, and Technopoly.

Tools and technology are the chief instruments of progress. They improved our material circumstances. They were supposed to bring superstition and suffering to an end. Technologies depend on information. They also control information. New technologies advance to compete with old ones for dominance in a world view.

Under what conditions can the effects of technology not be contained? Does technology always escape and produce its effects on the nonhuman environment as well as on other cultures? If so, then technology is always ecological, that is, it is always part of the environment. It always generates some change in the environment that it is part of. It is not the same environment after the introduction of the extension of technology. Any addition (or subtraction) is a change, and any change has many effects on everything. The conditions of survival for all change.

Technologies change institutions and the relationships between them. As an institution adapts a technology, its view changes. A new technology threatens other institutions, which have competing technologies, as the whole mass has considerable momentum and investment. As the nature of institutions changes, the nature of communities, cities and cultures, changes. Thus, technologies change the form of tools and whatever they shape, then change the quality of response to those changes.

Technologies change the structure of things with new metaphors. As metaphors (and totems) have been and are good things to think with, technologies become things to think with. Technologies based on organic metaphors have the potential to change the quality of the built environment and it supporting wilderness.

5.4.2. *Tools Small Things & Electronics*

We started by domesticating our world with tools. Tools promoted greater autonomy and decentralization. New tools have been developed to record images of everything from the planet to atoms; for instance, the Pancam creates billion-pixel images. The Gigapan system scans a landscape and stitches it together in 100 Gigapixels. The great resolution increases the interactive potential. Citizens can get tools and have knowledge for monitoring pollution (including mobile phone applications). All that is needed now a well-organized network of citizens to provide independent monitoring.

We begin using and mass-producing these tools before we finish testing them for durability and dangers, much less for their influences on our psychological and social states. These things have to be considered and monitored.

We can have better and smaller communication devices, as long as they do not cause cancers. We can incorporate biodegradable or reused materials, as long as they do not degrade into dangerous molecules. We can rely on autodiagnostic medical tools, as long as we are willing to accept machine and human error. We can produce long-lasting, high-quality, upgradeable appliances, as long as we understand that we have to change design practices and possibly the entire economic system.

5.4.3. *Houses & Buildings*

Most early buildings were pit houses, pole houses (covered with thatch, animal hide or shingles), or post and beam (covered with thatch or boards). The balloon frame house, developed in Chicago in the 1800s, made houses easier to build and more affordable. Lighter standardized studs held together with nails were nevertheless very strong. The balloon frame is still used with prefabricated materials on assembly-lines for manufactured houses.

Large buildings were built with masonry (bricks or blocks) for thousands of years. The height was limited by the strength of the masonry (and the ability to climb stairs). In the late 1800s in Chicago, steel was used to build a skeleton that could support a larger building. Glass 'curtains' were used to wall the buildings. Most skyscrapers are now built with (steel) reinforced concrete. The tube structural system is also used (trussed or bundled). Skyscraper heights were limited before elevators. The use of imported raw materials and cheap fossil fuels allowed corporations to build more of them. Beyond 100 stories, the costs exceed efficiency, although many skyscrapers are symbols of identity and economic success. Skyscrapers do have large costs associated with their extra height; the costs for foundation and support are large, as are the costs for multiple elevators and services such as water, lighting, and air.

Some designers claim that skyscrapers can be made 'green' because of the concentration of people reduces per capita needs for space and waste. Modular mass housing can be more efficient than the labor-intensive methods for building balloon frame houses. Buckminster Fuller suggested using aircraft manufacturing techniques to mass-produce houses. New styles of houses are reinventing post and beam construction with hay bale walls. Some companies promote self-designed houses, made of logs or other construction. Take-apart technology may allow kit or custom houses designed for disassembly (although the author remembers reassembling a grain elevator from Viola, Idaho into a studio on skids).

5.4.3.1. Material systems for Building

(Being edited)

It is likely that reinforced concrete building and balloon frame houses will dominate the number of constructions, although wood is getting certified and concrete is incorporating recycled materials and wastes to reduce its impacts. Salvaged materials can be reused in many buildings, saving good materials and avoiding additional impacts on the environment. Biodegradable materials are appropriate in some cases.

5.4.3.2. Walls

Walls are useful for the isolation of plants, animals and people. Cities became walled, originally for protection, although they led to separation and the distinction between outside and inside. Walls were one response to attacks. Further centralization, bringing local farmers under the protection of cities, was another. In Italy, did changes in class and equity lead to first private gardens, then public gardens, to areas walled off from others? The walls failed to lock out nature or drought. Walls are an example of the Principle of Separation. Limited flow, or nonflow, can be critical to the design of buildings.

More diverse patterns may result in more diverse social interactions, such as nurseries located near administrative offices or a bar located near a work area. The network of working and living can become much more dense if it includes pedestrian paths gardens and public spaces. Within each building walls and floors could be movable so that people could customize their own living or working spaces. As part of the participatory process, workers and builders can be simply given design principles and constraints rather than final blueprints.

5.4.3.3. Floors & Ceilings

All buildings over one story have floors and ceilings to protect the interior of the building. The floor provides a stable base for walking and for furniture; it can also house the heating system. The ceiling often holds insulation and ventilation systems, which reduce noise and guarantee air circulation. The height of a ceiling can make the space more or less comfortable and inviting. The texture of a floor can make it more or less comfortable for long periods.

5.4.3.4. Air & Water Material Systems

In many buildings water is pumped up and air is pumped down. Smarter buildings collect water on the roof and let it flow down; air is collected blow ground and allowed to rise. Water systems have always been a challenge as clay pipes leaked, iron and steel pipes rusted, plastic pipes decayed and broke, and copper pipes were very expensive.

5.4.3.5. Furniture

Buildings contain different sets of furnishings depending on the purposes of the buildings. Office buildings need chairs and desks, filing cabinets and computer stands. Apartment buildings have couches, beds, table, and chairs. Many larger buildings dedicate the first story or two to local businesses, such as restaurants or clothing stores. A large percentage of the floors may house apartments, with offices taking up the remaining floors. Of course a significant part of the building is given over to elevators, stairways, storage, insulation, and air and water systems. In his books on design, Victor Papanek shows examples of good design, such as a Secretarial posture chair by 'Team Design' in Stuttgart (1960s), or the Director chair (from 1860s), manufactured by Telescope Folding Furniture Co. in Granville, New York. Papanek emphasizes that furniture, from chairs and tables to bookcase and computer boxes, could be built to last, rather than built cheaply to become obsolete.

5.5. *Design Factors: Adaptive Patterns—Urbanization & Civilization*

A city is a place where a large number of people live permanently. The *polis*, the classical city-state, was a place where the citizens depended on and maintained the whole, which cared for and outlasted the individual. The Stoics declared the cosmos to be the great city "of gods and men." The citizen became related to the cosmos as a whole, in the same way. For those in Mesopotamia, the whole city, and not just the temple, was conceived as an earthly imitation of the cosmic order, a sociological middle cosmos, established between the macrocosm of the universe and the microcosm of the individual. Through the priesthood, the one essential form of all was made visible. The early Sumerian temple tower, with a hieratically organized city surrounding it, became the model for the Hindu world mountain Sumeru, for the Greek Olympus, for Aztec temples, and even for Dante's Purgatory. The Sumerian city was organized in the design of a quartered circle. This design has been a favorite for cities and utopias, as well as for many cosmologies.

The whole city, and not just the temple, was conceived as an earthly imitation of the cosmic order, a sociological middle cosmos, established between the macrocosm of the universe and the microcosm of the individual. Through the priesthood, the one essential form of all was made visible. The ideal landscape became the middle region, the garden between the complete order of the city and the complete chaos of the wilderness. The garden is cultivated from the wilderness as a middle landscape. The garden is a human order, but not usually sacred.

Cityscapes still contain gardens. Gardens and parks have been designed to express an idealized view of natural and agricultural scenes, since Sumerian times, at least, over 5000 years. Many names for the landscape—grove, lawn—are drawn from the imagery of the garden. The complementary aspects of landscape planning are the invariants of a given area and the artistic imagination of a planner. Conrad Aiken recognized that "The language and the landscape are the same, for we ourselves are the landscape and are the land." The city expands at the expense of gardens, fields, and then wilderness.

Many other cities were placed at the intersection of rivers or trade routes. Cities have distinguishing characteristics, such as permanent buildings, specialized buildings, monumental buildings, and large populations, which have increased over the millennia, from 2000 people to 50,000 or millions of people. Another characteristic is the density of the population, as a result of the shape, size and number of buildings. Cities in general are not self-sufficient; they rely on outlying areas for food and resources. Cities are places with surpluses that can be traded for things from other groups or cities; cities become central places in an area that provide services for outlying areas. The specialization of occupations in a city promotes trade of special goods. Specialization and unequal distribution allows people to become stratified in distinct classes, based on specialization and differential rewards. A city is also a place characterized by organizational complexity, with universal rules and central institutions.

As people are attracted by the advantages and excitements of cities, the size and number of cities increases. The spread of cities increases. Their areas are added to that of the vast agricultural lands, which together make up the physical impact of human habitation on the surface of the planet. The characteristics of cities change also, as their sizes and shapes are formed by desire and consumption. They become more "ideal" and more uniform, and with less and less to make them different and unique, they start to resemble ideal noplaces.

5.5.1. *City Patterns*

The earliest villages and possibly the first cities were established near water, near hunting and gathering grounds. In general, cities formed in areas where there was surplus and where larger groups of people were needed to deal with surpluses. Many of the first larger cities (500 or more) were located on waterways needed to irrigate crops. Permanent special buildings were needed to store surplus. Permanent buildings were built as homes, since people were living in one place all the time. Buildings began to increase in size to express faith in the place and the gods of the place, but also to express new differences in status or wealth as the labor system became more hierarchical.

Some cities were built as religious centers, others on trade routes to facilitate trading. Some cities built walls to protect their surpluses, wealth or people. Soldiering became a specialty. Cities came to resemble artificial land forms, such as mountains or canyons. Only certain plants or animals were kept in the cities. Almost the entire land surface was covered with buildings, paths or roads.

5.5.1.1. The City as an Ecosystem

The city was an adaptive exploitation of shifting environments. It was an adaptation to an ecological niche and a cultural niche. At first cities allowed for the coexistence of hunters and herders with the local farmers and residents of the city. The city is exploitation of a geographic site, which has a multilevel history from geology to plants and humans. The structure of a city might be like a cell, a specialized structure with a transmissible memory in the nucleus. Or like a sponge; the city cannot use sun to get energy but must use surrounding environment (plants) and other organisms in the water. Sponges must circulate food brought in, using energy from that food. A city has own metabolism, that is a network of circulatory structures used for exchanges. Water and wood (energy) are carried by channels to cells of organism (homes in this case). Channels also carry wastes.

As an ecosystem, a city has fewer plant and animal species. Many factors influence its balance with its environment. Water flow is a factor affecting the landscape. Household wastes are another factor. A network of corridors perforate the landscape; the number of small patches increases, and there is a reduction in other kinds of patches and corridors. Flows include energy, information, people, materials, and pollution. The net productivity of the city ecosystem is negative, due to massive imports of food and energy.

The size of a city grows, apparently without limit. During the mass migration from 1800 to 1991, urban population in developed countries rose from five percent to seventy-three percent. A mass migration in less-developed countries, from 1940 to 2004, the city population rose from three percent to over fifty percent. As cities increase in size, there is less flexibility for change. Land use planning and design becomes more important.

This size expansion has caused problems, effects, and side effects—that is, main effects that are unwanted or unanticipated. Villages lower connectivity to the landscape by increasing patches and corridors. Agriculture gets more homogenous, decreases fallow areas. Stream corridors are destroyed by environmental degradation. Connectivity is lowered, the matrix is minimalized. The nutrient mineral cycles are disturbed. The atmosphere is disrupted, resulting in drought and storms. Microclimates change, with heat islands and dust domes. There is lower photosynthesis and productivity, with lower diversity leading to homogenization (with cosmopolitan species) and inefficiency of use of energy and materials, and decaying infrastructure (roads, sewers, buildings, houses).

Urban ecosystems and agroecosystems, however, are less resilient than natural ones, due to the constant expenditure of human and fossil energy to maintain them. Cities need the environment for resources, food and water. People in cities tend to organize the environment by controlling markets and transportation on which the agricultural systems depend. Cities also produce large quantities of waste that have to be absorbed. Cities pollute air and modify climate. They are heat traps by absorption of solar and the production of heat.

5.5.1.2. Changes from City Living

The city is a consolidated area, a permanent part of the landscape, with stable home sites and larger, permanent houses. There are impressive public buildings and monuments that can lead to civic pride and use. The city offered opportunities for specialization, for choosing a mate from a wider pool of candidates, for excitement and stimulation, and for wealth. The city has a critical mass for inspiration and invention (in art and science). Of course, the city offered better protection from invading groups and it offered a more stable food supply

Living in a city also had disadvantages. It meant one had a smaller living space, lower standards of living, and a less varied diet. There was a greater chance of disease, nutritional deficiencies, and crime. The larger population meant an increase in competition and possibly a decrease in personal rewards. Public art styles could be limited. Due to the large populations, cities tended to ecologically degrade the surrounding areas. Although cities were fine adaptations to the climatic and environmental changes as the ice age was ending, they were vulnerable to larger groups of enemies and to longer environmental problems such as long droughts, over ten years, or series of changes.

5.5.1.3. How did Civilization rise from Cities?

A society based on cities, with a complex social organization, engages in a whole series of changes. Rather than reciprocity, economies are based on a centralized accumulation of materials. Social status is changed through tribute and taxation. Formal records-keeping arises, from knotted strings to cuneiform. A state religion develops, where the leader takes an important role, as for instance as god-king.

Table 5513-1. Urbiculture Summary

Change	Leads to	Which creates	And changes
Permanent crops	Irrigation	Salinization, exhaustion	Kinds of crops
Permanent buildings	Storage, property	Possessions, greed	Movement
Massive monuments	Status	Competition	Relationships
Trade	Common value	Standards	Shortages
Managers	Rise of elite	Taxes, power skew	Hierarchy
Transportation	Roads	Crowding	Perception
Water/Baths	Increased use	New reservoirs	Shortages
Walls	Protection	Separation	Inside/outside
Centralization	Concentration	Intensification	Obedience
Culture mixes	Violence	Laws	Behavior
Ecosystem replacement	Degradation	Protection	Invasions, disease
Unsustainable use of water and wood	Distribution	Drawdown, money	Money
Specialists: Artists, clothiers masons	Luxuries	Redistribution	Production patterns
Crowding density	Disease	Immunities	Health

This table (5513-1) is not able to show multiple causal chains or how changes lead to or influence other changes.

5.5.1.4. Intensification & Civilization

Cities are an intensification of trade and agriculture, and the things that surround them. Gravity a fruitful metaphor for intensification, for desire or crowding. Gravity is a universal long-range force. It is like centripetal, where the center of gravity is the center of the city. For the earth, the center is the sun. But, the sun is part of another gravitational sphere. As a metaphor gravity might explain why people are attracted to and move to cities: Intensity. Opportunity may increase due to the concentration of people. The number of links dramatically improves the possibility of better or more communications. Although gravity may explain the intensity of compression or miniaturization, it has trouble explaining the intensity of expansion. The metaphor raises some questions: How is danger related to intensity? Is it being exhilarated by the closeness of death? If the populations increased, would intensification have been necessary?

Cities are where most people live, where most resources and energy are consumed, and where most wastes are produced. The city becomes a center for intensification and excitement, but it also causes environmental challenges to appear faster and larger. A city changes the local environment around it, pulling in medium size cities, as it did in Mesopotamia, and reducing the number of small satellite cities, which are made continuous with the influence and attraction of the center city.

Urbanization was the result of permanent settlement and population growth. The growth of the settlement and intensification followed. There was specialization and engineering of a more complex infrastructure, with canals, larger buildings and massive monuments, and fortifications. Religion and trade enlarged in scale. Specialization As result of surplus food and larger population, special people can create a flow of specialized objects. The population has to be large enough to support a market.

Jane Jacobs asserts that cities are at the root of all economic growth, from agricultural, manufacturing, and technology growth to the information explosion, and therefore import replacement is the cause to all economic growth. This idea challenges one of the fundamental assumptions of Classical and Neoclassical economists, who consider the nation-state to be the main player in macro-economics. Jacobs argues that it is not the nation-state, but the city that is true player of this world-wide game. She speculates on the further ramifications of considering the city first and the nation second, or not at all.

The advantages of urbanization included consolidated area and stable home sites, with larger, permanent houses. Public buildings and monuments became more impressive. This formed a critical mass for inspiration and invention, including art as well. Measuring and writing, inventories and commercial dealings followed.

The disadvantages, however, included population increase, an increase in diseases and the danger of violence. There was a limit to public styles of art, ecological degradation of surrounding areas, and increased vulnerability to collapse. There seem to be limits to our personal space and levels of tolerance to human intensification, also. Urban intensification leads to the question: Is there a limit to human numbers? Perhaps space, but is there a psychological limit? People in cities seem to do well with high-contact, high-proximity living. What happens when people are crowded or feel crowded? Physical complaints, emotional complaints, sexual dysfunction, or feelings of fear, seem to be expressed often. There may be limits of crowding. Are there social limits, in terms of the number of people one can tolerate?

We may have requirements for personal space, home space, and wild space. Psychological limits may be the basis for some of the great failures of human life, for instance, the "failure of perception." We cannot see slow change or anticipate it. No one really sees the incredible interdependence of humanity and nature, of diversity and success. We do not seem to be able to see others as feeling human beings.

5.5.1.5. The Complex of Civilization (or Neolithic Revolution)

Many phenomena interacted during the Neolithic revolution. Not simply agriculture or urbanization, but profound economic and social shifts occurred. Agriculture, by taking over an ecosystem, allowed a short-term increase in biodiversity followed by a long-term decrease. To increase control, new tools such as hoes and plows were developed; the fields themselves were engineered with added irrigation systems. Gradually other ecosystems were taken over. Land use became a formal ownership to reflect the investment in all of the above. Land ownership was a logical step with permanence, as individuals dedicated themselves to a relatively small area. The technology of tools advanced, as did the engineering of fields and water systems.

Permanent settlement resulted in more permanent buildings, as well as in massive buildings and monuments, fortifications and religious centers. This led to an increased infrastructure to supply water, access, transportation and waste. Sedentation and surplus allowed larger populations and increased fertility. The subsequent trade in foods required better paths and transportation. Food storage and trade become concentrated in cities.

There were dramatic economic shifts that included trade, craft specialization, professional art, record-keeping, writing, direct engineering (pottery), and standardization. Personal property increased dramatically; one no longer needed to carry everything one owned. Records were needed to supplement the memories of individuals, since many individuals were required to keep track of stores, trade and rations. This required specializations in new trades, from potters to smiths and artists. Managers were required for recruit labor for fields and hydraulic engineering. Labor shortages on public projects required recruitment of labor from within the city or from outlying areas. Animal labor was applied to some activities.

This lead to many personal advantages. People had more property. It was not necessary to make everything for one's self, or know how to. One could trade for needs or luxury items. There was a new kind of leisure time, for doing nothing, as opposed to joining societies and telling stories. Ale was on tap for the entire year, to permit voluntary dizziness as a recreation; one no longer had to wait for trees to provide fermenting berries.

Unfortunately, the advantages lead to disadvantages. For instance, more work was required, especially to contribute to public projects and taxes. There was less overall leisure time. Trade for things required trust, which was a more difficult commodity between strangers. Records had to be kept of all the new kinds of transactions. Extra property had to be stored somewhere and protected. The distribution of luxuries and necessities was skewed, so some people had much more than others, in terms of things and respect.

Social and political changes were equally dramatic. With the distribution of power and materials, a political religious elite arose. And, to protect that, there was a rise of military organization, which allowed royalty and kings to replace or diminish religious leaders. Public ideologies replaced tribal or personal beliefs. Social hierarchies became more pronounced, and wealth was redistributed according to rank in them. Warfare became necessary for protection, or to acquire needed resources, or to control the social hierarchy.

There were advantages to these changes as well. People could expect uniform laws and protection from theft or violence. The stratification provided stability and identity with place or specializations. There was a greater diversity of jobs and places than ever before.

And, these advantages lead to disadvantages. A worker could be conscripted to work on public projects or to serve in the army. Workers had to pay taxes in the form of percentages of food or wealth. Social stratification could deny one access to luxuries and other things, and the stratification was not always fair. Bureaucrats, as well as medical workers or farmers, might have higher or lower status at different times. Power was also distributed differentially in society. In the larger population of the city, people were more vulnerable to violence.

Labor shortages to harvest crops lead to recruitment and some specialization, especially with tools and irrigation (engineering). The surplus of crops, with its necessity for allocation and storage lead to record-keeping and perhaps writing. With labor specialization and records came further specialization, especially with crafts (art) to make standard containers for distribution and professional art to provide necessary images, decoration and luxuries. Economic interaction shifted from reciprocity to formal exchanges. Trade expanded to include special products, tools and artworks.

Concentration in cities permitted and promoted these changes. Permanent settlements contained working houses. Supplies and control required a new infrastructure of massive buildings, better roads, and fortifications. This paralleled the rise of a religious elite to control the unknowns of weather and crops. And, this was followed by a military elite, which took over protection and combined it with universal gods, which lead to the rise of royalty and kings. Heroic buildings increased to provide palaces for royalty. Warfare was necessary to procure foods in times of famine or to protect the stores from less fortunate cities.

The increase in personal property, combined with the redistribution of wealth, lead to social hierarchies and classes. Public ideologies were developed to attune the people. Civilization provided immediate advantages, including supplemental animal labor, more property for everyone, especially the opportunity to have luxury items, the use of records instead of memory, which allowed verification and less misunderstanding, some leisure time for doing nothing, and the production of ale, as a mind-altering substance. Unfortunately, there were disadvantages, such as an increase in the amount of work, not only for production, but also for public projects. Leisure was reduced overall. Living required more work and materials for records, as well as trade and trust with a larger group of people. Personal property had to be stored and protected.

Social political changes include the rise of political religious elite, then the rise of military organization, then the rise of royalty and kings. Public ideologies followed, as did social hierarchies and warfare. Increases in personal property paralleled rationing and the redistribution of wealth.

5.5.1.6. Discussion of Cities: Ideal or Trap

Is the city a human ideal? An environmental ideal? It can offer ideal environments, as well as different kinds of physical environments, from streets, and squares, to religious monuments and parks. What is it about a city that commands awe? Increasing populations can lead to intensification of production, through labor or mechanization. Is the city a trigger for intensification? Interdependence becomes overconnected and then a trap. First cities were of bricks and concrete. John Thackera thinks that cities are held together now by human attention spans, which may be more gaslike than solid. The technology that dominates attention, such as the wireless infrastructure, just adds a new layer, as people still live in brickworks. Certainly, the city has always been an incubator of new forms and ideas. And, the medium for cultural transmission has been ideas, more than genes or bodies. And, there may even be standard units of cultural transmission by imitation, such as memes (Richard Dawkin's phrase for the unit of transmission; see section 7.8 for a more detailed discussion), of which cities, agriculture, and fashions are examples.

How did cities start? Was a town formed by neighboring villages? Thucydides used the word 'synoecismus' to refer to the union of several towns and villages under one capital city. Was the city a result of technology—The wheel, cart, or metallurgy? According to Rod Brooks, cities started from a subsumption architecture of organization, where higher levels of behavior subsume the roles of lower levels to take control (also called bottom-up organization). So, people start with villages, and get the kinks worked out; who lives where? Who does what? This fits with the idea that cities were adaptations to permanent siting, slow overpopulation, and changing environmental conditions.

When the villages are working, you can make a few towns. Coordinate the logistics of streets, sewers, water, lights, and law. When the towns are successful and reliable, you can make a capital city, adding a layer of law, taxes, schools, maybe trading and heroic institutions. The cities can be combined into a state or empire, which has new responsibilities, such as taxes and international affairs and defense. The empire subsumes the other levels, but lets them operate independently. Of course, there are advantages in the group. And there are further emergent structures. Especially international trade and forms of education. Of course, at some point, the cities are no longer adaptive to the environment.

How do cities end? Abandonment? Collapse? If the top level collapses, the others can continue to function. This is what happens with certain kinds of collapse. How does civilization steer? Does it push or pull? Pushing can result in a backlash or revolt. What is best for the individual is not always best for the species. What is best for the person is not always best for society.

Is it possible to have cities without states or vice versa? A city is an adaptation to keep economic surplus in one place. A state is a political unit governed by a central authority. Like a city, a state has to delineate rights and responsibilities for citizens, regulate social relations, e.g., marriage and family, support a religion or ideology, integrate networks of communication and transportation, control redistribution, control punishment, and have a monopoly of police, military forces, and weapons.

How is size important? Is it centrality important? Ancient big cities included Uruk (10,000 to 40,000) and Harrappa (20,000-25,000). Mohenjo-daro of the Indus Valley Civilization was one of the largest, with an estimated population of 41,250. Later, Nineveh and Carthage each had 700,000. Alexandria was large (~400,000 by 32 PE). Rome had an estimated population of 1 million by 5 BPE. Baghdad exceeded a population of one million by the 8th century PE. The largest cities now include Tokyo, Japan (28,025,000), Mexico City, Mexico (18,131,000), Mumbai, India (18,042,000), São Paulo, Brazil (17, 711,000), New York City, USA (16,626,000), Shanghai, China (14,173,000), and Lagos, Nigeria (13,488,000).

The best cities are no longer the largest. The best are considered Vancouver, Copenhagen, Zurich, which are usually less than 2 million. For a long time cities required larger numbers of people to be intense enough for stimulation and creativity. But, now new technology provides the stimulation and could allow cities to be significantly smaller, perhaps 10 or 20 thousand. A growing population creates environmental resistance to itself, in the form of reduction in the reproduction rate as the population approaches carrying capacity. Due to various time lags, e.g., to increase when conditions favor or to react to unfavorable crowding, the density can overshoot the capacity. Human population is controlled to some extent by self-crowding. The overshoot has to be on a local scale, never on a global scale. The Russian geochemist S. Vernadsky concluded that the property of maximum expansion is inherent in living matter as it is for gas expansion or heat distribution (and ideas in cities?). The pressure of life can be measured in terms of velocity. For cities, immigration can increase fast and cause overshoot and subsequent oscillations. What if all cities overshot simultaneously?

Is there a trend to larger cities? Should it be continued? Should arcologies replace unplanned cities? What about limits or nesting into optimum size groups? The degree of aggregation, as well as overall density, which results in optimum population growth and survival, varies with species and conditions (this is Allee's Principle); undercrowding, as well as overcrowding may be limiting. Aggregation can enhance group survival. Fish in a school may tolerate higher doses of poison than individuals. Bees can create more heat to survive than individuals. Applied to humans in cities, aggregation is beneficial, up to a point. But, bee or termite colonies can get too big.

How important is connectivity? Cities are unavoidably entangled in global nets, now. Cities used to be limited by the local carrying capacity. With global nets, self-correcting feedback can take too long. Localization makes feedback visible and more immediate. It would make more self-reliant cities.

How can we look at humanity? First we were wolves, catching animals and eating them. For thirty thousand years. Then we were cows, standing in places eating grasses, for ten thousand years. Now we are termites, swarming over everything in furious dances of labor and status, for the past five thousand years.

Are we urban by nature? Have we changed from hunting to urbaning? Of course, we may be preadapted for cities. We are clever social animals. We prefer edge habitat so as to move between other habitats. Urbanization is a characteristic of an edge species. Civilization produces edges that people like. Wilderness is fragmented into islands and patches. So, there is no more deepness, no more interior to wilderness. What are other edge species? Raccoons, coyotes, crows, and rats.

We are foragers and predators. What are the needs in human nature that a city answers? Perhaps the city is what wilderness was, according to Eisenberg, a place of passage, a place to be brave and test yourself. Humans have always worked to abstract their specialness. The city seems to be another myth of separateness and independence from nature. The city is the laboratory of human creativity, kept apart from the mother of nature. Lewis Mumford saw city life as a compromise between the hunting stage and the farming stage. The female principle of home is wed to the male principle of predation. Perhaps the city is a throwback to hunting. People are more likely to wander, and are less attached to a place?

Are cities unnatural? Is urban living a preadaption? We are of course social animals. We prefer edge habitat. We are hunters and foragers. Why is the city growing everywhere? Especially Africa, South America and Asia. Parts of Manchester and Detroit are being abandoned.

Are cities declining? Certainly, there has been a decline in the quality of living spaces, irrespective of energy or resource use. Some cities have declined in population as areas of a city are deserted. The decline of cities due to the replacement of creative architecture by architecture that lacks organic order and connection, may be due to a declining interest in planning, as a result of economic values, suggests Eliel Saarinen.

Are cities the ultimate creation of civilization where people can enjoy culture free from want or physical extremes? Or, are cities a gross alteration of nature that destroys human life and dignity—a gross ecological error. As long as cities grow without negative feedback, the second will occur. When ecology helps the city ecosystem fit in the surrounding ones, then the achievement is worthwhile. The city depends on its environment.

5.5.2. *Large Buildings*

The first large buildings were probably for multi-family homes, like longhouses or malacas. In cities, large buildings were built for storage, then for religious ceremonies, and palaces. Over centuries large buildings were built for other reasons, also, for parliaments and courthouses, for sporting and political arenas, for factory manufacturing and storage, then for offices and malls. Large buildings not only covered more ground, but also were built much higher, often 20, 50 or more stories.

The change in the scale of buildings required changes in lighting, air and water circulation, and movement corridors (horizontal, diagonal or vertical). Stronger or thicker materials had to be used. Building methods advanced to make it easier to build higher. The Boeing Everett Plant has a tremendous volume and floor area (4.3 million square feet); other plants and warehouses are also large. The Dubai International airport has a tremendous floor area. The planned Abraj Al Bait towers in Mecca, Saudi Arabia are expected to have over 16 million sq. ft. of floor space, in addition to being over 1600 feet tall (By comparison, the Pentagon in Washington has 6.6 million sq. ft.). Larger buildings required more materials and capital, as well as labor and social skills.

5.5.3. *Small Place Design*

(Being edited)

As cities enlongated over miles, many people tried to limit the sizes of neighborhoods. Some cities redesigned parts of their downtown as small public places (for instance, Portland, Oregon). Some cities created a public center, for daily social activities as well as ceremonial social events. These centers were sometimes combined with parks or fountain, and paths too and from other buildings or parks.

5.5.4. *Zones: Neighborhoods & Campuses*

Neighborhoods encourage identity formation with the place. In Sarasota, Florida, neighborhoods acquired their own identities as art areas or quiet places to live. A neighborhood seems to be the largest social unit in which tools and things could be shared safely. Or to which residents have shared responsibility. A neighborhood is an environment in which the patterns of responsibility are all known by all.

Over time, there have been different models of a college, from cloister to campus, business, partnership, intentional community, on-line community (internet node), and participatory network. A college is a place where classes of people—students, faculty and staff—work together to design, construct, live, and learn in a sustainable setting. The campus is an ongoing, experiential community, an intentional neighborhood in which residents are joined by educational purpose. Campuses are often set apart from other communities, regularly on high hills. Some campuses are designed now as neighborhoods, with the functions and amenities of one.

5.5.5. *Corridors & Roads*

Landscape ecology focuses on spatial relationships, fluxes and changes in species, energy, and materials, across large land mosaics. A structural approach shows how these objects (species, energy and materials) are distributed in relation to the sizes, shapes, numbers, kinds, and configurations of ecosystems present. Patch, corridor, and background matrix analyses have proved fruitful.

Streams and animal trails form natural corridors, often with unique boundaries and vegetation, following and changing the contours of the land slowly. Human corridors, such as roads and railways tend to recontour the land quickly with flat linear routes, often with strict boundaries and invasive vegetation. The aesthetics has changed as the landscape is less continuous and the roads are more.

The landscape is perforated. In cities and suburbs the networks of corridors and roads thickens. The number of patches increases. Physical connections increase, yet, biological connections decrease. The artificial support ecosystem of roads is imposed on a humanized ecosystem of urban agriculture (with its system of domestic animals and pets), which is imposed on a natural ecosystem of primary productivity that has been simplified with fewer layers and species. The three systems are linked by material and energy flows, and by specialized predators, including fleas and lice, and by decomposers, such as bacteria and fungi. Towns and cities lower the connections to wild ecosystems. The infrastructure of roads starts to break down, in spite of increased investments in repairs.

5.6. *Design Factors: Adaptive Patterns—Civilization & Industrialization*

By the end of the 1200s, the Mongols brought China into direct trade with Europe, and they developed paper money to make the trade easier. Their empire had its own transcontinental highway, as well as sea trade routes with the Chinese. At that time, the European cities were being changed by guilds, trade and gothic architecture. Colleges were beginning to replace monasteries as centers of knowledge; local languages were replacing Latin. By the mid-1300s, there was an economic downturn in Europe and China, which was related to plagues across Asia and then Europe. The regions were linked if not interdependent enough to influence each other. After a series of local disasters and political changes, China unplugged from the trade system. Did population growth prevent a Chinese industrial revolution? China had many advancements around 1000 BC: Papermaking, moveable type printing, paper-money, guns, gun powder, compasses, underwater mining, umbrellas, hot-air balloons, rockets, and laws of motion.

Then European commerce began to flow past local boundaries. But, the commerce had to be protected from pirates, so laws and agreements were needed. There was a slow transition to regional commerce. Specialization went beyond nations. The Dutch East India Company followed economic goals rather than political ones, although it was closely linked to political and may have become a tool for political objectives.

In the 1300s in Italy, most cities did not know how many people lived in them. With the plague, people began to know by counting dead bodies. How strong is order and morality when there is a plague or conquest? In the 1300s, Florence had a large industrial group of people. A third of the population was employed by the textile industry, as beaters, washers, carders, combers and laborers. Another third was classified as paupers. Companies in Italy started banks, so that there was capital for long-distance trade and development. Did changes in class and equity lead to first private gardens, then public gardens, to areas walled off from others? In the 1400s, the printing press (1455) was part of another transformation that included the Renaissance (1470), the formation of the Spanish nation (1492), and global exploration (1492-1550s).

The English encouraged trade to increase wealth. This led to Mercantilism, government regulation of the economy to ensure growth by granting monopolies. And, that led to the accumulation of capital. Capital required more energy and new technologies. Capitalism changed production (goods are privately owned yet there is separation of people from land

and resources—England 1650 slave, sugar, cloth triangle). Other factors leading England over France or other countries: lower internal tariffs, weakened guilds, intercourse between aristocracy and middle class, relatively less conflict and war (the isolation of the island), and a geography conducive to commerce (access to the ocean).

The impulse for better sources of energy, and the challenges from using coal, led to many new inventions, such as the water pump and steam engine. To produce these things on a larger scale required capital investment, special designs, regulated labor, free materials, directional flow, and sufficient environmental services for renewal. These inventions also transformed society, along with the modern university in Berlin, the institution of science, and the American Revolution.

Table 560-1. Agroecological Sequences Pointing to Industrialization (Read down)

Fewer species	Settlement	Human labor	Deforestation
Control of land	Population increase	Tools/technology	Erosion
Competition	Specialization	Animal labor	Water pollution
Conflict	Big religion	Machine labor	Salinization
War against humans	Big trade	Slave labor	Soil loss, wood scarcity
War against nature	Urbanization	Energy slaves	Species losses

Early cultures obtained energy from food or from a few domestic animals. Cultures practicing intensive agriculture were also limited by energy, but were able to obtain more food by mixed crops on engineered terraces. With the advent of machines, more energy, from wood or coal, was required. And, with the change of scale in industrial production, more concentrated fossil fuels were found and consumed. Industrial societies used triple the energy of intensive agricultural societies. Another watershed of change occurred in the 1900s, with the ascendance of Japan and the United States, the invention of the computer, and the US GI Bill of Rights, which made a university education available to many more people. Some countries, such as the USA and Japan, were able to use double the energy of other nations and 50 times what typical hunter-gatherer societies used.

Industrialization changed many cultures. The culture of consumption has developed from the industrial large-scale production of commodities. It is the psychological switch from being to having. The packages and advertising become part of the commodity. Consumption has private and public aspects. As the production of art became commoditized, the mass production of entertainment by a culture industry, it also became a tool for pacification of a relatively poor population conditioned to 'having' the surplus products of capitalism dumped on them. People, however, regard consumption as an 'empowering' act. And, of course, they personalize things.

Humanity was calculated by Norman Myers and others to be using over 40% of the ecosystem productivity for the entire earth before 1990. Later calculations, by Stuart Pimm and others, indicate over 55 percent a decade later. Humanity influences virtually every ecosystem to some extent, destroying some, interfering directly with many, and exposing the rest to exotic chemicals and materials. Species normally use a percentage of system productivity without disrupting the processes of production. Humanity interferes with the processes.

Based on limited scientific and cultural perspectives, humanity fails to value those beings and communities for which no use is known. But, as Aldo Leopold (1949) notes, the majority of the beings in nature have no human uses. Even some ecologists cannot think of uses for many large birds and mammals. The real danger is genetic loss, which is grossly underestimated. As wild areas grow smaller, even wild species interbreed. As species are lost, the ecosystems become simpler or start to collapse. Our perspectives seem to be limited by metaphors, and metaphors can drive the direction of activities.

Table 560-2. Changes in Metaphor by Societies

Social Type	Metaphor	Psychological case
Foragers/Gardeners/Chiefdoms	Animals	Otherness nonhuman
Pastoralists	Father	Competition
Agriculturalists	Mother	Infantile (land=woman)
Industrialists	Machine	Take-over
Post-industrial?	Ecosystem?	Partnership?

The new possible post-industrial metaphor might be more useful in designing changes on a global scale.

5.6.1. *Historical Patterns of Industry*

Industry is a specialized approach to producing things by a division of labor combined with a division of time and the use of special tools. The first industries of hunters and gatherers were woodworking, pottery, or other trade items, such as axes or toy cedar crest poles. Many industries were winter activities. The scale of industry increased with the advent of agriculture. Catholic monks used the ideas of industry to run monasteries and to produce publications. Industry as a process soon came to mean the use of energy from fossil fuels to increase the scale of production, its speed, and its form, as assembly lines became more formal. In fact, the by-products of industrial mining—steam engines—became the basis for further industrialization. The quantity of energy increased, from human labor with slaves, to animal and machine labor with energy slaves, but the variety of energy resources decreased.

Fashion industries exploited a single resource, such as beaver hats. Art industries produce status and luxury. The function of art in industrial society is to train people's perception of the industrial environment.

Industry is a form of intensification of culture, of miniaturizing nature and focusing consciousness on industry. Industry increases goods and the rates of consumption. It requires more energy and nonrenewable resources.

The key machine of the industrial age is the clock, from mass production to computers. The clock was invented in Benedictine monasteries on the twelfth-century to make daily routines more regular. Thus a thing invented by religious people dedicated to devotion, the clock, became more useful to those devoted to the accumulation of money. The clock made capitalism possible. It made mass production possible. It made wage slavery possible. The clock was one of the first metaphors for the operation of the universe. They were part of the process of mechanization. The clock sets the same time for everyone. It allows the units to be the same. So that people will do the right thing at the right time (of course, biological clocks did the same for animals in a changing world, to set the organisms rhythm to the natural world, in terms of temperature, rain and other conditions). Culture and industry have become driven by clocks.

Industry required free materials and 'externalities' (ecosystem services), as well as regulated labor, capital investment, and special designs and tools. Industry imposed more costs on society, as people had to act like machines, had to be regulated by time and specialties; pollutions and health-related costs accumulated as environments were processed into flatscapes.

Industry processes resources and dumped the wastes back unprocessed. It forces one-way flows. It does so at an unnatural speed. Unlike a pioneer stage of succession, industry is, in Evan Eisenberg's words, at a "fetal" stage. It relies on the maternal biosphere, or the biospheric matrix, to feed it and lick its wastes. Alas, this fetus is very large and could kill the mother soon, unless the fetus matures or the mother reacts faster.

Industry has grown in a very linear, then exponential way. In the 1800 and 1900s, industry became 'heavier.' Human populations rely more on the industrial production of food, which requires fewer people. Industry has resulted in factory farms, which handle insane numbers of animals in smaller areas, where increasing production means treating animals as machines, which is a form of cruelty. Soils are depleted. Human health is compromised by animal stresses and diseases. The industrial system destroys plants and animals, poisons water and soil, as well as soils and human communities in its quest for size and efficiency.

Industrial methods have also dominated manufacturing and sales, also. Industry eats away at cultural images and values. It focuses on discovery and use, rather than need or recovery. Industrial practices are sold as producing 'better quality' and more volume. Its myth is 'bigger is better.' The other myths of industry are familiar: Industrial agriculture is necessary to feed the world, to provide us with safe, nutritious, cheap food, to produce food more efficiently, to offer us more choices, and, of all things, to save the environment. Yet, industry implicitly recognizes the myths and lies. In fact, it points to advances in biotechnology and nanotechnology to solve the negative impacts. Its claims are broadcast through every communication channels. Industry is related to habitat destruction and extinctions. Industrial society is constantly mobilized for emergencies, in the battles against, noneducation, poverty, diseases, terrorism. Industrial development has never been nonviolent or respectful to people.

The scale of industry increases the scale of cultures. Increases the vulnerability of culture, as there are higher incidences of starvation and homelessness. The industrial revolution decreased contact with the natural world and objectified what was left. As a result of drastic changes in the production of economic goods, other political, social and even psychological changes occurred. Other kinds of order were de-emphasized. Human relationships became based on economic allegiances instead of kinship, and were formed in societies, not communities. Money became a symbolic representation for the value of labor and land. Similarly, labor was seen as minimum value. For example, the idea that "labor is a resource" implies that, like any common resource defined by industrial society, labor is cheap and can be used up. The real costs of free goods and externalities have had to be accounted for, yet—this often influences the selection of corporate priorities and growth. Furthermore, the production and distribution system for most economies is linear (straight throughput) and not circular (complete recycling), although this is logical economically, given our frontier resource-use accounting. Why is everything an industry now? The industrial revolution restored humanity to a place of grandeur and importance on the planet.

Despite a few paths of plastics recycling, aluminum or steel, most industry is a one-way path from energy and materials to waste energy and materials. Industry has not yet matured to fit ecosystems and recycle elements back into the systems at a proper scale.

5.6.1.1. Industrial Growth

Mesarovic and Pestel stated that "the issue for the economy is not to grow or not to grow; it is how to grow, and for what purpose." They claim that if a workable world system is to emerge, it must be after the establishment of an organic pattern of growth. Due care is devoted to describing such a pattern and contrasting it with other, tragically inapplicable patterns of growth. Their treatment of the world system itself was regionalized and multileveled.

They recommend that the establishment of organic growth was necessary with no need for special no-growth policies for populations or economies. They assume that further industrial growth will continue, that economic growth is good, and that this growth solves human problems as long as it is organic. Exponential growth is said to be bad, and organic growth is said to be good. In fact, although organic growth is better, there is little difference during a world crisis; both reach asymptotes of suffering. One need only regard the population crashes of lemmings and others to see how organic growth can go wrong.

Economics became enamored of growth during a critical time in history. Rapid European expansion occurred at rates rarely exceeding a growth of one percent per year, and with unparalleled opportunities for expansion into sparsely settled areas, such as North America, Australia, South America, and South Africa. Many cultures now do not have these opportunities; the continents are claimed, and violent population growth may have wrecked their hope for development by ravaging every resource.

The economy has been growing almost constantly since it has been studied. We have been trying to force it to grow, rather than let it stabilize or contract. Even if it stopped growing, the economy could still develop. Mesarovic and Pestel, as well as many others, confuse growth with development. J.S. Mill did not. In the organic world, growth is healthy only when the rate of change is decelerative in the long run; cancer and population are constant or accelerative. Mesarovic and Pestle fail to realize that continued economic growth in any form is a threat to the stability of the biosphere.

5.6.1.2. Industrial Intensification

There has been an intensification of tool-making in the past 3000 years. Population increased and new areas were exploited. The agricultural revolution was a technological revolution. The industrial revolution, by definition, was a technological revolution, which resulted ultimately in men being replaced by machines. Only the Luddites faced this limitation, unsuccessfully. Human labor is increasingly dispensable. Industrial culture confuses mechanical with personal power. According to W. I. Thompson, industrialization is really an intensification of civilization.

Table 5612-1. Important Cultural Revolutions

Date	Development	Effect
9000-8000 BC	Domestication	Animal things & labor, weaving, animal diseases
7000-4000 BC	Agriculturization	Sedentization specialization, accumulation
3000-0 BC	Urbanization	Cities, architecture, trade, metallurgy, writing, armies, calendars, disease concentration
500 BC-300 AD	Cosmologization Thought systems	Religion, ethics
500-1500 AD	Mechanization Science / Banking	Machines, wind/water power, literacy
1700-1800 AD	Industrialization	Steam power, assembly rules

Industrialization is really an intensification of civilization, which is still an ektropic process. In each case of cultural absorption, there was an attendant process of miniaturization. The forest was miniaturized in clumps of trees; animals were miniaturized in artistic image; time on a lunar tally stick; plants in a garden; women in a household; nature by culture in 1800, under the glass roof of the Crystal Palace; and now by the new consciousness surrounding the old. What is intensity?

Intensification brings cultures together faster. Conversion increases, technology and industrialization increase. Intensification is a gigatrend that could lead to violence. For instance, intensification leads to urban living, which increases creativity, as well as some kinds of illness, which leads to physical weakness, which leads to changes in population, which leads to intensification of resource use, which leads to the potential for conflict, which can lead to dislocation and violence.

Table 5612-2. Dates of Human Revolutions (Generally after W. I. Thompson)

To /-ization	Homin-	Art-	Agricult-	Urban-	Industrial-	Global-
Year BP	325000	90000	11000	5500	500	160
Yrs between	1675000000	235000	79000	5500	5000	340
Strengths	reproduction	power	energy	concentration	materials	trade
Weaknesses		secret	fragile	disease	suffering	cost
Ends		power	food	drought		luxuries
Effects		image	surplus	permanence		spread
Problems			Famine	inequity	waste	inequity
Solutions			ecology	ecology	ecology	politics
Strategies			fitness	incorporation	linkages	profit

Urbanization concentrates people in a smaller area. Does that leave more room for wilderness? Just squeezing in does not mean better conditions. There has to be enough energy, sanitation, water and other things. In cities, industrialization, in the form of increased wealth, can bring about a demographic transition. As health and wealth increase, people have fewer children. What is the actual trigger? The desire or opportunity for self-actualization after reproducing heirs?

The industrialization process has led to a dramatic extension of cities and buildings, as well as growing extents of machines and increases in energy use. Furthermore, people, property and machines are much more mobile, which requires more pathways and more energy. This has increased the impacts and pressures on ecosystem productivities. Design has not been able to effectively deal with this kind of runaway process.

5.6.1.3. Problems of Industrial Civilization

Our modern problems reflect an unbalanced and immature image of the earth, the earth as a machine, for instance. People sometimes constructed their worlds from preconceived notions, and many of these worlds did not survive, because they could not adapt to the environment. Our modern cultures are defective for this reason. The modern attitude toward nature as a resource has resulted in pollution and depletion of resources. It has allowed humans to overpopulate their habitats. Recent productivity studies indicate that the optimum sustainable human population is far below the current world population.

Even worse, decisions regarding resources are still made exclusively on short-term economic rationalizations and lead to material shortages and environmental degradation. The crises of environmental degradations are crises of cultures. Monocultures of the industrial kind lead to 'dedifferentiation,' that is, the decomposition and destabilization of complex structures. A species or culture that destabilizes its ecosystem through misbehavior risks its own extinction. Human beings make changes to ecosystems that endanger themselves.

Industrial society is constantly mobilized for emergencies, in the battles against, noneducation, poverty, diseases, and terrorism. Industrial development has never been nonviolent or respectful to people. Industrial production has its own unique style, shape, and scale.

Is the industrial city a jungle? Herbert Spencer saw the world as a jungle, where life was nasty, brutish and short. Spencer coined the phrase 'survival of the fitter'. He based it on the competition in industry, where entrepreneurs fought for money and power. Spencer thought the weak should be eliminated, so he opposed poor laws, charity, sanitation, education, clean water, and pure food, among other things. Rioting is lashing out at the cage, when people cannot leave cities that do not meet their needs.

One problem in industrial culture is the production of flatscapes. Our attempts at social improvements have proceeded without order, without sufficient insight and perspective, without sufficient confidence, without a comprehensive plan, and without a great dream. Our politics has been corrupted by special interests. The structure of our civilization comes from anonymous builders and mediocre designers, minimal engineers and rapacious financiers. We work within the rules as they have been for decades, rejecting any alternatives as too utopian. The rules themselves have been shaped by centuries of social metaphors and utopian ideals. They do not exist in place, either a human place or an ultrahuman place. They are designed to be no-place, without weeds, storms, or hard ground. Because they are nowhere, flatscapes, like demented unplanned utopias, lack any reference point.

Speed, so beloved by industrial cultures, can be a problem in archaic cultures. The Kelantese people in the Malay consider haste a breach of etiquette and ethics; slowness is important. George Beard, in American Nervousness, coined neurasthenia, to describe mental illness caused by increased tempo of life.

Information pervades society, but it is for an information market. Between uncoordinated information and the acceleration of activity, the social fabric gets fragmented. Projects are fragments of work. Industrial information work is bad for physical and psychological health, from bad physical conditions, high pressure, and low control. The information economy can lead to disconnection, loss of identity, and loneliness.

Physical laws create patterns in space, as well as in human history. There are simple kinds of patterns. A linear pattern tends to be interpreted as progress or regress. This is the dominant concept of modern history, unending progress. Despite the chaos of individual events, there does seem to be a direction of gradual improvement. Marquis de Condorcet suggested that civilization will always move in a desirable direction (1795). People become frustrated when it does not, and there have been collapses.

5.6.2. *Connections*
(Being edited)

Paths create connections. Human life is linked in an intricate web of connections. There are always unintended consequences. Humans have the ability to recognize or make connections, although often we do not pay sufficient attention to them. Analytic science and the compartmentalization of scientific fields has exposed the complex connections of the subjects. Philosophy has to be positive and constructive, to ask the questions that highlight patterns and rules for living, to contribute to healthier images and myths, to rethink the terms of debates, for instance, of tree farms versus wilderness, so that as many connections as possible are made visible and meaningful, and to provide good metaphors. Ecological metaphors, such as gardening or tending a forest, are more fruitful than images of machines or war.

Cultures and ecosystems can have too many connections. Overconnection functions like a trap. The process occurs too close to limit and it cannot expand away. In an ecosystem, individuals and species are connected to some degree, in terms of quality and quantity of connections. The connectivity of a component of a system is a measure of the number of direct connections between it and the rest of the system. Connectance is a percentage of the number of connections through predation or exploitation as a percentage of the total number of possible interconnections. A system is more connected if the absolute number increases and the percentage, and the strength, increases.

Agricultural fields are good examples of overconnectedness. This is even more true with corn than with wheat due to the changes from domestication. Overconnection causes greater lag times. But, in a city, feed bins damp food variations so the lag is less important. By being overconnected a system can lose stability. Overtconnectedness creates a lag in signals. This can lead to system collapse.

Underconnection can also create instability. Parts go their own way and there is no communication between them. If a system is underconnected, it may become unstable or lose resilience, because there is no pathway to allow a signal between significant parts.

5.7. *Problems: Ecosystem & Cultural Collapse*

Ecosystems and cultures share many properties, perhaps since cultures emerged from being in place in ecosystems or perhaps because they are both complex systems. They also share stages of development, which always end in collapse.

5.7.1. *Collapse*
Collapse is the rapid, significant loss of an established level of complexity after some catastrophic event. By this definition, ecosystems can collapse during climate changes, and cultures can collapse after social or environmental changes. For an ecosystem, collapse can mean less complex food webs, fewer species and decreased diversity. Ecosystem collapse is a regular process, often precipitated by environmental change. For a culture, collapse can entail the loss of centralized control and regulation by elites, decreased flow of information between center and periphery, less trading and redistribution of resources, losses of complexity in organization and stratification, drops in economic specialization, and reduced investment in the complex physical structures. Many cultures have collapsed and disappeared, often without being remembered or documented. The known collapses range from the Old Kingdom of Egypt (2181 BCE) and the Harappan in the Indus Valley (1750 BCE) to Rome (476 CE) and the Khmer in Cambodia (1431 CE). Recent collapses include the Kachin of Burma (1950 CE) and the Ik in Uganda (1970 CE).

5.7.1.1. Kinds of Ecosystem Collapse
An ecosystem can mature through a series of stages, described as succession, to old-age and dissolution. Pierre Dansereau delineated four general stages of succession (modified here): pioneer, consolidation, mature (Dansereau's subclimax), and closed (Dansereau's climax). A fifth unmentioned stage is collapse, where the system reacts to a disturbance by dropping to an earlier stage or by completely collapsing. Eugene Odum characterized the pioneer stage as having a linear food chain. A typical pioneer stage might have lichens, mosses, and annuals. The consolidation phase may have perennial forbs, grasses, mixed herbaceous plants, shrubs, and a few shade-intolerant trees. The mature stage might have shade-tolerant trees. The closed stage would have emergent trees and a mixture of plants and animals that resist exotic species, in effect closing the system to further colonization or invasion. In the mature

and closed stages, complex food webs are dominated by decomposers. For animals, body sizes increase and life-cycles and strategies become more complex. At this stage, the system has learned the cycles of the environment. Throughout this process of maturity, according to R. Margalef, there are changes in productivity, efficiency and nutrient cycles. Diversity increases and trophic structure becomes more complex, with added numbers of consumers and decomposers, and more specialized niches. During a collapse the networks and webs fall apart, the system can drop to any earlier stage, and individuals and species disappear or disperse to another system.

Catastrophes (used in the mathematical sense meaning a down-turning), such as storms, fires, glaciers, and clearcuts, can set back stages of succession or change its direction. Stages do not necessarily begin in order or progress uniformly; they can begin anywhere and overlap. Nor are they exclusive, as life forms in one can occur in another, for instance, when lichens appear in all stages in a coniferous forest.

A mature ecosystem can collapse when internal overconnections between species make it fragile or incapable of being flexible enough to respond to environmental change, either large changes or new patterns. Climate change, especially drought, long or short-term, has caused the collapse of many natural ecosystems. For instance, the change in temperatures and moisture patterns at the end of the last ice age replaced many forests with grasslands, and many species could not adjust to the change and were extinguished.

5.7.1.1.1. Local Ecosystem Collapse: Mt. St. Helens

During the eruption of Mt. St. Helens in Western Washington State in May 1980, thousands of animals (at least 5000 deer and 1500 elk) and millions of fish (probably 12 million Chinook and Coho salmon) were killed; over a hundred square kilometers of forests were destroyed by a super-heated wall of gas and ash. In the direct blast zone, everything was obliterated. In an intermediate zone the blast was channeled by the hills, and trees were flattened. In the outer zone, the trees were seared and killed, but left standing. Mudslides flowed down creeks and rivers. The landscape was described as 'scorched and lifeless.' Although natural disturbances, such as fire and drought, drive the direction of many forest ecosystems, this large-scale disturbance destroyed many systems, sterilizing a plain north of the crater and leaving a 'forest' of dead snags to the east.

Ecosystem recovery started within months after the eruption, in different stages at different rates. Communities of plants and animals under the remaining snowpack were little affected. Many burrowing animals and insects survived in the intermediate and outer zones, as did numbers of ground dwelling insects and spiders. The rate of recovery depended on proximity to remaining original forest, that is, the biological legacy. Areas close to surviving forests experienced faster recovery. In the blast zone pumice plain, as seeds blew in, willow and alder shrubs have rooted and other tree species are evident. Along the blast zone, seeds have created an alder forest with some conifers. Birds and mammals have returned to many areas, followed by their predators, although the species mixes are unusual. Along some streams, coniferous forests have been replaced by a deciduous woodland, with alder, cottonwood, and willows, inhabited by neotropical migrant bird species. Other coniferous ecosystems are reestablishing themselves. Salvage crews took out over a billion board feet of downed trees and planted stands of Douglas-fir, which should also be subject to succession.

5.7.1.1.2. Regional & Local Ecosystem Collapse: Mesopotamia

By the end of the last ice age, the storm tracks had shifted north from Africa to Mesopotamia, and brought moist air masses. The ecosystems switched from being boreal to being temperate. Grasslands and deciduous woodlands expanded.

Around 8200 BP, there was a sudden cooling that may have been caused by the collapse of Lake Agassiz, which had formed over 4000 years of melting, after the ice wall collapsed and the lake suddenly discharged through Hudson's Bay or the St. Lawrence Seaway (and may have changed the salinity of the north Atlantic and caused the thermohaline circulation to alter or shut down). Temperatures and snow accumulation dropped dramatically. It affected North America, Greenland, and Europe, as well as western Asia. The regional climate adjusted, and things became colder, drier, and windier. Lakes in Africa dried up, including a large one in the Sahara, which never reformed. In western Asia and Mesopotamia, ecosystems dried up and were replaced by desert ecosystems on desert soils. Some systems in Turkey and the Zagros mountains survived.

There were subsequent coolings in 5200 BP and 4200 BP. The first event caused a drought that coincided with cultural collapses in southern Mesopotamia along the Euphrates and Tigris rivers. Although irrigation canals brought water to fields, a weakened river flow would have reduced irrigation and yields, so people may have had to switch to drier crops or abandon the effort entirely. Settlements in the interior of Arabia were abandoned around this time. In the second event, the westerly winds bringing moisture to the Mediterranean and Mesopotamia failed. Precipitation fell 30 percent. Hot winds contributed to drought. Some of the cities, such as Tell Leilan, were deserted; many villages were abandoned. Between 4200 and 4100 BP, Mohenjo-Daro and Harappa, around the Indus River, declined suddenly.

5.7.1.1.3. Global Ecological Collapse: Chicxulub Asteroid Strike

Dust, asteroids and comets are part of the environment of the planet. Although collisions have become more rare as the system ages, there have been large strikes that have pushed life in new directions. Notable strikes occurred at these junctions and seemed to have precipitated large extinction events: Precambrian-Cambrian (570 million YBP), Ordovician-Silurian (438 MYBP), Devonian-Carboniferous (360 MYBP), Permian-Triassic (245 MYBP), and Cretaceous-Tertiary (65 MYBP). This last event allowed mammals to claim dominance over reptiles in the hundreds of thousands of years that followed the event.

Very large-scale disturbances, such as volcanic eruptions or asteroid impacts, can destroy entire ecosystems or disrupt global biogeochemical cycles. However, such very large-scale disturbances are rare, and the ecosystems often have thousands or millions of years to become reestablished, although changed. Disturbances that are not part of the history of an ecosystem may cause irreversible changes to the system, because the system has not evolved a defense or response mechanism to such a rogue disturbance. An asteroid strike would be such a disturbance, especially if the landform was altered by a crater. The Chicxulub asteroid strike 65 million years ago destroyed a large area, broiled the surface and ignited global firestorms, created a series of massive tsunami perhaps 300 meters high, and created thick dust and particles that blocked sunlight and covered the surface of the planet for 2-10 years. Species that depended on sunlight for photosynthesis declined or became extinct. Plant-based food chains collapsed. Organisms that fed on dead matter, omnivores and insectivores, survived. The climate cooled and was affected for thousands of years. Species that depended on a stable climate declined. Acid rain fell. Volcanic activity, for instance the Deccan Traps, may have increased and further destroyed ecosystems, changed the sea level and altered climate.

The Chicxulub crater is the largest impact structure still existing on the planet. Nonavian dinosaurs are found only below the K-T boundary. This event caused global extinction patterns and shifts in biomes. There was a global disruption of plant communities. Fungi, however, which live on organic matter, proliferated for a while. In general, aquatic and semi-aquatic species survived better than terrestrial species—perhaps due to available organic debris for food and the ability to swim or burrow. Species that had been stressed before the event may have been pushed to extinction by the effects of the impact. The biodiversity of the planet required a long time to recover.

5.7.1.1.4. Discussion of Ecosystem Collapse

If a catastrophic event is total but localized, plants and animals can recolonize the ecosystem. If it is total and regional, ecosystems may disappear. If it is partial and regional, many ecosystems can recover over ecological time. Certainly the subtropical ecosystems on the northern Yucatan peninsula disappeared immediately at impact, which not only changed the rock substrate, but also dramatically altered the soils and topography.

5.7.1.2. *Kinds of Cultural Collapse*

Cultures seem to mirror human growth, development and aging. A fragile or old culture may collapse at the end of its cycle. The stages of a culture could be described as: Origin or Renewal, growth, maturity, collapse, and dispersal; this might be a rough parallel to the stages of an ecosystem: pioneer, consolidation, maturity, closure, and collapse. After closure, the ecosystem either collapses or is bumped to a lower stage by some disturbance or interference. Lance Gunderson and C.S. Holling, in their book *Panarchy* (2002) present a stylized representation of ecosystem functions—exploitation, conservation, release, and reorganization—that they use also as a four-phase heuristic model of institutional dynamics and other applications. This representation will be related in a different section.

Joseph Tainter uses several models to describe cultures: In the Runaway train model, a culture is impelled on a certain path without the ability to change directions; although it is dynamic, its course is fixed, as when the Aztec continued to ensure the rising of the sun with sacrificial blood. The House-of-cards model stresses that a culture is inherently fragile, with little ability to respond to stress; perhaps the dramatic decline of Rapa Nui is an example. The Dinosaur model suggests that a culture is a lumbering colossus with a big investment in one environment, but too static and maladaptive to adapt to any change. One might suggest other models that still might result in collapse. The Fungus model presents culture as an adaptive, problem-solving, intensive social structure that, although it can respond to many environmental challenges, can still fall apart with scale or be overwhelmed by large events.

Cultural collapse can entail any set of many factors. The rapid loss of an established level of sociocultural complexity may be first expressed as a breakdown of central authority. Provincial provinces or villages may break away. Revenues to government may decline. There may be foreign challenges over territory or resources. The upper hierarchy may start to claim resources for themselves. Ordinary people may become disaffected. The military may be ineffective. Continued central direction may not possible. The center may lose power. Distribution of goods may suffer. Trade may decline. The city center may be ransacked and abandoned. Small states may start to emerge in the same territory. People may lose the protection of the law. Public art and monument construction may cease. Literacy may be lost. The small remaining urban populations may reuse the architecture and subdivide rooms. Palaces, government buildings, churches, and storage facilities may be abandoned. Technology may revert to simpler forms. There may be a great reduction in population size and density. Peripheral ecosystems and animals may recover, although not always.

Generally four concepts leading to collapse can be identified (J. Tainter 1989 and J. Diamond 2003): Cultural systems need energy to be maintained; increased social complexity results in increased costs per person; the investment in complexity reaches a point of diminishing returns; and, the cultural system fails to react to very large, slow or invisible catastrophes. Of course the culture could respond to those challenges. At some point, all energy income is used for maintenance and none is left for other problems, such as starvation of classes of people. Cultures often know how to deal with resource uncertainties, but do not know how to reduce populations to reduce pressures on scarce resources (although they allow disease and war to lower populations for a time). Cultures also seem to know how to deal with regular catastrophes, but do not seem to know how to minimize them or avoid them; catastrophic fires or floods could be minimized by relocating buildings or replanting native trees—perhaps the people grew tired of the increased costs. Cultures also seem to know how to provide and control labor, but they cannot reduce management levels (and this applies to royal classes as well).

Collapse can occur as a result of other reasons or combinations of circumstances, from invasion to social dysfunction. Such reasons include: Resource depletion, establishing new resources, catastrophe, insufficient response to environment or change, intruders, conflict mismanagement by class, social dysfunction, religious or mystical factors, chance series of events, and economic failure. An example of social dysfunction is class conflict. In the Mayan lowlands, class conflict was complicated by militarism, overtaxation, and land degradation. These contributing factors may have been solved, except that three long droughts weakened the Mayan cities further until they collapsed. Group conflict, deforestation, isolation, and population increase contributed to the collapse of Rapa Nui.

Table 5712-2. Internal Kinds of Cultural Collapse

Social dysfunction	Class conflict	Militarism, over-taxation, degradation	Mayan lowland
	Peasant revolt		Huari, Peru
	Group Conflict	Agricultural collapse deforestation	Rapa Nui
Economic inefficiency	By-passed by trade routes. Agrarian system unproductive	Spanish gold ruined economy. Expansion of government/ army	Ottoman 1500s
	Increase scale for control	No support for surplus population. Contrast rulers	Chinese dynasties
Political dysfunction	Loss of provinces	Increased taxation	Babylon
	Roman Withdrawal	Regional tribalism	England 411
	Maladaptive ideology		Aztecs Rapa Nui
	No inheritance from old rulers	Continued conquest to enrich new ruler	Inca
	Bad management	Lack of economic development	Spain 1700s
	High taxation		Netherlands 1700s

Economic inefficiency was a contributing factor to the collapse of the Ottoman Empire, which had to support an expanding population and an expanding government and military. Their agrarian system was unproductive, the influx of Spanish gold hurt the economy, and the economy was also by-passed by new oceanic trade routes.

Bad luck was a critical factor in the fall of Rome. Weak emperors and their mismanagement of the Huns and Visigoths contributed. Other, mystical factors may have been important; perhaps the final challenge was a loss of virtue or a spiritual or physical exhaustion. Other cultures also have been ruined by bad luck or chance events. Byzantium was unsuccessful in its competition with Venice for sea-lanes and trading, and they also suffered the loss of many agricultural lands. More recently, the Ik people of Africa lost their territory after European conquest and had to switch to farming, which they did not like.

Political dysfunction was a major factor in the collapse of Babylon; the loss of provinces and high taxation made things worse. Political dysfunction was also contributing factor to the collapse the Inca and the Aztecs, who also had a very maladaptive cosmology (the constant thirst of the sun for blood; see Table 5712-2).

5.7.1.2.1. Local: City Taxes and Collapse in Mesopotamia

City-states on the Tigris and Euphrates flourished around 2800 BCE. Irrigation systems and arts developed. The Akkadian empire formed. Two hundred years later, the city-states rebelled and became independent again. Shortly afterwards, the Third Dynasty of Ur set up a larger bureaucracy to collect taxes and tribute. They rapidly expanded the irrigation system, settlements and the population. This led to a more rapid and more complete collapse. In the following hundreds of years there was a 40 percent reduction in settlements and a 77 percent reduction in area. Political power shifted north to Babylon and Hammurabi. But that empire only lasted 80 years to the death of Samsuiluna in 1712 BCE. Succeeding kings ruled smaller realms. The last resurgence of Babylon was ended by Cyrus the Great. Mesopotamia was incorporated into successive empires. Under these empires, agriculture and city-building was expanded. Then after the 7th century CE the river systems changed courses and alluvial deposits. By 1100 CE the total area was only 6 percent of what it was 500 years earlier. Population was at its lowest point. Tax revenue was down 90 percent. People rebelled. The only remaining irrigation weirs were in the vicinity of Baghdad. The region was claimed by nomads until modern times.

5.7.1.2.2. Regional & Local Collapse: Diseases in North & South America

When European people started to explore and colonize North and South America, they came in contact with large native populations, many of which welcomed and helped them. Almost immediately, native people started dying. Entire villages and tribes were wiped out. Survivors often moved in with neighboring tribes, mixing belongings and cultures. Epidemics spread from tribe to tribe, often well in advance of the Europeans themselves. European explorers commented on how empty the land was. Some disease-devastated or abandoned villages were overgrown by vegetation; a few were noticed by the explorers. The pre-Columbian native populations were reduced to 5-10 percent of their original levels by the epidemics. In North America the original population may have been 10 million; in Central America, over 17 million; and, in South America 25 million—even these estimates are being revised by research.

Insects, diseases, and animals, more than being simply agents of mortality, are native components of complex food webs in ecosystems, and they contribute to the selection of species. The domestication of animals from Mesopotamia to Asia and Europe exposed people to animal diseases, which transferred to the human populations. Epidemic diseases originated in domesticated animals. Measles, smallpox and tuberculosis came to humans from cattle, flu came from pigs and ducks, and whooping cough came from pigs and dogs.

In North and South America, only the dog and lama were domesticated, with less disease transfer. American peoples did not have epidemic diseases or immunities because they did not have the domesticated animals that gave rise to the diseases. Europeans, however, were exposed to syphilis for the first time.

Disease was even more important than horses or guns in the European subjugation of the Americas, according to Jared Diamond. Disease not only reduced the numbers of native peoples, but it wrecked havoc on their cultures, so that the enormous competitive advantages enjoyed by societies with horses and guns were more effective when they were used. Diamond recounts how Spanish conquistador Francisco Pizarro used 62 horsemen and 106 foot-soldiers to destroy thousands of Inca soldiers on one day, November 16 1532. In hours, Pizarro's small band captured the Inca emperor Atahualpa, leader of South America's largest and most advanced state, by panicking the emperor's 80,000 guards.

5.7.1.2.3. Global Cultural Collapse
Although many local cultures and various regions, from the Mediterranean to the Indus River or Eastern Asia, have collapsed, there has not been an event that caused every culture on the planet to collapse within a short time.

5.7.1.2.4. Discussion of Cultural Collapse
In the Sahel and Mesopotamia, the argument was that overgrazing and human population caused the droughts. Exposed soil contributed to a changed albedo and hot air, which did not permit rain-forming clouds. Recent research indicates that, for the Sahel, a single variable made most of the difference—rising sea surface temperatures of the Indian Ocean, from greenhouse gases, was responsible for most rainfall decline. Despite the fact that we often focus on political and behavior causes, environmental changes are critical factors.

5.7.1.3. *Coterminal Ecosystem & Culture Collapse*
Cultures and ecosystems produce co-effects on each other. In some areas, such as Mesopotamia and Central America, the synchronization of plants used for crops can increase conflicts over food. The simplification of complex ecosystems for agriculture usually causes the collapse of the native system. Overuse of a wild ecosystem can cause it to collapse. Even targeted removal of keystone species can cause the system to collapse into a simpler system. English forests, for instance, changed after large predators, such as wolves and lions, were eliminated; deer and goats reduced the germination of trees, so grass or scrub lands expanded.

5.7.1.3.1. Local: Maya
The Lowland Maya formed around 1100 BCE. By 200 BCE, massive public architecture was rising. Temples and palaces were built. Vast public works, such as aqueducts, were undertaken. The arts flourished. The entire landscape was modified for planting. The zenith of organization and population occurred between 700 and 900 AD. Perhaps 75 percent of the region had been cleared for agriculture at that time. The Maya had a high density, stressed population, with intensive agriculture, complex hydraulic systems, in large centers, with an elite class, calendars and rituals. Population growth triggered competition and conflict, which lead to positive feedback that caused more stress on populations, which led to disease and stress on the surrounding ecosystems.

Drought was a major consideration for the Mayans, who had built so far from water. Maybe the southern centers were sited for water. Much of the culture was devoted to collecting and distributing water. The water lily had iconographic significance because it was indicated good water quality. Between 810 and 910, Mayans had three droughts of three, six, and nine years, in 810, 860, and 910. They entered the drought with a maximum population and limited flexibility.

People suffered from the stress. Analysis of skeletons shows that in the Late Classic they became more fragile. Although people became 7 centimeters taller by the Early Classic, by the Late Classic the stature of men had declined markedly; they also exhibited degenerative bone conditions, bad teeth, scurvy and other pathologies. There was a collapse and depopulation, from 3,000,000 to 450,000, which never completely recovered. The current population is 1,250,000.

Collapse probably improved things for the Mayan peasants, without the burden of rulers, elite, priests, and artists. In the long run, however, even the peasants were decimated, perhaps due to environmental deterioration, stress, and in-fighting. By the time of the Spanish, the area seemed to be unbroken forest. The Spanish introduced malaria and hookworm, which made the forest worse to live in. In some cases, such as Maya, environmental degradation did play a role in collapse of the civilization, either as a contributing cause or effect. Complex societies put harsher demands on local environments when political regimes set production demands too high.

Table 5713-1. Kinds of Environment-driven Collapse

Reason	Main factor	Related factors	Examples
Catastrophe	Maize mosaic virus	Overpopulation Droughts, lack of flexibility	Maya lowlands
	Earthquakes, plagues		Teotihuacan
	Eruption of Thera	Competition with Mycenaea	Minoan Crete
	Malaria	Lead poisoning, exhaustion, political corruption	Roman state
	Climate change, drought	Salinization	Hohokam
	Climate change	Agriculture-base	Hopewell
	Nile flood failure	Destruction, elite rule, over-taxation, parasites	Egypt
	Flooding	Collapse of trade	Harappan, Indus
	Climate, drought	Famine, migration	Mycenaean
	Climate change	Agricultural collapse, erosion, Deforestation, invasion	Roman

5.7.1.3.2. Regional & Local Collapse: Mediterranean

In the Minoan civilization of Crete, the earliest palaces were built after 2000 BCE. Although they were regularly ruined or destroyed by earthquakes, each was rebuilt more splendidly, with a sophisticated knowledge of architecture, engineering, hydraulics, and drainage. The Palace of Knossos was more luxurious than anything in Egypt, especially with water-flushing latrines.

The palaces were administrative and trade centers, but also warehouses. Knossos for instance had the capacity to hold 240,000 gallons of olive oil. There were craft production rooms for potters, weavers, and metal workers. Written records show how goods were directed to the palaces and redistributed from there. Scenes on widespread frescoes on the walls were generally peaceful. The palaces were not fortified.

In 1500 BCE, a powerful earthquake precipitated major destruction and widespread changes. Knossos started to dominate the other palaces. The Mycenaean Greeks competed with Crete for sea trade. They introduced new kinds of warfare, with new weapons and horses. About 1380 BCE the palaces were all destroyed, and the civilization collapsed. Parts of Knossos were rebuilt and a reduced administration carried on for a while but even that ended by 1200 BCE. The early script Linear A was replaced with Linear B, which was Greek.

The Mycenaean Greeks themselves inhabited a hilly topography with good forests, rugged semi-arid regions with small alluvial valleys. Villages dependent on domesticated plants and animals appeared about 6000 BCE. The core of Greek subsistence was small scale farming. Agrarian success led to increased soil erosion. This civilization began to develop around 1650 BC. It reached its height about 1400, around the Minoan collapse. Central and southern Greece was divided into independent city-states. Mycenae seemed to be the most powerful. It had 16 administrative districts, each controlled by a governor. Each was an economic center for distributing goods and foods. Aqueducts were built to carry water and roads to carry people and goods. The cities had massive walls with major structures. Palaces had frescoes and bathrooms. Under supervision of palace authorities artisans cut gems, worked metals, and threw pots.

After 1200 BCE, palace after palace was destroyed. The artisans seem to have vanished. Writing disappeared, and art became much simpler. Cities became more fortified and dug into their rock bases for water. The number of settlements dropped from 320 to 40. Athens survived but suffered a political collapse. Overall population declined from 75-90 percent. Some people migrated to the southwest Peloponnesus. Elsewhere, around the same time, after 1200 BCE, the Hittites in Anatolia, and the Vera basin people in the southeast Iberian peninsula collapsed—deforestation of uplands, and a shift to barley monocropping, were made worse by the desiccation and many Vera basin villages were abandoned. Egypt experienced troubles. There were also local collapses in the western Pacific, for instance Kangaroo island off Australia.

A shift in westerly winds brought long droughts to the entire eastern Mediterranean. Some hypothesize that this abrupt climate change may be related to a supernova, bolide activity, comet or asteroid impact, volcanic activity, El Nino event, or earthquakes. Cities were not able to survive on reduced crops. Populations dropped; people returned to herding or foraging when possible.

5.7.1.3.3. Global Collapse

No combination of ecological catastrophes and cultural disintegrations has resulted in a global collapse of linked cultures and ecosystems around the planet.

5.7.2. *Discussion of Collapse*

Combined cultural-ecological collapses are probably the most common kind of collapse. That is, every cultural collapse seems to have an ecological component, and every recent ecological collapse seems to have a cultural component. This should not be surprising, as people are embedded in ecosystems and tend to use them and transform them. Combined with the human propensity to multiply to fill in every niche, the sum of human activities puts strong pressures on ecosystems.

There seem to be two basic problems in living in ecosystems: Perceptual and managerial. Often, we do not perceive the system in its complexity. We depend on initial observations and ignore a wealth of details. We tend to make general responses to complex, detailed challenges. We may make general models of the environment for thinking about it. We also tend to transfer generalities from one ecosystem to another, without being attentive to critical differences. Assumptions by immigrants on entering new ecological zones shape their perception. The earliest aborigines must have though the herds of diprotodons stretched forever. The first Americans may have thought mammoths were numberless. The Maori may have thought the moa were limitless. Hunting would have been easy and successful. Deception is also a problem. Europeans thought that America and Australia were *terra nullus*, empty lands, or under-utilized ones. Deception can turn to disillusion as resources disappear.

Managerial problems can arise from difficulties with communications and assignment for responsibilities. With more than one level of management, the response to a challenge might be delayed or entirely inappropriate. Furthermore, with each level of management, there is a degree of detachment from the participant. These two problems can lead to poor recognition of challenges and to poor, untimely responses.

Of course, our perception of challenges is improving with advancing science and increased communications. Climate change is only one of the challenges to ecosystem or cultural health. Other threats can be summarized briefly as global or local problems. The local problems include such things as the removal of key elements, the addition of novel elements, and the losses of species and habitats. The global problems include such global things as global warming, ozone depletion, wide-spread contamination, and the disruption of global cycles. Any one of these threats, or some combination, can cause ecosystem collapse. Ecosystem breakdown happens as a result of stresses, singly or grouped, that relate to interference patterns in the system, most of which are caused by the human species now, although the potential for asteroids, volcanic eruptions or prolonged droughts remains.

Collapse is often part of physical, biological and cultural cycles. Collapse can happen when a culture is too static in a changing environment. It can happen a culture selects growth to try to overcome certain environmental obstacles. Sometimes the collapse is thought to be the result of internal moral or technological failings, as well as conquest or some other external influence. Cultures are adaptive systems that have to integrate a number of challenges, opportunities and pressures, from drought to invasion, and then reconcile them to changing economic and political situations. Collapse can happen by bad luck, chance that is, or just a coincidence of bad leadership, bad images, cultural problems, external social pressures, and environmental degradation.

5.7.3. *Patterns of Collapse: Recovery & Renewal*

There are many ranges of collapse. In some cases, the population collapses, but there is not much loss of cultural complexity. Ireland, for instance, lost half of its population, during and after the potato famine, because of its reliance on that one crop; due to English rule, the government continued. The population stabilized at a new lower level, under three million. After 1960, with the development of industry and then membership in the European Union, the population has been increasing to six million (still less than the eight million in the 1800s).

In other cases, there has been a tremendous population loss, along with a political crash and complete loss of organization. The Maya collapse eventually led to the complete abandonment of the cities and the urban way of life. The population dropped dramatically. Some parts of a culture may collapse, such as the Southern lowland Maya, while the Northern upland culture continued to survive for several generations before collapsing. As part of a cycle, after people disperse to return to older styles of living, the local ecosystems can regenerate, without the pressure of use or overuse, and recover.

In still other cases, local city states may collapse for a generation or two, and then be rebuilt in place, as the local ecosystems have experienced some level of renewal, perhaps at a lower level of complexity themselves, where the forests and woodlands may be gone, but the grasslands have recovered. Mesopotamia civilizations recovered numerous times, as new states were able to benefit from the natural process of desalinization of soils after the hundred years of rest for the soils.

Of course, some of culture may survive and be used by new or other cultures. The ideas produced by a culture may spread and reproduce. The culture of ideas continues through people leaping to other places. Lower levels may survive with many of the ideas of the larger complex. So, the Roman Empire collapsed but not parts of Italy or France. The Germans were able to return to chiefdoms.

5.7.4. *Potential Collapse in the Future*

Elman Service uses a biological analogy, where social organizations are modeled as plant or animal species that are initially successful because they are adapted to niche, but later become overadapted and less flexible. In this model collapse is part of an unalterable natural cycle. Adaptation denotes a systemic homeorhesis in which a range of variability is activated in response to various environmental and social perturbations. Service believed that complex systems were profoundly maladaptive since their responses to stress became less flexible.

This argument does not consider weed species or the maturity of the system. In a systems model, collapse is part of stochastic process, which implies that civilizations will die, but not necessarily within a definite time frame or for specific reasons, such as overconnection. Perhaps one reason for cultural collapse is due to overconnections, between trading groups and social classes. This could be a problem with modern globalization, in the 1990s, as well as with simpler regional connections, such as in the 1300s in Asia and Europe. In the general system model, complex systems are hierarchically composed of many stable lower and intermediate orders, strongly connected horizontally, but less so vertically. The problem may be more one of scale than complexity.

Human cultures tend to fill all available space, carrying capacity space, even though some of the spaces are occupied by other species or other cultures. This makes them prone to crash after sudden changes. They have adapted to marginal environments with behavior and technology, such as that for water storage and grain storage, to buffer themselves against the known changes of the environment, but when an unknown change happens, such as a nine-year drought, they fall apart. Few people have the luxury of moving to an open area.

Global climate change has been established by many scientists as a real threat to the public health and safety of cultures. Examination of Paleolithic records establishes that climate, far from being benign and uneventful, has been dramatically unstable for local or regional patterns. Droughts can start abruptly and continue for tens or hundreds of years, far beyond the ability of the adaptations to them, such as urban living, to cope; agriculture is disrupted and the technological innovations, such as immense grain storage buildings, cannot continue for the duration or amplitude of change.

Now, global warming is related to human activities. Human contributions of carbon dioxide, methane, and water vapor contribute to a much faster temperature rise. Natural factors, such as the eruption of Mt. Pinatubo, may temporarily mask the effect of global warming, as may technological factors, like jet contrails. Still, if the trend continues, global atmospheric temperature increase could develop deserts where croplands exist, dramatic sea-level changes would flood low-lying areas, and shifting rainfall patterns would affect crops and fisheries. Global warming could lead to reversal of ocean currents, unpredictable redistribution of rainfall, and other unpleasant, chaotic, deadly effects.

Any one strategy, such as reducing the human populations—without making changes in distribution, flexibility, ecosystem health, complexity a many other factors—might not let us avoid collapse, but merely postpone it. Reducing the population would relieve stress of ecosystems and reduce the destruction of species and systems. But, if we continued growing economically and connecting tightly globally, then the catastrophe of a single key resource, such as cheap oil, could set off cascading pulses of contraction and collapse. Developing and relying on relatively limitless resources, such as solar power, would reduce pollutions and the energy disruption of systems. But if we keep enforcing historical inequities while relying on a lifeboat approach to saving civilization, that could set off havoc and warfare that would lead to collapse. Reducing and refitting our massive appropriation of ecosystem productivities and services would allow those systems to recover and diversify again. But, if we continue to rely on externalities for free services, without restoring and setting side a majority of the planet for wild systems, the support system may break down and trigger ecological collapses. By becoming less complex and connected, by deciding on a satisfactory level of sophisticated culture, we might have the flexibility to recover from a collapse if we could not avoid it.

Lewis Mumford noted that each civilization began with a living urban core, but ended in a common graveyard of bones and broken pottery. Every complex culture so far has failed, although some like the Romans or Chinese, were able to renew and rebuild subsequent cultures. Collapse has been history, but it is not necessarily fate. Cultures do not choose to collapse, despite what Jared Diamond says; they choose to continue successful behavior in inappropriate circumstances, in spite of the unsustainable costs—the Greeks called this the operation of tragedy. We believe that we can avoid collapse by growing and expanding, even to the point of stealing and destroying natural capital—that just postpones the inevitable.

We can learn to understand the interdependency and scales of nature. We can change our strategies by reducing our size and impact, by decomplexifying our societies and governments. We can accept that our labors will be more intensive and that our luxuries will be smaller. We can balance our global push with a greater isolation and insulation, with self-reliance and self-control.

Because people live in groups having unique cultures, they react differentially to diseases and stresses, even war. In crisis, people tend to pull back to smaller groups. Smaller populations can adapt faster to smaller resources. This was an appropriate response to local catastrophes, but it may not be the best way to deal with global ecological threats or coterminal collapses. Dispersal to other healthier ecosystems is no longer an option for most cultures. The problem is scale and it has to be addressed on a larger scale.

Science can create models that can be used to anticipate future changes—not exactly or definitely, but in terms of probability and eventuality. Models of climate change can map climate change and perhaps even accommodate unexpected nonlinearities. And, models can provide the information to design strategies to give cultures more flexibility in responding to change and challenges. With international cooperation, the risks can be shared and vulnerable cultures can be helped with minimal disruptions.

5.8. *Problems: Culture & Bad Images*

No culture has developed a perfect balance of human and situational needs. Some do better than others, but all cultures change and age. As a culture ages, it may become abstract, indifferent, self-centered, and forgetful, suffering rigid rituals and cultural amnesia. If the contradictions become too great and maladaptive, however, then the culture dies.

5.8.1. *Cultural Transformations*

Cultures emerge from groups of human beings living in places. The variables of the environment influence cultures in a number of ways. For instance, A high Net Community Productivity (NCP) can allow larger annual crops, or, a low number of sunny days can contribute to psychological depression. Cultures have been unique programs, using local materials and ingenuity, for satisfying basic human material and spiritual needs.

Cultures develop over time as the groups change and always seem to grow larger. Cultures are influenced by climate, resources, and of course by human ideas about their places and themselves. There may also be larger patterns in human culture. For instance, reproductive success and overshoot of resources may always occur in the development of a culture. The asymmetry of sex and violence, ecosystem conversion, and limited time horizons are also things that seem to develop.

How does this happen? Culture does not fit together into a perfectly integrated whole. A culture is a loose-fitting patchwork of ideas, relationships and things. In that sense it is parallel to species adaptation to an environment. There are discontinuities and contradictions. Humans can tolerate inconsistency and operate with contradictory beliefs: soldiers fight for peace; ministers save the unborn for starvation. If the contradictions become too great and maladaptive, then the culture disappears.

The mode of operation of nature consists of a rhythm of dissolution and reformation. Often the elements of a culture will simply be rearranged by a succeeding culture. A new culture can only be made from the heritage of the old. The International Workers of the World urged its members to make the new world in shell of old one (cf. genetic recombination) Our survival depends on the capacity to remake the image of the world from within, phoenix-like. The planet is wild beyond our imagining, and we need a wild image that can capture that wildness.

According to most theories of cultural adaptation (or integration or evolution), resistance to change is normal as a cultural process. Groups like the pygmies have specialized to fit the requirements of the environment, successfully. This makes it difficult to adopt other cultural arrangements.

On the other hand, resistance to change itself is an adaptive mechanism. According to Betty Meggers, it works as a successful "cultural isolating mechanism." Isolation remember is what allows a culture to develop in the first place. But, then does it force a culture to become stagnant?

Stability is a characteristic of cultures. We can even state it in Newtonian terms: A culture at rest tends to remain at rest. According to Harding (1960), when forced to act on changes, a culture will only accommodate those changes that preserve its fundamental character. Since cultures have been traditionally self-reliant, resistance is a positive act.

Thought experiment: If we found a way to live in the forest, using solar power and eating only farmed algae in small arcologies, would the pygmies adjust to that?

As place changes, a culture changes. As people change, under the influence of each other and other cultures, a culture changes. Culture does not fit together into a perfectly integrated whole; it is a loose-fitting patchwork of ideas, relationships and things. It is parallel to species adaptation to an environment. There are discontinuities and contradictions.

William Thompson identifies six major shifts in the transformation of human cultures. The first he calls the feminization of primates; females abandoned estrus and became open sexually (200,000 YBP). In observing synchronicity between bodies and nature, women established system of symbols and notation (art 50,000 YBP). Then women discovered that they could collect enough cereals in three weeks to last the entire year, more than a hunter could kill; but it required storage. Agriculture gave a surplus; that and crop failure and excess property lead to excitement of war (Agriculture 9000 YBP). Civilization became the domestication of women, according to Thompson. This emphasized a patriarchal structure (Civilization 5500 YBP).

Industrialization (200 YBP) is really an intensification of civilization, but still an ektropic process. In each case of cultural absorption, there was an attendant process of miniaturization. The forest was miniaturized in clumps of trees; animals were miniaturized in artistic image; time on a lunar tally stick; plants in a garden; women in a household; nature by culture in 1800 (206 YBP) under the glass roof of the Crystal Palace; and now by the new consciousness surrounding the old. Ratio becomes *logos*. Measuring becomes pulling in, and perhaps shaping (*morphos*). Planetization, along with miniaturization (48 YBP), contributed to the idea that the planet was an organism, that it could maintain itself in the environment of the solar system, and that it could balance its atmosphere with living communities. The photograph, from the Moon, of the earth in space, became the symbol of the change.

5.8.1.1. *Patterns & Renewal*

Nature consists of moving patterns whose movement is essential to their being. As a rope makes the knot visible, so the body is a pattern made visible. The body is a movement that maintains a topologically stable pattern; it is a vortex but not the water. The thing, the pattern, is a cross section cut through the movement. The mind is an invisible knot that is capable of recognizing both visible and invisible patterns, that is to say, a rope is not always necessary for the demonstration of a knot. Culture is also this kind of pattern. Culture can be analyzed into smaller blocks; the pattern of the whole organization is reflected at every division in differences of organization on either side of boundary. The wholeness of the character of a culture is reflected at every level. Patterning relates symbolic meanings in the context of a cultural system as a whole. The patterns form another level of meaning that has to be addressed in understanding a culture.

There are patterns of interactions of cultures, which arise out of several possibilities: Indifference, trade, competition, cooperation, conquest, or respect. Some archaic cultures seemed to be limited to indifference, that is, they ignored one another, and to trade. Competition and conquest may have accelerated with the acquisition of territory for agriculture. Cooperation and respect seem to have occurred under some circumstances of trade or unification.

The mode of operation of nature consists of a rhythm of dissolution and reformation. Perhaps this process applies to cultures. Often the elements of a culture will simply be rearranged by a succeeding culture. A new culture can only be made from the heritage of the old. The International Workers of the World urged its members to make the new world in shell of old one (in a way similar to genetic recombination perhaps). Our survival depends on the capacity to remake the image of the world from within, phoenix-like.

5.8.1.2. *Weaknesses of Culture*

No culture has developed a perfect balance of human and situational needs. Some do better than others. But all cultures change and age. As a culture ages, it may become abstract, indifferent, self-centered, and forgetful, suffering rigid rituals and cultural amnesia. Even a culture that fits its adherents and place changes and may become unfit even as the adherents and place also change. Cultures seem to have no limitations of size or kind, although declining mental health may be indicative of some limit exceeded in industrial culture; a culture can grow beyond ecological limits.

5.8.1.2.1. Holding Arbitrary ideas

People sometimes construct their worlds from preconceived notions. Success in one area may become associated with a chance happening, an event that is repeated to continue the success. In this way a maladaptive image of nature can be built. The culture on Easter Island developed over 9-13 centuries, but extinguished itself before contact with Europeans. The people devoted enormous amounts of energy and materials to building heroic statues; other spectacular public works projects included a road network and agricultural terraces. Their culture allowed them to exceed the carrying capacity of the island. Even converting almost all land to human use was not enough, and the population dropped from about 12 thousand to 111. Some primary traditions may work against the conservation of a place; for instance, the Algonquian notion that game animals spontaneously regenerate after death means that there is no reason not to over hunt.

A powerful arbitrary idea, such as the Christian principle of plenitude, can influence many cultures over centuries. The principle states that an intelligent creator gave an earth of unlimited bounty to humanity for their use; this seemed to be confirmed in the Renaissance with the discovery of the richness of stars, microscopic life, and unexplored continents. Modern ideologies have even been shaped by the principle of endless wealth; the economist Adam Smith believed that the real price of anything was just the toil spent acquiring it. Many European cultures would have vanished if they had not been able to leave their exhausted fields for new lands.

5.8.1.2.2. Remaining Indifferent

Industrial cultures desacralize nature. Since the advent of the machine image, the concept of the sacred has been reversed. In the primary view, the familiar was sacred. When modern cultures made the familiar trivial, it became profane. The quality of sacredness was bestowed on the unknown, wilderness, or children. Modern cultures show reverence toward that which cannot be dominated. So, reverence for nature diminishes as control escalates. In industrial culture, all aspects of life become interchangeable artificial units, including soil, water, and land. This view impoverishes humans by claiming all consciousness for humanity. It claims that nature offers no joy, love, peace, or certitude. Emphasis on the emptiness of nature creates a gap between humans and their environment; there is no room for the intrinsic worth of nature. By granting human sovereignty over the entire earth, industrial culture justifies usurping the habitats of plants and animals for coffee plantations and recreational boating.

Many ecological problems are a consequence of populations in excess of those that can be indefinitely supported. Many different peoples have deforested their lands and poisoned their waters, regardless of their religious ideals as Buddhists, Taoists, Moslems, or Christians. Industrial cultures are indifferent to the limits of a natural carrying capacity. Some cultures suffer from a collective amnesia regarding what life was like in earlier times; much of the richness of life is simply not known by later generations. The Ik, for instance, thought their dietary habits were fine, although less than two generations earlier, they ate a greater variety of foods that were more nutritious.

5.8.1.2.3. Overexploiting Nature

All peoples want some power over the natural order. Primary peoples rely on ritual acts instead of machinery. As technology supplies power to primary peoples, rituals decline. Power increases exploitation and interference. Exploitation can become pathological, when it interferes with the natural processes that maintain an ecosystem. The intrinsic worth of beings can become supplanted by monetary value. For example, some North American Indians were seduced into the fur trade by the lure of manufactured materials. The spread of power has two other effects. The natural order becomes simplified, the human world becomes increasingly complex—and both orders become unstable. Human society acts from ignorance of the bonds of living and nonliving beings. Applying culture beyond a small scale gives rise to behaviors that are nonecological and unsustainable.

Distinctions may not be made that could encourage survival in a long-term context. For example, modern cultures have not learned to distinguish between renewable and nonrenewable resources, or between the temporary and permanent carrying capacity of a habitat. Human carrying capacity is part of the natural system. Cultures may also be indifferent to long-term catastrophes, such as species extinctions, or to short-term hazards, such as volcanic eruptions or flooding; people always resettle floodplains and volcanic lowlands.

5.8.1.2.4. Being Incomplete

The very circumstance that makes each culture unique—being in a unique place—ensures that each culture is limited. All cultures produce destruction and waste, all of them produce at least some of the opposite of the good intended. A culture rarely meshes perfectly with the natural order or even its own social order. That a culture includes so many patterns and dimensions makes its fitness less. To the degree that it is effective, any ideology can fit the order of nature. But the total mix of ideologies makes the overall fit very sloppy. As long as nature can be dominated, without catastrophe, the importance of the fit is not critical. But we do not know enough of nature to know when catastrophes occur, nor how to avoid them or minimize them. Nature is unpredictable.

Cultures rarely have long-range plans; they do not concern themselves with global problems. They rarely consider any cultures other than their immediate neighbors; they do not have policies to help them. They are rarely conscious of their activities. Many cultures have little interest in gaining new knowledge on how to exploit their areas more effectively and efficiently. Many cultures have no way to cope with their own expansion or contraction.

5.8.1.2.5. Staying Inflexible

It was thought that cultures could vary infinitely and change rapidly. This is an exaggeration. Change is not always easy or adaptive. The inertia of cultural practices makes change painful. People may become fixed in permanent roles and personalities. Even if cultural attitudes are appropriate, they can trap a people if there are no longer functional reasons for the practices. The Nembi of Papua New Guinea may be trapped in their system; making stone axes is difficult when thousands of steel ones are available. Although the ritual of making axes can bond people together.

Cultures can determine inappropriate attitudes towards nature. The Ik had a string of misfortunes after their hunting ground was turned into a national park. The difficulty of farming and adverse social conditions made their situation worse. The Ik acquired an attitude as victims, characterized by a cluster of new beliefs, including: Nature as alien, unjust, violent, or vengeful; things being better in the past; humans beings are out of place. By contrast, the English treated tropical lands as enemies to be defeated, then enslaved them in plantations. Their cultural attitude as conqueror of nature led them to treat biogeochemical cycles and soil requirements as temporary obstacles in a world where everything had its price. This is the prevailing mode in institutions dealing with land use today: nature is a beast to be tamed, controlled, and exploited. Despite proof of the importance of tropical forests and knowledge of their destruction, corporations still mine them for short-term profits.

5.8.1.2.6. Keeping Exclusive
When the largest social unit was the tribe or nation, it was possible for the local mythology to represent other people outside its bounds as inferior, and the local inflection of human mythology as the one true mythology. The young were trained to respond positively to tribal members, to love their home, and to project hatred outward. But, there is no outward.

5.8.1.2.7. Being Aggressive
Are humans innately aggressive, or does the nature of the culture of civilizations promote aggression? Cultures allow more aggression against people outside the home culture. The size of a local population increased the likelihood of its success. For cultures, size was important. More important cultures were larger and more aggressive. Aggression may be encourage to protect a culture trapped in low food outputs or scarce resources. Aggression would be important to protect a culture, but it can become a prime way of relating to other cultures, especially very different culture, or neighboring cultures that may hinder the expansion of the home culture. Cultures need not be limited to aggression. They can also divert aggression from violence and war into propaganda and art.

5.8.1.2.8. Ignoring the Limits of Culture
A culture will often develop without concern for limits of complexity or scale. How large or small can a culture be? How simple or complex? There are human cultural limits in numbers. For instance, a minimum might be genetic, at only 2000 people; but for ideomass, the minimum might be one million. A maximum, based on wilderness, might be ten million, or there might be a social maximum of forty million. Each limit must be worked out, depending on place and the structure of the culture, but they should not be ignored.

5.8.2. *Challenges & Traps*
Sometimes, rearrangement leads to a position of not being able to rearrange further. In many cases it is hard to tell if the destructive use of land preceded ecological problems or followed from efforts to maintain production after an ecological challenge. Cause and effect are hard to separate. The same environment that challenges a culture with some kind of change, also offers opportunities with the change. New resources can stimulate economic activity and increase the level of living.

Cycles that do not operate with the right kind of feedback function as traps. Thus, on an elemental level, phosphorus becomes trapped in an ocean sink, and can only be recycled by long geological processes or by specific harvests through human activity.

Karl Marx contended that humans live in cages, partly natural and partly of their own making. However, human actions can modify the situation. The word cage is a metaphor; it implies being trapped. It is however, a metaphor that can be expanded with a description of space as a four-dimensional box. Perhaps the trap is a better metaphor, since we depend on nature and society as a foundation for life.

Traps function in different ways. The use of resources by a people, where the replenishment rate is constant and the rate of use exceeds it, is a serial trap. This trap results in ecosystem degradation that is less reversible. The industrial age mistakes the rate of discovery for the rate of recovery.

Agriculture is an energy trap, because it allows a higher concentration of energy, that is, higher yields, but then it requires more energy be put into the system to maintain it. The system has to produce more energy than it uses to be sustainable, with a surplus for trade.

Sedentism is a trap. As the population of sedentary communities increased, the wildlife numbers decreased. The productivity and narrowness of food increased. Thus, there was less possibility of returning to the foraging lifestyle. People became committed to the new lifestyle. Intensity was no longer an option either; it had to be pursued. Habits were set. One problem of sedentism is that the individual cannot simply move away to avoid conflict. People are tied to a particular place and have to communicate to adjust to sharing places.

The city is a different kind of trap, that offers intensity and opportunity, but requires massive imports of supplies to survive. The size and scale of cities create the dual centers of attraction and despair.

The cultural trap is that: One cannot transcend culture unless one knows the hidden structure, axioms and unstated assumptions about how life is lived, viewed, analyzed or changed. Cultures are systematic wholes composed of dynamic interrelated wholes. They are more easily described from outside with comparison to another culture, although transmitting culture to youth and watching the culture collapse expose hidden structures.

Language and art are also traps. Language because one is limited by words. Art because one is limited by styles or demand. Even if cultural attitudes are appropriate, they can trap a people if there are no longer functional reasons for the practices. The Nembi of Papua New Guinea may be trapped in their system, making stone axes with difficulty when thousands of steel ones are available.

Global capitalism can lead to a consumption trap. Capitalism claims to serve the wants of the people, but it spends half its income creating more wants in people. Not many of those wants are real, or as real as cereal and roofs. Few of the soft services satisfy real psychological needs. Markets advance individual desires and not social goals, by offering running shoes, not inner city restoration. Instead of being free from economic want to develop their potential as creative human beings, people are trapped in a consumer cycle. Self-actualization is postponed for self-gratification. Furthermore, capitalism can undermine traditional cultures by offering consumerism in the place of guides for behavior. Social roles seem irrelevant by comparison, if the good life can be bought without effort.

Being in a trap means much-reduced flexibility and fewer choices. Climate can drown whatever is in the trap. That is, being in a trap makes one vulnerable to many other changes that could be avoided if you were not in a trap. When the weather got colder, then hunters and gatherers could move south. Cities could not. Civilizations are more fragile and more vulnerable to smaller climactic changes.

Addictions, such as those to foods or oil or money, make it difficult to escape from a trap, a trap being a kind of energy well or gravity well. Addictions can amplify some emotions, such as fear or hate, especially as they relate to the possible end of the addiction or the threat of that end. Addictions can justify illegal behavior, especially those that seem necessary to continue the addiction. Of course, many cultures are addicted to the illusion of control and power. The U.S. is trapped in the belief that only it, among nations, can bring prosperity and peace to other nations, with trade or violence. Eventually the trap is escaped, or more likely destroyed or collapses with its victims. Paul Shepard suggests that the entire Neolithic revolution has trapped us in behaviors that only end in madness. The feedback is inevitable.

Table 582-1. Agriculture and Population Feedback Sequences

Direction	*of*	*Progress*	——>	——>	|
Concentration	Intensification	Disease	Stress	Decline	Madness
Simplification	Instability	Famine	Drawdown	Destruction	Madness
Territory	Defense	Male dominance	Military	Take-over	Madness
Surplus	Specialization	Technology	Novelty	Stress	Madness
Distribution	Taxation	Inequality	Insecurity	Slavery	Madness
Human order	Abstraction	Isolation	Stress	Drift	Madness
Knowledge	Habit/tradition	Manipulation	Control	Laziness	Madness

Now we have to ask, what happens after madness? Do we die? Do we change and get better? Do we stay the same and destroy everything. How would we act if we were mad? Better than consumers? Can we analyze our way out of the trap?

The cultural trap: One cannot transcend culture unless one knows the hidden structure, axioms and unstated assumptions about how life is lived, viewed analyzed or changed. Cultures are systematic wholes composed of dynamic interrelated parts or wholes. They are more easily described from outside with comparison to another culture. Although transmitting culture to the young and watching the culture collapse also exposes the hidden structures.

5.8.3. *Problems Facing Individual Cultures & a Regional Culture*
Problems of individual cultures include: Lack of Resources; environmental change; conflict and violent relations; environmental degradation; species habitat diversity loss; introduction of exotics species and artificial elements; cultural exhaustion; human limits; and human inequity. Many of these problems arise from other activities, such as: Competition; take-over; cooperation; trade; exchange; clash of images and metaphors; differing ethos or ethics; trends or gigatrends; and random changes or changes in luck. And these activities and problems can result in the problems of overshoot, reproductive success, ecosystem conversion, stupidity, violence, or stagnation.

In Mesopotamia by 3000 BCE, for instance, a stratified class society had developed. Slaves at the bottom. Peasant farmers, craftsmen, then the elites of administrators, religious and military. As problems developed, temporary military leaders became permanent hereditary kings. Palaces were built and staffs numbered thousands. But, the fields could not keep producing wheat and barley. Many of these problems like salinization are long-term problems and do not become evident for several generations. They are also very difficult to reverse. For a society than needs surpluses to continue, with growing dependents and growing people, there is little flexibility to change. The only way to avoid the problems was to let the land be fallow for long periods until the water table fell. This was impossible due to food demands. So, the land and the society collapsed.

Our cultural images, our worlds, do not have a close match to the organization and complexity of nature. Our image of the world has failed; our myths, from progress to science, are no longer effective in dealing with regional or local changes. There is no formal global culture. No global image has been formed.

A regional culture would have emergent problems far more complex than any one individual culture. It could be crippled by hyperadaptivity (success and overshoot), since humans can adapt to poverty, bad diets, crowding, stress, and other unhealthy practices. There might be homogeneity of forms, from a loss of diversity, which could make the global culture more fragile. The global culture might exhibit a lack of flexibility, which could lead to stagnation and collapse. Other possible problems of a global culture include: Hyperpersistence of error (stupidity or violence); lack of consciousness and planning (ecosystem destruction); the limits of capitalism, or any single system, as a global economic strategy; the acceleration of interactions beyond the human social or political ability to keep up with them; and, a breakdown of boundaries and limits, especially ones that control productivity.

5.9. Problems: Erasing Cultures

Our dream of civilization, in modern industrial culture, is the dream of order and beauty. But, as Aldous Huxley notes, the dream of order begets growth and tyranny, and the dream of beauty ends in monsters and violence. Striving for the good life for many has left us with crowded roads and regimented jobs; trying to build beautiful cities has given us gigantic boxes and neighborhood violence. Trying to fulfill our dreams of comfort and security has provoked global threats and local nightmares.

5.9.1. Begetting Problems & Erasing Cultures

To ensure the success of our species, we have appropriated the places of other species. To extend our families, we have increased our numbers and our rate of increase exponentially. Our overpopulation has led to aggression against other cultures and species, then to indifference at their suffering. Even low levels of food and fulfillment can be maintained only through theft from other species, from future human generations, and through the degradation of billions of humans, as well as the ecosystems on which they depend.

To provide for the needs of many and the extravagant luxuries of some, we have produced waste and pollution on a geological scale, from islands of garbage to acid rain. Manufacturing processes result in the production of new dangers, such as recombined genes, and new substances, which are not easily incorporated into natural cycles. The overuse of ecosystems results in deforestation, devegetation, and desertification, then in depletion of raw materials and depletion of agricultural land. Economic and political pressures, derived ultimately from population pressures, force farmers to intensify their efforts to increase crop production, instigating a dismal cycle of population expansion, environmental deterioration, and poverty.

To provide efficiently, we have increased the scale of our activities. But we have decreased the diversity of habitats by filling in wetlands, felling forests, plowing grasslands, and irrigating deserts. Agribusiness has caused widespread landlessness; people who try to grow their own food are forced onto marginal lands. Acquiring fossil fuels also creates landlessness; coal mining in the Black Mesa mountains of the United States, for instance, may force the resettlement of 20 thousand Hopi and Navajo people. Without land, and the economic independence it allows, cultures are more likely to disintegrate.

To achieve even greater efficiency, we have increased the speed of our activities, converting materials and cultures without consideration of the meaning of or need for efficiency. The speed of our economy is too great for many cultures to adjust to; and the thoughtless transformation of cultures may result in great mistakes. The speed of our conversion of wild habitats to domesticated lands is too great for many species to adapt.

To increase trade and claim resources, states worked with business corporations to enclose entire areas and stabilize them with physical force. Now, every piece of land on earth is enclosed and claimed by major states. States have also incorporated traditional tribal territories into their artificial boundaries with little regard for representation. The Oromo in Ethiopia, for instance, comprise half the nation's population, but do not have a voice in government or even title to their lands. In sub-Saharan Africa alone, 450 million people speaking 1800 dialects have been compressed into 50 states.

To increase our economic wealth, we have created a 'global marketplace' in a 'global village.' We have tied together people with millions of telephones, hundreds of millions of televisions and billions of radios. More and more people eat the same foods, wear the same style clothing, and read, watch, and listen to the same entertainment. People are pressured to give up their ethnic identity and kinship for the 'global unity' of humanity. This global culture suffocates local cultures; unique dialects and ways of life are diminished.

Population pressures, resource shortages, and manufacturing "side-effects" cause instability in many societies; militarism, intolerance, crimes, and health problems are symptoms of that instability. Confusion and misinformation contribute further to the destruction of cultures. The instability of cultures, as well as stress, insecurity, and insufficient diets, results in psychological problems for people. Individual powerlessness and disillusion provokes the further disintegration of cultures.

Progress is erasing archaic cultures. There have been great cultural transformations over the past 20 thousand years. Primary cultures, previously called 'primitive or archaic,' regard the relationship of human beings to nature as one of kinship; all neighboring beings fall within moral consideration. These cultures observe the synchronicity between their bodies and nature and understand their culture through mythical explanations. They employ hunting, gathering, and shifting agriculture.

Secondary cultures analyze and deduce the operations of nature; rituals become more stylized. Cultural innovations permit larger human populations; ecological limits are raised by agriculture, although they are not eliminated. Moral consideration is reserved for human beings and sometimes other conscious beings.

Tertiary cultures (in fact, the real meaning of a third world: twice removed from nature), are based on mechanical images that objectify nature. Drastic changes in the production of goods forces other psychological and social changes; human relationships are based on economic allegiances instead of kinship and exist in societies instead of communities. Money becomes a symbolic representation for the value of labor and land, which are considered mere commodities. Social stratification and the specialization of labor become fundamental characteristics. Orders are rearranged during the process of urbanization. Moral order, for example, becomes subordinate to technical order.

This last transformation of culture is not the only one in existence, however. There are hundreds of others, although around 1900 there were over one thousand unique cultures and 3000 languages (roughly equivalent to the number of natural biogeographical provinces, subprovinces, and habitats on earth).

Cultures fail for many reasons. Early cultures had little understanding of their impact on ecosystems. Mesopotamians silted their water supplies and salted their soils; the Greeks overcut and overgrazed the Mediterranean hills. Other cultures were not able to adjust to a qualitative change in size. The Mayan culture probably became too large to grow and distribute food. The Marajo Island culture probably collapsed due to population pressure.

Some cultures, stagnant or senile like Rome, only avoided failure by expansion into new areas—for instance, the European expansion into Africa and America. Often the elements of a culture will simply be rearranged by a succeeding culture into a new pattern.

The images and diseases of secondary and tertiary cultures had immense repercussions on primary cultures. American cultures had no resistance to diseases bred in European cities. Many cultures could not compete with more aggressive groups. Many primary cultures have lost 60 to 99 percent of their populations. The Tasmanians, for instance, lost over 98 percent of their population; this much stress on a culture usually results in extinction, as happened to the Ona and Yahgan in Tierra del Fuego. About one third of the known groups in Brazil were gone by 1957.

Some cultures are simply wiped out. The Herero people in southwest Africa were exterminated as a culture by German forces. The Yanomami and others in Brazil are facing threats from prospectors and ranchers, now. Other cultures subside or intermarry out of existence. The Birale people in southwest Ethiopia have only 89 remaining members—and only 19 of them speak the tribal language, Ongota.

Industrial culture is wrongly considered to be the evolutionary successor to primary cultures and is displacing them rapidly. Scholars once plotted an evolutionary trend of cultural types, from primitive through historic, modernizing, and modern; they speculated that later developments were more adaptive than earlier ones and should replace them. It was assumed that the modern view culminated from earlier ages; thus, the 'superior' modern cultures were justified in exploiting or removing 'primitive' cultures.

There is no evolutionary trend of cultural types from the primitive to modern. Later developments have not proven to be more adaptive than earlier ones; nor do they necessarily replace them. Ethnic groups are not anachronistic stages that point to Switzerland or Japan; they are equally valid ways of living. Any culture is only one of many possibilities, one way of living in a unique place—there is no single correct way.

Industrial culture, depending on its expanding market system, is becoming unstable—worse, it is attempting to become a global system at the same time. We know that cultures can destroy their ecological basis, but we do not know how to extend their existence.

5.9.2. *Redefining Culture*

Much is known about the dimensions of cultures. Culture is everything created by a group, society, or nation, physical or ideal, past or present. Thus, culture embraces cookware, arrows, and turbines; art, books, and legal codes; images, values, and social structures.

Culture is a codification of reality, a symbolic system that transforms physical reality into experienced reality, which can be preserved and transmitted through many generations by language. The uniqueness of the place of a culture is exhibited through language. The Inuit may have over 17 different words for snow, while the Taureg may have 11 words distinguishing the kinds of blown sand—and Americans recognize 94 shades of lipstick and 22 kinds of road surfaces. Language is not only a medium to express thought, it is a major element in the formation of thought. Different languages recognize events differently, therefore no culture can be considered apart from language or entirely apart from place.

The difference between cultures is not due to the number of phenomena taken into account; it is due to a difference in the basic postulates of thought. Nor is the difference really a matter of truth or falsity. Truth and falsity are less important than relevance. Primordial water is no less true than six-dimensional phase space or primordial *ylem*. Each group views and reconstructs the world through their experiences and values. Codifications are true to reality; they represent different facets, not exhaustive catalogues. The inconvenient truths of one culture, like the germ theory of sickness or the vengeance of Coyote, are usually disregarded when they conflict with the direct beliefs of another culture.

Although the capacity for culture is innate, culture itself is learned. Culture provides the framework and boundaries of reasoning as well as the basic ideas of discourse. Culture provides kinship rules, obligations, and entitlements, the values that determine behavior. It also defines the means of production, kinds of livelihood, and the ways of distribution. Culturally determined values and ideas are the basis of political behavior; these ideas determine the kind of institutions and their ability to change.

Culture is an active way of living for a group of people in a unique place, a set of behaviors that allow survival in one place. Culture is not simply a matter of territory or human ancestry or names; it is a living organ, like human skin, that allows the interpenetration between the human and the natural. Every culture has strengths and weaknesses, however, that may color or distort human perception, or limit or expand human adaptation to the environment.

5.9.2.1. The Strengths of Culture

A culture has social and ecological functions. The image a people have of their world makes sense of the overwhelming confusion of nature; it gives people a unique identity and justifies their behavior. Culture ties them to a place, whose qualities are known and preserved. The strengths of a culture have to do with a common image, which orders experience, personalizes a place, offers identity and security, and justifies human actions. Culture can adapt to place and preserve its health.

5.9.2.2. The Weaknesses of Cultures

Culture does not fit together into a perfectly integrated whole. The balance of freedom and necessity ensures that no culture will ever fit its environment or its members perfectly. A culture is an incomplete, loose-fitting patchwork of ideas, relationships, and things. There are discontinuities and contradictions. A culture can have an arbitrary idea, that may let it overexploit some resources or ignore a long-term trend. Due to habit or tradition, a culture can stay inflexible or exclusive—or remain indifferent to serious problems. Many cultures have strict hierarchies that limit the mobility and development of their members (see Figure 5922-1). Humans can tolerate inconsistency and operate with contradictory beliefs: soldiers fight for peace or ministers save the unborn for starvation. If the contradictions become too great and maladaptive, however, then the culture dies.

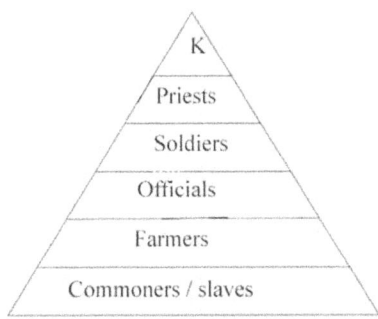

Figure 5922-1. A common hierarchy pyramid, from king to slaves

5.9.4. *Summary: Nightmares & Wisdom*

Our dreams have become nightmares, which are symptoms of rotten images and unbalanced ways of living. Our ecological, social, and political problems do not have simple technological solutions; a single industrial culture cannot solve all problems in every place. The problems are cultural and polycentric.

Industrial cultures have two great myths, progress and nationalism, originally successful, that are reducing our fitness to the environment. Progress has been described by Aldous Huxley as the theory that one can get something for nothing, that the gain in one field is not even paid for in another. Progress assumes that all consequences could be foreseen and that ideal ends justify any means. Primary groups, it was thought, only obstructed the march toward paradise and could be murdered or assimilated if necessary. Nationalism is the theory that one state is the only true god; all other states are considered false. Conflicts over prestige or power, as crusades for the true state, still lead to human and environmental destruction. The cure for progress lies in the responsibility of small nations for their own environments; the cure for nationalism is a holocultural framework of small nations.

Nations have embraced a dominant industrial culture for the benefits they perceive that it gives them. But this monoculture tends to displace local cultures, such that the local knowledge and traditions—the ways of living in unique places—are lost. Nations have tried to assimilate their indigenous cultures. Are nations going to be able to continue under catastrophic changes?

6.0. Designing Domiture

The early discoveries of science seemed to confirm that there was a real dualism between humanity and nature. New scientific research in quantum physics, psychology and ecology, focusing on anthropogenic effects on nature, suggest that the perceived dualism is erroneous. Nature is no longer a separate realm that we can look at from outside. Although the old dualism is a common misconception, human actions are embedded within human-dominated matrices, which are embedded with massive biogeochemical cycles and wild matrices.

Our level of interactions in many places suggests that human processes are significant drivers that affect the function, organization, and composition of many ecosystems (Turner et al. 1990). As the degree of our influence on the entire biosphere increases, especially interference with global cycles, we need to work to fit our actions into the limits of those cycles and systems, or risk having them collapse and require human restoration and control.

6.0.1. *Nature Itself.* G. Spencer Brown understands a much wider concept of the self. In describing the conception of form, Brown notes that the self constructed boundaries in order "to see itself". But, in order to do so, it must divide into one state that sees and another that is seen—it must become distinct from itself. In this sense, the world has divided and subdivided itself. Whenever another division is made, a self—Brown says a "universe"—comes into being. The skin of an organism only cuts off an inside from an outside. But, the skin is permeable.

The earth has innumerable modes of being that are not human modes. Our direct intuitions of nature tell us that the earth is infinitely strange; it is alien, even when gentle and beautiful. It seems often mysteriously impersonal, unconscious, immoral, hostile, and awesome. J.B.S. Haldane recognized the strangeness of nature. "I have no doubt that in reality the future will be vastly more surprising than anything I can imagine. Now my own suspicion is that the universe is not only queerer than we suppose, but queerer than we can suppose." Perhaps the queerness results from sheer complexity. George Perkins Marsh believed that the equation of animal and vegetable life was "too complicated a problem for human intelligence to solve, and we can never know how wide a circle of disturbance we produce in the harmonies of nature..." Barry Commoner echoes them both: "not only is nature more complex than we think, but perhaps more complex than we can ever think." In its immense complexity, nature seems wholly other, nonhuman, or ultrahuman. Nature consists of moving patterns whose movement is essential to their being. The holomovement enfolds and unfolds in a multidimensional order that is undefinable. Nature seems distant and unknowable, so it is feared as unfathomable and uncontrollable. Nature seems contradictory and sinister, shaped by death, which we fear. We fear to understand, to be compassionate. So, we try to dominate and control nature, to overwhelm it before it can overwhelm us.

6.0.2. *Culture Itself.* Culture as a filter to keep the details of nature from overwhelming us. Culture feeds back into what nature is, as ecosystems and species. As people learn a language, and as they learn from the collective store of memories and experiences, they contribute to their own change and the historical change of a culture. People, like termites, beavers and birds, modify the environment to improve their chances of survival. The modifications and changes become cumulative, so that houses change, ways of generating energy change, without having to be reinvented. Our general cultural ability, rather than specific biological improvements or specific adaptations, give us a greater survival potential. The accumulated cultural knowledge and their meanings as they are transmitted, is a semantic environment. The limits of this environment limits how we can design future environments. This environment needs to be redesigned before the overall environment can be redesigned. When experience is distilled in norms or laws, these secretions, as Anatol Rapoport calls them, change the semantic environment, in a way parallel to toxins or nutrients in the biological environment, encouraging or inhibiting actions.

Most cultures are multiracial and multilingual. Culture forms an integrated design. Elements are fit into the design and related. A traditional culture integrates these relationships in a coherent whole, according to cultural values. By virtue of its integrative potential, culture provides an ideal framework for public and private decision-making. Order provides stability and security. Cultural order is necessary to deal with the redistribution of wealth and power. Furthermore, justice, ethics, freedom, and truth are based in culture. The study of cultural ecology—human adaptation to nature—has been neglected. The dilemma of ecosystems in the planet reveals the extent of this neglect.

Our minds are not only nature dependent, but culture dependent as well. As the wind, trees, and birds are sources of signals and symbols, so are gestures and words. A community is a place where a group of residents share ways of experiencing or the same experiences. This enables an individual to go beyond a finite view, to see the embedded culture as one of many ways of relating the self to the universe. Culture evolves from the interactions of humans with nature; both are in a constant state of flux. The development of cultures may be thought of as parallel to the development of species, where memes or ideas in a culture have the similar function of genes in a body.

Nature needs a place to play, without being controlled or converted. Culture needs a place to play, without being forced by nature. Nature is an extension of the field, Culture is an extension of nature through human existence.

6.1. *Domiture: The Coevolution of Nature & Culture*

Domiture is the entire field of extensions as a unit of study, rather than the reification of something new, perhaps divided among the studies of natural history, ecology, human ecology, and cultural ecology. Domiture, a neologism made from the Greek word fragments for 'home' and 'again,' could do that. The system of culture is embedded in nature; domiture is a larger term to enclose the previous nature/culture dualism. It has to include human reason and human emotion, rather than simply putting them on opposite columns, as with order and chaos, higher and lower, linear and cyclic, as well as agriculture and wilderness. Nature and culture need places to play. Domiture envelops both concepts, giving them room to play.

Nature and Culture are systems. Domiture is the combination of those two systems. Culture was once called a "Second Nature," but human culture has expanded so dramatically that the two systems are better identified as one hybrid wild/developed system, now. The fitness of human systems are intimately related to the fitness of species and natural ecosystems. The human attachment to place is critical to understanding why people live where they do.

Domiture is the system of culture embedded in nature; it has to be a larger term to enclose the previous nature-culture dualism. It has to include human reason and human emotion, rather than simply putting them on opposite columns, as with order and chaos, higher and lower, linear and cyclic, as well as agriculture and wilderness.

Even where the culture may be good, the scale of the culture results in conversion of wild ecosystems into limited human ecosystems. This is part of a large-scale problem of interference, not just in artificial systems, but also in wild ones. Landscape change is about intertwining ecological and cultural processes. Because we are humans and have cultural ideas and images, and because we have impacted ecological systems for ten thousand years, culture is part of any restoration. Environment is the natural; the landscape is cultural. But, it's the same thing. It is not environment or landscape—It is domiture, which includes all agriculture, cities, geospheres, and wilderness.

Table 610-1. Changes between Culture and Nature over Time

Humans as part of environment	Second nature is equal to environment	Second nature exceeds natural
Wild ecosystems	Agroecosystems	Urban ecosystems
Natural environments	Cultural landscapes	Built environment
Dependency on natural environment	Modification of the environment	Major impact and control of environment

Stewart Brand thinks that the levels of a healthy society move at different rates. Each operates at its own pace, with the lowest and slowest sustaining the others. Culture moves at the pace of the "long now," according to Brand, at the pace of language and religion. Nature moves slower than culture. Culture is the work of whole peoples.

The study of domiture requires a 'Pan Ecology,' that is human ecology combined with ecosystem ecology. Design has to consider the rates of change of culture and nature, as well as those of emotion and technology. Nature evokes feelings of beauty and terror, joy and sadness—we can observe the wild or feel rewarded by the experience of nature. Nature is fun. It invites play. Ecology can be fun. It invites play. No matter how tiresome or frustrating ecology as a science can become, there is always the potential to enjoy it. Who cares if nature is not natural anymore? It is only human words and ideas. Nature is healthy, and that is what makes us healthy. Who cares if culture is claims dominance? Hurricanes, tsunamis and other events are agents of cultural humility.

Nature becomes more complex through filtering, combining, dividing, and mixing. B. Mandelbrot suggests that nature is not made of three simple dimensions. Things can have 1.63 or 2.17 dimensions. Regular simple dimensions are imposed on nature by culture, which has its own odd dimensions of complexity. Euclidian nature is a fantasy. But, nature has gotten more playful with fractals. But, fractals too are necessary fictions. Thus, every form of knowledge becomes a human fiction. Still, that fiction is based on the 'wisdom' of the wild planet, and we are compelled to ask questions: Does nature exist? What is the role of humanity in the destiny of the planet? Who knows? Let's play.

6.2. *Practicing Ecosystem Medicine*

Medicine has made great progress in the past hundred years, especially with the invention of machines that allow noninvasive examination of virtually every part of the body and the discovery of drugs that can control moods or modify diseases. It is possible to transplant malfunctioning organs or sets of organs. Some diseases seem to have been extinguished; others have been controlled. The understanding of diseases has been extended to psychological dimensions and to social contexts. Some medical practitioners and researchers even talk about removing death—as if it were a disease—or cloning healthy replacement body parts or whole bodies. Medicine, in combination with advances in hygiene and food technology, has consistently extended the average life span.

But, medicine is not perfect. Some diseases are making a comeback, and new diseases are starting to appear regularly. Some of the new diseases transfer from wild animals to humans, and a few from humanity to the planet, such as greenhouse fever or ozone loss. Some cures, such as chemotherapy, seem to cause more damage in the long run than no cure at all. Some medicines, or at least their effects, including packaging, chemicals, and testing cycles, are degrading human and wild environments; medical toxins, such as mercury, are killing plants and animals. Some treatments drain individuals and entire insurance companies with their costs—and some of the extreme costs are passed on to all individuals, healthy or sick. Some procedures, especially many kinds of elective plastic surgery, have no medical purpose at all. The distribution some medical procedures is centered on those rich enough to pay the recognized costs. Modern medicine has seemed to reach some invisible limit.

Alternative kinds of health care are attempting to address these problems and limitations, but they also seem to be limited to the exclusively human dimension. Public health emphasizes prevention for individual and communities. Other crisis environmental sciences, such as conservation biology and ecoforestry, try to address the unwanted effects of the medicine, but are unable to influence the causes.

A new field of inquiry is needed to focus on the health of the entire system, and to reconcile the care and health of ecosystems, populations, communities, and individuals. Using the term first used by D. J. Rapport, this field can be designated Ecosystem Medicine. This field will be contrasted with traditional medicine in the following pages.

6.2.1. *Medicine and Its Brief History*
The Greek physician Hippocrates theorized that illness was a natural biological event and that the body could heal itself naturally, in time. The Hippocratic corpus focused on prognosis rather than diagnosis but, it insisted on observation and reason as the basis for any actions. Hippocrates suggested that there was a relationship between the occurrence of disease and the physical environment. Hippocrates' insight was that health can be produced and maintained by natural elements, such as hygiene, diet, mental balance, physical conditioning, and a supportive home. Health depended on being in harmony with these things.

Greek medicine was typified by the notion of well-being. The goal of medicine, like Greek philosophy, was to establish well-being (*eukrasia*) or happiness (*sophrosyne*). Happiness was conceived as harmony with nature and self. Democritus wrote that by using the principle of harmony (balance) we can attain calm of body (health) and calm of soul (happiness).

Over the thousands of years since then, medicine has changed and developed. The entire history of medicine can be compressed into six stages.

1. Traditional medicine, which is common-sense and faith-based. Things have to be bal-

anced in nature. Sickness results when things get out of balance. A shaman or physician can return the balance with actions and medicines. People are expected to limit themselves as the shaman or physician tells them.

2. Rational medicine, as the result of observation. The Egyptians noticed that brain swelling could be reduced with trepanning. The Romans noted that people living near swamps got sick, so they drained swamps. The reason for the engineering was common sense, without medical or ecological knowledge of the cycle of mosquitoes and microbes (and with the belief that the environment is something outside that can attack the body).

3. Scientific medicine (from the 1860s), where the body as a machine can be fixed with surgery, amputation, pressure release, balanced with medication, stopped with radiation, or modified with other tools of the trade. It is based on the chemistry and anatomy of the body.

4. Psychological medicine. The body can be harmed by bad thoughts and improved by good thoughts, that is, the mind shapes the body as they both age. Some diseases are psychosomatic and some cures are psychological, based on the functioning of the brain within the body.

5. Social medicine. The body can be harmed or helped by the thoughts of other people. Shunning can kill, as peoples as diverse as the Inupiat and Amish have found. The love of family and friends can heal. But, of course, prayers can help, that is the thoughts of others can help individuals recover, sometimes, regardless if the individuals even know about the prayers. The effectiveness of this is based on the functioning of human consciousness, e.g., the use of the brains, in groups. There are scientific hospital studies that show the efficacy of prayer on distinct groups of patients.

6. Ecological medicine. These first five areas are defined by different levels of human consciousness or understanding. The environment is still considered an outer thing that fires its insults in towards innocent humans. The environment is not outside, however; we are embedded in the environment, which can direct feedback into physical, mental, and social dimensions. But, our consciousness of the earth still can manifest itself in our modified and artificial ecosystems, and through the degradation and destruction of wild ecosystems. Our love or hate for specific places can influence the state of those places. This is what many call spiritual growth, where the consciousness transcends its limits. People can create ways to avoid painful personal or ecological experiences. People can create a harmony that unites living with the living of every other being in the network. People can cooperate with other people and flow with natural processes.

Despite the history of medical thought, many forms of medicine still exist together and still work simultaneously within a culture. General medicine, as per Hippocrates, addresses the healthy person. Reactive, or scientific, medicine addresses diseases and illnesses. Preventative or alternative medicine is concerned with environmental causes that can unbalance a healthy person. Holistic medicine considers the spiritual and community effects on disease and health. And, ecosystem medicine recognizes that human health ultimately depends on the health of the environment.

Medicine exists because people become ill. Illness is a subjective state where a person detects a change in harmony, and suffers an altered state of existence. There may or may not be demonstrable pathology. In medicine, a disease is identified through a syndrome of patterns. Medicine has a long history of trial and error in identifying patterns, but it has built a foundation of science (hierarchical knowledge), now, with advances in cellular and molecular biology.

Medicine, as what doctors do, has been called a science, humanely practiced. It has also been called an art. Yet, it is neither an art nor science, but an intermediate discipline that combines theory and practice; the theoretical knowledge of science and symbols, as well as the practice of individual experience, is combined with an altruistic commitment to the patient.

According to Pellegrino and Thomasma, medicine is the cognitive art applying science and persuasion through a complex interaction in which a mutually satisfactory state of well being is sought, and in which the uniqueness of values and disease, in the institution in which care is delivered, determine the judgments made.

Clinical judgment is the central element of a theory of medicine. The clinical event is a key feature showing why medicine is neither science nor art. Medicine is a discipline of practical experience arising from clinical interactions, e.g., interhuman events. The ingredients of a medical event include: etiology, responsibility, trust, and decision. Medicine bridges the scientific study of symptoms and diseases as well as human values. It creates a tension between the abstract and the concrete.

Ivan Illich pointed out that people assume something is wrong inside them when they respond to a difficult, intolerable environment. Self-care and modern technology can be mutually supporting. But, people have to work to resolve their problems in the whole network and not just give themselves over to machines. Illich suggested that convivial institutions are better than manipulative ones for education and healthcare.

6.2.2. *Extensions to Medicine*

By 1997 almost half of adult Americans had visited practitioners of alternative medicine at least once, if not regularly. This social trend is occurring because alternative practitioners are willing to acknowledge the necessity of treating illness within a larger context, one of meaning and spirituality, and not just the symptoms.

Alternative medicines seem to be better at managing disease prevention. C. Everett Koop, in the 1988 Report on Nutrition and Health concludes that dietary imbalances, which are preventable, are the leading contributors to premature death in the US. The Center for Disease Control calculates that 54% of heart disease, 49% of atherosclerosis, 37% of cancer, and 50% of cerebrovascular disease could be eliminated by lifestyle modification.

Alternative medicine recognizes an intuitive dimension for healing systems and people. Everyone who is good in their field works sometimes by intuition, which is the integration of knowledge at a subconscious level, involving more than just senses and technical expertise.

6.2.2.1. Definitions of Health

Human health has been studied for thousands of years. Our first clue to a lack of health in a person is abnormal behavior—not moving or breathing, for instance. The first thing a doctor does is to classify the symptoms and measure vital signs: heart, blood pressure, temperature, and maybe blood chemistry. Then she makes a diagnosis and verifies it with more measurements. Doctors, who often rely on their long experience and learning, then discuss a prognosis and prescribe a treatment. So, a first definition of health is the continuity of normal behavior, of living.

There are problems with definitions, however. The problem with health is that it is such a general concept—it is a judgment without an operational definition. Since the human body is complex, it is hard to determine its health. Stress and pathology are normal parts of living systems, as are their agents, such as bacteria or viruses. Furthermore, not enough data exists about what levels of stress or pathology threaten the health of individuals. Since extremely sick individuals display continuity, for extended periods of time, further consideration is necessary to define health.

Health can be redefined as the absence of disease. Health has an exact scientific definition in medical science—it is defined as the absence of disease. In this traditional way, *via negativa*, health is not having cancer, infections, high blood pressure, diabetes, or other ailments. The opposite of health is disease. But, what is a disease? Many diseases, at some stage or concentration, obviously compromise health. But some human diseases, such as sickle cell anemia, make one healthier in the sense of being able to resist another worse disease, G-6-PD deficiency (glucose-6-phosphate dehydrogenase), a genetic disease that causes red blood cells to dissolve.

There are problems with redefinitions, also. Absence of disease is not adequate as a definition, unless we can measure every kind of pathogen. People who seem not to have diseases are termed healthy, yet, there are some people who do not seem healthy, although no disease can be found. The definition of health, according to the World Health Organization (WHO), is "a state of complete physical, mental, and social well-being and not merely the absence of disease and infirmity." This definition of health is the condition of being sound in body or being well. The human body is a dynamically changing being. Human medicine itself is changing from being disease-driven to wellness oriented, from focusing on symptoms to describing the properties of health, such as interconnectedness and self-realization. We could define health as the potential for recovery after collapse or attack.

Health can be redefined again in terms of vigor and resilience. Too much stress, for example, leads to unsustainable patterns of behavior; continuous stress leads to a breakdown of processes that becomes irreversible—the system dies. The imbalance is a form of stress. There is a direct link between the reactions to stress and the body's ability to regain balance. Hans Selye identified three stages of stress: Alarm, resistance, and exhaustion. Each stage causes a different response in the body.

The basic medical definition of health used to be freedom from disease. Part of a new definition of health is resilience to stress—of course, not all stress is harmful; there is good stress (eustress) as well as bad stress. A system may require a certain amount of eustress to be healthy (athletic competition, lovemaking, and childbearing all can cause eustress for instance). Stress may be related to the rate of change for the system, in addition to loss or gain of components or changes in structure.

After Illich and Dubos, the British Medical Journal redefined good health as "the autonomous personal capacity to master one's conditions of life, to adapt one's self to the accidental modifications of one's surroundings, and to refuse if necessary environments that are not tolerable." To be considered healthy, a body has to be stable, to maintain its structure and metabolism (rate of energy use) despite occasional stresses. Health is a dynamic measure of system organization, vigor, and resilience. Organization is described by diversity and connectivity; vigor is related to the amount and speed of productivity; and resilience is a measure of reaction to stress. Since resilience is only one form of stability, that component should be an additive function.

There are further reconsiderations of redefinitions: Until we understand how people change and move around their physical and imaginary landscapes, we will not know which changes are important and inevitable and which are the unhealthy result of human imbalance. Until we understand the changes, we will not be able to adjust our needs to the limits of supporting ecosystems.

Health can be redefined again as harmony and wholeness. The body reacts to as well as shapes its world. The living body selects harmony, that is, health, as a goal. Health is a value of living bodies. Health is the coherence of the pattern of living in other words. If the pattern is disrupted, the local entity dies. Many local patterns flow together through time interdependently, sharing materials. The death of one pattern sometimes leads to the death of other patterns. The body of a human or any entity is a dynamic pattern supported by dynamic processes that include other entities.

The disease event throws one's perception of health out of balance or harmony. The physician helps to restore the whole. In Chinese medical tradition, the highest good is harmony, especially social harmony, or good relations. A good person is one who creates and maintains harmony. Perhaps this is a good working definition of health. But, it requires inclusion of a healthy environment.

Health can be further qualified as harmony in a whole context. Health is a quality that is grounded in the total order of the environment—or implicate order. Health is a dynamic quality of the entire movement of the environment (holoverse) as it flows. As organisms sometimes interfere with others or with the flow of change, the harmony breaks down—we call that disease. Health is the dance of bodies that interpenetrate (in Paul Shepard's image). None of the bodies are completely independent or completely bounded; they are interdependent and open systems. A body is only maintained by a flow of energy and materials from its environment—much of this flow is in the form of other entities, usually much smaller, such as prey, insects, bacteria, and viruses. Harmony is a constraint of the whole system.

Human health has been the focus for many thousands of years. The concept of health has been applied to domestic and wild animals more recently, and now to any complex system, such as a forest or ocean, or even to the planet according to James Lovelock). Applied to large whole systems, health is an integrative concept that can address the well being and functioning of life forms and environments. It also highlights the dependence and intertwining of living beings at home in living ecosystems.

Signs of system health include the homeorhesis of the system (a similarity of flow, rather than the stable state of homeostasis, after Conrad Waddington), the stability of the system (that is, its resilience after stress, such as infections), the diversity of its components, the continuous recycling of elements, and flourishing. Health is the overall ability of a system to maintain itself under a normal range of environmental conditions, which may include extreme temperatures or lack of water.

The relationship between environment and health is integral. The environment is a critical factor in human health and disease. Health, being a complex harmony of the environment, society and individuals, requires that the ethic of medicine must consider individual, social, and environmental good—all are rooted in living bodies.

6.2.2.2. Extensions of Medicine to the Community

Medicine already has to consider the whole spectrum. In disease, the harmony of the body is disrupted by some event through the environment. The disruption causes a symptom, of some kind, which becomes objectified as it, the disease. But, the disease is a conceptualization of the disharmony of the patient's world, that is the self-image and human image, and that image is partially defined by the human community and the larger environmental community, that includes movements of animals, plants, and microorganisms. Medicine has to consider multiple etiologies of disease, from environmental causes, to human social, genetic, somatic, and psychological aspects.

Modern medicine already considers epidemiology (or population medicine) "the study of the distribution and determinants of disease and injuries." The goal of epidemiology is to limit disease through early intervention, controls and treatments. These actions are performed through a community—a group of people with a shared location, environment, and fate. Community health is the private and public efforts of individuals, groups, and organizations to preserve the health of those in the community.

Community health education emphasizes the importance of preventative health care at the community level. Their basic function is to give people facts about health, causes of disease, and methods of prevention, so that they can act for their own well being (and for that of their families). There are established strategies for community health: Establish a board of health, collect statistics, perform research, implement sanitary measures, then preventive services, health protection, and health promotion.

There are shifts in roles and continuity. Physicians offer advice for individual patients in an office. Community medical doctors are advocates for individual patients by communicating with employers, public health agencies or other agencies; they may advise and educate citizens, colleagues and leaders about environmental health.

6.2.2.3. Extension to Environment: Environmental & Conservation Medicine

Environmental medicine usually refers to—for instance, in the 1988 Institute of Medicine report—the diagnosing of and caring for people exposed to chemical and physical hazards in their homes and workplaces through media such as contaminated soil, water and air. This definition emphasizes nonlife-style environmental factors, but does not omit traditional lifestyle factors such as diet, smoking and drinking alcohol. Environmental medicine has been considered as dealing with the impacts of air, water, or food contamination on the well being of people and not much more (see Table 6233-1).

Emerging infectious diseases cause novel complex problems, especially if diverse organisms are perceived as threats to humans. Emerging diseases are either infections newly expanding their territories or new creations. Viruses can exist as latent or persistent sources of infection in a population.

Table 6223-1. Possible hazards to the whole system

Site	Earthquake, flood, wind, drought
Physical	Radiation, vibration, forces, abrasion, humidity
Chemical	Toxins, poisons, allergens, irritants
Biological	Microbiological, vegetative, insect, animal
Psychological	Stress, discomfort, depression
Sociological	Overcrowding, isolation

To some extent human disease has always been the result of the reaction to environmental stress, certainly from microbes and other things, but also from the failures of agriculture, sanitation in cities, deforestation, and the use of exotic chemicals. Population size and structure are also critical factors in infection susceptibility. Medicine is more sophisticated and accomplished, but environmental degradation and the decrease in biological diversity—and the stresses from them—are increasing rapidly, resulting in new diseases and new problems, that is, new stress on individuals and populations.

There are already in place city, county, state, and federal laws regarding sanitary standards (also for the disposal of hazardous materials). These laws are responsible for five human resources: air, water, food, shelter, and waste. Human sanitarians are responsible for controlling, preserving or improving environmental conditions so that human health, safety and well being are maintained. Environmental medicine is an anthropocentric medicine to help people overcome problems from environment. It has a bias for human welfare without regard for cost to other species.

There are examples of relationships of environmental conditions to human health (and the discovery or classification of diseases that probably existed previously but were misdiagnosed). Health has started to take account of the health patterns that result from ecosystem changes. New infectious diseases and cancers occur. For example:

6.2.3.3.1. *Lyme disease.* The white-tailed deer and white-footed mouse thrive in an impoverished landscape where large predators no longer select out sick animals. The disease gets magnified when the vectors, ticks, hitch up the spirochete to humans.

6.2.3.3.2. *West Nile Virus* (WNV). WNV is a virus that is maintained in nature through biological transmission between susceptible vertebrate hosts by blood-feeding arthropods, such as mosquitoes or ticks. The virus is maintained in a complex life cycles involving a non-human primary vertebrate host and a primary arthropod vector. These cycles typically remain unknown to humans until humans interact with the natural focus or the virus chooses human hosts as the result of some ecological change.

WNV was first isolated in the West Nile District of Uganda in 1937. The ecology was characterized in Egypt in the 1950s; it was first noted in Egypt and France in the early 1960s, and it first appeared in North America in 1999.

Humans and domestic animals can develop clinical illness. The most serious manifestation of WNV infection is fatal encephalitis (inflammation of the spinal cord and brain) or severe meningitis in humans and horses, as well as mortality in some domestic and wild birds. WNV has been a significant cause of human illness in the United States in 2002 and 2003.

6.2.3.3.3. *HIV/AIDS.* Human Immunodeficiency Virus (HIV infection) causes an Acquired Immune Deficiency Syndrome (AIDS) in monkeys and human beings. The virus appeared to be resident in wild African monkeys. Possibly after decades of forest cutting, the virus was able to use humans as hosts. The first attested date for AIDS was 1959. In order to trick the human immune system, HIV can evolve new molecular pieces of the microbe, called antigens, within an individual patient.

6.2.3.3.4. *Hantavirus.* Several people died during the Hantavirus pulmonary syndrome outbreak in 1993 in the southwestern United States. The summer of 1992 was rainy, leading to an increase in the crop of pinon nuts, which lead to an increase in the rodent population. Various species of rodents have carried (presumably for a long time) this strain of Hantavirus; cases may have occurred previously but been unrecognized medically. The larger reservoir population of rodents led to closer contact with human populations and the emergence of the Hantavirus in 1993 in the Southwest.

Subtle environmental changes, after a delay, affect the spread of disease in humans. This spread may be accelerated not only by exploration and destruction of tropical rainforests (where HIV may have emerged), but also by the increase in the amount and speed of travel and commerce (which aided the Asian tiger mosquito, a recurrence of cholera, and others). Also, large human populations allow a disease to shift to another local area and to avoid dying out.

Culture is intimately intertwined with health. Culture provides the mode for pain and suffering. It provides rules for sickness and death, as well as compassion for others suffering. African customs of polygamy and extended families provide a larger pool for HIV and hastens its spread.

The health of ecosystems is also intertwined with human health. Since 1996 a group of Canadian and Peruvian researchers has been developing an adaptive ecosystem approach to human health. They used two phases: A conceptual framework that combined ecosystems as complex systems with secondary data and fieldwork, and an application to understanding and improving a problematic situation in the Peruvian Amazon. The goal, however, seems focused on human health. They included natural resource use.

The term "conservation medicine" was introduced by Koch to describe the broad context of human health. Tabor says that conservation medicine "studies the multiple two-way interactions between pathogens and disease on the one hand, and between species and ecosystems on the other." He says it combines human health, animal health, and ecosystem health (shown as 3 intersecting circles equally). But, the application seems to narrow to human health. Michael Soule says "conservation medicine is the right medicine." It has more expertise, more knowledge, more research, and more cures—and possibly more understanding and prevention.

6.2.4. *Ecosystem Medicine*
Medicine has gradually extended its attention to wider domains, from the personal health of individuals to the health of human communities and the health of those communities in an ecological matrix. It will be concerned eventually with the health of environment (ecosystems).

The next step is to integrate ecosystem medicine with environmental medicine and ecosystem science. This means starting by creating a body of cases, then, evaluating the responses to the cases, and finally using that body of experience as a predictive for future treatments. A short discussion of ecosystems is necessary first.

6.2.4.1. What is an Ecosystem? Characteristics and Definitions
An ecosystem is a set of communities of species, using the flux of energy and matter in the same place and time, that is the smallest unit capable of recycling the elements of its membership. An ecosystem is also a topographic unit, a volume of land, occupied by organic beings, extended over an area and through an extended time, with connections to larger mineral, chemical, water and air cycles.

The new paradigm in ecology is the "flux of nature," which replaced the balance of nature, a homeostatic balance, with a homeorhetic balance, the flux. The ecosystems, in general are: open to external forces, do not maintain stable point equilibrium, have directional change as a result of stochastic forces and partial intention by itself, are influenced by both history and current composition, are adapted to natural disturbances, and have adjusted to small scale human influences for thousands of years.

6.2.4.2. Ecosystem Health
Concepts of health applied to organisms can also be extended to communities and ecosystems. Both are complex whole systems with parts and functions. Of course, ecosystems are more complex than humans, which is also why we have trouble measuring social or cultural health (there is no standard society or standard ecosystem as there is a "standard human"). That complexity means we that should be measuring a large number of variables, starting with soil depth (richness, compaction), then annual nitrogen uptake (often related to leaf litter), trophic flows, species counts, and patterns of activity.

Norms for humans and animals are more definable due to the large numbers and the long history of observation. Ecosystems often have only one sample, if they are very unique. This means that there is no population and no possibility for replication. Therefore, the use of analogy for individual systems may not be very productive.

Life energy, according to the physicist David Bohm, belongs to the implicate order, the unseen totality that underlies our reality. Health is the harmonious interaction of every pattern/flow that is part of the implicate order. But, since flow and order (in the holomovement) are imperfect and uncertain, perfect harmony or perfect health is not possible. Breaks in harmony appear as disease. Blockages of flow or stagnant flow result in disease. C. S. Holling considers that local pockets of chaos keep ecosystems stable by forcing the evolution of new forms to create new niches. Ecosystems reach their fitness near the edge of chaos. So crashes and explosions occur. In human medicine, that chaos seems to be a feature of ill health.

Health is not a final state, as modern medicine often regards it. Furthermore, the flow has channels or limits. The harmony of motion is more than just parts. Health is a description of the kind of harmony, not the opposite of disease. The health of any wildlife species in a system is tied to the entire system, as well as to changes at the boundaries. This is especially true of preserved places. Gray wolves in parks have been shown to carry antibodies to canine distemper, canine parvovirus, and infectious canine hepatitis. It is likely that coyotes and fox in the same range also have been exposed to these diseases.

Ecosystem health must become a scientific discipline, addressing losses of biodiversity, habitat degradation, and disease explosions. With an understanding of ecosystem health, patterns of mortality in species may be more be able to be predicted and interdicted.

It is possible to examine ecological health from various perspectives: The emergence and resurgence of infectious diseases of humans and others; the effects of hazardous substances; the health effects of unwise actions, such as fragmentation or other alterations of systems; and, the interdependence of species and the connection of health, also. There was a link between natural history (descriptive ecology) and medicine in Thales and others. Ecosystem medicine reexplores the connections, which have never disappeared.

Aldo Leopold started to describe a science of land health: "Health is the capacity of the land for self-renewal. ... A science of land health needs, first of all, a base datum of normality, a picture of how healthy land maintains itself as an organism." Notice that Leopold anchored this concept in a theory of organism rather than an ecological theory of community. Ecosystem medicine uses the broader concept of community to explore ecosystem health.

6.2.4.3. Local or Global Threats to Ecosystem Health

Threats to ecosystem health are well known. They can be summarized briefly as local or global problems. The local problems include: Removal of key elements, species, resources, and productivity; the disruption of natural cycles; the introduction of novel elements, as the result of inappropriate technologies; human take-over of habitats for human purposes, often the result of simple population pressures, but also of greed and sloppiness; and, extinction spasms in general, but specifically, for example, the decline of amphibians worldwide, due to habitat loss, pollution, and fungal infections (exacerbated by global warming). The global problems include global warming, ozone depletion (chemical causes), the disruption of global cycles, and contaminations (e.g., nitrates, mercury).

These threats can cause ecosystem collapse. Ecosystem breakdown happens as a result of stresses, singly or grouped, that relate to interference patterns in the system, most of which are caused by the human species now, although the potential for asteroids or volcanic eruptions remains.

As with people, health is related primarily to lifestyle (or life habits) and not to intervention by a doctor. Alas, the lifestyle of an ecosystem is often determined by its keystone species, and in most cases now humans are that species. There are specific issues in ecosystem health: Biomagnification of pollutants; diseases crossing species boundaries (and boundaries in general); the relationship between biodiversity and ecosystem health; between habitat destruction and health; and, plants or animals that cross boundaries (requiring us to track health in several ecosystems). There are also specific pathways for infection of ecosystems: Any kind of simplification, from monoculture to habitat degradation or simplification can allow disease to spread faster and farther; the decline in predators; a lack of competitors (to which pathogens are not as adjusted); and, dominance by generalists or specialists, which allows higher pathogen levels. This is not a comprehensive list of threats to ecosystems. But, it might be adequate to use to develop the concept of ecosystem medicine.

6.2.4.4. What is Ecosystem Medicine?

Traditional medicine has developed traditional medical specialties, including: Anesthesiology; Immunology (allergy, reaction to foreign substances); Dermatology; Family; Gynecology/Obstetrics; Internal medicine (further divided into specialties: cardiovascular, metabolism/endocrinology, gastroenterology, hematology, infectious disease, nephrology, pulmonary, rheumatology); Neurology; Pathology (chemical, neurological, derma); Pediatrics; Physical/Rehabilitation; Preventive (Public Health, occupational health, and aerospace health); Psychiatry; Radiology; and Surgery (which deals with diseases that require operative procedures to restore or preserve function or to cure the disease). There are rough parallels with ecosystem care. For instance, ecosystem medicine might include:

- Examination by infrared or other parts of the spectrum (the equivalent to radiology)
- Study of boundaries, or ecotones (similar to dermatology)
- Immunology (reaction to exotic species or foreign substances)
- Pathology (chemical, virology)
- Internal functioning (metabolism, element cycles, infection)
- Surgery (from the French word for "hand-work;" in this case, the removal of exotics from invaded ecosystems)
- Rehabilitation (restoration of structure or function); and
- Preventive efforts (ecosystem health).

Ecosystem medicine covers the spectrum from human organs to the complete ecosystems of the planet and maybe the planet itself. Everything could be considered nested ecosystems, from a body to the planet. Ecosystem medicine is not exactly humanitarian or ecocentric. Although it addresses ethical concerns and the focus is on ecosystems, it is more ecoperipheral; it approaches its subject sideways, cautiously, slightly out of focus, and aware of the frame.

Ecosystem medicine asks a much larger question that is integrative and contextually sensitive: Is the physiology of the base system healthy enough for self-renewal? Focus is shifted from the symptoms of a disease to the entire functioning process of existence within ecological limits. Health is embedded in the context of the system, with all of its constraints, limits, and opportunities for development.

Our knowledge of the world is returning again to a unified image, so it seems every field is converging. Alas, every field is also diverging and looks fragmented—perhaps because we are using the wrong lens.

6.2.4.4.1. Organic Metaphor

What we have learned from physics and psychology is that the body is a pattern in dynamic relationship with its surroundings. Health is an extension into the relationships of bodies. The breakdown occurs in the larger pattern, the body in context with other bodies. Health and disease, harmony and motion, are all connected, so none of them is negative or positive as local events in context. In most cases the causes of disease, bacteria and viruses, are only competing in the life base. A metaphor needs to reflect the reality. Machine metaphors or war metaphors are not adequate.

Lovelock used the human body as an analogy for the self-regulating planet. After all, the planet has recovered from several major traumas, such as oxygen wastes and strikes by planetesimals (up to 10 miles in diameter). Each time, although the time may have been a million years, life bounced back. The biota of the planet appear to regulate the surface temperature, atmospheric composition, and ocean chemistry, for a start (perhaps like the human body regulates its temperature, blood chemistry, and other vital signs).

But, the organism is only a metaphor. With scientific study, the earth appears more regulated. Organisms, however, seem less regulated, considering symbiosis and neural competition and selection. Nevertheless, organisms are more a functional whole than the planet, which is whole more because of limits. Most organisms are limited by their original designs, that is, the body cannot replace the heart and liver with one purifying pump. The planet is not limited to one set of designs for its composition (biologically anyway).

The central metaphor of the Gaia hypothesis is that the earth is an organism. But, we consider that it is more loosely organized and less integrated than organisms. But, then we find that organisms are less integrated than we thought. Organisms seem like the ultimate bricoleurs, which gather up patterns, only some of which are then selected in living. Still, organisms are more functional wholes, with genetic programs that orchestrate growth and development.

The organism might not be adequate as a metaphor. As a metaphor, the web offers connection. Perhaps a nested set of boxes would be more fruitful. As the nested series of embedded centers, it is the largest center, the environment, that is the most crucial. Without a healthy environment, community and individual health with only be a temporary situation. Healthy ecosystems are the foundation of healthy communities, which are the foundation for healthy individuals.

6.2.4.4.2. Ecological Perspective

A germ tries to live well and to reproduce. If it is quick, and spreads to more than one new victim, then it does not matter if the first host (often human) dies. If it is too fast, the host dies before the germ can spread. Human beings and germs all evolve tricks and counter-tricks to stay alive. Medicine is one of our tricks.

The many breakdowns happening—in government, in education, in health care, in the family, and in business—may be negative or positive forms of rearrangement. In the larger, evolutionary context in which these social changes are taking place, the disintegration of our social structures may be a positive phenomenon. Without understanding the larger context, it is easy to mistake these breakdowns as pathologies. From an ecological perspective, most change is neutral.

Our planet is friendly, because it was a humaning world (or composed of "peopling rocks" as Gary Snyder put it). We have been afraid of the planet and depressed by it, and we have responded by degrading it. Human impact in the form of overuse and inappropriate technologies is rapidly degrading the environment. This syndrome creates new patterns of human and ecosystem disease and instability. But, we can be rejuvenated by understanding, and we can work to rejuvenate the systems we live in.

An ecological perspective recognizes humanity's deep connection with our supporting ecosystems. Individuals cannot live healthy lives in unhealthy ecosystems. Due to the large human population, healthy communities and biological systems depend on human restraint and responsibility in technologies, population, production, and consumption.

This ecological perspective is a way of examining the whole context of medicine, to see how it might be redirected to examine its context within community and ecosystem health. The ecosystem approach requires a knowledge or list of actual and potential pathogens in the system. Consciousness is an important emergent property of physical health. The common goal is a healthy environment.

This means transforming our global and local economic and social structures, not totally, but within limits of the cultures that furnish with our identities and rules. Both can work. Current economic structures are locked into exploiting the maximum for the minimum, or for controlling scarce resources for a favored few. Profit with emotional and spiritual impoverishment is not healthy, but support with emotional commitment is. We need to develop a sense of enoughness within the real natural abundance, within the ecological limits. Security is found in working within the limits.

6.2.4.4.3. Ecosystem Medicine and Ecological Principles

In medicine a false positive diagnosis is an inconvenience; a false negative diagnosis can be catastrophic. This is true in ecology and conservation biology, as well. Making a diagnosis is a form of gambling, so it should be done on the side of caution.

The practical ecology of a scientist needs to be grounded in a theoretical formulation. The science of ecology provides principles that can guide actions. These include:

- The importance of scale cannot be ignored; things that work at a small scale do not always work at a large one; and, large scale rules can not be predicted from small ones.
- Patterns of interdependence (connection and embeddedness) are as real as bodies, and have to be considered.
- Resilience, the ability to recover from disruption, is a characteristic of healthy ecosystems.
- Appropriate measure, the Greek ideal emphasizing minimal intervention, is a proper approach to less healthy systems.
- Diversity is crucial for stability and interest. Approaches must be matched to the type of ecosystem and culture.
- Cooperation is necessary. For patients, partnership. For communities, the interaction of patients and professionals. A cooperative framework for healing within the depths of communities and ecosystems.
- Reconciliation (equality of opportunity for treatment) is necessary.
- Conviviality, basically living together in harmony, according to Ivan Illich, is a necessary strategy.

These ecological principles, and many more not presented here, can guide our actions to make good, healthy places.

6.2.4.4.4. Actions and Treatments

Medicine can be destructive to individuals, land use can be destructive to landscapes, and ecology can be destructive to animals and ecosystems. The concept of ecosystem health, by bringing the pieces together into a conversation, could avoid much of the destructive behavior of semi-autonomous departments that are not questioned or contained as part of a larger picture.

There are limitations for our treatment of ecosystems. We do not have a list of diseases for ecosystems. It is difficult to see the effects of ecosystem change or disease from the air or from a distance. Ecosystem illness is similar to chronic illness in humans. The appearance of symptoms indicate that the disease began long before the symptoms became apparent. So, the exact date of onset cannot be pinpointed. If so, then prevention becomes more important. And that is related to genetics, lifestyles (of forest entities, for instance), stress, and environmental effects. Furthermore, prevention is not a short-term, one step solution.

Some patterns of resource use need to be understood on the landscape level (and now on the global level). Others need to be examined over a year, decade or life of the system (especially in the Amazonian forest). There has to be a landscape level mechanism (and also a global one now) for regulating common resources. Local diverse patterns are good, but they must be coordinated at the landscape or higher level. It is the normal range of variability that has to be defined. Models need to be created for a wide range of habitat ecosystems.

There must be an institutional framework, with legal and social mechanisms, for monitoring and managing common resources. This is necessary for the health of the system, as well as the long-term health of humans. That means all interest have to be represented, even global corporations trying to profit from disease or rareness (brought about by overuse).

Public health specialists and population biologists look at disease in populations, whereas, doctors and veterinarians consider disease in individuals. The latter intervene while the former observe. The ecosystem doctor has to intervene in ecosystems. Ecosystem doctors care for the basic systems and communities.

Surveys need to be made of every kind of ecosystem. We need to have an inventory of kinds of systems and kinds of changes. How is conserving terrestrial animals part of conserving ecosystem health? We do not know whether animal declines were caused by disease or some other factor (competition or predation). We need to find that out.

Monitoring is the key to understanding changes. Disease needs to be monitored as an important indicator of integrity. Other indicators are surveys of key species, habitat mapping and human impacts monitoring. Complex interactions have to be monitored, using a range of indicators at levels from behavioral to ecological. There may be limitations of the bioindicators of ecosystem health. Perhaps we need to find common and endangered indigenous species and monitor them, hoping that would reflect the health of the system.

Many systems that have been overused to collapse need to be restored. Especially in forests, many small changes have a cumulative effect. Climate change can lead to loss of forest biomass. Change of rainfall patterns lead to water stress in adapted species. Deforestation results in other patterns of change, especially in vectors. Reforestation is associated with rapid changes in vectors, which can adapt to nonindigenous species of vegetation. Restoration is a necessary part of reforestation. Maintenance of remnant populations may not be enough to prevent loss of key species on a landscape level.

6.2.4.4.5. New Goals for Human Medicine as Part of Ecosystem Medicine
Medicine can continue to advance and confer benefits to societies, but it has to have a larger perspective. The physician had traditional obligations to the patient, to herself, and to her society, but now there are obligations to the environment. All these are partial because all of them are inescapable and unavoidable.

Medicine needs to monitor disease and health, as well as the human use of the environment and its consequences. Medicine needs to improve research for the prevention and treatment of new diseases related to ecosystem changes or degradation.

The standard for our efforts should be health—of communities, ecosystems, and corporations—not profits, technological sophistication (although that is sometimes a good thing in use), or fame. The common goal is a healthy environment as a basic human right.

Medical advertising needs to change. Rather than promote the unquestioned use of drugs, advertising needs to promote awareness of consequences of wildlife diseases (and of course the causes, as a result of change or human intervention), as well as paths to human and ecosystem health.

6.2.4.5. What is Global Ecosystem Medicine?

Is the planet really in peril from human activities? Or can it work with new changes? The planet has a concept of health, or rather we have the concept, and we can describe it by contrast with an organism, but we have too little experience to try to make the health of the planet a real issue.

James Lovelock launched a medical model of Gaia (in his book *Healing the Planet*). The earth can be considered a patient that humans can address with some medical training. A few planetary diseases are evident now, such as greenhouse fever, ozone loss, or acid rain 'indigestion.' And there may be potentially fatal diseases at early stages, such as nuclear winter or extinction spasms. As a patient, Gaia is less threatening than a ruthless pagan goddess. Lovelock suggests a practical medicine for the planet, which grows from guesses and empiricism to practical solutions and good hygiene. Of course, this is what ecosystem medicine is—the application to the special cells and organs of Gaia.

Medicine is the identification of problems, such as stress; and, we are aware that humans put stress on global systems. We need to monitor the health of the system, checking temperature and pressure, breathing, biochemical tests and biopsies from ecosystems. We need to increase our understanding of decay and healing. Medicine has shifted responsibility from individual, to be healthy, to doctors and medicines, after decades of dramatic success with some medicines and treatments. This way shifts the burden to the intervener. The way out is avoidance, or awareness of long-term restructuring. Gaian medicine returns the responsibility with the planet, not with an incomplete flawed human management.

This form of global thinking is a critical correction for the machine world view, which traps us in a narrow atomic, context-free extreme individualism. Some forms of biological thought, such as Neo-Darwinism, have converted Darwin's organic view to a noisy market of manipulative machinery. The metaphor of the machine justifies our indifference to environmental disasters, since nature is considered a lifeless aggregate of atoms.

We can figure out if the planet is healthy—it is relatively stable. But, we are not sure what the planet requires to stay healthy. Does it require asteroid collisions? Does it require the increase in solar radiation? Does it require the ellipse of the planet orbit, with refreshing distances? Lovelock suggested that the galaxy was a giant warehouse containing spare parts needed for life.

Lovelock suggests humans be the stewards, as representatives of the varieties of life, for the planet—not managers, with full responsibility, or masters that make it provide for us. As stewards, we would address species in communities and ecosystems. Gaia as a living planet is a self-regulating system that can correct itself, under most circumstances, and without too much interference.

6.2.4.6. *Summary: A New Category of Medicine*

Less than 30 years ago, the environment was of little concern to most people. Now it is the primary issue for most people. Calvin Coolidge once said that the "business of America of business." The biggest single business in most of the world, now, is the environment. All farming, most pharmacology, most tourism, and many other "industries" have their basis in the health and beauty of the environment. The environment contributes to the largest share of most gross national products directly or indirectly.

Historically, we have used ecosystems without regard to their continuity or to their health. Partial knowledge and technology has allowed us to exploit our environment beyond what is desirable for us or for other species. While moderate exploitation is necessary to live, too much exploitation is unwise. A wise use of resources would not make the world less habitable. We are part of the system and must protect its health as a whole.

A new category of medical professional is needed: People who address the health of ecosystem themselves. Human physicians may need to be able to identify critical environmental conditions that affect human health, but others are needed to identify the health of those systems themselves. Human physicians need to know the basic principles of diseases related to environmental change or chemical exposure; others need to know the principles of ecosystem health and how that is related to human health.

Conventional medicine has great strengths, as do alternative medicine or traditional cultural medicines. Ecosystem medicine must develop such strengths. Ecosystem medicine incorporates all other kinds of medicine as special cases. It can use the best and most appropriate of any procedures. No single approach works best for every instance.

Using a model of ecosystem medicine offers a comprehensive framework for investigating the problems of health, from individuals to ecosystems, especially the interactions between individuals, social groups, place, and environmental change.

Ecosystem medicine is a medical discipline, aimed at restoring ecosystems to health. As with any medicine, the patient actually does most of the work to become healthy, although the doctor gets the credit and the payment. This leads to respect for the practitioners, but also to responsibility and more rules. The first rule, which we might take to be basic, is identical to the first vow of the Hippocratic oath, "Do no harm." Noninterference is a basic ecological principle—do not interfere with the health and stability of the ecosystem—the health, diversity, and stability of the ecosystem are a first consideration.

Ecosystem medicine bases itself in a community context and limits the use of the ecosystem to that which the ecosystem can afford to provide and remain healthy over indefinite time. Ecosystem Medicine undertakes the responsibility to preserve the healthy functioning of the ecosystems under its domain. It also has a responsibility for the ecological production of goods from an ecosystem.

This is the local application of ideas for specific ecosystems, which may have a direct link to human health. The systems still have to be self-sustaining and self-renewing, Of course, the planet can also be linked directly to human health. In traditional medicine the organs are more integrated than in ecosystems, which are looser and more complex. James Lovelock has emphasized the health of the planet as a single system.

The health of ecosystems and human institutions should be measured with a holistic index. We have not developed qualitative indicators of ecological health or quantitative measures of social health, much less an ecocentric view that would value preserves of nature for themselves.

To address the health of ecosystems, ecosystem medicine would be a temporary medicine, not a constant intervention or even a continuous diet. We have already tried to gain complete control over ecosystems through scientific methods and technological applications. We have not been able to control them successfully. We regard modern medicine as a fool-proof system that tried to eliminate weakness, disease, and mistakes. It has not. Ecosystem medicine must limit its goals.

One goal is the pursuit of the health of ecosystems and their inhabitants. Their goal is good health. Individual health is in the context of community health, which is in the context of ecosystem health.

Another goal is security. Symptoms of insecurity are poor human health, migrations, and conflicts (territorial or religious). Environmental security is more than the abundance of natural resources and having a stable social and economic situation. It is the health of the whole system resulting from balanced exploitation by all beings.

Rather than telling the ecosystems what to do, rather than controlling their growth, we need to watch ecosystems to see what they do (this used to be the function of natural history), and we need to let them do it (this requires patience and temperance), with a minimum of interference. Abraham Maslow regards this attitude as taoistic, and the way to ecosystem health is letting the ecosystem do most of the choosing and working.

Our response to the ecosystem, being concerned with its health (as ecosystem doctors or nurses perhaps), is not benign neglect or complete anticipatory stewardship, it is participation in the process of the ecosystem as a harmonious system, with mutually restrained conflicts and constrained influences.

The goodness of our lives reflects an imperfect balance of love and selfishness, reason and passion, sensuous materiality and spirituality. We have the responsibility to be healthy, to contribute to the health of our community, and to contribute to the health of natural ecosystem communities. Good intentions have to be combined with ecological knowledge and ethical behavior for the discipline to be meaningful. Aristotelian ethics emphasized justice and fairness; Epicurean ethics, gentleness and kindness; Stoics, duty; and the later Stoa, love, compassion, and mercy. Our rules for living together have to be a compassionate participation in the whole planetary process.

6.3. *Problems: Conversion & Interference*

We exploit resources, plants and animals too fast for natural renewal. We exploit minerals and fossil fuels too fast for the wastes to be integrated in cycles. We create novel substances and use them too fast for the substances to be integrated in natural cycles. There is a conversion of planet from wild to urban ecosystems. System used to be wild with wild humans, now going towards one domestic urban system.

6.3.1. *Cultural Conversion of Places*

Agriculture converts ecosystems into simplified ecosystems that have to be maintained with large amounts of human labor to keep the system from diversifying and maturing. Even modern forestry tries to convert wild forests into single-species managed plantations.

6.3.1.1. Conversion of Land to Fields

In North America, which held the largest extent of forests on earth, forests were considered wilderness, to be tamed and converted to agriculture—although no one really expected the conversion to be so rapid or dramatic. Since the first agriculture, 10,000 years ago, forests have been cut and burned to create fields for food crops. The invention of iron plows speeded up the process because they could break up soil too heavy for wooden plows. Much later, the invention of the horse collar allowed horses to pull plows and loads with greater strength. Technological innovation, combined with accelerating population growth, lead to clearing of many forests for agriculture.

Traditional agriculture proceeds by substituting selected domesticates for wild species in equivalent niches, or by manipulation of the ecosystem, whereas modern agriculture replaces the biota, or transforms the ecosystem into an artificial system. Margalef pointed out that all agricultural systems are laid out for low maturity, to increase production per unit. Traditional agriculture changed the pattern of ecosystems. Modern agriculture simplified those systems dramatically. From the perspective of a balloon or airplane, agricultural fields are easily identified, as circles, squares, rectangles or parallelograms, with a more monotone color and texture than surrounding wild areas.

To achieve even greater efficiency, we have increased the speed of our activities, converting materials and cultures into new designs without consideration of the meaning of, or need for, efficiency. The speed of our economy is too great for many cultures to adjust to; and the thoughtless transformation of cultures may result in great, irreversible mistakes. The speed of our conversion of wild habitats to domesticated lands is too great for many species to adapt to.

Although our behavior may not be qualitatively different from our remote ancestors and the worst pathologies of wild animals, which usually result from miscommunications under certain circumstances, it is quantitatively different. More and more activities affect larger and larger parts of the planet; many problems have global effects now.

In tropical areas in the past, people practiced land use that permitted the vegetation to maintain itself despite human exploitation and the constraints of soil and climate. With population and production pressures, tropical areas everywhere are rapidly being exploited. The consequences may be disastrous.

Size is almost always the greatest threat. In spite of the fact that many Buddhists planted trees regularly, the Buddhist traditions of wooden temples and funeral fires contributed to the denuding of large areas of forest. It is our misfortune to live at a time when the accumulated effects of the conversion of nature for human ends are becoming obvious and cutting into the survival potentials of many other species.

Modern agriculture replaces the biota, or transforms the ecosystem into an artificial system. The application of modern agricultural techniques in tropical areas disregards the local realities, and is socially and ecologically disastrous. Traditional ways have operated successfully for thousands of years, although not without serious problems of their own. But, these problems may be overcome, unlike industrial problems, which have fundamental flaws and cause rapid conversion.

We exploit resources, plants and animals too fast for natural renewal. We exploit minerals and fossil fuels too fast for the wastes to be integrated in cycles. We create novel substances and use them too fast for the substances to be integrated in natural cycles.

The myth of a return to earlier rural times is only a myth; the soils have been destroyed. We must learn to create bioshelters that sustain us and provide us with shelter and more. The soils are starving for organic matter. Waste and manure need to be returned to the soil. Soils, like animals and plants, can be domesticated. European soils bear little resemblance to wild soils. American soils bear even less. Smaller farms could fertilize by crop rotation and manure. Build up soil, choose better seed stock; make hedgerows for birds to control pests.

Agriculture works best where it is least perfect: that is, the farmer still hunts and gathers, keeps many kinds of plants and animals, produces little surplus, uses no chemicals, grows adapted crops, and has wild lands. That situation may be ecological and durable. Successful agriculture depends on artificial climax or sustained successional state. A pioneer is self-sufficient in more than material.

6.3.1.2. Conversion of Land to Cities

There is a conversion of planet from wild to urban ecosystems. System used to be wild with wild humans, now going towards one domestic urban system. Cities convert ecosystems into cityscapes, a special artificial ecosystem with fewer plant and animal species, disrupted water flow as factor, household wastes as factor affecting the landscape, a network of corridors that perforate landscape and cut off wild flows, an increase in the number of small patches, and a reduction in other kinds of patches and corridors, woodlot, stream, golf. Flows include energy, materials, information, people, and pollution.

In a converted ecosystem, there are three imposed ecosystems with minor linkages: (1) a natural system of primary productivity based on depauperate system with few trees and layers, a simple trophic structure of birds, squirrels; (2) a human system, with imported food and water, smaller system of carnivore predators, such as fleas, lice, and bedbugs, and decomposers, such as bacteria, fungi, and gulls; and (3) a distant support systems for food, materials, and energy, but also smaller emotional support system of pets, such as cats, canaries, and fish.

The daily inputs for city, in terms of sunlight, atmospheric deposits, water, food, fuel, manufactured goods, new buildings, roads, and infrastructural support, is tremendous. The daily outputs, including heat, water, sewage, solid waste, and pollutants, are likewise tremendous, although they are skewed to unusable wastes, including heat.

These extra inputs and outputs have effects, side effects, or rather main effects that are unwanted or unanticipated. By thinking in terms of side-effects, we have been unable to manage the side-effects of our massive technologies. They are in fact unmanageable in a mindset that thinks of them as side-effects. A holistic rationality might be better to use, where everything is an equal effect, and we manage out technology to have all the effects.

Land is taken over and covered, but then idle land may total 9-28 percent of urban area.

Villages lower connectivity to the landscape by increasing patches and corridors. Agriculture gets more homogenous, decreased fallow areas. Stream corridors are destroyed, environmental degradation. Connectivity is lowered, matrix is minimized. And, these things happen: Disturbance of nutrient mineral cycles. Disruption of atmosphere, drought, storms. Changed microclimate, with heat islands and dust domes. Lower photosynthesis/ productivity. Lower diversity leading to homogenization, with cosmopolitan species. Inefficiency of use of energy and materials. Decaying infrastructure, roads, sewers, buildings, and houses. Disruption of supplies, due to social and political actions, e.g., strikes, breakdowns, attacks. Budget crisis from loss of tax revenue. Increase in violence, sickness, drug use, crime, and impoverishment. Increase in mental and physicals illness from stress. It is almost as though human beings are major agents in changing the wild planet into a dust planet.

6.3.1.3. Industry & the Artificial

Artificially created environments are cultural environments. Nature thus acquires characteristics from agriculture and social institutions as well as from geology and climate. The practices of environmental conservation must be complemented by careful policies of environmental creation. Societies have images of the future that influence their policies, and those images can be shaped by knowledge and persuasion. Human beings can and have created new ecological values by collaborating with, or following the laws of, nature.

Many ecosystems in the temperate zone are artificial. The tame character of ecosystems in England, for instance, results from human intervention. English parks are based on an imperfect understanding of natural systems, and on their potentialities, as well as on a series of happy accidents. Much native vegetation is a social artifact. In Scotland, for instance, forest cover was reduced from 55% of the total area to 5% by primitive stock-keeping and agriculture; moors decreased by half, but meads increased eight-fold. The soil was more fragile, so the accidents were not happy there and the forests did not grow back. Some forest plantations have been started. Rene Dubos mentioned Kentucky, Western Europe and Japan as examples of radical conversion from the native fauna without disorganization.

6.3.2. *Interference as a Problem*

Humanity is exploiting nature recklessly, by converting ecosystems without paying minimal attention to their health. Many ecologists, such as Eugene Odum, have observed that complex communities have existed for thousands of years in relatively stable environments, even though these environments are characterized by regular disturbance and constant exploitation. As a result of our growing population and increased use of technology, these environments are now vulnerable to human interference, which is a different thing from disturbance or exploitation. Disturbance, by definition, is an event that can be caused by climate, biological entities, or other actors. Exploitation is the normal use of a resource or of a species by another species, including the human species (this ecological definition differs from a sociological definition, which means 'selfish or unethical use,' although it may suffer from negative connotations due to the latter); in fact, ecological exploitation has a rejuvenating effect on populations.

Interference is an activity that can degrade, destabilize, or destroy entire ecosystems. Interference is not a form of disturbance, exploitation, or competition; it is destruction without gain to any species. Sometimes it is caused by planetary events, such as volcanic eruptions, but in the case of human interference, it is the destruction of the structures and processes of evolution for a large-scale, one-species, short-term gain. Interference behavior that characterizes the nonecological activities of the dominant human, industrial culture. The pandominance of ecosystems is related to the biological and cultural characteristics of the species. Ignorance and indifference are identified as major reasons for continued interference.

Although rare large-scale or novel disturbances can interfere with ecosystem processes, the term 'interference' is reserved for constant destructive or novel effects. The destruction of ecosystem processes in nature by the action of one or more species is rare; any species that did so would become extirpated or extinct, unless it was not dependent on a single ecosystem, as is sometimes the case with wolves or human beings.

Human populations have increased exponentially, with billions of people in giant urban ecosystems. Agriculture has produced monumental yields, but only at the cost of tremendous erosion and great subsidies of fertilizers and pesticides. Dams have been built all along rivers, and riverine forests have been cut, altering rivers and fishing grounds. Changes have been made without regard to the long-term impact on the ecosystem or on its human population. We dominate as many systems as we can.

6.3.2.1. Pandominance of Systems

This pandominance—that is, dominance of every system—has major effects on ecosystems. Human beings have contributed to the extinction of species and to the destruction of ecosystems. Human hunters are hypothesized to have wiped out the most of the large mammalian species of the Pleistocene through overhunting—not for future food, but rather from the style of hunting—by driving herds over a cliff. It is possible these populations were vulnerable due to changes in climate and food sources and may have become extinct without help. There are other instances. In the 1880s, soldiers and cowboys slaughtered buffalo as a political strategy to reduce the resources of native peoples. Farmers and loggers destroyed the dense forests of Ohio and other states. Settlers and industrialists in the Amazon are destroying vast tracts of rainforest, as part of a political strategy to move peasants out of cities. Industrial forestry in the Northwest America is content to take a high percentage (well over 90 percent) of a forest for wood and pulp, destroying the basis for the continuity of the forest, as well as all beings that depend on the old-growth, fungi, and physical properties of the forest to live.

Human exploitation at the tremendous physical scale that occurs in industrial states is different from exploitation by other species, because it results in the destruction of the entire system, the very basis for renewal of a system that human beings, as well as other species, need for life. Human actions are damaging global biogeochemical cycles, such as the carbon or nitrogen cycle. For instance, deforestation, burning, wetland loss, and industrial processes are releasing massive quantities of carbon dioxide into the atmosphere, which disrupts the carbon cycle. Although the destruction of large species, from whales to frogs, has a dramatic effect on ecosystems, the destruction of microbes, which generate oxygen and recycle nutrients, has a critical impact on the entire food web. These actions are global, like a large volcanic eruption, but, unlike a volcanic eruption, they are constant and hourly. These pandominant human activities are best referred to as interference, and may be related to our industrial style.

6.3.2.2. Industrial Style in Systems

Interference has been a rare phenomenon on earthly ecosystems; it has happened in the past as the result of global catastrophes, such as meteor impacts. Now, interference, as opposed to more limited and predictable disturbances or exploitations, is threatening the stability of all ecosystems. It is dangerous to interfere with the processes of ecosystems because it disrupts the communities on which other species, and ultimately human communities, depend.

Furthermore, in the deepest sense, it violates the idea of living together with other species on the planet. The proper relationship of humanity with nature should include competition and exploitation and mutualism, but not interference.

We kill millions of animals in laboratories to insure our safety, we kill billions of plants and animals for food and clothing and products, while indulging in the sentimental preservation of some individuals of other species. Animals do not need to be saved from natural death, a great regulator of life, but from unnecessary suffering, experimentation, and premature extinction. The world would not be a better place without sharks, silverfish, rats, cockroaches, or hyenas. They need their own places. The places, entire ecosystems, need to be saved. If we diminish variety in nature, we debase its stability and wholeness. To save ourselves, we must preserve and promote the variety of nature. Perhaps we should reconsider our unconscious role in nature as agents.

6.3.3. *Humanity as an Agent of Change in History*

What are the forces of history? Certainly, we should include geological processes, as well as solar system effects, such as the output of the sun or meteorites. Certainly, we should include the environment, such as climate and the distribution of resources. Especially important are human impacts on ecosystems, such as deforestation, identified by John Perlin as very important, perhaps the reason for the decline of the Hittites and Babylonians. Another impact is desertification, according to Uwe George and others. Disease patterns, according to W. H. McNeill, are crucial. To disease, or germs, Jared Diamond adds steel and guns as forces that have shaped human history and societies. Then, there is simply luck, the position of a culture in the stochastic chaos of nature.

By their activities, human beings change the places they live. Much of the change is easily incorporated in the cycles of renewability of the ecosystems. However, humans often change the directions of such systems by simplifying or degrading the systems. In this case humans act as agents of interference.

6.3.3.1. Humanity as Biological Agents

Humans have had a great impact on nature, and should be considered themselves as a force of nature and history. One could consider humans as special agents. One analogy of humans as special agents is as a parasite: A consumer feeding on another living organism, usually inside, drawing nourishment and weakening the host. States acted like macroparasites, according to William McNeill, but becoming less violent or unpredictable over time, as they adjusted to their host populations.

As crowding increases competition, there is more pressure on remaining reserves. The system parasitizes humanity and nature. Humanity becomes an autoparasite, a new pseudo-species. Technology enlarges the number of niches for us; tools fit humans to different habitats, displacing other species. We steal from animals and plants, from the earth, from our own descendants. Hobbes foresaw this war of each against all. The systematic destruction of human beings and animals is not an isolated peculiarity. A fat parasite often kills it host and then dies itself. Perhaps, humanity is an agent of a different sort, a systems agent that encourages only positive feedback.

Perhaps human expansion is like a cancer. Alan Gregg (1955) compared the world to a living organism and the explosion in human numbers to the proliferation of cancer cells. He sketched other parallels between cancer in humans and humans' cancer-like impact on the world. Cancer cells proliferate rapidly and uncontrollably in the body; humans continue to proliferate rapidly and uncontrollably in the world. Crowded cancer cells harden into tumors; humans crowd into hardened cities. Cancer cells infiltrate and destroy adjacent normal tissues; urban sprawl devours normal open land. Malignant tumors shed cells that migrate to distant parts of the body and set up secondary tumors; humans have colonized just about every habitable part of the globe. Cancer cells lose their natural appearance and distinctive functions; humans homogenize diverse natural ecosystems into artificial monocultures. Malignant tumors excrete enzymes and other chemicals that adversely affect remote parts of the body; humans' motor vehicles, power plants, factories and farms emit toxins that pollute environments far from the point of origin.

It is not in a tumor's self-interest to steal nutrients to the point where the host starves to death, for this kills the tumor as well. Yet tumors commonly continue growing while the victim wastes away. A malignant tumor usually goes undetected until the number of cells in it has doubled at least thirty times from a single cell. The number of humans on Earth has already doubled thirty two times, reaching that mark in 1978 when world population passed 4.3 billion. It is over six and a half billion now. After thirty-seven to forty doublings, at which point a tumor weighs about one kilogram, the condition is usually fatal—that would be the population equivalent of 5.4 billion people. We have exceeded that; the question is if it has been fatal—large complex systems may take a long time to collapse—or if the system has more flexibility than an organic body.

The metaphor of cancer may be more appropriate than a footprint. After all, a footprint can stimulate some kinds of ecosystems, such as short grass prairie. What humanity does is transform the ground under the footprint into a new human system.

6.3.3.2. Humans as Ecological Agents

Every species exploits its environment to the extent that it can, with no regard to consequences. Usually, each species is checked by another, because there are so many competing for the same food, and equilibrium is maintained.

Partial knowledge and technology has allowed us to exploit our environment beyond what is desirable for us or for other species. While continued, moderate exploitation is necessary to live, massive, unbalanced exploitation is unwise. Wise use of resources would not make the world less habitable. We are part of the system and must protect its health.

By distorting the equilibrium, we have destroyed whole species and favored many others, many wild as well as domesticated. Rats and mice have been carried to all parts of the world and live in direct competition with humanity, invading our buildings for food and shelter. Crows and coyotes have also profited from their human association.

Basic principles need to be examined in relation to human ecology. Natural populations are maintained at ecosystem limits; this maintenance is achieved by the production of excess young and the elimination of the weaker; mating and parenthood are denied to the young by instinct in most species, until a territory sufficient to raise, feed and protect offspring has been acquired. If the population becomes excessive, glandular conditions are activated to induce stress and complications, which reduce the population.

6.3.3.3. Humanity as a Geological or Climactic Force

No single change is exclusive to humans as a species, but they are excessive, rapid, compounded, and large-scale. There is movement of soil, but also massive erosion. There is movement of minerals, but also disruption of mineral cycles. There is the addition of novel elements into the atmosphere, but there is also a massive release of carbon.

When people use more of the earth's supplies in a certain period than can be replenished in the same period by the sun, they are eating into the natural capital. Humans have caused the extinction of hundreds of species. Rhinoceros, buffalo and crocodiles are disappearing. Getting timber for fuel and construction, clearing land for agriculture, has destroyed whole habitats. Demand for timber has been insatiable, for houses, ships, paper, and fuel. Trees have been cut from vulnerable watershed sites, with resulting floods, erosion, and diminution of rainfall and water table. Domestic animals inimical to growth were introduced, such as rabbits and goats. The introduction and maintenance of sheep in Spain, Italy and Cyprus has changed whole ecotypes.

Sixteenth century Spain, under Philip II, experienced unprecedented growth; shipyards flourished; gold was brought from South America; wars were fought. But, every year it cost thousands of hectares of centuries-old forests to build and maintain the fleets. By the end of the seventeenth century, the rich Iberian forests had disappeared. After the soils eroded away, there were famines.

Vegetation holds soil in place, reduces wind speed at the soil surface, and improves water absorption and transport in the soil. Erosion destroys soil and makes it difficult for plants to be reestablished. Recovery, if it occurs, may take decades. Erosion is an ecological catastrophe on a planetary scale, causing thousands of higher plant and animal species, and countless lower species, to be lost forever. This is what is planned in Brazil and South America. It is not known how massive deforestation through overgrazing, firewood collection, and timber exploitation will affect terrestrial and atmospheric systems. Perhaps we should just accept erosion. Erosion is picturesque. Cezanne's paintings of France are striking. The abstract terrain of Greece is pleasing to many.

Israeli scientists doubted the theory that Arab methods of cultivation were responsible for spread of desert in North Africa and Middle East; desiccation on that scale is beyond even modern technology. They concluded that deserts are usually created by a relatively small change in climate. However, grazing practices could cause a small change in climate and thus contribute to desertification.

Only in the nineteenth century beginning with G. P. Marsh, did people start to realize that humanity has done as much to change the environment as the environment has done to mold human history. Marsh, the first U.S. ambassador to Italy, was one of the first to study the role of humans in changing the face of the earth. When he visited the near east, he was shocked to find deserted cities, silted harbors and wastelands instead of flourishing civilizations. He concluded that ecological errors had led to the deterioration of agriculture in Mediterranean countries. He advocated agricultural conservation practices.

Environmental factors have shaped the course of human history to a greater extent than had been realized. The decline of Rome is a study in forest ecology. There were previous and later catastrophes in the Tigris and Euphrates valley, Greece, Khmer, Maya, Midwest United States, and the Australian outback. Many people did not change their behavior in time to solve the problems. Worse, the current civilization is global, not local; so, there will not be a migration to unaffected lands. Ecologists have not unraveled most mysteries of ecosystems, so the long-term consequences of most human interaction cannot be predicted.

6.3.4. *Practicing Noninterference*

Exploitation, in the ecological sense, is necessary and beneficial to biological populations. A machine metaphor approach, with its assumptions of interchangeability and quantity, apparently has difficulty distinguishing between exploitation and interference. An ecological metaphor, that is more receptive and reverential, may be more appropriate to understanding organisms and nature in general. Such an approach would stress noninterfering observation rather than controlling manipulation.

Applied to nature, human intelligence could discover the significance of natural rules of interaction and exploitation. The reverence for beings as they are could result in a rule of noninterference (Wittbecker 1984). A rule of noninterference states that human beings ought to avoid behavior that disrupts essential ecological processes or destroys biotic communities. As Paul Taylor states his rule of noninterference, it requires a "hands-off policy" for whole ecosystems and biotic communities; the rule stated here is concerned with limited and sustainable exploitation of ecosystems already shaped to some extent by human activities. Many other ecosystems, perhaps covering 50 percent of the land area of the planet, would be reserved by law for predominately natural ecosystems or adapted first nations. Noninterference also means "letting be" (after Martin Heidegger), or "letting alone" in the words of E. O. Wilson. Noninterference is not indifference, which is diffuse. It is caring. Noninterference will not lead to chaos, poverty, or stagnation. It permits the rational exploitation of resources.

We need to practice the rule of noninterference so that all beings can enhance their lives and habitats. Noninterference can be derived from nonviolence or even from English Common Law, which is well-established in Western law; it includes a precept: "Use what belongs to you in such a way as not to interfere with the interests of others" (*Sic utere tuo ut alienum non laedas*). This rule could be defined by positive laws and by negative restraints on behavior. This attitude would entail using what is necessary, exploiting some ecosystems completely, changing a place to fit human aspirations, and killing plants and animals for sustenance. But, it would also mean limiting humanity and its technological effects, limiting human use to local impacts, and letting other beings live without interference. It is not necessary to dominate or to terraform the earth completely. Humanity could contain itself to a small percentage of the planet's ecosystems and only visit or ignore the remainder.

6.4. *Problems: Technology*

Problems can be subdivided into physical and cultural problems. Although it is tempting to divide physical problems into natural and human generated, given the scale of human influence, it is too difficult to separate them definitively; humans contribute to natural trends, and some trends add to human problems. Humans themselves create carbon dioxide and wastes, by breathing and excreting, and also by concentrating, burning, recombining, and synthesizing. Our activities, from hunting and growing to deforestation, transportation, urbanization, and industrialization, produce more wastes, on a larger scale than most other animals. For instance, 25-50 percent of all aerosols, that is anything ejected into the atmosphere, are put out by human activities, from deodorants to coal-powered energy plants; volcanic action adds a significant percentage.

The purpose of technology is to create a tool to solve a specific physical problem. But, this can create other problems, if the tool is inappropriate or has deleterious effects. There may be a disjunction between the effects of technology on human activities and its effects on the ecosystem or atmosphere.

6.4.1. *Physical Problems*
Is pollution a problem? What is pollution? Usually pollution thought of as materials, but what kind of materials? Things that are useless? Side-effects? Things that do not fit? Resources out of place? Things that cannot be cycled by a system? Contaminants? Poisons? Ugliness? Anything not wanted, such as light at night or noise during the day? In many cases pollution is a difficulty with density or distribution. The activity of the atmosphere generally keeps gases diluted and distributed, although particles like dust can cause problems. To some extent the activity of the ocean dilutes and distributes particles and chemicals. Land-based particles and compounds have a greater likelihood of clumping, although wind and water can spread them somewhat, for instance, when runoff dilutes salt from evaporation.

The planet has a long history of concentrating or diluting particles, as well as changing the composition of its spheres. Volcanic eruptions have added carbon dioxide, sulfur and water to the atmosphere, and molten stone to land and sea. The 1991 eruption of Mt. Pinatubo in the Philippines ejected 20 million tons of sulfur dioxide, which caused 0.3C (0.5F) degrees of atmospheric cooling. Through photosynthesis, plants have added enough oxygen, often considered one of the first 'pollutants,' to change the operation of atmospheric processes, such as oxidation. Do ecosystems cause or have pollution? Or, do living organisms eventually use the 'pollution' as a nutrient? Animals add carbon dioxide to the atmosphere and chemical wastes to their habitats. Inadequate recycling of plant and animal matter, over a long time, can lead to peat, coal and oil deposits. Although these materials can be poisonous, because we regard them as resources, we do not consider them as pollution. Pollution seems to have to be a nonresource and nonneutral to be pollution. Perhaps it is the dose, as Paracelsus said about poison.

6.4.1.1. Nature-dominated Changes
The dynamics of the planet, from tectonic motion to volcanic activity and storms, creates change and challenges for individual organisms and ecosystems. Climate change can be 'caused' by changes in orbit of the planet. Milutin Milankovitch (finally translated in 1969) identified three principle cycles that cause variability in the planet climate. The longest is the elliptical orbit of the earth around the sun, a 100,000 year cycle. Thus, intensity of the rays

varies during parts of the year. In 2004 only a 6% difference between January and July. The second cycle is the tilt of the earth on the axis, which varies from 18 to 24.4 degrees. The cycle is 42,000 years. The third is the wobble of the earth on its axis. The cycle is 22,000 years. At extreme of cycles, variation in sunlight is less than $1/10^{th}$ of 1 %. But that can change temperature by 9 degrees F. These cycles can cause an ice age only when continental drift puts land near the poles.

The meteor event about 65 million YBP changed the conditions for dominant dinosaur species and permitted smaller organisms to eventually dominate. The 55 million YBP event heated the surface of the planet from 9- to 18 degrees F. Oceans had become much more acidic. There were massive extinctions of forminifera. In 2004 researchers found that 1650 to 3300 gigatons of carbon had been injected into the atmosphere by ocean gas, from 500 PPM to 2000 PPM, from a nature gas release under the ocean. It took 20,000 years to reabsorb the CO_2 mostly by surface plankton blooms.

The Maunder minimum of sunspots affects weather. During the period 1645-1715, Europe's temperatures plummeted. Changes in the galactic environment, such as the solar system passing through a dust lane, can contribute to the increased in solar radiation, as the sun attracts more matter.

When the changes are gradual, most life forms can adapt to them or contribute to them. When the changes are sudden or rapid, many organisms and systems cannot adjust and collapse.

6.4.1.1.1. Climate Change
The climate has changed in large, millennial pushes. Some of the pushes, which seem to come from communities of living organisms, can contribute to instability and change, as when bacteria and algae essentially pumped large quantities of oxygen into the atmosphere. Other organic pushes can contribute to climate stability. For instance, about 540 million YBP, living things began building skeletons of carbonate, as they absorbed CO_2 from sea water. This factor alone has made ice ages more rare.

Andy Ridgwell and colleagues argue that the shell-forming plankton of 300 million YBP stabilized the planetary thermostat. Planktonic calcifiers changed the positive freezing loop because they were not tied to continental shelves; they floated in the ocean, which prevented too much CO_2 absorption and further cooling. About that time forests were starting to cover the land. This tied the carbon cycle down somewhat. In another example, modern coral reefs of 55 million YBP drew volumes of CO_2 from the atmosphere. About 1 million YBP, the ratio of forests to grasslands affected the climate, especially water regimes. Even now, in some African savannas, elephants work to keep forests back. Removing elephants will likely shift the ecosystem towards trees.

In the past 12,000 years, climate change, especially after the last ice age, has caused the collapse of many agricultural civilizations. As areas became more open and grassy, there were fewer wild animals. In Europe the tundra was replaced by forests that reindeer did not like.

A shift in the orbit between 10,000 and 4000 BC brought 7-8% more sunlight to the northern hemisphere. Rainfall in Mesopotamia increased 25-30% and increased available moisture by about 700%. But, after 3800 BC the orbit reverted and the rainfall pattern changed. By 3100 BC this forced changes in tilling fields and double-cropping, as well as with specialization and irrigation canals. The city became a key adaptation to drier climates.

6.4.1.1.2. Global Warming

Global warming is an unfortunate term, as it might imply a gentle tanning. It also implies a comfortable, warm future where plants grow more and people need less heat from coal. Global 'burning' or 'suffocation' might raise more alarms. How we use language influences our priorities. People in glass houses should not throw rocks, and people in greenhouses not throw coal on the fire.

Did the megafauna in North America and Australia die out about 13,000 years ago due to global warming, or from stress or overhunting? In earlier periods, such as the Jurassic and Cretaceous, was it warmer than now? Was that because of global warming? How stable is the atmosphere? Do coral reefs and forests make it more stable? Do industrial processes make it less stable? Do the characteristics of life cause things to fluctuate more or less? Including human actions? At what point does climate become global warming? Is it a symptom, problem, or nothing? Are green houses gases increasing? Is the average annual temperature increasing? Is it a positive feedback system? Is it too complex for us to understand?

Is global warming bad? For us? For our companions plants and animals? For the Planet? Should we try to control it? Will it reverse on its own and head towards another ice age? Can we control green house changes? How could we try? How do we know if we can? Should we try anyway? What kind of behavioral changes could we make? What kind of technical things could we do? Mirrors? Ocean Doping? How would they work? What other effects would they cause? What is the worst that could happen?

We can measure overall warming of the atmosphere; it is warming. We can measure the increase in temperature, as well as heat content, in the ocean and on land. Drought, fires, and air pollution have contributed to global warming. Global warming makes ecosystems more vulnerable to long-term changes in precipitation patterns. Inadequate rainfall may lead to more droughts and fires and then to more global warming.

6.4.1.1.3. Carbon Dioxide Increase

The amount of carbon dioxide in the air has changed dramatically over millions of years. In the past, concentrations of CO_2 were much higher, perhaps 12 percent, rather than the 0.00038 percent today. It was quite high at one time, due to volcanic action. It has been much lower as systems matured during and after the ice age. We now can measure human additions to carbon dioxide in the atmosphere. Carbon dioxide is a molecule that can hold heat in the atmosphere, due to its chemical shape and characteristics. CO_2 is also a trigger for a more powerful greenhouse gas, water vapor, since heating allows the atmosphere to take up even more water. Heat travels where CO_2 is effective. If the earth were a black billiard ball, then doubling the CO_2 would raise temperature 1.8 F. However, because it is wrinkled, wet, plant and cloud covered, doubling the CO_2 would increase the temperature exponentially.

CO_2 is very long-lived, over 100 years, and 56% of it from industrial processes is still aloft. CO_2 interacts with water vapor, methane, and sulfur, affecting how those gases retain heat or contribute to cloud formation. CO_2 is good for some plants, mostly trees. But, plants in extra CO_2 have tougher leaves, less nutritional value, and higher amounts of defensive chemicals. One species that will benefit from changes will be parasites. Mosquitoes will spread malaria in some places.

The CO_2 is already in the air. We have no easy way of getting it out. So, the course of climate change is set for next several decades. Significant CO_2 is removed through natural processes, which also generate quantities of CO_2. We could remove quantities, but it would be expensive, and may not be at a scale large enough to be effective. We do not know the threshold for anthropogenic change, but it can only be 2.3 F according to Steven Schneider.

6.4.1.1.4. Ocean Acidification, Nitrogen Pumping & Phosphorus Loss

Oceans are becoming more acidic. This has influences on CO_2 uptake rates and on the health of animals and plants. Applied fertilizer, usually anhydrous ammonia, which is oxidized to nitrates, can acidify the soil. Erosion from these soils can increase ocean acidification.

Although the bulk of the atmosphere is nitrogen, only a few physical processes, such as lightning, and biological processes, such as leguminous bacteria, can make it available for plants to use. The rate of removal from atmosphere is many millions of tons per year. The nitrogen cycle pushes nitrogen through the compartments of the vegetation and soil, from the actions of precipitation, leaching, fixation and other processes. The cycle is relatively slow (See Section 3.1.3.1 and 3.2.4). It has stayed within certain limits for many millions of years. Although we probably will not significantly alter the nitrogen cycle, we can increase nitrogen losses through activities such as clearcutting. Synthetic fertilizers and fuel combustion adds another 100% of reactive nitrogen to the biosphere.

Phosphorus, like nitrogen, is an element closely associated with life, and living processes preferentially concentrate it. Phosphorus runoff is a problem. The phosphorus cycle essentially runs into the ocean, as a sink, at a relatively high rate, and is returned to land surface by geological events or smaller events such as fish predation by birds and animals (See Section 3.2.5). Another smaller sink for phosphorus is guano deposits. Organisms keep phosphorus in a relatively closed system, which keeps it from entering sinks. Phosphorus is a limiting nutrient for life. Too much of it in one place, or too little, can cause problems. We are increasing phosphorus 12-fold in systems. Waterborne phosphorus and nitrates cause eutrophication of streams and lakes, and even some estuaries and seas (and this can affect the albedo of the planet).

6.4.1.2. Human-dominated Changes

Human activities, especially from agriculture and industry, have the effect of precipitating other changes. Our farms, fields and mines can dominate an entire landscape. Our roads dominate landscape patterns. Even new technologies of wind and solar farms, because of the scale of energy use, are starting to dominate coastal or sunny landscapes.

6.4.1.2.1. Land Conversion

We have converted a large percent of wild forests and grasslands to cropland (currently about 12 percent by area). We have converted wild, old-growth forests to modified forests and tree plantations. We are converting rich estuaries and shorelines into fish farms. Much of the destruction of land can be traced to our ignorance of ecological connections and to our inconsiderate use of gigantic machines and tools, using unlimited amounts of fossil fuels.

6.4.1.2.2. Biodiversity Loss & Extinction Rises

Currently, 100 species per million per year are entering extinction. The extinction of species is certain, and there is a high degree of certainty. Temperatures of mountain habitats are easily measured, as are conditions that mountain species can tolerate. As climate warms cold-tolerant species have nowhere to go but up, and we know the height of the mountains. We know the planet will heat by 2 F this century, or 5 F, if business and industry are conducted as usual. Chris Tomas et al. found that at the lowest degree of global warming (1.4-3 F) about 18% of species are 'committed' to extinction. At higher rates the number goes up: 3.2-3.6 F then 25%, over 3.6 F then 33%.

6.4.1.2.3. Chemical Pollution

Chemical processes driven by solar energy let complex molecules build up. Chemical orders underlie the biological order of life; one self-replicating molecule allows life to continue. The science of chemistry has been able to make wonderful advances in materials, from nylon stockings and parachutes, to replace silk and hemp. Many new materials were based on hydrocarbons from oil. The new molecular patterns, however, did not react by degrading in natural ecosystems; there were no predators capable of consuming the molecules as food.

We are not sure what the minimum or maximum values are for various chemical elements in biogeochemical cycles. If the cycles continue, the elements should be enough. Elements, however, exist in nested environments. Smaller pools of elements are more vulnerable to interactions with pollutants. The pollutants themselves often have longer life spans than complex molecules. Amounts of many pollutants get concentrated in the environment.

Sulfur dioxide (from blast furnaces) causes tissue degeneration and interferes with enzymes in plants. Sulfur pollution can kill entire forests; the Norilsk Nickel Combine in Russia has killed over 4000 square kilometers of larch forest and serious affected another area of equal size with its sulfur emissions. Sulfur pollution from burning coal and oil in factories in the Midwestern United States combines with rain, fog or snow to make acid rain in New England and Canada (only recognized since 1972), 100 times as acidic as normal, causing declines in crops, sugar maples, and forests.

Although there are a few chemical responses to pollution, such as liming the ground around pines against acid rain, they are expensive and difficult to apply. The prevention of pollution remains the best option.

6.4.1.2.3.1. *Stratospheric Ozone Depletion.* Ozone has an annual metabolic flux of millions of tons. Ozone, with three oxygen molecules, is created from the action of lightning discharges, making a lucky coincidence for life on land; before an ozone layer started forming, the solar emissions like ultraviolet light were too dangerous for animals to move to the surface of the planet. On the ground, however, ozone is a pollutant that can cause tree death by reducing root and leaf biomass. With sulfur dioxide, ozone can influence ecosystem health. Low-level ozone, mostly from electrical and machine discharges, can damage crops like wheat and soybeans (about $40 billion in 1988).

Stratospheric ozone can be depleted by a variety of pollutants, such as methyl bromide or chlorofluorocarbons (CFCs). Methyl bromide is used for fumigating logs, as well as tomatoes and strawberries. Lighter than air, it drifts into the upper reaches of the atmosphere. CFCs, originally used as refrigerants, pose another danger, a long-term, broad-spectrum, invisible danger. They accumulate in the stratosphere, where they are broken apart by sunlight and react with ozone. Worse, the CFCs absorb wavelengths that carbon dioxide does not. Far worse, a single molecule of CFC-12 traps 20,000-times more heat than one molecule of carbon dioxide. The marvelous science of chemistry cannot seem to be dissociated from the dangerous of its applications. Although banned in most nations for the past decades, CFCs can last 75-110 years in the atmosphere. Since the ban, the ozone layer is stabilizing.

6.4.1.2.3.2. *Aerosol Loading.* An aerosol is a suspension of particles in a gas. The particles can be molecules, pollutants, or viruses. Insecticides, paints, hairsprays, and drugs are spread with aerosols. Often the propellants, hydrocarbons, are or more dangerous than the particles, creating Local particulate concentrations in atmosphere. Other chemicals used for cleaning medical instruments or electronic parts, especially terpenes, end up in the higher atmosphere. Maximum quantities for these have not been set, although we can correlate it with ozone depletion.

6.4.1.2.4. Water pollution

Water is a universal solvent; it is such a great solvent that absolute purity is only a theoretical goal; even highly distilled water contains gases and solids. The fresh water we drink contains about 1% solution of carbonates, with various nitrates, silicates, minerals, compounds, and trace elements. These elements in normally pure water are absorbed by plants. Obviously, pollution is the fault of water.

6.4.1.2.4.1. *Biological.* Although many animals, such as cattle and sheep are raised on ranges, they often spend months in feedlots being fattened with grain for human consumption. About 95 percent of this food goes for respiration or ends up as manure, which overwhelms natural forms of recycling. Many domestic animals, especially kept in intensive conditions in feedlots, produce many tons of feces, which piles up in mounds, before leaching into the water table and polluting water. Free range domestic animals can compact soil and seedlings, reducing grass or forest reproduction. Normally, dung beetles or termites remove the feces, although they often cannot cope with the wastes of that scale of production.

6.4.1.2.4.2. *Runaway Plastic Nurdle Concentrations.* Charles Moore, sailing in the North pacific subtropical gyre (one of 5 high pressure areas in the world) found a floating trash island that went on for thousands of miles (perhaps twice the size of Texas). The gyres cover 40% of the ocean or 25% of the entire planet, and all of them are attracting islands of trash. Much of Moore's 'island' was made of plastic, from fully formed pieces down to small nurdles. Plastic is a petroleum-based mix of monomers shaped into polymers. Other chemicals are added for inflammability or suppleness. Plastic is replacing iron and glass as containers; it is lighter and more easily molded. Every year we produce 450 million kilograms of 'phthalates' used to make plastic soft and pliable (known to be toxic to human reproduction systems). They can leach from packaging and coatings. In some food containers and plastic bottles, phthalates are found with a compound bisphenol (BPA). We produce 3 billion kilograms of BPA every year.

Only 3-5 percent of plastics get recycled. Glass and iron are more easily recycled. PET and HDPE (numbers 1 and 2) can be recycled. Plastic retains pollutant and gives off deadly vapors. Products made from plastic recycling are limited to carpet and boards and jacket linings. Except for some incineration every piece of plastic made still exists. Recycling also uses resources and energy and creates pollution. But it does reuse resources and it is wiser use.

Plastic does not biodegrade, it crumbles into smaller fragments. Plastic can decompose in seawater. And it can contaminate marine live at the molecular level. Samples contain styrene monomers, dimmers and trimers (which seem carcinogenic in mice). Plastic is moving into the food chain. The danger is eating it or becoming entangled in it. There are miniscule pieces of plastic, called nurdles (lentil-size pellets of plastic in raw form), in the water. By weight, it can total 6 times more than plankton. The pollution seems invisible and ubiquitous. They are easily mistaken for food, can be ingested, and can screw up genes. They disrupt the endocrine system, so that some male fish and gulls have female sex organs.

Like sand on a beach, the entire biosphere gets mixed with plastic particles. These particles change the properties of water and soil. Plastics pollution at this level and scale is almost completely unrecoverable for recycling or breakdown (from burning or solution).

6.4.1.2.5. Freshwater Drawdown

Rates of human consumption of water are tremendous. Aquifers many thousands of years old are being drained. We are using almost all of peak water in every system. Peak renewable water limits are the total renewable flows in a watershed. Peak ecological water is where, in any hydrological system, increasing withdrawals reaches a point where any economic benefit is outweighed by the ecological destruction caused by the action. Furthermore, the efficiency of use is poor and needs to be improved; in parts of the Balkans, over 60% of the water is lost through broken pipes.

6.4.2 *Cultural problems*

Part of the problem with technology is the lack of limits in a culture. We rarely deny a new technology, before we come to depend on it, and thus become trapped by it. Technology has greatly increased the kind and quality of materials used for buildings and machines, especially plastics, aluminum and other light metals and silicon constructs. Yet, the scale of technology produces pollution that reduces the productivity of natural and agricultural systems.

6.4.2.1. Risks & Threats

Effects become problems, if they are too large, cause interference, or are not understood. Problems become threats if they are not solved within a required time frame. Threats to our survival rise from a disjunction between our powerful technologies and the wisdom to understand the effects and limit their use. Threats to human survival include: Environmental degradation, extinctions, climate destabilization, nuclear weapons, terrorists, and use of un-tested technologies. They seem to become threats when we underestimate the risks of apply-ing industrial processes to global scales. We are too insulated from small real danger, although not from global slow invisible ones. We just do not recognize them. We have problems with scale. We have a human scale, based on our body and images, but this scale is not favored by much of the universe. The problem of using metaphors between scales is that we can miss the significant changes between scales.

6.4.2.1.1. Risks

Natural processes pose risks to every culture or nation that depends on the constancy and sta-bility of the environment. Flood, earthquakes, volcanic eruptions, and droughts for example are not regular or predictable. In the Sahel and Mesopotamia, the argument was that over-grazing and human population caused the droughts. Exposed soil contributed to hot air and changed albedo and no rain-forming clouds. For the Sahel a single variable made most of the difference. Rising sea surface temperatures of the Indian Ocean, from greenhouse gases, was responsible for most rainfall decline. We focus on political, religious and behavior first, but find that sometimes environmental changes are culpable.

Agriculture and permanent settlement increased the risks of drought. The inventive agriculture entailed a transformation of how people change their mode of life to one that entailed a greater risk and vulnerability. Larger sedentary communities subsisted on a fused staple crops but these became prone to long-term fluctuations in rainfall floods and weather. These communities also began to impact the surrounding areas through the intensive use of wildlife and would. This lead to waves of intensification, extensive occasion and the current expansion of the labor force.

Because of an increase in risk and uncertainty, religious ideologies and management strategies achieved a reduction of that risk and uncertainties some extent by linking com-munities in a network. This also led to a disparity in wealth and power between workers and those affiliated with power. This situation has only gotten worse in six thousand years.

Urban centers became linked with religious establishments and administrative institu-tions quite early in history. The current environmental crisis is bonded to urban agglomera-tions that were unimaginable in earlier times; these cancerous expansions of urbanization have placed humanity at unprecedented risk. Although the system is capable of serving humanity through judicious integration of food resources and distributions to ensure equity, it has in fact led to disruptive, disastrous consequences as certain urban centers, especially in industrial areas, use the world for ruin the short-term gain.

Every civilization encounters a risk spectrum, which may push the civilization to find new increases in the area food or energy. The need to solve such risks pushes the system further from its original state sometimes at increasing speeds. The conjunction of long-term risks can be called a crisis. When European colonists settled in the Americas, they recognized the similarities of the environment, but not the difference in patterns. Cutting only familiar species of trees changed the forest dynamics and led to destructive impacts. The spectrum of risks shifted and human exploitation eventually had to be restructured.

Human beings often begin by adapting themselves to the dynamic of the environment. Over the long term, however, they modify these dynamics to suit themselves. They appropriate the environment by reducing its complexity in exchange for the increasing complexity of their societies. There does not seem to be a return from complexity. In this sense complexity is a trap. People cannot stop investing in knowledge and the system they modified. The current crisis as well as many in the past, is primarily a function of how people trap themselves by coming as close as they can to the limits of the environment. This of course is a form of trap, made worse as risks enter a spiral.

A risk spiral results from the transformation of environmental mental complexity into social complexity. Human actions therefore can create new risks and risk spirals to occur. The deep-time perspective can reveal proximate and ultimate causes of the collapse of systems. Social and annotations or cultural traditions that may appear in efficient or illogical in the short-term may reduce risk and increase resilience in the long term. The risk is driven by human cultures attempts to cope with risks or to exploit opportunity. The process involves more management of the environment and although different parts of the environment operate in a range of scales, most of the natural dynamics and landscape occur slowly by comparison with human dynamics. As a result humans adapt themselves to the immediate dynamics of the environment at the beginning, but over time cultures serve their own needs by modifying the environmental dynamics. Human cultures thus become dependent on colonized systems, which require certain social institutions, especially those involved in organized production and storage.

Machine technology can reduce risk, as can management expansion. However, if the risks are perceived as smaller, then people take more risks (sometimes a challenge or danger is more fun than a safe, optimum environment). If every hot or sharp surface is labeled or protected, people can be surprised when they get hurt. In one sense, knowing dangerous conditions is enough to encourage people to act in a safer manner. Is technology then a trap also?

6.4.2.1.2. Threats

What are the greatest threats at the moment? Climate change or species extinction? Social equity or levels of luxury? Humanity, that is most cultures, seems to be treating the threats as problems and pushing them forward to the next generation to solve. Although we act as though we cannot tolerate risk, we are willing to accept a very high level risk for the next generation and the planet itself. Threats to our survival rise from disjunction between our powerful technologies and the wisdom to understand the effects and limit their use. Threats to human survival include: Growth in demand for resources; exponential growth in population; transformation of the atmosphere, geosphere, hydrosphere, and biosphere; conversion of wild ecosystems into domestic; deforestation; release of pollution into land, air and water;

environmental degradation through use; extinctions; destabilization of systems and cycles; the release of novel substances into historical systems; the use of untested technologies; the use of highly destructive technologies from giant earth movers to nuclear weapons; uncontrolled genetic changes; economic failures; social failures; cultural error or inflexibility; the use of violence or terrorism to achieve social objectives—obviously, this is not a complete list of threats.

Some of these threats are the result on unconsidered growth of population, demands, and conversions; the scale is far more damaging than just the activities. Other threats have to do with the unconsidered use and scale of technology. Many have to do with our personal and cultural failures to understand the system in which we are embedded. Humanity seems to be amazingly poor at predicting what important issues are going to be important with in the next 30 years. For example, the 1972 Stockholm conference on the human environment did not mention major threats such as mass extinction of species, tropical deforestation, desertification, ozone depletion, or climate change.

6.4.2.2. Problems of Civilization
Civilization has become so beneficial and desirable that we do not question its direction, size or momentum. But, internal problems have major effects on civilization as a whole, not just its technological brilliance. The external problems affect regions and the planet.

6.4.2.2.1. Growth & Momentum
Unending economic growth is emphasized in some economies and cultures. Yet, growth can cause a community to fall out of balance or scale with its surroundings. And, being out of balance can lead to massive disruptions.

Population growth, even if not combined with technological innovation, can lead to greater deforestation and land conversion for agriculture and urban areas—which pushes the regional systems further out of balance. Economically, we reason that growth is necessary for living standards to rise, and individual self-interest has proved to be a stronger motivation than patriotism, altruism, or recognition. Modern economics depends on economic growth to avoid crisis. The major premises assume that the population will grow, social good is related to equitable distribution of material products, and if resources are limited, technology will erase the limits. A large literature has treated perpetual growth as the only conceivable state of affairs. Kenneth Boulding suggests that it is a short-lived 'cowboy economy.'

Capitalism depends on growth for stability. There is some analogy with plants (to be elaborated on later). Some stability can be gotten from growth in early stages; later stability must result from limits and metabolism. Growth in plants can delay the onset of senility by ridding the plant of waste products in more diluted form. However, too much growth produces a strain on tissues and early decay. In fact, one herbicide promotes excess growth as a means to kill plants labeled as weeds.

The production of wealth from growth depends on technology. The technological perspective is oriented toward materials and not humans or forest processes. Nature is considered to be a resource to be exploited. The immediate objective of technology is to create wealth through knowledge. Technological activities are justified on humanitarian grounds, scientific discovery increases the well-being of human society, yet the social consequences of scientific activity are ignored; short-term suffering will be offset by long-term benefits, it is claimed. But because the long-term view is not taken, long-term benefits will be worse.

Economic growth can produce great wealth for some. Inequality is more the result of differential development than of exploitation. According to Boulding the greatest source of the differential is different rates of accumulation of knowledge, capital and organization; the rates are essentially internal properties of cultures. Although a minor element in terms of transfers, it is a large psychological perception, which may need to be compensated for in a global community.

However, gross production may not be as desirable as thought. The world prices of food and industrial raw materials have increased far more rapidly than those of manufactured goods. It can be seen that economic growth is not equal to progress.

6.4.2.2.1.1. *Growth & Stability.* Mesarovic and Pestel stated that "the issue for the economy is not to grow or not to grow; it is how to grow, and for what purpose." They claim that if a workable world system is to emerge, it must be after the establishment of an organic pattern of growth. Due care is devoted to describing such a pattern and contrasting it with other, tragically inapplicable patterns of growth. Their treatment of the world system itself was regionalized and multileveled. They recommend that the establishment of organic growth was necessary with no need for special no-growth policies for populations or econo-mies. They assume that further industrial growth will continue, that economic growth is good, and that this growth solves human problems as long as it is organic. Kinds of growth, linear or exponential. The J-curve graphically represents exponential growth, such as by human populations. A positive linear increase is generally steady, such as food production. Exponential growth is said to be bad, and organic growth is said to be good. In fact, although organic growth is better, there is little difference during a world crisis—both reach asymp-totes of suffering. One need only regard the population crashes of lemmings and others to see how organic growth can go wrong.

In the organic world, growth is healthy only when the rate of change is decelerative in the long run; cancer and population are constant or accelerative. Mesarovic and Pestle fail to realize that continued economic growth in any form is a threat to the stability of the bio-sphere.

Economics became enamored of growth during a critical time in history. Rapid Euro-pean expansion occurred at rates rarely exceeding a growth of 1% per year, and with unparal-leled opportunities for expansion into sparsely settled areas (North America, Australia, South America, South Africa). Many cultures now do not have these opportunities; the continents are claimed, and violent population growth may have wrecked their hope for development by ravaging every resource.

The economy has been growing almost constantly since it has been studied. We have been trying to force it to grow, rather than let it stabilize or contract. Some have argued that contraction causes losses and suffering; yet, growth has caused exponentially more losses and suffering. Mesarovic and Pestel, as well as many others, confuse growth with development. Some theorists, like Samuelson, have concluded that growth is necessary to rid the economy of disparities. Even if it stopped growing, the economy could still develop.

6.4.2.2.1.2. *Size & Impetus.* The larger a moving thing is, the more momentum it has. Stars have far more momentum than snowballs. Perhaps big science and big technology have too much momentum. Theodore Roszak acknowledges its schizoid attraction and repulsion, with the twin promises of glorious accomplishment and hideous death. Who could escape being torn between yes and no, if even our end would shine with radioactive, Promethean grandeur? Our image of big science—the scientist as tragic hero, isolated in chaotic nature, but strong in his proud individuality, perhaps driven to research by hubris and madness—is a barrier to any new vision, especially a small vision.

We cannot imagine beauty in the old and messy nature, and we are afraid to try to do without luxuries or to try to sacrifice anything to try to change the momentum of industrial civilization. The most likely result will be the total destruction of the combatants, nations, and natural habitats. Sadly, the only people who do not know this, or admit it, are those in decision-making positions, who are compelled to prepare for what they subconsciously know will be a terrible disaster. Their power has trapped them in the momentum of their nation, afraid to be caught in any criticism. Yet, they direct the money, skill, and knowledge of their citizens into projects that lead to misery, servitude and hideous death, and not to life, liberty, and happiness. Our civilization is based on its early momentums. We pray it never stops.

6.4.2.2.2. Technology & Technopoly

Technology has reduced the globe to a single, closed system, which humans can share according to their financial powers. Our direct experience of the world has become shallow, in spite of faster travel. Travel used to broaden the mind, but now it narrows it. We travel in sealed corridors like boxed goods, comforted by homogenized foods and a few common languages. Technology or social structure can mask the internal stress from fast economic growth.

New technologies compete with old ones for dominance in a world view. The medium of technology contains an ideological bias. Tool attacks tool, according to Neil Postman; printing attacks manuscripts, television attacks printing, painting attacks rock art, and photography attacks painting. Postman refers to this fight for dominance as technopoly, and defines technopoly as a form of cultural AIDS; the cultural immune system is inoperable. The immune system protects the body against invasions and uncontrolled growth of cells.

The change to Technopoly may have started with James Watt's invention of the steam engine in 1765. Adam Smith in 1776 justified the transformation from small-scale, personalized, skilled labor to large-scale, depersonalized, mechanical mass-production. He argued that money, not land, was the key to wealth. In a technocracy, an unseen hand would eliminate the incompetent and reward the efficient. Several years later, Richard Arkwright, a barber, developed the factory system in cotton-spinning mills, where he trained children and others to conform to the regularity of the machine. It was the first mechanization of production. Twenty-seven years later, the power loom eliminated skilled workers altogether. Every ten years a new invention changed industry. Whitehead thought that the greatest invention was the idea of invention.

Technologies, in a technopoly, make other things invisible. This has the effect of eliminating them from consideration. Now, instead of being the most important of things in a culture, the invisible things, like the environment, become the irrelevant and useless. Other things, like religion art and history are redefined. August Compte argued that things that could not be seen were unreal and undeserving of attention.

Technopoly is dismissing philosophies and traditions, as well as moral democracy and cultural beliefs. Not having a transcendent narrative or moral order, technopoly has to depend on techniques for control the information from technology. A bureaucracy is one filter. Bureaucracy is a technical solution to the crisis of control. It is administrative not governmental, as de Tocqueville recognized them. A bureaucratic form restricts information to what is asked or can be put in boxes. It is to make the use of the information efficient. Bureaucracy is independent of culture. It is a very low-context extension. It also has no intellectual political or moral theory. Detachment allows escape of responsibility. Not responsible for human consequences of decisions, just efficiency of that part of the bureaucracy.

Humanity becomes an autoparasite, a new pseudo-species. Technology enlarges the number of niches for us; tools fit humans to different habitats, displacing other species. We steal from animals and plants, from the earth, from our own descendants. But, science and technology are spiritually impoverished, divorced from awareness of values and purposes.

Technology must be placed in perspective. The analysis of complex problems is beyond the specialist as is the synthesis.

By the time of the Frederick Taylor system of scientific management, in 1911, the primary goal of human labor and thought was efficiency, not the production of goods. Technical calculation is superior to human judgment. Humans are at the disposal of technique and technology.

Why did it work in America? Perhaps the American character of wonder and frontier life. Or, the audacity of American capitalists in robbing America's past and heritage. Or, by providing people with sufficient abundance and comfort. Old beliefs were discarded and history was disconnected from change and the future. Technology was something to believe in. It never made mistakes (only humans did). Antibiotics always cured. Airplanes always flew. Well, maybe in our memories.

Science and technology are the chief instruments of progress. They were supposed to bring superstition and suffering to an end. Technologies depend on information. They also have to control information. In technopoly, according to Postman, cultural symbols are trivialized by corporate enterprise. Symbols become common place by use in television and movies. The promiscuous use of images may seem like irreverence, but it is worse; it is trivial overexposure. Should all symbols be used for commerce? Is that okay? Technopoly tries to fill the void of dead narratives with its idea of progress without limits, improvement without costs, and rights without responsibilities. Society can be engineered is a thing of technopoly.

Science and technology are spiritually impoverished, divorced from awareness of values and purposes. Technology must be placed in perspective. The analysis of complex problems is beyond the specialist as is the synthesis.

6.4.2.3. *Economic Problems*

Economics posits rational actors in economically efficient optimization policies who would spend only as much money averting global warming as would be lost to costs. Efficiency here is not about death or destruction but finding the least cost and highest return solutions for the business of life. Saving the commons however is a serious market failure because it depends on morals not profits. Institutionalists argue that more is needed, such as better organizations and norms, which go beyond the market. Climate change could slow progress, and science is uncertain. More cooperation, on things like global warming are needed. Better organization is needed.

6.4.2.3.1. Consequences and Costs of Pollution

The industry gadfly Dixie Lee Ray claims that 'side effects' cause us to worry to the exclusion of considering the benefits of a new technology. Her example of the internal combustion engine is ambiguous. In fact, it has had great undesirable effects on society, regions, and the planetary atmosphere. Pollution is not a side effect here; it is an equal effect, along with mechanical power. The costs of air pollution are staggering: $40,000,000,000 in health care and lost productivity in the U.S.; $4,000,000,000 from ozone damage to wheat, soybean, and peanut crops; $5,000,000,000 from acid rain damage to agriculture, forests, and aquatic systems; and destruction of 20 percent of European forests (figures from the Worldwatch Institute, 1988). Other pollutions are as bad. The oil pollution of the oceans from spills as well as the continuous discharge of poisonous sludge (up to 17,000 gallons per month per supertanker, including the BTX compounds—benzene, toluene, and xylene) and toxin-contaminated water kills thousands of animals and fish every month (including salmon, ducks, and sea birds with concentrations of metals—zinc, chromium, and cadmium).

Airborne pollutants have increased world-wide, according to a 1982 UN report by Dr. M. Tolba, Executive Director of the UN Environmental Program. Although some forms of pollution may have lessened in some industrial countries, the U.S. and Britain, for example, due to lawful control measures, other forms of pollution, such as acid rain, have increased dramatically everywhere, threatening fish, trees, crops, and buildings.

Air pollution, with acidification and toxic substances, has been implicated as a primary factor in 'forest death syndrome, first noticed in Germany and other parts of Europe. There is overwhelming evidence of pollution-caused death in Europe and eastern United States. The fact that there are other causes that work independently or with pollution does not invalidate the other evidence. To argue so, as some scientists do, is based on a logical fallacy, the semantic fallacy of complexity.

The costs of modern industrial farming in England, for instance, are applicable to China and other nations: Two billion dollars for removing pesticides from drinking water, damage from soil erosion, medical costs of poisoning and mad cow disease (90% of what farmers earn); $0.4 billion for subsidies to farmers (180%); $0.1 billion for healthcare costs for poor choices (45%); and, $3 billion at least in loss of productive land.

The costs of industry are more subtle, but equally expensive: Unhappy people act as identical replaceable machines; people are tightly regulated (leading to disrespect and violence); cleanup of pollution is prohibitively expensive (especially compared to prevention); Cleanup; health-related problems are almost too expensive (only the complexity of the effects of mixed pollutions prevents a violent revolution against industries); and, the boredom of sameness is impossible to put a price tag on.

6.4.2.3.2. Effects on Human Health & Ecosystem Health

Pollutants can contribute to or cause diseases, including cancer, lupus, immune diseases, allergies, and asthma. Pollutants can cause levels of irritation leading to death. Higher levels of background radiation have led to an increased incidence of cancer and mortality associated with it worldwide. Some illnesses are named for the places where specific pollutants were first formally implicated. One example is Minamata disease, which is caused by organic mercury compounds.

Bad air quality can kill. Ozone pollution can cause sore throats, inflammation, chest pain, and congestion. Oil spills can cause skin irritations and rashes. Noise pollution induces hearing loss, high blood pressure, stress, and sleep disturbance. Contamination caused by pollution can have damaging effects in the brain and central nervous system. Studies have shown that brain of animals actually shrink from prolonged exposure to contaminants in the environment. Humans can become stupid.

Changes to the environment can lead to disasters and collapse. A few human environmental disasters include: Mesopotamian soil salinization, 3300 BC; Chinese soil erosion, 3000 BC; Deforestation of the Mediterranean, 500 BC; Deforestation of Rapa Nui, 1400 AD; and, the US Midwest Dust bowl, 1930s. These are signs of a decline in the health of these ecosystems.

The consequences of land degradation are many. For instance, productivity has been reduced by one-third on half of India's soils. Salinization in the Middle East and on most irrigated lands means that many crops can no longer be grown; for instance, 34% of land in Bangladesh affected, and this has quadrupled since 1990. On-farm expenses have risen 100%. Over 7% of the agricultural productivity in SE Asia has been lost. Off-farm expenses have ballooned. The costs from air and water pollution, road damage, desertification, cleanup, and health are steadily increasing. The loss of biodiversity and ecological services continues. The increased uses and costs of energy affect both personal lives and industrial improvements.

6.4.3. *Conclusion: Questioning a Machine Image*

Modern industrial technological cosmology, beyond being another kind of order, more linear and abstract, is wrongly considered the evolutionary successor to traditional cosmologies, and is displacing them rapidly. Using the metaphor of the machine, this cosmology tries to render other images as incomplete. The machine image has allowed tremendous advances in understanding mechanical systems, and even some biological systems, by treating everything as a machine, which can be repaired with replacement parts or improved with better design.

We need to question the image of the machine. Questions widen the narrow field. Hardin points out that concerns about narrow issues, such as pollution, can cause a deep examination of the process, such as distribution theory, that cause the issue. Human activity simply produces things that we want and things that we do not want, such as pollution. As we ask questions about who pays and who benefits, we are able to think or rethink about these things and keep them in our consciousness.

6.5. *Cultural Changes from Technology?*
(Being edited)

Figure 65-1. Remains of Mohenjo-Daro on the Indus

6.6. *Rethinking Technology: Integrating & Limiting*
   (Being edited)

Before designing new technologies, we can break down technology into purposes and strategies. We can then enlarge the scope of inquiry to include the fitness and appropriate application of technology.

6.6.1. *Purposes of Technology*
What does technology do, basically? It creates extensions, as tools, to our bodies, letting us access new resources and create new things. Many of our basic tools have very basic functions: To cut, burn, mend, collect or move, for instance.

6.6.1.1. Cutting (knives, shears)
Pulling down a tree for shelter or slaughtering a large animal for food was difficult for fingers and muscles. Sharpening a piece of certain kinds of rock made cutting wood or flesh easier. The first tools were stone-flaked, then ground stone, used with boomerangs and clubs. Ground stone axes appear first in Australia. Tools are simple, but effective, easily manufactured and maintained, for example, the spear-thrower, or woomera (atlatl), had hook on one end and an adze on other; it could be used for a shovel, fire-starter, or percussion instrument.

The earliest lodges may have been pit houses, covered by up to 3 feet of earth over grass mats on a frame of wooden poles, with a central opening for smoke. After the Indians acquired horses in the mid 1700s, wild horses from the southwest, and became more mobile, the shape of their buildings changed. During the winter, they lived in A-framed or conical-shaped mat lodges. Mat lodges on wooden-pole frames were easier to dismantle and move, although they were usually just stored in rock piles. In the summer, mat or deer-hide lean-tos were used on hunting trips. Later, deerskin tipis were favored for trips.

6.6.1.2. Burning
Something as simple as a knife can replace the muscular effort of chewing large pieces of food, which can be rendered more tender by another tool, fire.

The biggest energy contribution from cows and bulls is their dung. Most of the dung could be used for fertilizer at no cost to the farmer or to the world's fossil fuel reserves. The remainder could be used for fuel. It is odorless and burns without scorching, giving a slow, even heat.

By burning fossil fuel every year, we are burning light from previous ancient years. Most energy is generated by variants of the steam turbine, which relies on burning something, not much different from dung and wood, just centralized. Burning fuel to produce fertilizer to grow feed, to produce meat and to transport it produces 9% of all emissions of carbon dioxide, the most common greenhouse gas.

Burning allows us to ignore inefficient practices. For example, in 2000 fishers burned 13 billion gallons of fuel to catch 80 million tons of fish. Without calculating the real value of carbon fuel, it takes over 12.5 times as much energy to catch fish as they provide.

Later technologies tried to burn waste to break it down into component parts that could be sifted out of ashes; the burning, however, resulted in air pollution and some toxic wastes that had to be isolated. In the USA, a third of municipal waste is burned that includes the percentage recycled though).

Derrick Jensen and Aric McBay note that US paper industry burns more fossil carbon than the entire chemical industry uses as raw material. Our transportation is supported with energy from burning. Burning fossil fuels lets bacteria form new routes for mercury contamination. Burning of things for energy is bad? Yes, we could be eating it or using it to replenish soil. Is burning waste bad? Yes, for similar reasons, but also for the release of toxins.

6.6.1.3. Mending Gluing & Weaving

As people had the need to separate things with tools, so they had needs to put things together, for instance two pieces of hide or a sharp rock and piece of wood to handle it. Sticky substances from the environment were used to glue things together—mud, asphalt, animal fat or tallow. Thorns and later needles were used to pierce hide and cloth. Cloth itself was woven from threads of fibers, cedar, wool, cotton.

Then, they make clothing, for making life easier. Clothing and tool-making is a universal in all human cultures. Polynesian cultures used 3-way weaving (at 60 degree angles) to keep the weave from spreading under pressure. By the way, the words weave, woof and web are derived from the Indo-European word *webh*. On a loom for weaving, the upright threads (vertical) are called the warp; this means 'throwing.' The fabric or web is made by weaving new threads horizontally; the crosswise threads are the woof. Warp means to 'twist' or bind.

6.6.1.4. Collection & Storage

For aborigines, the whole environment was a storehouse for gathering. Sun-dried fruit on the ground, eggs. Most personal belongings could be carried with people, on their backs or in bags or on horseback. In homes, it could be stored in bedding, baskets or wooden boxes. Food was stored in different kinds of woven baskets, made from bear grass, wild hemp, and cedar and spruce roots. Mat lodges on wooden-pole frames were easier to dismantle and move, although they were usually just stored in rock piles.

With the surpluses from agriculture came the needs not only for larger forms of storage—large pots or buildings—but also for standard measures to distribute earnings measured in crops. Salish people often built separate storage longhouses and storage boxes for fish and oils. Canoes were stored in special canoe houses. Fresh and curdled milk, carried and stored in long, decorated gourds, is the basic item of the Masai diet. Inuit could store food in ice caches or smoke it.

Agriculture had new requirements to get materials to and from the fields, preparation sites and storage sites. The surplus grain had to be transported, stored, reallocated. This required organization. Information is stored, as data. Biological symbiosis results in a greater store of genetic information—a new species. Information is stored in various media or on computer disks.

6.6.1.5. Moving (Wheels & Vehicles)

For many lifetimes, things were dragged from place to place. Even after horses and oxen were domesticated and carried a few things, most things were dragged or strapped to poles and dragged. Eventually wheels were invented for moving water or toys. The Southwest Asian system, involved clearance of savanna, swamps, and fringe forests. It led to changes in the water regime by irrigation, with systems of embankment dams, canals, buckets, and water wheels. Pottery wheels mass-produced bowls for measuring grain and rations that were distributed to outer villages.

The wheel and the cart, for moving larger loads, was invented about the same time as the city. As paths become wider more permanent structures that can accommodate wheeled vehicles, their edges interact with human places. The wheel allowed a much greater mechanical advantage for human and animal effort. Because of its relatively low coefficient of friction, it is useful in a variety of circumstances. The bicycle is still one of the most efficient ways of allowing human mobility.

6.6.1.6. Production
The simplest form of production was gathering ripe foods where they grew, then carrying them home to be prepared. As people started to tend plants, they had to being water and soil or night soil to them.

6.6.1.6.1. Food Production
Some desired foods grew well far from their native habitat, but they required care to get them water or keep them safe from their other predators.
As tools became more sophisticated the scale of production increased.

6.6.1.6.2. Industrial Production
As things were made on larger scales, new ways of making them were tried. Much human labor was replaced by energy from wood, coal, and then fossil fuels. Tools were put in sequences to make complex machines in a linear order. Tools and machines required more metals and materials to be dug up, which exposed living beings to the toxic wastes.

Peasant society featured a plow, with a domesticated animal on dedicated farmland, plus a wide range of artifacts and tools. The transition to the industrial society began in England with steam engines, textile machines and tractors, which forced the peasants off the land and into the cities. Machines began to dominate production.

At the beginning of the Industrial Revolution, the sudden change in production and distribution of goods benefited many people. For example, fabrics became much more common, and women who have never had more than two outfits in a lifetime, including one for ceremonial purposes, now had several. As the number of goods rose dramatically, they became less expensive and more widely available. By working in specialized factories people's income also rose. Science discovered many new patterns. Educational opportunities were increased by employers who had to educate their staff.

The Industrial Revolution is credited with ending struggles against famine, pandemic disease, and extreme poverty (compounded by endless cycles of despotism and war). Industrial technology is good at solving certain problems, but it creates new problems and depends on the specific characteristics and scale of a centralized society. And, these new problems included starvation, diseases, and poverty from the distribution of new wealth.

Industry did not solve the problem of war, which seems to inhabit the core of industrial culture, also, especially in terms of expenditures. Conflicts over identity and resources occurred when people saw that their needs were being ignored or threatened. Industrial culture values competition, violence and domination more than it does cooperation, peace and equality. This seems to be a historical trend.

In the last century, humanity doubled the amount of its crop land use; the number of people increased by four times; water use increased by eight times, energy use increased by a factor of 16, and industrial output increased by a factor of 40. Understandably, the quality of human life also increased, especially in terms of life expectancy.

Industrialism is a special case of production, with special characteristics, not the least of which is cheap energy. Using the benefits of science, industry has been able to produce items that aid and enhance human efforts to understand nature, as well as make luxurious places within nature.

6.6.1.7. Enhancing

Human beings were clever enough to find ways to enhance their senses, for instance, certain chemicals from plants. Someone discovered that a lens of water enlarged an object underneath it. Lenses of glass allowed things far away to seem close-up and small things close up seem larger. This allowed us to see deeper into the structure of things, as well as see new things with microscopes and telescopes.

6.6.1.8. Repair/Replacement

Tools are also used to replace parts of or repair the human body. We can use chemicals to repair imbalances of acids or bases. Other chemicals can relax or stimulate the body. Medical tools can be used to remove pathologies or repair breaks.

6.6.1.9. Communication & Information

Clay markings, the abacus, and now the computer allow exchanges to be recorded. Computers permit us to wildly exceed the limitations of our brains.

Technology is a long historical development that started during hunting and expanded with agriculture, and it has ushered in tremendous changes. Ivan Illich points out that in the 1930s, nine of ten words a person heard were spoken directly, one to one. By the 1970s, the proportion was reversed, and nine came out of a speaker. He noted: "Computers are doing to communication what fences did to pastures."

Any of the newer tools, such as computers, can be used just as well to bolster profits as to save forests. The government has learned to use the same technology, although the nature of the technology may in general undercut authoritarian use. Lewis Mumford thought that authoritarian rule encouraged technological progress. It makes it easier to manipulate and control. When the system promoted some of the ideals of democracy, that everyone should have a share of goods, the system gave the largest share to the rich. And, the poor took the bribe, and can wish for more and be happy. Those were 'magnificent bribes' as Mumford noted.

The introduction of silicon and the fabrication of the microprocessor is now the heart of information technologies. Computers dominate industrial production. Some computers may radically increase our ability to understand and shape the future.

With expanding computer capabilities, virtual worlds can be created that can embody many of the values of the material world. The generation of these worlds would require fewer resources than if the activities had occurred in physical space. The only problem is that it takes a lot of energy and space to create those virtual worlds. The virtual world is a model of reality, but not having the origins of reality, its simulations could meet needs where tangible material products no longer are necessary to satisfy some of those needs, which may resemble Maslow's hierarchy needs. Obviously the lower needs, such as food or comfort, could not be met in by virtualization.

As tools increase in complexity, from knives and levers to computers and space stations, so does the knowledge needed to support them. Complex modern tools require libraries of information that has to be continually increased and improved, then spread and understood. Technology first simplifies life, then complicates it. Some technologies, considered network technologies, increase benefits when others also use them: Mobile phones, internet, and computer operating systems.

6.8. *Rethinking Cities*

Cities developed as adaptations to a changing climatic conditions, first so food could be stored against droughts and hard times. Then cities started to change mobility and habits, as well as work styles and possessions.

6.8.3. *Cities*

Cities are the chosen domiciles of most human beings. Yet, we rarely think of planning them. We need to think about cities as a form of ecosystem that can be made more self-sustaining and complex. The size of cities can range from 25,000 to 20 million.

6.8.3.1. What is a city?

This city is a complex structure characterized by buildings, roads, residents, density, division of labor, air-conditioning, domestication, partnerships, patterns of movement, numbers, concentrations, intensity, and miniaturization. Who could produce a city just described—over a hundred million years ago? Termites.

Termites have a division of labor. Several million photophobic insects have special jobs. Workers, with no wings to cool themselves, build the tower. Structure of building with microclimates and air-conditioning due to the tower effect or chimney effect—hot air rises and is replaced with cool air from the subterranean levels. Mounds are created particle by particle, over years, and get to 10-12 feet high, but may last centuries. They can reach 120 feet deep into the soil. The principle cost is the tower. No other costs associated with cooling. There can be interior farming of fungus, like leaf cutter ants. The fungus grows without weeds or diseases. Termite mounds in west Africa have cooling fins. Some mounds use mud, or wood, some use excrement. Termites also make roads on tree trunks and on the ground; some of these are covered, where different species come together they make traffic circles. Termites develop domestication and partnerships (with flagellates to digest cellulose, and fungus). Termites do not consciously design and plan their city mounds. The structure is assembled by many millions of workers following simple individual rules.

Human cities have many similarities to termites and to other natural structures. The development of cities from encampments and villages is like that of tidal lagoons, where groups of algae ended with communities. Unorganized, homogenous ensemble transforms to an organized structure that cycles energy, materials, and information, and provides benefits to its residents. Human beings have characteristics, however, that make their cities more complex, diverse and mobile. We can design cities consciously.

6.8.3.2. Why have a city?

Why should termites or human beings live in a city? To band together to survive droughts? To trade in one place? To worship together? To promote specialization? To increase excitement and inspiration? To defend wealth? To change the planet? Because it's genetic? Because it's inevitable? Why continue to have cities, now that technology can provide the excitement and intensification in rural communities, or now that urban problems seem insurmountable?

6.8.3.3. Problems with Cities

With cities come unanticipated or unwanted effects, sometimes erroneously called 'side-effects,' although they are equivalent to the intended main-effects. The new city ecosystem, that replaced a native ecosystem, is greatly simplified; there are fewer plant and animal species, and lower diversity leading to homogenization, especially when cosmopolitan or favorite species are present. Plant productivity declines, due to fewer plants and lower rates of photosynthesis. Connectivity to the surrounding matrix is lowered—the matrix itself is minimized by increased patches and corridors, such as roads, which perforate the landscape and also act as barriers to plant and animal movement. The number of patches increases, which decreases interior species and increases edge species, such as squirrels, raccoons and coyotes. Native patches and corridors, such as woodlots or streams, are reduced or eliminated. The agriculture around cities is more homogenous, with decreased fallow areas. Stream corridors are degraded or destroyed, making a city more vulnerable to floods. Sewer systems route more water faster to an ultimate sink, lakes or the ocean. Household wastes affect the landscape directly, if they are buried or burned. Nutrient and mineral cycles can be disrupted. Local weather can be disturbed. And, the environmental can become degraded.

Cities change their own microclimate, with heat islands and dust domes. The masses of materials used to build the city are used inefficiently. The energy to run the city is greatly concentrated—thousands of times more than in a native ecosystem—and the energy is used inefficiently. Unless the infrastructure, from buildings and power lines to sewers and roads, is kept up, it can decay and cause greater inefficiencies and problems. A budget crisis, from loss of tax revenue or improper use of funds, can affect the infrastructure and most amenities of a city. The disruption of supplies, due to social and political actions, such strikes or attacks, can lead to other problems, from displacement to disease. Even in small, well-managed cities, there is an increase in violence, sickness, drug use, crime, and impoverishment, all of which require some response from the political structure of a city. Ian McHarg noted that with crowding, there was an increase in mental and physicals illness from stress.

The human system in the city, with imported food and water, has a smaller system of carnivore predators, such as fleas, lice and bedbugs, and decomposers, such as bacteria, fungi and gulls. The surrounding native system of primary productivity is based on depauperate system with few trees and layers and a simple food chain of rats, birds and squirrels.

6.8.3.5. Redesigning & Designing the City

Much design has to do with solving one problem or a set of problems. Very rarely is a city redesigned to solve problems. And, very few cities are designed to avoid recognized problems.

One challenge for a city is to balance its inputs and outputs to minimize the effects on the residents and the environment. For example, G.T. Miller notes that for a city of a million people, there are daily inputs: Sunlight (higher or lower depending on size and shape), atmospheric deposits (), water (625,000 tons), food (2,000 tons), fuel (9,500 tons), manufactured goods (), and buildings (high or low depending on area and height and very importantly the success of the city attracting new people or businesses). The outputs do not always balance inputs: Heat leaving is often greater because of the use of fossil fuels; water may be less due to evaporation or use in processes; sewage may be high (500,000 tons) and include water; solid waste/refuse may be high (9,500 tons); and, pollutants may be significant (for air 950 tons).

Because cities are so concentrated, and occupy a relatively circumscribed area, they require distant support systems for food, materials, and energy, as well as a smaller system for pets (monkeys, cats, canaries, or fish) and support machines, such as automobiles. This requires constantly increasing inputs and outputs.

6.8.3.6.1. Goals & Enantiodromia

The goal is not only to create good places, and be connected with other good places, but to express meaningful ways to live. Unfortunately, goals are shifting patterns that are different for different images in different times, and also become responses to the fitness of past goals (see Tables 68361-1 and 68361-2).

Table 68361-1. Goals of modern individuals (or the small picture)

In	Neighbor-hoods	Housing	Transporta-tion	Jobs	Leisure
Goal	Separation	Size	Personal	Money	Fun/rest
What's Missing?	Community	Comfort	Usefulness	Meaning/ plea-sure	Fun/rest
What's the Result?	Isolation	Waste	Spread	Dissatisfaction	Fatigue
What's the Re-sponse?	Violence	Unhappiness	Pavement	Sabotage	Ennui
What's replacing that?	Ecovillages	Small homes	Light rail	Self-employ-ment	Hobbies
And, what's Replacing it?	Villages in arcologies	Self-built	Walking	Satisfaction	Watching

For instance, in neighborhoods, the goal of separation resulted from older traditional urban neighborhoods with some crowding. The separation as it was achieved, however, lost the feel of community, as people had their own quarter-acres and personal cars to reach the local malls and supermarket complexes. They felt more isolated and more vulnerable to fear and violence (especially the middle classes). Many of these people decided to intentionally bring back the sense of community as well as reduce waste and vulnerability to violence by creating intentional communities, such as ecovillages. But, even these did not always have the density that inspired creativity and invention.

Table 68361-2. Goals of modern institutions (or the big picture)

In	Science	Technology	Management	Corporations	Religion
Goal	Data	Technique	Efficiency	Money	Truth attendance
What's Missing?	Knowledge	Appropriateness	Humaneness	Real cost	Meaning
What's the Result?	Irrelevance	Mass produced	Bureaucracy	Conversion	Rock & roll in services
What's the Response?	Personal search	Simplicity	Law suits	Activism	Personal religion
What replaces that?	Astrology	Wood shop	Computers	Limited Liability Cos.	Scientology
And, what replaces it?	Ecology	Personal	Work effort	Nonprofits	Loose associations

Goals are obviously imperfect. There is always some element missing that results in a suboptimal condition. And, the resulting dissonance prompts a response, which may result in an improvement or a less satisfactory element. Perhaps with continued experiments on ways of living, combined with an ecological perspective and forms of ecological design, communities can optimize human happiness or at least satisfy it minimally. What do we design for?

Maximum numbers? Maximum luxuries? Maximum happiness? Based on our knowledge of systems and maxima, it is dangerous to design for maxima of any kind, or even for optima—perhaps it is best to aim for a satisficing level. New local village and city patterns have suggested that we can design for safety, design for peace, and design for health.

6.8.5.3.1. Deep Ecology Movement

When Arne Naess presented his deep ecology alternative, starting in 1972, he contrasted it with the shallow ecology movement, which he characterized as fighting against resource depletion and pollution—its objective is the health and affluence of people in developed countries. Deep ecology is a movement that goes beyond a concern with pollution and resource use, with its humanity-in-the-environment image, to consider humanity in a relational, total-field image. The movement promotes human equality, conservation, and local autonomy. In principle, it proposes a biospherical egalitarianism, that is, the equal right of all beings to live in place.

Table 68531-1. Contrast of Industrial and Deep views.

Industrial view	Deep ecology
Nature is a human resource	Nature has intrinsic worth
Human pan-dominance of nature	Human harmony with nature
Unrestricted growth	Population fit to nature
Unlimited resources/substitution	Limits to use of resources
One-way progressive technology	Appropriate technology
Unrestrained consumption	Intelligent frugality
Unsustainable economic growth	Homeorhetic economy
Global management	Local/regional control
Competitive, destructive relations	Cooperative relations
Material goals	Spiritual/material goals

Naess characterized deep ecology in his article thus: It is a relational total-field image (knots in a field of intrinsic relations), with biospherical egalitarianism (key words: respect, understanding, right to live and blossom, space and crowding). It is based on principles of diversity and symbiosis (where diversity means live and let live rather than either/or, and symbiosis means coexistence). It takes an anti-class posture (key words: exploitation), in the fight against resource pollution and resource depletion (ecologists as informants). It seeks to create and use complexity, not complication (there is multiplicity and lawful factors; and division of labor, not fragmentation). For effectiveness, it depends on local autonomy and decentralization (to strengthen local regions and encourage self-sufficiency). The movement is based on a philosophy of ecology, ecosophy.

6.8.5.3.2. Technology Design Related to Properties of Culture and Life

We can use the properties of ecosystems to consider design concerns with technology, although those concerns might be better linked with the properties of a culture: Conduct, wholeness, flexibility, adaptation, endurance, and vitality. Each of these properties should help us understand the limits and alternatives to design technology.

6.8.5.3.2.1. Conduct Related to Design Technology. Conduct is the course of cultural behaviors through a behavioral landscape. The concept of the epigenetic landscape can be used to explain why people become trapped in the use of a technology. The use-need path or chreod is deeper than the cognitive path that relates technology to history and need. So, we have to use the technology even though we may be aware of simpler solutions, such as bicycles or evaporative cooling. If the need chreod is too deep with investment or profit, we cannot explore the cognitive chreod. This may explains why necessity cannot be the mother of invention; the necessity path, eating to avoid starvation for instance, is too deep to allow exploration of a shallower path of daydreaming or design. The broader behavioral landscape of leisure is needed. Conduct can be described as the stable path of culture through a landscape of possibilities. Once the course has been set, it is most likely to channel most subsequent behaviors, unless some event or catastrophe triggers a deeper course.

6.8.5.3.2.2. *Wholeness Related to Design Technology*. Culture provides an identity for its members. It tells them who they are, where they came from, and why they are special. Identity is basic to human existence. People are identified by their roles. A person is an incarnation of his and her group—even in industrial culture, one is identified as an astronomer or farmer. Specific technologies are often part of an identity. Identity is that persistent quality that can be described apart from its performance in interactions, but not isolated. The relationship between identity and wholeness is a rhythm, with unique patterns. In fact, culture is concerned with all things and beings in a whole. If the rhythm is whole enough, it could incorporate appropriate technologies into it.

6.8.5.3.2.3. *Flexibility Related to Design Technology*. The rules of a culture lend order and stability to the whole, as well as flexibility. Flexibility means not being over connected, or not being too rigid or too efficient. Than means that culture is able to slough off older technologies for newer ones—or new ones for appropriate technologies that fit better into a sustainable pattern. That way culture is able to keep some options and unused connections open. Some of the flexibility comes from different ways of establishing connections in specific places. Culture is bounded by ideas and habits, but is open to flows of energy and materials. It is a loose-fitting patchwork of ideas, relationships and things. It requires order, but not too much order. It is tolerant of discontinuities and contradictions, and this gives it flexibility. If the contradictions between technology and moral behavior become too great and maladaptive, however, the culture can collapse. Flexibility can allow the choice of technologies regardless of novelty or fad-value.

6.8.5.3.2.4. *Adaptation Related to Design Technology*. The patchiness of culture is parallel to the co-constrained construction of a species and its environment. Co-constrained construction enforces coevolution, the emergence of a highly ordered complexity to full structuration. Culture has to balance between embracing change and resisting change. People show a desire for new technologies, tools and products, but often fear and resist change. Resistance to change is a normal part of a cultural process. Groups like the pygmies have specialized to fit the requirements of the environment, successfully. This makes it difficult to adopt other cultural arrangements. A culture modifies and is modified by the environment, and both can adjust and survive. Technology is part of a culture and it aids survival in the wider environment. Each technology is a way of exceeding the limits of the environment. This changes the environment, which pushes new changes in human thought and technology. Technology itself, however, has not been used for improving ecological balance. It could and it needs to be used that way to aid survival.

6.8.5.3.2.5. *Endurance Related to Design Technology*. A cultural system is stable and persistent in time. It is a general property of some systems that acquired information is used to close the door to further inflow. A mature culture needs less information, since it works toward preservation, and closes itself off to information that does not fit the shape of the culture. The effect of maturity is to allow a maximum variability between systems with slight external differences, such as place or initial conditions. Cultural shapes need to be loose enough and diverse enough to change when new information is required. Cultural stability is the ability to maintain cultural identity under the flow of external forces and disturbances. A culture has to be able to resist disturbances that are too disruptive. Sometimes those disturbances result from an inappropriate technology that then needs to be modified or rescaled. If the technology cannot be modified, it has to be resisted. Resistance is a positive act for a self-reliant culture. Culture also has to be resilient enough to recover from intermediate and small disturbances from the environment or technology.

6.8.5.3.2.6. *Vitality Related to Design Technology.* To be constant and stable, a culture has to be vital, that is, it has to be productive, to be able to convert energy and materials into foods and structures for survival. With order and integration come stability and security, without which no one can survive. Primordial security comes from a physical, knowledgeable relationship with nature. Stability can also be related to compartmentalization, communications, and the richness of interactions and connections. The system, with all of its technologies, is self-creating. It renews itself as its contents change, as disturbances change the parameters of the system. But, technology that does not fit the culture or its environment is a barrier to cultural renewal. If a culture is vital enough to be productive with less complex technologies, then it can judge and choose technologies according to its best longer-term interests and not be trapped in the depth of habit or investment. When human societies were small, the amount of control and security required was small. Although societies have grown to immense sizes, human security has not, despite incredible increases in the sophistication of technologies. Perhaps that is because vitality does not always require sophisticated technologies beyond a certain point to achieve goals of a good and stimulating existence.

6.8.5.3.3. Technology Design Related to the Properties of Life.
We can use the properties of life to consider design concerns with technology. These properties are: Movement, cellular structure, irritability, growth and development, adaptation, and reproduction. Each of these properties should help us understand the limits and alternatives to design technology. Because of the emergence of technology from human creativity, many properties of technology seem similar to living.

6.8.5.4. What is Necessary?
Everyone agrees, to some extent, that we need science and technology to meet human needs. And, these systems have to be integrated systems of production, consumption, and distribution. They have to be postindustrial and ecological.

We tend to think of civilizations or nations as metaphors for behavior. Primasck and Abrams suggest that identity with 'generations' might be more meaningful than civilizations. But, generations are shaped by cultures and now by global influences and trends.

Advanced communities have to adapt biologically sound processes. Pollution is only a symptom of imbalance and improper resource utilization. The wars on pollution and on poverty are as ineffective as the wars on people. A serious problem is our lack of understanding of the extensive, long-term effects of pollution on the atmosphere. Once it is determined that materials and wastes, especially nuclear or chemical are dangerous and long-lived, they should be minimized. There are ways to reduce generation of wastes by acting on the flow: Reduce number of products; reduce the quantity of waste in each product; increase the durability of each product; and make sure that the cycle is a spiral. The best results might come from a mix of these strategies.

The greatest threats facing ecosystems and biomes, and the organisms and species adapted with them, are not just disease organisms or pollution, but the synergistic effects of fragmentation, pollution, and climate change. And these are best addressed all at once.

One problem with an industrial ecosystem might be how it fits with natural ecosystems. A related problem might be the total energy and material in the industrial system—that is, the scale might overwhelm natural ecosystems accustomed to solar energy budgets. Although we can expect many species to invade or exploit an industrial ecosystem, we need to choose them carefully so that they do not overwhelm the system.

We can use the properties of ecosystems to consider design concerns with industry, starting with 'Course.' Living systems do not display homeostasis—constant value—so much as a particular course of change in time—homeorhesis (from the Greek meaning 'same flow'), according to Conrad Waddington. The course is stabilized, not the constancy. Changes to a system are symbolized by trajectories in a multidimensional phase space or landscape. Homeorhesis is a significant phenomenon in evolution. Waddington applies it to the tendency of a process to continue in its original pattern, even if disturbed. Homeorhetic mechanisms protect the system from many disruptions. Negative feedback counteracts the effects of change to maintain the system in a steady state or homeorhetic state. A mature community is self-perpetuating and homeorhetic, with a dynamic balanced energy-matter budget.

The course exists in a natural topology with valleys and hills. The valleys occur in various shapes. Some could be very narrow canyons, others large meadows, similar to old mature earth forms, with meandering rivers. The name for a characteristic of an attractor surface in multidimensional space is a chreod, not a valley. The cross-sectional shape of the chreod describes the reaction of the system to fluctuations. With a steep slope, for example, it is difficult to divert the developing system from the bottom of valley; even with a strong force the system will return immediately as soon as the influence stops. With a shallow slope, like a flood plain chreod, it is easy to divert the system; it meanders before returning. But actually the perturbations alter the landscape itself, making a steep valley again, not just shifting the river. As the environment changes, the system changes with it.

Ecosystems have many properties and are affected by many environmental conditions. Their changes are symbolized by trajectories in multidimensional phase space; orderliness can be described in terms of constraints on trajectory courses, and these constraints are visualized as attractor surfaces. If the system starts from any condition, represented by a point in multidimensional phase space, the trajectory will move to nearest attractor surface and then move along it. If industrial design can be fit, with proper scale and energy impacts, into the order of ecosystems, then the whole system could function with minimal disruption

Another property is identity. An ecosystem has an identity as a whole. The ontology of any living system is the history of the maintenance of its identity as a whole through continuous self-making, or autopoesis. If there were no identity, there would be no differences and so no relationships. Ecosystems are part of an unending, imperfect process, without any final state. Furthermore, the human attempt at perfectibility through self-improvement causes disharmony, which is part of the same imperfect process. Each system is a practical application to place. Unknown factors determine a large part of the operation of any system. Furthermore, there is chaos in every system; there are plagues and random frenzies. Industrial ecological design has to accommodate the uncertainty and imperfection. In grassland, industry has to be part of that grassland.

Diversity is an emergent system property. The environment has been constant enough for organic evolution, but variable enough for natural selection to be challenged. Variability challenges organisms to adjust and thrive. Variability, even in small ways, leads to diversity.

Diversity, as a measure of genetic variability in ecosystem, enlarges information. A mature system needs less information, since it works toward preservation. The limit of maturity allows maximum variability between systems with slight external differences, like temperature.

The industry has to enlarge the system, and increase diversity, but help the system stay within the limits of variability to preserve its basic structures and functions. The variation of climate, for instance, an external disturbance, can change the physical and chemical parameters of the forest. Acid rain causes biochemical and soil chemical reactions, which in turn provide in-form-ation to other organisms, which use it to regulate their physiological processes that can affect the forest, e.g., insect damage changes the level of seasonal growth of trees. Damage to trees from pathogens, or disease agents, is an internal disturbance. Industry can produce some disturbance, like any natural phenomenon, but it has to be limited to the stability and resilience of the system. The use of energy has to keep pace with the process of maturation of the system.

The property of Coconstruction suggests that the organism and environment are co-implicative, co-defining, and co-constructing. They engage in a process of self-assembly, where the complete self is the organism-environment system. Construction requires participation, complexity, and development. The process of construction involves a self-presentation offering new symbiotic relations and novelty. Novelty always enters with environmental change, which serves to maintain the openness of the system. Novelty enters with fluctuations. The "strategy" of ecosystem development is increased control of, or homeorhesis with, the physical environment and novelty—probably to protect itself from perturbations. There is a fundamental shift in energy flows, as increasing amounts of energy are used for maintenance. As more and more energy is used for maintenance, the net community production (NCP) approaches zero. The mature system becomes more efficient, as it supports a larger biomass with the same amount of energy. The food chains become more weblike, dominated by detritus chains as opposed to linear grazing. Construction depends on diversity for the reciprocal constraint. Local context allows for more rapid construction. The constraint forces species to change. Industry, when young, resembled pioneer ecosystems—as it matures, and is shaped by ecological design, it may more resemble a more mature system, with less gross productivity and a more scaled and directed productivity that fits within the limits of the ecological framework.

The property of stability is the ability to maintain the identity of a system under the flow of external forces and disturbances. Stability can be refined through the specifics of constancy, resistance, resilience, and accommodation. Stability can be related to ideas of compartmentalization, communications, richness of interactions, and connections. Ulanowicz suggests that stability might be explained by diversity flow topologies, where flow topology is a descriptor of how ecosystems develop. The stability of ecosystems, as originally proposed by Eugene Odum, becomes the result of regular flows of energy and materials. Growth and development are characterized by a qualitative formalism of increasing ascendancy, which explains the drive towards coherence, specialization, and self-containment. Industry cannot kick any of these properties out their topological courses.

Productivity, the final property, is the ability to convert energy into living forms and the ability to incorporate materials into living forms. Productivity, in general, depends on the vigor, or strength or vitality or health, of the system. Health is the overall ability of a system to maintain itself under a normal range of environmental conditions (which may include hurricanes, volcanic eruptions, or fire). Health is a dynamic measure of ecosystem organization, vigor, and resilience. Organization is described by diversity and connectivity; vigor is related to the amount and speed of productivity; and resilience is a measure of reaction to stress. Too much stress, for example, leads to unsustainable patterns of behavior; continuous stress leads to a breakdown of processes that becomes irreversible—the system dies. To relate health to growth and productivity, we could say that the capital of an ecosystem would be its physical environment and its gross primary productivity; interest would be the net ecosystem productivity. Our measurements of productivity, however, are not adequate. We are measuring over a year or two only to establish a growth rate or productivity. We should be measuring over centuries.

The production percentage would be the amount necessary to keep the ecosystem healthy. Obviously, a pioneer community may change the conditions to favor a new level of the system with new components. With a goal of maturity, an industrial ecosystem could furnish some of those new components, as long as they fit into the processes that lead to productivity. This may limit the scale of industry, so that economic scales may become less important than the ecosystems scale, with its limits and long-term changes.

An ecological industrial system would be concerned with the interrelated design of the movement of things and energy, and with the creation of power and waste. If resources stayed in place then the culture in each place could specialize in value-added economics, instead of moving the raw resources around. The system could optimize the scale and style of mass production, so that some needs, such as clothing and transport, could be made efficiently and inexpensively, but to be high-quality and long-lasting (within the constraints of the need for change and fashion). The system would emphasize repair, reuse, and reassembly; in fact, this part of the system could be extended to local shops and crafts people or engineers, which would involve local populations.

6.8.7. *What About Civilization?*

The real question is whether any civilization that human groups have formed is sustainable. We know that foraging groups (hunter gatherers), and some herding, horticultural and agrarian societies are sustainable, mostly because of their size and limitations. But, we do know if this is or will be true for modern civilization, especially an urban, industrial civilization. Amory Lovins suggests that modern civilization is imperiled by three main problems: The dissolution of civil society, weak support systems, and dwindling public wealth to address problems and suffering. Ignoring these problems could lead to worse violence and terrorism that would further undermine communities and nations. Perhaps the problems are unavoidable and emerge from our commitments to science, growth, and flirtation (commitment) to edges of chaos.

The direction of modern science seems imperative. Francis Bacon thought that we had the power to remake nature for human needs, and that history was moving to a complete human mastery of nature. Many other thinkers state that nature is moving towards a goal, of consciousness or completion. This might be the controlling idea of modern civilization: Progress leads to a perfect heaven on earth. Progress cannot be stopped, and the suffering and bad ideas in its name are tolerable side effects. The atomistic aspect of science concludes that nature is composed of indifferent atoms and selfish genes, and not a holistic structure with integrity. Nature is simple, orderly and predictable; she can be dominated and controlled. Therefore, we can do anything, or try anything we want, and see what happens.

In a similar way, growth has become an imperative, also. Civilization, with population and energy use, has to grow to be efficient and successful. But, growth can affect the extent and kind of problems. Growth can lead to an unsustainable pattern of civilization were over energization and overuse can lead to inappropriate sinks and explosions in ecosystems, starting with human ones.

We can think of the complex of civilization as rational response to climate change; we can think of cities and centralized religions and governments, as responses to drought, in a different way than nomadism. This alternate direction became more complex. Government became important to coordinate larger numbers of people engaged in irrigation of plants and storage of surplus food, to distribute rewards, settle disputes, and defend against aggressive neighbors, whose crops may have failed. Religion became more important as the climate became less predictable, as people looked for ultimate rewards.

The direction of modern civilization is based on myths and fallacies that are not rational or sustainable. For instance, one American myth holds the contradictory views that civilized processes cannot rip the fabric of nature, but if it could, it could repair nature. Another myth is that the US is doing everything it can to help the disadvantaged and poor in its nation as well as around the world—this is ridiculously not so. The poverty of other nations affects any one nation. Ignoring international diplomacy, military decisions to protect the nation are based on the fallacy that military actions will provide security for one nation even if the rest of the nations are unstable. Extremism and violence threats affect every nation eventually. This is more so, when nations are linked in global travel and trade.

After traveling in Italy, as American Ambassador there, G.P. Marsh pointed out that civilizations have collapsed due to ignorance of natural laws. He recognized that climate change could contribute to collapse, as well as that human actions could result in large-scale harm to the environment. At first humans responded to and adapted to environmental changes; then they started changing the environment in all its diversity and dynamics to suit our immediate needs. Sometimes, civilizations collapsed, if the there was imbalance or excess. When earlier civilizations collapsed, however, they were isolated from groups by natural barriers as well as by looser links of trade. The drivers of the collapse were local and maybe regional. In general, groups were self-sufficient. When Maya civilization collapsed, at least the urban political structures, the people were able to return to simpler forms of foraging or agriculture.

In an interconnected global civilization, massive ecological or social failure in one region might threaten the stability of the entire global system, if people cannot go back to a state of relative isolation and self-reliance. Furthermore, the role of chance or contingent events becomes larger, as droughts, shortages or diseases affect different groups or areas. Geoffrey Chew suggests that the collapse of civilizations is not really a problem for learning, but it is a problem of not learning, as when cultures choose not to learn from past mistakes. Of course, like death, collapse is hard to learn for the dying or collapsing entity—in fact the lesson cannot be learned if the collapse is complete. In order to avoid collapse, a culture has to learn from the past. But, nothing in our past has prepared us for the number and strength of global connections. During any collapse, many cultures have been able to adapt and evolve by expanding into other areas, but this is no longer possible in a completely occupied, territorialized, owned planet.

Every civilization faces a spectrum of risks, which may push the civilization to find new sources of land, food or energy. The need to solve such risks pushes the system further from its original state and can accelerate the operation of the system. Acceleration makes decision-making, especially weighing alternatives, much more difficult. The conjunction of long-term risks can lead to a management or governmental crisis. We have never planned or designed our civilization to respond to impacts and the dynamic changes of the entire planet. We never designed civilization to respond to complex challenges or to slow or invisible catastrophes. Thus, we are not responding to many kinds of catastrophes, and when we do, we only shore up a quick fix, regardless of the regularity of the catastrophe. Sometimes we may not recognize that the catastrophe is an effect of our designs and activities. Sometimes when we do recognize that, we think that we can use the same mindset and approach to correct the problems later, when it might be more economically or politically feasible, yet it rarely is.

Furthermore, we are engaging in several large-scale experiments, which can conceivably impact large areas of, or possible the entire planet. These experiments are not just our civilization, but also our way of growing food, traveling, and converting landscapes. They are experiments on the change in the makeup of the planet, because we select so many different kinds of organisms that we need and we transformed so many ecosystems. Civilization and cities are large parts of that transformation, now. Albert Einstein noted that problems cannot be solved at the same level of awareness that created them. Because the global system creates problems, such as war, poverty and environmental destruction, it cannot solve them by continuing that behavior.

But problems can be solved at a different level, by new type of planetary civilization with different values and social institutions. Many problems are not global or regional or ubiquitous. Change can results from either random mutation or accidental behavior. But, conscious invention can produce new structures with new capabilities. If a global civilization is to form and to survive, it must develop into a completely new type of social system. A blind, consumer society cannot be transformed into a conserver society without constructional change.

Because modern civilization is dynamic and developing, it has the potential to change itself to be more stable and less intrusive. Individuals, groups, and nations have the ability to design social and living systems. We do not require more data or knowledge before we start redesigning systems. Buckminster Fuller said that to change something you have to build a model that makes the existing model obsolete—it is not necessary to fight existing reality.

In this approach, a eutopian strategy needs to dedestruct the destructive impacts of civilization. Maybe we can design a general self-regulating feature to lessen environmental stresses when they become apparent. But, we need to be able to see and understand them first. Another thing that a eutopian strategy can do is to dewesternize civilization. We also need to deglobalize those things that should not be globalized, such as traditional cultures. Perhaps what we need is a form of cultural anarchy that could be coordinated through the United Nations or new framework. Culture is the process that gives meaning to the lives of people who live in the culture—it is not decoration or artifice, it is a body of knowledge that allows people to make sense of the world. We need the eutopian framework also to dehomogenize the world. We tend to be drifting toward a more homogenous world since our weak global culture tends to be bland and generic. Yet, if it wipes out all other cultures, there will be no rivals or successors. Traditional cultures are not failed attempts to be industrial; they are manifestations of the human experience of specific places. They allow us to be aware of our place in this area of the planet.

One serious question is what to do with those nonurban or nonindustrial cultures that choose to continue to be nonurban or nonindustrial. Should we make a park for them? Should we isolate them in some way or have some kind of boundaries that sort the technology that they can use? Maybe the word Park is not the best way the best word, since they are not zoological specimens. Maybe we cannot save them that way, but what we could do is allow them to create the boundaries they want.

Vaclav Havel has suggested that a fundamental shift is needed to change the direction of the current civilization. His deep conviction is that the only option is to enlarge the sphere of the spirit. It may not be enough to invent new machines and new institutions. We must develop a new understanding of existence on earth. Only by making a fundamental shift will we be able to create new models of behavior and new sets of values.

Morris Berman considered that as our civilization matures, we could recognize that visual communicators sometimes manufacture misleading memories, which can be as dangerous as 'hot steel.' One solution is that everyone has to participate in design, to become a designer, perhaps as everyone may have been in a foraging society, where we made everything and built everything ourselves. The future of civilization is our common design project, Berman stated. He suggested that designers have more power than they realize, especially in critical situations, and ours is critical. One great threat to human future is consumption or overconsumption. At the present, design itself fuels mass overconsumption. We could leave a better legacy by using our best ideas immediately, and not rely on copying our chromosomes. We have made many major errors in human civilization, from our agricultural conversion of ecosystems to the production of plastics and synthetics and to our unplanned growth. Human civilization cannot afford major mistakes. We have to try balancing civilization as a local and global phenomenon.

Figure 687-1. A once successful city in Mesopotamia

7.0. Managing Design Levels

Even after heroic designs have been composed and applied, they have to be maintained and managed. Management is a continuous requirement. Management is necessary to keep the designs from being abandoned or subverted. Human settlements can be managed in small-scale ecovillages and communities. Large human settlements, or urban areas need to be managed in watersheds and regions.

The history of human use has resulted in numerous problems, which could overwhelm any attempts at frameworks or designed landscapes. For instance, the continual growth of industrial societies, not just populations, but the entire infrastructures, could overwhelm any design. The historical inequities between cultures, and even within most cultures, unless it is addressed, could tear apart the society before it could improve circumstances with designs. Thus, design has to include such things as poverty and unfairness, bad distributions and human greed. Corporations are examples of situations where legal changes and scale have released an entity from any normal ethical or legal restraints. Corporations have to be limited before designs can be effective.

Many design factors have to be considered, from economics and city shapes to religion and the power of religion to bind people into voluntary limited arrangements. These factors could distort a design, so design must account for them, also.

7.1. *Problems: Growth, Inequality, Poverty, Dominance & Slavery*

Growth serves as a mechanism of evolutionary adaptation, by carrying out genetic instructions in the environment; but growth is also conservative and stabilizing rather than innovative and reorganizing. Growth is homeorhetic; in a homeorhetic process, the flow is constant, not a stationary state. Flow processes follow fixed trajectories, called chreods. Growth, from fertilization, embryo states, birth, youth, and maturity, represents a homeorhetic process following more or less fixed chreods, programmed genetically and conditioned environmentally. Chreods seem to be like electron paths, that is, probabilistic. Sometimes, growth can be problematic, when a physical body grows too much flesh or a population keeps growing past the carrying capacity. Other things, such as ideas, and wants have no natural size to exceed, but can cause other kinds of problems at inappropriate scales or locations.

7.1.1. *Growth of Populations & Economies*
Variability in resources and weather, as well as dynamics with other populations, cause changes in a population. As a result of these influences and interactions, a population can grow or shrink or die out. It can stay in place or disperse. Natural growth of a population is related to development and maturity. Population growth occurs when the birthrate exceeds the death rate, or when immigration from other populations exceeds emigration. Under many circumstances growth confers a survival advantage. At some point a population stabilizes. Economic growth can track population growth, although economies can grow with stable populations.

7.1.1.1. Brief History of Human Populations
A population of individuals grows, or overgrows, like a population of cells in a body. And, individuals, like cells, receive signals from the environment that tell them when to die. Unlike cells, however, which rely on chemical signals, these other signals are in the form of food sup-

ply, climate, and predators. Some animals try to avoid such signals by moving; some plants try to avoid those signals by growing larger or producing chemical defenses. Human beings have used movement and technology to overcome those signals. And, we have become very good at converting ecosystems to food systems and using technology to control our home environment.

Regarding the success of human population growth, it is possible to relate it to explicit changes in ideas and technology. Agriculture, for instance, sparked an acceleration in population growth. It offered more food in one place. It allowed a sedentary lifestyle, which permitted larger families, especially for increased labor requirements. This increased population density in small areas. And, it triggered a demand for more food to feed more people in a permanent area.

The ideas of storage and urban settlement allowed more people to live together. But, large populations allowed diseases to survive, also, including those that would die if human populations were isolated. Other changes that permitted growth included improvements in sanitation, water and waste flows, and building.

Urban cultures responded to population growth by intensifying resource extraction (which still continues with the green revolution and genetic engineering), by relying on more technology (that can result in overfishing or overharvesting), by integrating support systems, e.g., transportation and banking, by increasing centralization, bureaucracy and stratification, and finally by migrating.

Cities grew and continued trading at larger scales. Early globalization affected populations by introducing diseases from other groups into isolated populations, resulting in catastrophic losses of peoples. This lead to changes in immune systems, also more symbiosis between hosts and agents, and much later, to public health changes like sanitation.

Traditionally, population growth has been a simple issue of the pressure of numbers against resources. However, there are many other factors that contribute to this, such as lifestyles, energy use, waste, and ecosystem conversion. Population is a complex issue, but many pundits try to simply it by saying that if populations continue grow, the economy will do well. That might be true if resources were infinite and ecosystems were invulnerable, but they are not. Others say that the danger is a confrontation between a pessimism and smugness. Neither extreme is a good approach to addressing the limits on or impacts of populations.

Design of populations through planning has to balance migrations, levels of luxuries, needs, inequities, and many other things. International agreements can balance emigrations and immigrations. Some violence erupts from uncontrolled migrations, but this can be ameliorated with the lessening of institutional inequities and the need to migrate. The United Nations control of common areas, especially the entire atmosphere and hydrosphere, can set limits for the unbridled expansion of economics and populations. Finally, a shift in human consciousness, that results from new images of the planet and human destiny, could make these changes agreeable and desired.

7.1.1.2. Growth of Populations

Agriculture, growing crops and raising animals, provided more food, but it was less nutritious and less palatable. By increasing the population, agriculture increased widespread hunger. The reduction in biodiversity by monocropping caused some ecological crises, that resulted in periodic and devastating famines.

Despite famines that caused hundreds of groups to wobble or collapse, the overall human population has kept increasing. The world population is increasing at an alarming rate. By 1900, the rate started to increase steeply. At a rate of increase of two per cent per year—

below the current rate—the earth's human population will reach 50 billion by 2100. By 2280 it would be 1.2 trillion. That is 300 years from now (1980).

Nations that seem to have adequate resources, may be growing too fast. Even the Union of Soviet Socialist Republics, with all its potential agricultural land, and its uncertainties of drought and frost, may be increasing their population too fast, as may Canada and Australia. Population growth can become a political weapon, in a democracy, when two groups coexist, with neither ecological differentiation nor geological separation, and only breeding control for coexistence. For instance, in Sri Lanka, the Sinhalese and the minority Tamils, breed for political control, ignoring population control. Voluntary controls fail; the population increases by almost two and a half percent per year. The problem of population control is a tribal problem, not a racial or economic one. It is a local problem, more than a global one. Even with trade and charity, in most areas population is limited by local resources and local limits; only in a few cases, such as the Netherlands and Hong Kong, can a population be supported by massive ghost acreage elsewhere.

Sir Charles Darwin's view of humanity was that it was not domesticated. Humans were still wild. Unmastered, they have no breeding control; therefore, they will eat to the limits of the natural food supply and press against social and biological limits. The global population explosion, an example of autocatalytic nonlinearity, is not considered a problem for many thinkers, such as Eric Jantsch, who believe that this growth constitutes an essential factor in the creative act of gestalt formation. Growth does lead to intensity at some stages of development in the life of an individual, group or species, but growth can stop, and development can also lead to the kind of creative intensity that is identified by Jantsch and others with mindless growth.

The principles of ecology can offer information on human societies. For instance, the principle of competitive exclusion states that two dissimilar races cannot occupy the same niche in the habitat. If humanity is wild, and not domesticated, this may help explain war and racism. Territoriality is no longer a rule in human ecology, though the instincts may still be operative. Human ecology defies maintaining its population at an optimum level. The young are overproduced and protected; the least fit are not eliminated. Medical science has also increased the number of old and maladapted dependents of society. Problems with populations bristle with social and political implications. A balance in birth and death rates is necessary for ecological stability. We have controlled epidemic death and infant death diseases. We need to correct the overbalance in births. Toughness and ingenuity might be required.

7.1.1.3. Growth of Economics

Although economics is a social science that studies human behavior, it considers itself a positive science, that examines "what is" with theories, as opposed to a normative science, which addresses "what ought to be." Keynesian economic theory, the predominant theory in industrial countries, holds that the full utilization of resources is necessary to ensure full employment and the maximum social good. This economics depends on economic growth to avoid crisis. The major premises assume that: Population will grow, that social good is related to the equitable distribution of material products, and that if resources are limited, technology can erase the limits. The economist Kenneth Boulding referred to this as a cowboy economy, an economy that has yet to bump against real limits.

M.D. Mesarovic and Eduard Pestel stated that "the issue for the economy is not to grow or not to grow; it is how to grow, and for what purpose." They claim that if a workable world system is to emerge, it must be after the establishment of an organic pattern of growth.

Due care is devoted to describing such a pattern and contrasting it with other, tragically inapplicable patterns of growth. Their treatment of the world system itself was regionalized and multileveled. They recommend that the establishment of organic growth was necessary with no need for special no-growth policies for populations or economies. They assume that further industrial growth will continue, that economic growth is good, and that this growth solves human problems as long as it is organic.

Exponential growth is said to be bad, and organic growth is said to be good. In fact, although organic growth is better, there is little difference during a world crisis; both reach asymptotes of suffering. One need only regard the population crashes of lemmings and others to see how organic growth can go wrong. In the organic world, growth is healthy only when the rate of change is decelerative in the long run; cancer and population are constant or accelerative. Mesarovic and Pestle fail to realize that continued economic growth in any form is a threat to the stability of the biosphere.

Economics became enamored of growth during a critical time in history. Rapid European expansion occurred at rates rarely exceeding a growth of one percent per year, and with unparalleled opportunities for expansion into sparsely settled areas, such as North America, Australia, South America, and South Africa. The economy has been growing almost constantly since it has been studied. We have been trying to force it to grow, rather than let it stabilize or contract. Even if it stopped growing, the economy could still develop. Mesarovic and Pestel, as well as many others, confuse growth with development. Mill did not.

7.1.1.4. The Assumptions of Growth

The metaphor for the economy used to be a simple mechanical model for turning resources into products. It was assumed that to be successful, the economy had to grow and turn a profit continually. Unfortunately, the assumptions of the model were also simple and failed to consider human needs and natural cycles, causing great suffering and great disruption. These assumptions resulted in overgrowth, with increases in complexities and costs. Overgrowth contributed to economic and ecological instability (refer to Section 6.6.2. for a longer discussion).

Continued growth of the "free market" is amoral and pathological, benefiting the elite of authoritarian regimes as much as the oligarchs of democratic ones. It refuses to recognize, much less to pay, all of its costs, such as depletion, loss of security—which may be most important—or extinction. The entire system perpetuates mass poverty and justifies it by blaming individuals, but the system itself fails to reduce inequity or poverty. This loss reduces effort; it is responsible for tiredness and low kinds of health, productivity, and esteem, things that are necessary for personal and systemic renewal.

Some theorists, like Paul Samuelson, have concluded that growth is necessary to rid the economy of disparities. The need to grow is intrinsic in this kind of economic system. A large literature has treated perpetual growth as the only conceivable state of affairs. Capitalism depends on growth for stability. In an organic system, however, growth contributes to stability, but it cannot continue beyond maturity, due to real physical and biological limits. Mature organisms grow to maturity, but then they can keep developing without growth during maturity.

7.1.1.5. Economic Growth or Development

Many physical structures, such as talus slopes, grow by addition, until some physical limit, such as gravity or the coefficient of friction, is passed, then they stop growing. Some stability can be gotten from growth in early stages; later stability must result from limits and metabolism. Growth in plants can delay the onset of senility by ridding the plant of waste products in more diluted form. However, too much growth produces a strain on tissues and early decay. In fact, one herbicide promotes excess growth as a means to kill weeds. A biological organism grows to maturity, which is a stopping point for size. The organism continues to develop, however, experiencing and learning the environmental complexities through mating and then to the end of life. Development may include growth at some stages, but development refers to the continued change after growth has stopped. Mechanisms for growth can become pathologies when central authorities meddle in stable lower orders.

Development instead of growth would equalize wealth more efficiently—after all, economies have been growing for at least 400 years and increasing the disparities. There is no necessary association between development and growth in economics, as Daly and others have shown. Growth means increase. A community is forced to accept an upper limit, beyond which it cannot grow any further. Further growth results in destruction or disruption of itself and nature.

There is another distinction between growth and development. The ecological social approach (or a redistributive environmental strategy) to development makes it irrelevant to discuss global limits to growth. Local limits are far more significant to majority of population. Regardless of how much food exists, people will starve unless they can get it. Redistribution of resources and improvement of environmental quality (in the home environment) are more important than increased production by sophisticated technology. The natural capacity of regional photosynthesis must be limiting factor in development, especially in tropical and subtropical areas.

We have economic growth; we can see the numbers. But, the growth is premised on saving costs by forcing down wages, or by reducing the number of workers in the name of efficiency, and forcing overqualified workers into service jobs. The growth promotes inequality, improvement for a few and impoverishment for most. It is the growth of a tumor, issuing a healthy glow from a fever and the false image of health. Profits go up, but public services decline for lack of funds. There is no money for schools, none for libraries or parks, little for private institutions, and little for national, state or local governments. Where did it go? Profits? Profits for individual corporations, profits for individuals? Could we find them, can we track the money? We should be able to, since the management revolution has made paper trails everywhere. Perhaps the trails are too complex.

It has been said that international goals are best realized through national self-interest. This must be the old wisdom. It might have been true, if nations were perfectly rational and knowledgeable; they do not seem to be. Nations now seem to be the handmaidens of corporations, whose interest is only in profits for shareholders. The views of shareholders are notoriously short-sighted. The UN is limited by images and ideals of progress. The UN's solution to economic problems is "sustainable development"—that is, "growth that respects environmental constraints," as if growth respects any constraints. The Bruntland Report indicates a five to ten-fold increase in world industrial output within the next hundred years before population stabilization occurs. While the appeal of growth is unarguable, it is really not likely to be sustainable in any meaning of that word, since sustainable growth does not recognize known ecological limits. Furthermore, the UN has no power to coerce its members when it does make good recommendations.

7.1.2. *Inequality—Economic & Social*

The assumption of inequality is new with agricultural civilization. At the time of Bacon, it was assumed that there was: An absolute, immutable, omnipotent God, everything was sorted into a great chain of being, economic subsistence was preferable, and social inequality was unavoidable. Later, Darwin incorporated a different set of assumptions into his theories: Absolute space-time; atoms as discrete units; economic discrimination, and continued social inequality. These assumptions contributed to the misuse of his theories to justify social and economic conditions at the time.

Many modern political ideologies and economics have been shaped by the principle of endless wealth. Adam Smith once calculated that the real price of anything was just the toil acquiring it. Inequality in a world of abundance could only exist through human suppression and exploitation of other humans. The invalidity of this principle of plenitude came with the recognition of limits.

With surplus, came redistribution and the beginning of inequality. By 4400 YBP in Mesopotamia, there was status differentiation, as evident in the sizes of homes.

Government has become subservient to economic actors, according to John B. Cobb, Jr., partly because the ideology of economics is so positive. It proclaims that continued growth will solve most of the problems of modern civilization, from poverty to conflict, although the promise has not been fulfilled. The problems have increased: From food shortages, housing shortages, and energy shortages, to unemployment, inequality of opportunity or goods, environmental deterioration, increase in weapons, and insecurity.

Growth has been touted as the answer to inequality. We have economic growth; we can see the numbers. But, the growth is premised on saving costs by forcing down wages, or by reducing the number of workers in the name of efficiency, and forcing overqualified workers into service jobs. The growth promotes inequality, improvement for a few and impoverishment for most. It is the growth of a tumor, issuing a healthy glow from a fever and the false image of health.

Economies are out of balance. Modern economies, embracing the idea that "nature is capital," draw on the accumulated "capital" of ecosystems for production. By ignoring the real cost of the capital, as well as the costs of natural services, such as nutrient recycling, soil building, and atmospheric renewal, these economics create a temporary wealth—similar to the healthy flush of a fever, perhaps—and a long-term imbalance. When an economy falls out of balance with its local environment, massive disruption often results; industrial economies have only avoided disruption by trading advantageously with other economies, by using fossil fuels, and by promoting general institutional inequality.

The economic government, according to Robert Reich, creates a corporate hierarchy, with pervasive inequality between all levels, rather than a democracy. To keep profit growing under any circumstances, corporations have been willing to accept damage and conflict.

Nations are basically exploitative of other nations and smaller cultural groups, which may not be considered nations because they lack a permanent military: The United States concentrates on Latin America; Europe on Africa; Japan over Southeast Asia; Russia over Eastern Europe; and China over Tibet. The reasons for this continued behavior include: The rapaciousness of society; the acceptance of war; and the economic advantages of large-scale operations. The cultures of industrial nations are based on unethical accumulations of materials. Inequality is maintained by power, not persuasion, and also by the assumption that solutions are extrinsic and external and have to be found by spreading out rather than intensifying efforts to find solutions at home.

The predominant value in small cultures, and then large, was harmony. This minimized conflict that might have resulted from inequality. Confucian concepts of ritual and etiquette helped to regulate social conduct and made people feel good about their station. For example: "Inequality is the nature of things" and "seek no happiness that does not pertain to your lot in life." This may not work at the scale of nations and global corporations.

7.1.2.1. Distribution of Wealth

In almost every size of human groups status is divided unequally. In hunting groups, better hunters become the hunting leaders. The success of a hunter allows the hunter to distribute the cuts of the game between others, often according to an understood set of rules. In larger groups, with a surplus of materials, the materials are distributed according to status. People have equal access to status positions in society. The number of positions of prestige is adjusted to the qualified candidates. Status is separate from wealth. An animal divided up according to ratios regardless of who kills it.

J. S. Mill narrowed the scope of economics to production and the scarcity of means; he considered distribution to be a political process, since it depended on laws and customs that varied widely in different cultures and ages. One function of culture is to distribute material goods or energy. Economic culture defines the means of production and livelihood, techniques of distribution, and values and norms underlying economic behavior (can be more closely related to kinship. Order is a cultural problem. Order provides stability and security. Cultural order is necessary to deal with the redistribution of wealth and power.

With the creation of more wealth or new wealth, the distribution becomes skewed. Accumulation increases. More kinds of wealth become invented, dangerous, and useless, and more skewed. But, as long as an economy does not reach the limits of wealth, it can keep growing; these new kinds of wealth allow that. Cultures encourage the unequal distribution of resources.

Wealth is based on production in capitalist countries. The production of wealth from growth depends on technology. The technological perspective is oriented toward materials and not humans or ecosystems. Nature is considered to be a resource to be exploited. The immediate objective of technology is to create wealth through knowledge. Technological activities are justified on humanitarian grounds, scientific discovery increases the well-being of human society, yet the social consequences of scientific activity are ignored; short-term suffering will be offset by long-term benefits, it is claimed. But, without a long-term view, long-term benefits will be worse.

7.1.2.2. From Difference to Inequity

Inequality is more the result of differential development than of exploitation. According to Boulding the greatest source of the differential is different rates of accumulation of knowledge, capital and organization; the rates are essentially internal properties of cultures. Although a minor element in terms of transfers, it is a large psychological perception, which may need to be compensated for in a global community.

Most of the wealth used by modern economies is nonrenewable. These resources are limited, interrelated and distributed unevenly. Forests are a special problem; although trees can grow to a good size in 30-40 years, forest ecosystems may take 300-600 years to develop and then last for thousands of years. Oil, coal, peat, and some woods are functionally nonrenewable. Geological time periods are required to produce them.

The distribution of symbolic and real wealth is very inequitable as the result of historical trends, old economic rules, and cultural confusion. Karl Marx considered that misery is caused by class conflict about distribution of material things. Some of the problems of distribution have to do with the size of society and the scale of its operations. For instance, in Hutterite communities, usually less than 150 people, the distribution of goods rarely failed; people got their share, rarely less or more. With a larger group of Hutterites, the distribution started to fail, and the group divided.

The modern market distributes some benefits to all, but its scale allows unfairness. Economics are the harmonious distribution of wealth among people. But, problems outside small-cell scales. Kohr points out that Marx failed to link misery to the scale of economics rather than the system. The problem is with overgrowth more than style, which is why socialism based on overgrowth looks the same as capitalism overgrown.

7.1.3. *Poverty & Consumption*

Poverty can be defined as a lack of needs, such as food and clothing. Affluence and inequity seem to be producing new forms of poverty, such as lack of use (use-value of feet in Los Angeles) or the potential poverty of not having what is possible elsewhere in NY or Paris. Modernizing needs creates these new dimensions of poverty, and this leads to new discrimination. A violent cargo cult that seizes traditional cultures. The opposite of this is new forms of wealth, such as useless wealth or harmful wealth that actively subtracts from the enjoyment of life.

Ecological analysis would force us to look at the obvious—generating nonmarketable use values occupies the center of every culture because it provides a satisfactory life to its members. Needs are almost completely defined in terms of commodities. Even relation needs a commodity, travel, to be fulfilled. Once the need is defined and partly satisfied, it becomes a right. From walking to the right for passenger miles at high speed in a mere fifty years. Stung by the suffering of people with new unsatisfiable needs, we have offered to produce more, safer goods. Instead of ecologically analyzing the relationships of needs to satisfactions.

Consumption is touted as the solution to poverty. We all live, thanks to modern advertising, in a dream world of mass consumption. The purpose of expositions changed gradually from instruction in wonder to simple entertainment and buying. The Crystal palace in 1849 shows the former and the 1855 Exposition in Paris, the latter, where they put price tags on things and charged admission. The first department store opened in 1853; this changed the tone of shopping from haggling to inspecting a fixed price item.

Consumers are trained to enjoy waste, violent speeds, harsh medical treatments to extend life, and standard rations (mcRations?). The state has to try to discourage the lack of patriotism that comes from deserting from the standard needs diet. The military is actually a symptom of the orientation of the state. Health and education use a military rationale. Industrial society is constantly mobilized for emergencies, in the battles against, noneducation, poverty, diseases, terrorism. Industrial development has never been nonviolent or respectful to people.

Specialists incapacitate people's autonomy by forcing them to become consumers of care, instead of learning to care for themselves. Half of our needs are made by armed bureaucracies that have been growing since Louis IVth Empire. Industrial services see state security the generator of society's production patterns, even if it is a spin-off of military needs.

7.1.4. *Cultural Dominance: Colonialization*

Dominance is a social phenomenon that describes social relationships. Dominance in animals is related to a social hierarchy that allows the dominant animal to feed first. Dominance in plants is often related to mere numbers or the ability to grow faster and reach the light first to spread more leaves to catch light. Dominance is humans is more complex. Like wolves, human beings interact with the individuals of other species or with entire species. Like other mammals, humans change their habitats to suit themselves. Humans have modified animal and plant associations in a different way from other mammals, simplifying patterns of energy and chemical exchange and solidifying themselves at the end of many food chains as a dominant species. A dominant is a species with greater influence than any other in its biotic community, changing the lives of other species and the character of the habitat.

Human populations have increased exponentially, with billions in giant urban ecosystems. Agriculture has produced monumental yields, but only at the cost of tremendous erosion and great subsidies of fertilizers and pesticides. Dams have been built all along rivers, and riverine forests have been cut, altering rivers and fishing grounds. Changes have been made without regard to the long-term impact on the ecosystem or on its human population. We dominate entire ecosystems, starting with our own species.

7.1.4.1. Cultural Status

Like many mammals, human beings create social hierarchies, which allows some individuals to get more food and better mates. Status is also assigned according to the hierarchy. Although, in primates, the largest, fastest and usually younger males often dominate, old males can sometimes dominate with the help or permission of females who may prefer an old predictable leader to a young unpredictable one. Status has to do with standing in human society, and with appearance and ownership. Status may come from longevity, of the self or ancestors, as well as from the results of good decisions, from hunting for example, or from owning more than others, or just more things or more people. Status is a powerful human need, and may drive the growth of goods or populations. Perhaps related to population growth and size, changes in cultural status become related to equity, distribution, and dominance. As distribution becomes more unequal and as conflict occurs more between groups, dominance and slavery appear, and are both related to status.

Darwin has been criticized by Malthus for extending the popular theory of economics to the natural world. It is true that the biological considerations of economics inspired an economic description of a biological law, but Darwin enlarged and supported the metaphor. He retained the idea of hierarchy, but status became a form of fitness, specifically in birds and mammals. It can be argued that status has fitness values in human groups also. For instance, a woman is more likely to mate with a man who has higher status in the community.

Status is a social need, which is justified by a culture. Status levels in traditional society included: Nobles, commoners, and slaves. Slaves were captured or purchased. In earlier days, people were sometimes captured by enemy tribes. The return home of the captives, either through payment of ransom or owing to a retaliatory raid, was a special return. Like other property, slaves were given away as gifts. Occasionally, they were killed and eaten, for instance during the Kwakwakawakw cannibal dance. A commoner was simply anyone without a chief's position, potlatch position, seat, or standing place. They are not a class with a function. At times, nobles may retire from potlatch positions and become commoners. Kinship was the main determinant of status. For the Kwakiutl, the display of status validated the social system. The redistribution of food validated the cultural status of the leader.

Inequalities of status and wealth became associated with permanent settlements.

Furthermore, other social things developed such as complex division of labor, with castes and slave labor. Trade increased and competition increased. This may have had to do with resources that were abundant and could be stored, and transformed into political prestige and power. Agriculture provides more status, more wives, and more things. It allows armed struggle also. Art, in addition to luxuries and profits, was tied to status.

7.1.4.1.1. Equity & Distribution—Having More

In almost every size of human groups status is divided unequally. In hunting groups, better hunters become the hunting leaders. The success of a hunter allows the hunter to distribute the cuts of the game between others, often according to an understood set of rules. In larger groups, with a surplus of materials, the materials are distributed according to status. People have equal access to status positions in society. The number of positions of prestige is adjusted to the qualified candidates. Status is separate from wealth. An animal divided up according to ratios regardless of who kills it.

J. S. Mill narrowed the scope of economics to production and the scarcity of means; he considered distribution to be a political process, since it depended on laws and customs that varied widely in different cultures and ages. One function of culture is to distribute material goods or energy. Economic culture defines the means of production and livelihood, techniques of distribution, and values and norms underlying economic behavior (can be more closely related to kinship. Order is a cultural problem. Order provides stability and security. Cultural order is necessary to deal with the redistribution of wealth and power.

With the creation of more wealth or new wealth, the distribution becomes skewed. Accumulation increases. More kinds of wealth become invented, dangerous, and useless, and more skewed. But, as long as an economy does not reach the limits of wealth, it can keep growing; these new kinds of wealth allow that. Cultures encourage the unequal distribution.

Inequality is more the result of differential development than of exploitation. According to K. Boulding, the greatest source of the differential is different rates of accumulation of knowledge, capital and organization; the rates are essentially internal properties of cultures. Although a minor element in terms of transfers, it is a large psychological perception, which may need to be compensated for in a global community.

The distribution of symbolic and real wealth is very inequitable as the result of historical trends, old economic rules, and cultural confusion. Karl Marx considered that misery is caused by class conflict about distribution of material things. Some of the problems of distribution have to do with the size of society and the scale of its operations. For instance, in Hutterite communities, usually less than 150 people, the distribution of goods rarely failed; people got their share, rarely less or more. With a larger group of Hutterites, the distribution started to fail, and the group divided.

The modern market distributes some benefits to all, but its scale allows unfairness. Economics are the harmonious distribution of wealth among people. But, problems outside small-cell scales. L. Kohr points out that Marx failed to link misery to the scale of economics rather than the system. The problem is with overgrowth more than style, which is why socialism based on overgrowth looks the same as capitalism overgrown. The change in scale drove the changes, from the transition of states, from surplus distributors, to tribute-driven to commercial exchanges. There were political transitions from leaders, to chiefs, to tribute systems, and to economic and political trade systems.

7.1.4.1.2. Dominance

Strength or status can give rise to dominance. A dominant is an animal or person with greater influence in the community. Dominant behavior may be biologically-based or cultural. Dominance gives priority of access to food, sex, or space. In a majority of primates, males are not dominant over females, except where there is a large size difference. Where sizes are roughly the same and where female coalitions occur, there is not much sexual dominance. In some human groups, males may attempt to dominate females with the threat of physical violence, for instance, among the Yanomamo. Among other Amazon peoples, such as Mundurucu, it may be due to control of ritual objects and rituals, for dominating ceremonial life. In spite of the cultural forces dominant at any moment, an individual has the potential to determine a different course of action.

7.1.4.2. Colonialism

What is colonialism? Who does it, who benefits? The history of colonialism. The Phoenicians sent out colonists to found Carthage. Settlers from Carthage founded Barcelona. The Greeks made colonies in Italy, Sicily, Turkey, Libya, and Spain. Norway sent people to Iceland and Greenland, France, and Ireland. India sent colonists to Sumatra, Borneo and Java. The Inca moved villagers to Ecuador from Bolivia. Southern Indians called Tamils, crossed over to Ceylon, which was occupied by the Sinhalese. And, of course the Europeans tried to control all of their global trading by controlling the people and the resources in Africa, the Americas, Asia and the pacific.

Colonial countries introduced new ethnic rivalries wherever they took over. English imported Indian people from India to work sugarcane fields. By 1970, half the Fijans were Indians. Fijans tended to eat strangers and castaways, sometimes enemies from raids (and some recipes were the same as for pigs). But, only men could eat human flesh because of the strong manna.

In Tasmania, aboriginal people had woolly hair rather than curly, were shorter, had brown skin rather than the matte-black of the continental Aborigines. In technology, they had digging sticks and spears, but not boomerangs or woomera (atlatl). They understood curating fire, but could not start it. Their 12,000 year separation for any other people or culture is the longest for any people, even Rapa Nui. The English wanted to create or recreate a European world, without savages. The native peoples did not fit. All of Tasmania was first made into a prison; all the native people killed off. The culture was destroyed. The environment was destroyed. The English were not tolerant of dissent, variety or diversity. Neither seems to be the modern civilization.

The Kuna women of the San Blas islands north of Panama, wear clothing with appliquéd materials in geometric designs with real and mythical beasts and random words or letters copied from travelers. Now popular with tourists. The Kuna, who were never conquered or subdued, even now must require ships to pay homage to the chief, aboard the ship, in terms of meal or gifts. The Kuna build simple, large houses and cultivate bananas, breadfruit and calabash trees. They are the only people visited by Columbus who are still alive and living on their islands. Others were displaced or killed.

Colonial administration required central government and strong armies. The Ottoman expansion into Europe annexed one food-producing region after another. Sustainable intensive farming was abandoned as colonialism gained. There is a direct connection between colonialism and the abandonment of land, having to do with urban immigration or slavery.

7.1.5. *Cultural Dominance: Slavery*

To be enslaved means to be owned as property and divested of freedom and rights, or it means to be completely dominated. The word slave comes from the old Slavic word for the Slavic people, who were among the first slaves to be held in Europe.

Slavery is the idea that people can be held in bondage to perform work. The Greeks thought that freedom depended on some slavery. Some societies thought that their economies depended on slavery. It was actually thought that slavery improved the lives of slaves, since they were often considered to be from subhuman groups. It was also thought that it was a personal decision or personal right, although the slave might feel differently. There are of course, different kinds of slavery now. For humans, there are living slaves, but future generations might become enslaved to the decisions of their parents, especially as regards losses and debts. Animals have been enslaved. Machine slaves were the next economic boost. Finally, industrial societies have energy slaves, ten to twenty for each person. Slavery is based on a number of assumptions, such as contempt for the enslaved, or denial that there is wrong.

Different kinds of slavery can be distinguished: Opportunistic, institutional, or comprehensive. Opportunistic slavery was the result of raids or conflicts with other bands in archaic societies. Institutional slavery included labor force collection, as practiced in Egypt in 4500 YBP, or economic slavery for labor collection, as run by the British 300 years ago. Comprehensive slavery refers to the use of wage earners, animals, and energy.

7.1.5.1. Opportunistic Slavery

Many archaic cultures had slaves, who were usually the victims of conflicts between bands. In Nez Perce gatherings and celebrations, slaves were exchanged, along with furs, roots, berries, and fish. Slaves could often intermarry with their captors and acquire status; their children were almost always free. Slaves became a separate social class in some archaic societies. By 5000 YBP a stratified class society had developed in places like China—slaves at the bottom, then peasant farmers, craftsmen, then the elites of administrators, religious and military. In Assyria, slaves could not wear veils. Slaves worked everywhere in Assyrian society, for instance, even running businesses. Some people entered slavery voluntarily to pay off a debt.

7.1.5.2. Institutional Slavery

For the Tukano of Peru, the Maku group was a slave class, that is, a source of slaves. The Kwakiutl also had slaves, who were captured or purchased. Like other property, slaves were given away as gifts. Usually, they were captives kept near the door to guard the house. Occasionally, they were killed and eaten during the cannibal dance. Viking raiders and colonists owned their own land and used Irish slaves for work and for wives specifically (based on genetic evidence). As the number of slaves diminished through death or marriage, some farmers lost their land to absentee landlords, who, without ecological or personal feedback, would make poor decisions about crops and practices.

Economic slavery occurred in Mesopotamia, Slaves worked in all levels. Many were prisoners of war; some were criminals being punished; some entered slavery to pay off a debt. But, slaves could also own a business. If a slave married a free person, their children would be free. Slavery became specialized with agriculture, in China and Egypt.

In China, by 1615, slavery was hereditary, for agricultural or household slaves. Slaves reflected the stratification of society as it became more complex. Slaves in the Americas were considered to be needed because the amount of acreage increased with the discovery of new lands; the acreage and the slaves were considered necessary to develop more wealth. Slavery kept labor artificially cheap. This dampened the incentive to develop new technologies, but, new technologies eventually surfaced because they proved to be cheaper than slaves.

The appearance of mechanical devices such as the sugar mill or Eli Whitney's cotton gin helped to support the system of large plantations based on a single crop. The Industrial Revolution after the late eighteenth century swelled the population of towns and cities and increasingly forced agriculture into greater integration with economic and financial patterns.

Internal slavery was a feature of Europe from the Romans to Middle Ages, but disappeared when feudalism disappeared, and labor was no longer the scarce factor in production. With the opening of America, however, land opened up—or rather was claimed and conquered—faster than it could be filled with European people. Therefore, colonists tried to force native Americans into labor. Native Americans, however, were still dying from European diseases, which made them unsatisfactory as slaves. Europeans looked to another tropical continent, Africa, where the people were resistant to old World diseases. Less than thirteen years after Columbus, in 1505, African slaves were introduced into Haiti. Over 350 years, ten million more people were brought over, in a large trans-Atlantic trading circuit that involved rum, sugar, cloth, and timber as well.

In the American colonies, the independent, more or less self-sufficient family farm became the norm in the North, while the plantation, using slave labor, was dominant, although not universal, in the South. With the British, Americans, Africans, and Spanish, slaves became a commodity in a trading empire.

7.1.5.3. Comprehensive Slavery

After the enslavement of peoples, some dominant cultures turned to the enslavement of nature and then of energy. The English treated tropical lands as enemies to be defeated, then enslaved them in plantations. Their cultural attitude as conqueror of nature led them to treat biogeochemical cycles and soil requirements as temporary obstacles in a world where everything had its price.

The new economics of the industrial age depends on wage slaves, that is, people who need jobs to get necessities in urban areas. An overeducated labor pool led to secretaries and then managers, and to the fulfillment of the managerial revolution. Now, industrial society is able to use energy slaves to accomplish work. Energy, in human work equivalents, from animals, natural and fossil fuels, is used to perform work.

The concept of slavery need not be limited to plants, animals or humans. If we consider our machines as "energy slaves," then Americans each use the equivalent of fifty slaves per day. The computer is an information slave and must be the equivalent of at least three other human slaves. Slave owners love the privilege. And the exploitation of inanimate slaves is more easily justified. In fact, to some extent, the culture of arts and sciences may not be possible without machine slaves. But, at some point, slavery corrupts the owners, making them physically or mentally soft. What is the exchange for summoning these information slaves so easily? The failure to develop intellectual ingenuity? Loss of imagination? Lack of trust in intuition? Let us consider an earlier exosomatic adaptation that helped humanity. Knives permitted hunting larger game animals or deeper roots, but the long-term anatomical result was partial degeneration of the human jaw. Garrett Hardin notes that all exosomatic adaptations bring about "a corresponding degeneration in the endosomatic function," i.e., knives change the function of teeth. The species is then more vulnerable to accidents, since external adaptations, like pacemakers or artificial kidneys, become required for the health and maintenance of civilization. Perhaps writing has changed the function of human memory. Perhaps computers will change the structure of thought and maybe reflective consciousness.

7.1.5.4. A Question of Consciousness

The institutions of slavery raise questions about our relations to each other and to nature. What is our human relation to reality—slave, master, participant, or partner? Cosmology describes the place of a culture in reality. Cultures can determine inappropriate attitudes towards nature and that can result in ruin for the air and land. The worldviews of many people defined other people as nonhumans, who could therefore be improved by being allowed to work for true humans.

Although slavery is officially condemned in virtually every nation, the inappropriate use of people, animals, machines and energy continues, as a necessary part of the agricultural and industrial economies. And, it will continue to be until there is a shift in consciousness that would let us consider the ethical implications of our uses of everything and everyone, especially how these uses perpetuate patterns of domination and inequity, as well as contribute to the degradation of ecosystems and geocycles.

The ideas of dominance and slavery have been integrated into many kinds of designs, from the design of housing and chains to clothing, roads and auditoriums. Changing some of the traditional designs may reduce feelings of inequity and perhaps the actual inequity.

Religion could help control disruptive forces, especially those that result from distribution and power. Religion coerces people into a social contract. Religion and story-telling reduce the variability of individuals wants in a group. The introduction of new stories and myths, concerned with partnership and equality and with the value of other places, species and people, could change our attitudes about them into a healthier context.

7.2. *Avoiding Adaptations & Traps*

Human groups have adapted to their environments, have changed them and have been changed by them. Some human groups were trapped by their environments and could not change. Sometimes the adaptations themselves can act as traps that limit human behavior. Perhaps education can be designed to afford understanding of adaptations and traps.

7.2.1. *Adaptation & Drift*

Rene Dubos suggested that humans can adapt to anything. Humans adapt rather than plan. Are we genetically hard-coded to do that? Overpopulation is adaptation to an uncertain environment (or is it an expression of expansion of all life?). Agriculture is adaptation to diminishing game and slow overpopulation; it is stable source of lower quality, lower diversity food. Irrigation is an adaptation to growing exotic crops under low moisture conditions. Permanent settlement is an adaptation to more abundant local food supplies. Large buildings and cities are an adaptation to permanence, slow overpopulation and uncertain environmental conditions, such as droughts. Coal is an adaptation to overuse of wood, oil to overuse of coal. Forest over cutting is an adaptation to environment and increased demands for houses and glass, but it changed the environment and made it more variable. Ecoforestry and conservation biology are adaptations to the overuse of ecosystems.

Our genetic make-up predisposes us to some things and pushes us in some directions. It makes limits on our plasticity. It could promote behaviors that damage the environment, and hence our long-term interests. If limited by stone-age genes, then some pro-gene behaviors may be ineffective, others effective. For example: We must have contact with natural environments where we evolved or suffer psychological and physical harm. Behavior is determined by immediate personal consequences (short-term egoism), regardless of its consequences in modern world.

Places that have been successful for a long time, or that are likely to continue to be successful, may well have another quality, which may not be immediately apparent—the places adapt easily to changing circumstances. Finally, places that are successful in the long term, and which contribute to the wider quality of life, will prove to make good use of scarce resources. They are sustainable.

Adaptation is a response to changes in the environment and the evolution of organisms. There seems to be no direction in evolution other than to live with a changing environment; genes become plans on how to live from experience; drift is selected to fit an environment that is evolving. The planet as a whole inclusive of many subsystems regulates the stability of the climate and environment. Plant diseases, plagues, floods, volcanoes cannot occur too often, or nature could not purify air and water or reseed devastated areas.

Evolution is an integrated process, partly open-ended, involving choices, and the selection of whole individuals in whole environments. The cost of evolution by selection is so heavy that most of the time most populations are not perfectly adapted to changing environment. For the evolutionary process, Varela suggests the metaphor of "bricolage," which is the putting together of parts in complicated arrays because they are possible (rather than an ideal design). He calls his alternative view evolution by natural drift.

Change is inevitable and we have to adapt to it, but we must make sure that we do not adapt to a low-level style of living, to drifting lower standards and performance, as Donella Meadows warns. We must not try to adapt, for instance, to an impoverished world without wilderness that we could preserve or keep separate. Some human systems get worse, that is, they drift to low performance. Rivers get dirtier, hospital service gets worse. Rather than adapt to these conditions, humans need to create goals and plans to avoid the drift to low performance. Similarly, our many individual enhancements to the human body, especially mechanical replacements of organs, without consideration of changed interactions or consequences may result in what William Thackera calls 'Borg drift' (from the word cyborg). Drift is a systems trap.

7.2.2. *Drifts & Traps*

Meadows calls the systems structures that produce common patterns of problematic behavior 'archetypes.' These archetypes are traps. A trap is a device for catching or holding animals (from the old English word 'to step'); it can also mean a stratagem for catching people. Many things can act as traps; many situations can act as traps. Other systems traps include: Addiction, policy resistance, arms races, and the tragedy of commons.

There are energy traps for instance. Agriculture traps energy for people, but it also an energy trap of people, because it allows a concentration of energy, that is, higher yields, but then it requires more energy be put into the system to maintain it. The system has to produce more energy than it uses to be useful. And be sustainable, with a surplus for trade. The use of energy in general creates an entropy trap, since most of the energy (84%) is wasted, and it creates disorder in the surrounding system, making it harder to get more energy.

Agriculture could be considered a population trap, also. Not being limited by having to move or carry children, also, being sedentary, farmers had more kids. More kids required more food, which could be provided by agriculture. But, because of the surplus, farmers could out-compete foragers on a smaller territory, so if agriculture collapsed they could not go back to foraging. Perhaps the trap was also a material trap. Things required storage or transportation. Storage was cheaper and easier.

There are serial traps, also. An example of a serial trap is the use of resources by a people, where the replenishment rate is constant, and the rate of use exceeds it. This trap results in ecosystem degradation that may be irreversible. Industrial strategies sometimes mistake the rate of discovery for the rate of recovery. The two are different.

The rate of population growth can act as a trap, and the shape of population pyramids can act so, since it can change the reproduction rate dramatically. There is nothing to constrain humans in their growth. We could even adapt to too many humans and too much suffering, perhaps through abstraction and art. But, art could also be survival technique. That is, it undermines the paradigm by showing the invisible effects of adaptation.

Design in west suffers from easy living and convenience, suggests Victor Papanek. He states that convenience is the enemy of excellence, fashion the enemy of integrity, and cuteness the enemy of beauty. Convenience is a trap. Can the new object, advertised as more convenient, do something more, in a cheaper cost? Is it more desirable and environmentally friendly? Papanek suggest that there are ways to question technological improvements that can lead to convenience traps: It is too small, like watches with a television screen? Is it too powerful for what it is, such as certain power tools or knife sharpeners? The fact that there are too many of them may indicate a trap—televisions, radios, and stereos. It is too much to fit with our other things. It is too complex to operate or repair, with too many buttons, switches, and options. It has been improved too much, with idiot lights and time-zone maps. It is 'state of the art,' but untested. And, finally, the package may be as or more valuable than the product, if appearance or newness is the primary selling point. We are trapped in the production-consumption-discard cycle. Conservation was a convenience during good economic times, but no one wants to really be inconvenienced. We recognize the damage, but do not want to sacrifice yet. Thus, the slaves are bought off for a fix. Ironically, our wasteful habits may force discipline on developing nations.

Language and art are also traps. Language because one is limited by words. Art because one is limited by styles or demand. If art is articulate, it may use silence; but if art eliminated all "speaking," does it not trap itself in a dumb stasis? Does it not reduce it self to one dimension? Silence is not enough to communicate. Science can act as a trap, when its logical structure limits approaches to nonhuman phenomena.

Culture can be a tool, an external set of memories and guides, but it can also function as a trap. The human culture that allowed humans to survive rapidly changing environments seems to create crises during more benign climates. In small scales, culture can be very flexible. But, with changes in scale, culture can be very inflexible and wrongheaded. Although culture is loose enough to incorporate medium-sized changes, such as automobiles, it may still be rigid enough to act as a self-perpetuating, inflexible monster when dealing with large challenges. Traditions, especially those reinforcing identity, can force people to keep to old ways that are inappropriate in new situations. For instance, the Vikings in Greenland tried to maintain a style that involved raising grasses for cows in a cooling climate, with disastrous results; they refused to learn from the Inuit style that was effective under those conditions. Culture can be a mental trap that limits choices and actions. People may become fixed in permanent roles and personalities. Even if cultural attitudes are appropriate, they can trap a people if there are no longer functional reasons for the practices. The Nembi of Papua New Guinea may be trapped in their system; making stone axes is difficult, when thousands of steel ones are available. There are other cultural traps, such as slavery, alcohol or drugs.

Humans developed as generalists, using creativity and flexibility to cope with unstable conditions following the glacial climate of the last ice age. If the future returns to the climatic variability of earlier interglacial times, this variability would impose difficult conditions on agriculture and cities, which developed in the relatively mild past 5,000 years. The most recent trends to interdependence, just-in-time supplies, and globalization are not adaptive to any instability in the system.

Global capitalism undermines traditional cultures by offering consumerism in the place of guides for behavior. Social roles seem irrelevant by comparison, if the good life can be bought without effort. Yet, it does not seem to be working in Europe and the US. Instead of being free from economic want to develop their potential as creative human beings, people are trapped in a consumer cycle. Self-actualization is postponed for self-gratification.

Capitalism is as massively adaptable as other human behaviors. Instead of being chastised for its weaknesses, it has incorporated the criticism into its advertising program. Protesters are shown drinking Pepsi. Aborigines are shown drinking Coke. The computer revolution, once thought to become the backbone of creative anarchy, is absorbed into corporate service. People become addicted to new commodities.

Addiction is a trap, since it requires higher rates to appease the need. Addiction can appear in human social systems as well, such as dependence on government subsidies or reliance on chemical fertilizers to improve crop yields. Cheating is a trap, that is avoiding the rules for personal or social gain. We are changing conditions so fast that new adaptation is necessary to fit into conditions we cause through growth and adaptability. What is the solution? Plan or adapt?

Violence might work. War has been a way to get out of traps. War can also destroy filters. But, there are other ways to change filters—by perception or thought. War can destroy cultures and ecosystems, so it may be too destructive to work as a solution.

Growth is a way to get out of some traps, but sometimes it leads to a larger trap.

Meadows notes that because of the reinforcing positive feedback loop in the market, or 'success' to the successful, for example, US automakers were reduced to third place. This is a trap also, It can be broken, in species as well as corporations, by diversifying—and how does one diversify? Find the differences in the environment through education, specifically through ecolacy. Another way is to legally level the playing field with antitrust laws. One solution to growth is mature development within limits of age and size.

Traps can be escaped by reformulating goals, or by weakening or straightening feedback loops, by adding new feedback loops, or by recognizing them or altering the structures. The way out of the trap of escalation is to avoid getting in it, refuse to compete, negotiate, or interrupt. This is where education can offer alternatives, through the filters of numeracy and literacy. The solution to cheating is to redesign rules in a direction to achieve the original purpose of the rules. In fact, design and redesign are ways to remove traps.

7.3. *Design Factors: Economics or Koinomics*

The formal study of how people use their surroundings is economics (from the Greek words meaning 'law of the house'—house is used as a metaphor for human society and nature). The word has come to mean the management of resources to supply human needs. It is basically concerned with sharing resources to meet physical needs of a people.

Although economics is a social science that studies human behavior, it considers itself a positive science. As a social science, economics addresses human problems. The acknowledged fundamental problem of economics is the contradiction between scarce resources and unlimited human wants. The kinds of resources and the possibilities of using them in production are considered in the scope of economics, as are the role of the government, business cycles, monetary details and policy, stabilization and growth, international trade, consumer behavior, production costs, pricing, and resource markets.

7.3.1. *Traditional Economics*
Natural economies are based on solar energy and plant productivity, as well as geological heat and energy. Ambihuman systems (those surrounding the human) operate under natural limitations; they are empirical assemblages whose properties have been shaped by chance, filtering, and selection over millions of years. They stress adaptation to the environment and the survival of a breeding society.

Traditional human economies, although basically dependent on the same energy and production, reverse the priorities. Human concern is for individual survival by modifying and managing the environment. Over the past 40,000 years humans have used many economic systems, which can be characterized by reciprocity, distribution, and redistribution. Variants of the last system have included tribute economies, command economies, and market economies, all of which are forms of redistribution.

7.3.1.1. Reciprocity
Foraging bands divided their labor by sex; this seems to be a cultural universal and a natural division. Ostensibly it was done to allow people to use their unique strengths. Men fished and hunted, while women gathered marine animals,worked in gardens, and women with children gathered plants and insects. Later, labor was divided by age (age grades) and class (social division by prestige and religious rights). The fruits of their labor were divided recipro-

cally, to family, friends, and all other members of the band.

Foraging bands in general moved their camps regularly, from three times per year to once every three years. There were some instances of permanent settlements. Soon associated with permanent settlements were inequalities of status and wealth. Furthermore, other social things developed such as complex divisions of labor, with castes and slave labor. Trade and competition increased. This may have had to do with resources that were abundant and could be stored, and thus transformed into political prestige and power.

Traditional reciprocal, or subsistence, economies are dismissed as providing only a slim margin between life and death, which Marshall Sahlins exposes in his book, Stone Age Economics. If we think about it, there are numerous advantages in the way of life of archaic peoples: Fewer working hours, perhaps three or four per day; more leisure to talk, sleep, engage in rituals, and make love; a diverse and healthy diet; deliberate underproduction, usually well below the maximum levels; deliberate control of population growth below maximum levels; and, deliberate under-use of resources, resulting in a small ratio of people to resources. Subsistence economics means simply that surpluses are not accumulated. This might make them more vulnerable to food shortages, although the low ratio reduces the possibility.

These economies still occur in many places in the world, although they are under pressure to adopt the form of capitalism used by the nation in which they operate. The mark of these economies can still be seen underlying capitalism; many people in capitalist countries volunteer for their community, swap, trade and give away many different products, as well as hours of labor, which are not accounted for by most forms of capitalism.

7.3.1.2. Distribution
The basic change in economics is from equal sharing to a form of collection and redistribution, which allowed for specialization. Big men distribute the products of their own activities. Headmen are more informal, in bands of 50 or villages of 150. Every one knows and understands others; reciprocity is the bank. Headmen are leaders without power. Big men give away extra wealth. Chiefs accumulate it.

This form of economics seems to be rare now in the Pacific areas, where it was practiced. Perhaps it is shadowed in certain kinds of industrial exchanges, for example, between bosses and underlings or laborers.

7.3.1.3. Redistribution
Redistribution involves a different kind of step in exchange. Things are given to a chief and then redistributed to whomever is favored by the chief. Almost every kind of subsequent form of economics involves some form of personal or abstract redistribution.

7.3.1.3.1. Chiefdoms and Redistribution
Traditionally Chiefs redistribute the fruits of other's labor. That is to say, production is kin-oriented and reciprocal. Tribute, in the form of food or luxury wealth, is mostly redistributed by a chief (as a kind of informal taxes). The family is still the unit of production. In general, supplies are not accumulated. The institution of bridewealth serves to redistribute wealth from the bridegroom's family to the bride's family.

7.3.1.3.2. Command Economies
Command economies try to control the entire economy. The state owns the basic means of production, as a historical response to capitalism and inequalities. The communist model provided many things for many people, while maintaining a large war and research establish-

ment. The philosopher Karl Marx thought that socialism was inevitable, that public owner-ship of the means of production would provide equality and social security for all people. But, in practice the distribution of wealth was very inequitable, as a result of historical trends, old economic rules, and cultural confusion. Therefore, some people were treated as more equal than others. As a result of central planning, the patterns of life under the communist model were pressured to be uniform and efficient.

Command economies, such as Marxism and Socialism, are underrated as having free market functions in some areas. Yet, the strengths of this kind of economy—especially the planning of production and the control of resources—are not admitted. Instead, the military competition that ruined many command economies is left out of the equation, and thus this form of economics becomes a distant also-ran. The collapse of Marxism and Communism was the end of one kind of history, the belief in salvation by society, and a secular religion foreseen by Rousseau. Peter Drucker says a new post-capitalist society will use the free market as a proven mechanism of economic integration—although it is not free! Market economies have begun exploiting the resources, such as the forests of Siberia, while at the same time trying to rehabilitate other resources, such as the forests of East Germany and Poland. This contradictory behavior is due to a combination of economic myths and practices.

7.3.1.3.2.1. *Tributary Economies.* Certain chiefdoms use command economies. Chiefs distribute the fruits of others labors, sometimes according to social need, but sometimes for the prestige of peers or nobles. Early Chinese empires up to the 1400s were tributary econo-mies. In fact the Chinese explorations in the 1420s were by fleets prepared to accept tributes from other peoples.

7.3.1.3.2.2. *Social Economies.* Socialism is where the state owns the basic means of production; this was sometimes a historical response to capitalism and inequalities. Hy-brid systems have existed, such as a tributary system in medieval China, or do exist, such as Democratic socialism in England.

7.3.1.3.3. Market Economies

Market economics exhibits the same symptoms as traditional and command economies. For instance, industrial cultures have far higher incidences of starvation—50 million children starved to death in countries with market and command economies in 1978—than archaic cultures with traditional economies,perhaps because of indifference or tolerance for suffering. Our industrial culture is approaching a form of subsistence now, in the sense of a minimum of necessities—but we will not be able to leave after we have eroded the soil. Perhaps we will be less tempted to exploit the land for short-term profit, since we have to remain after the profit leaves. Market economics avoided the same fate as command economies by having a larger store of trust and credit. But, they exhibit some of the same shortcomings, such as gross inequity from the leaders to the followers. Military expenditures may still ruin some market economies.

Modern market economics privatizes profits and commonizes costs. One consequence of this is the decline of common or private lands resulting in environmental degradation. The value of goods in monetary value is determined by supply and demand. At the very begin-ning of the Industrial revolution, the English encouraged trade to increase wealth, which led to Mercantilism, the government regulation of the economy to ensure growth by granting monopolies, which could be said to be the start of global unity. This led to an accumulation of capital from mercantilism. Capital required more energy and new technologies and thus led to capital production, where goods are privately owned, yet there is a separation of people from land and resources. Eventually, there was a shift to assembly-line factories using fossil fuels. Capitalism is disruptive of resources and culture; plus unequal exchange leads to accumulation. Thus Corporations as large individuals lead to monopoly capitalism.

As the only factor, the invisible hand is a failure. Combined with strict and fair regulations in some nations, it has allowed a more chaotic capitalism. Modern market economics has proven to be more slippery at avoiding any kind of regulations created and imposed by the populace or representatives.

7.3.2. *Modern Economics*

Modern economics is defined, by Roger Chisholm and Marilu McCarty, as "the way people make their living." Economics attempts to address the 18th-century concerns of Adam Smith by using a scientific method to collect and interpret information—in fact, economics considers itself a science like physics, chemistry, or biology. Smith had noticed the trend in England away from mercantilism (where the central government regulated the output of goods for trade for gold) towards a free-choice economy, where people would decide what to make and how much to sell it for. Smith thought that a free market of independent buyers and sellers would let the entire community prosper as if an "invisible hand" were guiding it. That is, competition would increase public well-being, or to say it in a different way, self-interest is linked with common interest (or "What's good for General Motors is good for the USA"). Many countries in the 18th and 19-century, including the United States and Canada, adopted the free market system, and their citizens did acquire more symbolic wealth.

Economics considers itself a positive science, that examines "what is" with theories, as opposed to a normative science, which addresses "what ought to be." Oddly though, Chisholm and McCarty establish a list of goals (thoughts) for an economic system that includes: Productivity and growth—growth is necessary for living standards to rise, and individual self-interest has proved to be a stronger motivation than patriotism, altruism, or recognition; stability and security—stability in the form of full employment and set prices; security as providing the material necessities to the elderly and poor; efficiency—where the maximum amount of needed or wanted goods is produced from scarce resources with minimum waste; and, personal freedom and equality—by enhancing the personal dignity of everyone with maximum freedom, that is, the elimination of discrimination and limitations on opportunity.

The focus of economics, however, is rather narrow, in that the concept of resources is very limited and the unlimited wants are not much discussed. There is no psychology or ethics; there is no ecology or aesthetics. There is no concern with the triviality of the free choice of a worthless doodad. There is no thought for beauty. There is no concern with the welfare of the other beings that share the ecological community. Like the old discarded physics, economics applies a rigid standard of objectivity to its analysis.

7.3.2.1. The Modern Model: Abstraction & Accumulation

The metaphor for the economy used to be a simple mechanical model for turning resources into products. To be successful, the economy had to grow and turn a profit continually. Unfortunately, the assumptions of the model were also simple and failed to consider human needs and natural cycles, causing great suffering and great disruption. The old analogy of the economy as a machine leads to bad assumptions: That everything is a resource; That resources are unlimited; That production must continue endlessly; That the economy has to keep growing to survive; That the purpose of the state was to legitimize exploitation; That the purpose of humanity was to multiply, produce, and consume; and That the purpose of the universe was to supply human needs.

The bad assumptions of the machine analogy lead to false economic beliefs: That mass production is most efficient; That obsolescence is necessary for successful growth; That people's needs and wants are fulfilled by advertised products; and, That quality does not matter very much. These bad assumptions and false beliefs lead to problems in the economy as a whole. These colossal problems include: Overgrowth, with an increase in complexity and costs (many of them social); Economic and ecological instability; Social burdens (from pressures on families from relocation and powerlessness); Misdirected efforts on ill-conceived, low-quality products; and, Slack consumer attitudes and employee performances.

Economics has not been unsuccessful with its models, for instance the model on buying behavior, but it has become a highly abstract academic discipline. All its abstractions are applied to the real world without acknowledgment of the high degree of abstraction involved. The philosopher A. N. Whitehead warned that the economic method would triumph if the abstractions were judicious, but even judicious abstractions had limits, and the neglect of those limits lead to disastrous oversights. Considering a fictitious human nature under imaginary circumstances and thinking it is real is the fallacy of "misplaced concreteness" according to Whitehead. Daly and Cobb suggest that the classic instance of the fallacy in economics is "money fetishism," where the characteristics of an abstract symbol, such as limitless growth, are applied to real commodities and values.

As an aside, Daly and Cobb point out that modern economics might better be called chrematistics, after the distinction made by Aristotle between chrematistics and economics; the former related to the manipulation of property and wealth to maximize the short-term abstract money value to the owner, whereas economics was the management of the household (community) to increase the concrete value for everyone in the community over the long-term.

Business as usual, with its inertial model of growth, could end in catastrophe for humanity and its environments. Industrial cultures, with their characteristics of simplification, naiveté, homogeneity, and incompleteness, turn wild landscapes into "flatscapes," where variety disappears and significance is ignored for the comfortable standards of meaningless continuity. Rapid growth might precipitate a catastrophe sooner, while modest efforts at environmental protection and increased efficiency may only postpone catastrophe a few years.

7.3.2.2. The Wealth of Nations

J. S. Mill and Jeremy Bentham used the Associationism of D. Hartley to develop a Felicific calculus in which individuals were social atoms (another ghost of the old physics) seeking pleasure and avoiding pain—and benevolently controllable through arithmetic. Since then, economics has become more scientific and mathematical. Economic "man" (used as the traditional term for *Homo economicus*) is still considered a kind of cash-register, however. There is also a strong relationship between Protestant religion and economic development (just the opposite of Marx, who regarded religion as a superstitious atavism).

The sociologist Pitirim Sorokin indicated that the wealth of an area was a function of its physical attributes and its culture. In fact, the attributes are only possibilities until appropriate perceptions and technologies exist (due to culture). Economics has always been concerned with measuring wealth. The basis of wealth has been variously described as labor, resources, production, net plant production, and information. Yet, no single basis is adequate. Perhaps wealth is just as simple as what we value. The basis for Smith was a basic resources considered by economics: Labor.

7.3.3. *Holeconomics*

John Stuart Mill wrote that beyond the progressive state lies the stationary state; each advance is to approach it. A stationary economy is not synonymous with a stagnant economy; there is always room for developing and increasing the scope of economic culture. It could be highly sophisticated, dynamic, and imaginative. Mill said: "It is scarcely necessary to remark that a stationary condition of capital and population implies no stationary state of human improvement. There would be as much scope as ever for all kinds of mental culture and moral and social progress; as much room for improving the art of living and more likelihood of its being improved." Mill's vision of a stable state was a response to the goal of an industrial society. Technological progress would abridge labor, not necessarily increase production. In a state of equity, persons would have room for solitude and leisure development. But Mill's idea was anthropocentric—"man" was not bound to protect nature.

7.3.3.1. Holeconomics (Steady State Economics)

Herman Daly concludes that the steady state economy is both necessary and desirable. He notes that in the definition of economics, as the study of the allocation of scarce means among competing ends, the entire ends-means spectrum is not considered. Only intermediate ends or means are considered, not ultimate ones. He anchors the ultimate means in physics and ultimate ends in religion. Economics falsely concludes that the middle ranges represent the entire spectrum. In another computer model, Laszlo's scenario calls for an across board reduction in population, investment, and resource usage, to create a steady state with the present disparities maintained.

Some theorists, like Samuelson, have concluded that growth is necessary to rid the economy of disparities. But, development instead of growth would equalize wealth more efficiently—after all, economies have been growing for at least 400 years and the disparities have increased. There is no necessary association between development and growth, as Daly and others have shown. Development means the introduction of an innovation. Economic development will require technology. Ecologically sound technologies will minimize stress to the environment. Economies could be modeled after climax vegetation and not successional vegetation, where diversity in scale is greater. A community is forced to accept an upper limit, beyond which it cannot grow any further. Further growth results in destruction or disruption of itself and nature. This is the law of the maximum. Production could be stabilized in a steady state economy, a mature economy, like a climax system, where processes and cycles are constant. A steady state economy must be based on natural laws and ethical principles. Natural laws include thermodynamics and ecological theories. Rules of economics, laws of nature, and ethical principles must be related.

There is another distinction between growth and development. The ecological social approach (or a redistributive environmental strategy) to development makes it irrelevant to discuss global limits to growth. Local limits are far more significant to the majority of population. Regardless of how much food exists, people will starve unless they can get it. Redistribution of resources and improvement of environmental quality (home environment) are more important than increased production by sophisticated technology. The natural capacity of regional photosynthesis must be the limiting factor in development, especially in tropical and subtropical areas.

Development calls for a social and educational organization, more than technological style. Styles of technology must be determined by culture and context. Such development requires a local authority working with suitable economic and ecological conditions. No authority can be affective without the participation of the populace. To stop growth, a strict regulation of the productive system is needed. States should be able to control which products are made, and with which technology. Practical implications of the steady state are that nonrenewable resources will be conserved as much as possible, by recycling, while erosion, depletion and pollution are minimized; energy sources may be greatly decentralized and diverse. Using a field concept for development emphasizes the dynamic transformation of the domain, rather than its structure as an individual. Selective advantage or cost benefit become adjectives or afterthoughts to the domain.

Jane Jacobs, in The Nature of Economies, argued that the same principles underlie both ecosystems and economies: "development and co-development through differentiations and their combinations; expansion through diverse, multiple uses of energy; and self-maintenance through self-refueling." *The Nature of Economies*, also in Platonic dialogue form, is based on the premise that "human beings exist wholly within nature as part of the natural order in every respect." Jacobs' characters then discuss the four methods by which "dynamically stable systems" may evade collapse: Bifurcations; positive-feedback loops; negative-feedback controls; and emergency adaptations. Their conversations also cover the "double nature of fitness for survival," traits to avoid destroying one's own habitat as well as success in competition to feed and breed, and unpredictability, including the butterfly effect characterized in terms of multiplicity of variables as well as disproportionally of response to cause, and self-organization where "a system can be making itself up as it goes along."

7.3.3.1.1. Holeconomic Wealth

Economists try to bronze the economy in its current structure, but it is a changing system. Since it is changing, strategies that are appropriate at one stage are totally inappropriate at another. This is the remorseless working of tragedy. Any lasting economy must have a dynamic approach. Wealth must be renewable.

Rich sensory experiences can be derived from direct contact with nature, but economists and planners rarely mention these values. Light, wind, dirt, plants, and birds, all perform during a walk, but not with the same meaning as crops, which is for their utility—they just are. People do not live without these things. All values are based on a healthy ecology. It must be kept healthy; arable land must be limited, and mineral exploitation must be limited.

There are two roads to wealth: producing a bigger pie (supply) or reducing each portion (demand). This assumes that wealth is defined as supply divided as demand. If supply is limited then wealth can still be increased two ways: reduce expectations of individuals or reduce the number of individuals. Supply may be mostly material things, but not status, for instance; demand has the more psychological dimension. Therefore, wealth will always have a psychological dimension. For this reason, Gregory Bateson thought that economics may be founded on a fallacy. Economists cannot account for intransitive preference: where a is preferable to b, b to c, but c is preferable to a; as in Money being preferable to resources and resources being preferable to wilderness but wilderness is preferable to money. Preference curves in economics should not intersect.

An individual's perception of environment is a semantic map that forms the context for decision making by delimiting possible acts. People respond to felt needs; economic growth occurs when these psychological needs are stimulated and satisfied. As soon as people think that they need clean air, like snowmobiles, these demands can be met somehow.

Values are based on knowledge, which is measured partly in terms of information, where information can be considered a source of wealth. Some business economies are based entirely on providing information. Information is apparently boundless. Yet it can be manipulated. It is information that defines the use of resources by people. For example, hydrogen is worthless unless technologies exist to transmute it to helium and manage the released energy. What is not limited is our use of information. A sophisticated technology needs fewer resources. The natural productivity of ecosystems is less important if food can be grown intensively in tanks using solar energy.

The Gross National Product (GNP) is the money value of the final goods and services produced by the economy. It includes all the goods and the bads. That is, it is a measure of throughput, the flows rather than the capital stock of wealth. Meadows suggest that capital stock could be maintained with the lowest possible throughput. That would require redoing the economy and its technologies and making them more efficient. The trap is seeking the wrong goal. The way out of the trap is identifying goals that reflect the real wealth, then focusing on result not effort.

7.3.3.1.2. Limits of Economics Revisited

Both ecology and economics attempt to understand and predict the behavior of complex, interconnected systems where individual behavior and flows of energy and material are important. There are many other common or similar processes: resource allocation, optimal behavior, and adaptation. For each ecosystem, and at each level of technology and kind of social structure, there is an optimum size of population that offers a high quality standard of living. The optimum size in this sense is a working one based on our knowledge of all of the factors—and we cannot know everything. No one has complete information about the current environment or the results of their actions. Complete knowledge is not necessary, however—only the knowledge of a law of the minimum. For humans, trade can ease the law, but not repeal it. Rachel Carson demonstrated that we lived in a world of limits.

Resources may not be absolutely limited, in the sense that history shows that advances in technology can expand the availability of resources; less is needed to produce more. But, the same history also shows that humans reproduce up to the new limit of misery allowed by the new technology. New advances are used to increase the size of humanity, not its happiness or wealth. Society is growing at the same time as needs are growing, resulting in lessening of natural systems.

The redefinition of wealth in an ecological framework would tie it to limits, which might increase human enrichment and natural preservation. Diversity is a form of wealth. Differences do not necessarily cause conflicts because each can fill the needs of others. Since nature is a non-zero-sum game, many groups can gain at the same time, human or not.

7.3.3.1.3. Poverty of Humanity & the Earth

In general, poverty is a lack of things that others have, especially necessary things. There are different forms of poverty. For instance, one form is the lack of needs, such as food or money. Another is the lack of natural or public services that would provide food and shelter. On the other hand, what is the use-value of feet in Los Angeles? The lack of potential is harder to define, but it means that people who suffer that poverty cannot improve him or herself, or change, or develop. Finally, the lack of luxuries has become a form of poverty. The idea of affluence in a global village creates a new modern kind of poverty, the potential poverty of not having what is possible somewhere in New York or Paris. Modernizing needs creates a new dimension of poverty. This adds a new discrimination by a violent cargo cult (what do you mean by violent cargo cult?) that seizes traditional cultures.

Industrial society is constantly mobilized for emergencies with, in the battles against noneducation, poverty, diseases, and terrorism. Industrial development has never been nonviolent or respectful to people or nature. Some human poverties are poverties of the local ecosystems. Loss of entitlement to a natural resource base needed for foraging or agriculture is a political poverty. Interference with or destruction of the processes of ecological balance and renewal, the loss of diversity, the loss of capacity for the renewal of ecological systems, are ecological poverties.

7.4. *Managing Multiscalar Interacting Technologies*

At one time, new technologies were adopted because herding animals and cultivating crops were more reliable ways of producing food than hunting and gathering. This may have been a choice between moving and staying in one place, between finding food in different places and increasing the food in one place. These waves of innovation started with digging sticks and horse harnesses, and continued to iron and water power, through steam powered cotton mills, to electricity and the combustion engine, and to petrochemicals and aviation, digital networks and biotechnology. We have to entertain the possibilities and directions of further human technological evolution. As humans continue to create extensions of their bodies through technology, using surgical or genetic enhancements, their patterns and behaviors change. Human cultural evolution has already resulted in more complex forms of culture, with technological modifications. If humans finally become domesticated, then sexual selection will become controlled and perhaps artificial. Evolution can be considered in Merleau-Ponty's term as a 'pattern mixed-upness' of styles of living, as beings radiate through time and place, unfolding into new, complex patterns. Evolution is less a hierarchical ladder or an up-escalator than the history of forms fitting and refitting changing environments. Human needs and goods get mixed up with the needs and goods of other beings. It is complex.

We already use computers for dazzling communications. We are starting to embed chips in our bodies and brains, to repair diseases or enhance memories. Perhaps smaller computers will be internalized, for better modeling and communication, or perhaps for dealing with the complexity of global design. Biotechnology and nanotechnology seem to be qualitatively different from external tools, such as knives or tractors. With nanotechnology, we think it might be possible to create a distributed energy plan with alternative energy networks. This would be a radical solution harnessing quantum foam and manipulating atoms. Nanotechnology is the emerging science of manipulating molecules at the atomic scale. One need only move a few atoms around to create wood or energy. Nanobiology is expected to push a convergence of biology and technology.

Several writers and futurists have predicted teleportation in 20-30 years. Of course, this would create a real problem with what to do with the excess matter and energy at each end. There are also predictions of space tourism to the Moon or Mars, designer genes for improved health, manipulating matter for smart products, using hydrogen engines for transportation, cybernetic health enhancement for humans, downloading memories and drugs, and domestic robots.

Some computers may radically increase our ability to understand the complexity of ecological systems as well as the shape of the future. There might be an infinite number of multiple universes or mere worlds that we may be able to visit physically or virtually. The interactive realities of gaming may become more immersive. Robots may become more common. Artificial life is projected to arise from new electronic beings, perhaps emerging from robots doing web searches or digitally engineered characters in games. If a network mind forms and gains self-awareness, we may need new forms of communication or influence (control?). Can we translate its communications if it operates at the quantum level?

Will these technologies use less resources or more? Maybe more in the short-term development. Combined with biomimicry, the argument is less energy, less materials, and less waste. By controlling matter on a scale smaller than 1 micrometer, waste may not be a factor. It is likely, however, that these technologies, independently or as hybrid systems, will create new problems and threats.

7.4.1. *Threats of Technologies & Scales*

Many of our current emergencies are a crisis of our image of the world as a machine and nature as a passive environment. This leads to excessive consumption and blind faith in technology and luck. We have had dramatic global changes, such as the ozone layer disappearances and increases in annual wildfires. But, we prefer to pursue comfort rather than adjust to the real actions of a dynamic and dangerous planet. We have made an enormous investment in technologies with little regard for the environment, related to our goal of growth at any cost. Our growth, pollution and terraforming have put us in a trap.

Advances in technology allowed cities to scale up. But, now the infrastructure investment is high and some parts of cities get abandoned. Aluminum and steel are being replaced with composites. This is transformative of the overall socioeconomic production process. The introduction of silicon and the fabrication of the microprocessor that is now the center of information technologies. The telegraph transformed technology first, so it is not just computers in the last twenty years, although communications technology makes a difference to the to the political structure.

7.4.1.1. Technological Dependence

Dependence is a problem with any technology. As we depend on a tool, we sometimes lose the ability to work without that tool, whether knives or writing. We are dependent when we cannot go back to living without it. For instance, now that cash registers in fast food restaurants have pictures of foods on the keys, and it might be hard to return to numbers.

7.4.1.2. Technological Stagnation

Stagnation becomes a problem when we cannot think of alternative forms of technology. For instance, with transportation and energy, we have been stuck in a rut of burning. Most energy is generated by variants of the steam turbine, which relies on burning something, not much different from dung and wood, just more concentrated and centralized. Transport is primitive burning also, for automobiles or trains. For the flight of aircraft, the gas turbine jet was developed in the 1930s and although it has been improved, it has not been replaced.

7.4.1.3. Heroic Scale

The unconnected actions of almost seven billion people are really a scale change that can result in environmental disasters. Of course there are other factors, including extinctions and poverty. The scale of industrial output is massive.

Furthermore, emergence, which describes the appearance of a macro scale form from the evolving interactions operating at microscale, has to do with local cultural landscapes social structures and organizational problem-solving.

Feedbacks continually emerge as products of evolution at the local level, and they can have a large influence at the global level. One question is why regulatory feedbacks predominate at the global scale to make and maintain habitable conditions for living beings on the planet. Perhaps because the role of life on influencing feedbacks means that when life is wrong it disappears, when species make mistakes they disappear. So far, no local mistakes have made the entire matrix collapse.

7.4.1.4. Heroic Collapse

External disturbances such as asteroid impacts or basalt eruptions could have triggered significant transitions to less stable states or to collapse of the system. However, most transitions appear to have been generated internally with evolutionary innovation playing a large role. There are many questions relating to transitions or collapse: To what extent is the Earth system self-regulating? How does the system develop? What is the contribution of life to maintaining habitable conditions for other life? Why does regulatory feedback predominate at the global scale?

One thing to consider about collapse and timescales is the presence of multiple stable states. We know that cultural systems can bounce to a lower state and yet survive for a long time in that particular state. We also know that change is neither continuous nor gradual, but is often chaotic. Change is better described as episodic with periods of slow accumulation of natural capital that are punctuated by sudden releases and reorganizations. Redmond and others suggest that scaling up from small to large is not a simple process of aggregation. Stabilizing forces are important in maintaining diversity, flexibility and opportunity. Stabilizing forces are also important in maintaining productivity, fixed capital and social memory. Political management that only applies fixed rules for achieving constant yields, independent of scale and changing contracts, can lead to systems that lose resilience.

Joseph Tainter notes that one factor in addressing these issues is the change of scale that occurs when local people become embedded in larger systems at national and international levels. This is now called globalization, but that term merely denotes the most dramatic changes of the past seventy years or the recent phase of the development of the past 500 years. When the periphery is incorporated into a larger economy, it experiences a change in scale of its economic and political relations. However, the information pool remains local. This disjuncture of scale can undermine the sustainability of local populations. For instance, the introduction of roads after WW II initiated many changes in remote villages in Europe.

Tainter concludes that the lesson is that, unless these and other local people become knowledgeable about the full range of factors that affect them, they will continue to be vulnerable to distant processes and to others who profit from their ignorance. One approach would be for systems theory, either alone or in conjunction with other research, to develop a body of knowledge of such fundamental importance that it would be incorporated along with other mainstream social sciences. Tainter concludes that world systems create disjuncture and scaling between the flow of materials and the flow of information. This disjuncture must be remedied. People must become cognizant of the factors that does affect them at all scales. Tainter wonders if his proposal is utopian and unrealistic, but suggests that globalization demands the attempt. One certainty is that failing, local places will eventually collapse from poverty, dependency and ecological deterioration.

7.4.1.5. Technological Trends

Trends, however, can be predicted. To understand trends well, we need an ecological perspective to look at the interacting factors. Some trends are long-term, such as climate change and urbanization; others are short-term, such as biotechnology and nuclear energy. Cascading problems can trigger the next problem. Trends can also be deflected or reversed, with changes in behavior.

Human beings have no complete guidelines to interacting with other species in an ecological context. Cultural ethics are usually restricted to some members in a local ecosystem; such ethics are assembled inductively, from experiences from living in specific places. Many ultrahuman cultures have standards (or codes) of behavior to regulate interactions. In birds and simple mammals, these rules may be very rigid and predictable. With increasing brain complexity, however, learning takes a larger role. Human learning can find meaning in absurd patterns of change.

7.4.2. *Extreme Technology & Scale*

Scientists have worked on reasoning holograms and on cloned sheep (although the first one did not live to a normal life span). Scientists might continue to clone humans. They might design specialty bodies for working in a vacuum, for eating cellulose or even for converting solar energy directly into calories. More complex bodies could perceive other wavelengths. Would those bodies replace the evolved organic bodies we use to dream and make art? What characteristics would be lost or added or improved? Of course, it might be possible to create new bodies with new brains that have self-awareness and consciousness. Technology and design could take great leaps, but they would need to be enhanced to deal with the unanticipated consequences of such radical changes.

Extreme technology could remotely combine new bodies with faster than light signals (possibly through quantum entanglement or wormholes). Communication would literally take a quantum leap. External storage would be unnecessary, since it could be done at the atomic or molecular level inside the body, although its permanence, as in the life of the universe, might be problematic. Nevertheless genes, memes, and wenes could be accessed and modified temporarily or for the long-term.

7.4.2.1. Problematic Extreme Fixes

The technological fixes, from the modern industrial response of geoengineering, are problematic themselves. Geoengineering is the deliberate manipulation of global metabolism to correct or adjust human interferences. This approach offers several options: Orbit mirrors to deflect sunlight, launch sulfate particles into the atmosphere, or fertilize the ocean with iron. It would consider burying CO_2 gas or solids in the ocean or land. They are simple symptomatic responses that ignore the system. If the problem is simple, pollution or fossil fuels, and the fix is simple, then it neglects the complex issues of human conversion and scale. It ignores the proper relation of humanity to the planet. Technofixes are based on an inappropriate logic, linear and cause-effect, instead of nonlinearity and multiple effects.

Americans and Russians made proposals in the 1950s and 1960s to control the weather and modify the environment, for instance, Mikhail Budyko's proposal for artificial volcanoes in 1977. How do we remove the 100 ppm excess CO_2? Atmospheric scrubbing on a heroic scale? Planting more trees on continental scales, and restoring more forests to Pleistocene levels?

7.4.2.2. Limits of Extreme Technologies

Although several national space programs have been able to keep people on ships and space stations for limited amounts of time, with bottled oxygen, scientists in Arizona spent $200 million to prove that we did not have the technology to keep eight people in oxygen for more than a week, using a balanced ecosystem approach. So far, our best technologies cannot substitute for natural processes, especially in terms of detoxifying wastes.

Friedman suggests that the limits of systemic growth can be understood as the threshold of exhaustion of the system. A limit would be related to technological and economic factors, as well as to contradictory tendencies within the system. One bad assumption about computer intelligence and manufacturing is that people and machines perform optimally at all times. Another is that we have anticipated all the consequences of new technologies.

7.4.2.3. Scales of Technologies

On the other hand, we experience problems with scale. We have a human scale, based on our body and images, but this scale does not seem to be favored by the universe. Scale is problematic with smaller technology. What works well at one level may not adjust well when scaled up. Scale changes the interaction. If the technology is too large, there are too many interactions to monitor or understand. Scale also changes the relation to the individual.

The problem of using metaphors between scales is that there are too many emergent properties that do not appear in the smaller, and better known, scale. What can translate as a metaphor between scales? Foam can be found at the smallest sizes imaginable smaller than subatomic particles, and at the largest scales imaginable of the entire universe. But, its predictive value might not be adequate. Seafoam, which is white, refracts a large fraction of sunlight than water. The interaction of light and foam has an effect on temperature and climate.

The problem of scaling is central to linking local case studies to global processes, as we know. And, as we know, ecological variability tends to increase as the scales become smaller in our understanding of the factors on variability is modified by the scale of observation. What does that mean? Global cycles of local exchanges link scales. But the larger the scale, the longer the lag time before any changes show.

Really large scales, bioregional or global, are not considered often at all. At large scales, other factors have to be considered. Many cosmic tendencies, such as disorder (entropy) and order (ektropy), as well as change, creativity and temporality, are relevant to these scales. But, managers tend to emphasize only those tendencies that seem positive, such as creativity, diversity, and order, and not those with negative connotations, such as entropy, death, and uniformity. The universe comes with the whole package, and we risk serious error by choosing only the tendencies we want.

7.4.2.4. Connection & Complexity

Global cycles of local exchanges link scales. But the larger the scale the longer the lag time before any changes show in interactions. All interactions are ecological interactions, from the modification of mammals in Australia to the transformation of grasslands. These are all dynamic interactions between human societies and their environments, from the perspective of long-term patterns and processes. The interactions become tightly coupled. Ecologists call the excessive integration of a system hypercoherence. According to Holling a too-highly connected system is an accident going to happen. Rigidity increases vulnerability to change. The number and strength of the connections is a problem if disruptions occur. Essentially because the different processes were so overly interconnected that all the flexibility disappeared from the system. A reduction in flexibility is not the only effect of interactions. Globalization suppresses many diversities. But, diversities are essential for human stability, capacity and resilience.

Even progress is not even. In the short run it has incredible dips and highs as we go from accomplishment to disaster. William McNeill suggests that escalating vulnerability is the price of growing local and short-term mastery, especially when the scale is increased to the global. Joseph Tainter argues that societies choose to increase their complexity to solve immediate problems, their social and political dynamics, but that leads to increased vulnerability in terms of returns on investments of energy in the complexity itself. Human complexity, like ecosystem complexity, requires demands for more energy. Globalization is a titanic effort to put us all in the same boat, more complex and advanced, but just as vulnerable to human error and natural catastrophes, and ultimately to a global collapse.

Table 7424-1. Models of Fragile (3) & Robust (1) Complex Cultures
(after J. Tainter 1988)

Dinosaur Model	Culture is a lumbering colossus incapable of change. Big investments often maladaptive. Cannot adjust to stresses. Static (e.g., Rome or China)
Runaway Train Model	Culture is impelled on a path without the ability to change directions. The course may be dynamic, but fixed (e.g., Aztecs or Inca)
House of Cards Model	Culture is inherently fragile with low reserves and low ability to meet stresses (e.g., Maya, as per B. Meggers, or Rapa Nui)
Fungus Model	Culture is an organic, adaptive social structure responding developmentally to environmental and social challenges (e.g., Aborigine, Desana or Masai)

Table 7424-2. Five Concepts Leading to Cultural Collapse
(after J. Tainter 1988, J. Diamond 2005)

Cultural systems are adaptive organizations that become more complex in response to challenges and problems
Cultural systems need (solar, animal, fossil) energy for their maintenance
Increased complexity carries increased energy/tax costs per capita
Investment in complexity reaches a point of declining marginal returns leading to dissatisfaction and stress
Cultural systems fail to react to slow or large extended catastrophes

8.0. Designing Good Places

Any single human group can make a good place for themselves. Many cultures have made good places. Often, the making is a slow mutual adjustment among life forms, environmental constraints, and human cultures. But, many stakeholders need to participate in the making. And certain problems have to be addressed. The levels of intolerance and conflict that are permitted between groups, communities and nations have to be reduced. This change can occur through actions of large design factors, such as recognized global commons, or through local factors like reducing inequity between stakeholders. Local corporations or local governments could take ownership of all commons, especially those that provided ecosystem services for cultural areas.

This could happen as a result of a framework to allow cultures to develop without unfair competition or dominance. But, the frameworks can only be viable if they are run on a local level. The framework could allow good places to be designed and managed for long periods of time, within the limits of the dynamicism of the region and planet as well as accidental events in the larger universe.

8.1. *Problems: Intolerance Conflict & War*

We tend to think of problems as unwanted 'side-effects' of the wanted main-effects, but all effects are equal, as Buckminster Fuller noted, and must be addressed as equal. Most things identified as problems are embedded in a network. Nothing is simple; there is not one truth, there is not one way, and there is not one goal. Problems could be considered also as challenges that we must respond to continuously, in the process of living, not as puzzles that have to be solved once for all time. It is about consciously choosing to see what can be done, rather than dismissing a conflict as terrible and unsolvable. When challenged by some situation, we react by habit, although this may be disconnected from other habits. Habits protect us from many problems. Addressing a problem often has to do with a power struggle, which becomes part of the problem of the problem. If problems are regarded as challenges that require a social response, then much of the conflict can be avoided.

8.1.1. *Intolerance*
One problem that typifies every culture is intolerance. Intolerance means a lack of toleration or the unwillingness to tolerate the contrary beliefs of persons or groups. Intolerance seems to be a fact between cultures and between individuals; it can lead to conflict and violence. It used to be understood that intolerance and violence occurred between 'irrational actors,' whether individuals or ethnic cultures were involved. Therefore, violence could not be explained rationally.

If cultures or individuals were rational actors, then the irrational actions of conflict and violence would have to be explained in some other way. M. D. Toft suggests that ethnic group settlement patterns, socially constructed identities, charismatic leaders, indivisible issue, and state concern with precedence could lead rational actors to escalate a dispute to violence, even when doing so was likely to leave contending groups much worse off.

The problem of intolerance and conflict can be more easily understood with reference to the cosmology of a culture. A cosmology is basically the complete set of ideas about the nature and composition of the universe used by an individual culture. This idea provides people with a collective orientation. It is a collective image. Ezra Pound stated that the image is an emotional and intellectual complex in an instant of time.

All cosmologies cause destruction and waste; all sometimes produce the opposite of the good intended. Once cosmological ideas are adopted, they may remain unquestioned and unquestionable, and they can influence a culture over centuries. The principle of otherness, where others are not true humans, is an example. This principle may have made perfect sense when cultures had to protect their territory, especially when the territory was rich but limited, or when there was a large investment in it, as with agriculture. By representing others as nonhuman or subhuman, they did not have to grant them consideration in the ethics of their culture. Many modern political ideologies and economics have been shaped by this principle of otherness. Intolerance justifies inequity, domination and conflict.

8.1.2. *Kinds of Conflict*
Conflict is sometimes perceived as the only response to intolerance or to other problems. However, conflict can be related to various losses or cultural norms. Understanding these conditions may allow us to resolve some kinds of conflict.

8.1.2.1. Conflict as Loss of Fitness
Fitness is the ability to function under normal environmental conditions. Stress, obesity, illness, toxic chemicals, social conflict, and other kinds of dysfunctions, can reduce fitness. A lifestyle dependent on physical and energy slaves, not to mention a diet riddled with addictions to cheap fats and sugars, decreases our fitness. All of the elements of a true addiction are present in modern lifestyles: We get a short term high, and we suffer long term health problems. Air pollution, sedentary life styles, exposure to stressful city environments makes things worse, even as we acquire more information about the dangers and options to deal with them. Conflict is an attempt to correct the problem by capturing more food and energy, or by acquiring clean air and water, or by conquering new land. The stress of conflict however, can lead to a declining cycle less fitness and more conflict.

8.1.2.2. Conflict from Loss of Equity
Although many resources are distributed unequally over areas, as a result of different kinds of historical geological processes, trade can allow access to those resources. However, as a result of long-term processes of inequity, from keeping people enslaved to cultural hoarding, many people have far less than others. This has resulted in permanent overclasses and underclasses, which are often maintained by physical force, as well as by the force of economic and religious myths. These myths tell all people that they participate in the "best possible economic system" regardless if it justifies the differences of inequity based on history, on luck or on perceived racial abilities. Most people are hungry; few people are fulfilled. Even low average levels of food and fulfillment can be maintained only through theft from other species and from future human generations, and through the degradation of billions of humans, as well as the ecosystems on which they depend. Conflict is one attempt to bring about more equity, although it can result in greater disparities and losses.

8.1.2.3. Conflict from Loss of Accord
As a result of the unequal distribution of natural resources, including unincorporated waste and pollution, and the unequal distribution of materials and wealth between people, economic conflicts arise, often becoming violent political conflicts. Accord, by definition, means agreement or the concurrence of will or action. As an agreement between the parties in a

controversy, by which satisfaction for an injury is stipulated, accord allows people to reconcile their injuries and interests. Because many political boundaries were drawn as a result of colonial expansion and contraction, many cultures are artificially combined in large territories or stretched across several traditional territories. This has created the conditions for continuing cultural and political conflict. In addition to the normal conflict between different cultures with different ways and values, usually resolved by trade or distance, this new conflict resembles an indefinite series of small permanent wars over large numbers of territories. This kind of hot conflict not only destroys habitats and resources, but it causes immense human suffering. As the rules change, and conflict includes noncombatants, as well as plants and animals, there is less accord between conflicting groups. Accord requires the ability to trust, which requires self-reliance and confidence.

8.1.2.4. Conflict as a Result of Bigness

Leopold Kohr identifies the basic conflict between man and mass, citizen and state, large and small communities, as being the result of size. Kohr's theory of size states that the cause of most forms of social misery is bigness. We have always tried to exceed the physical and biological limits of places rather than recognize them and be guided by them. Every advanced country is now over-technologized by its attempts at control. Past a certain point, the quality of life diminishes, not improves, with each advance. Big science serves big technology, which supports and is supported by big government. And there is no science like big science, and no administration like big administration. But this enthusiasm is misdirected. Scientific advances and technological changes result in unforeseen consequences, good and bad. They cannot be controlled or legislated against before the fact. But the investment seems too big to abandon.

8.1.2.4.1. Personal Conflicts

Individual conflicts often resulted from insults or broken promises. In Mesopotamia, people paid fines for hurting others. For instance, severing the bones of another man with a weapon resulted in a fine of one mina of silver.

When arguments could not be resolved, there was open conflict. Resolution of individual conflicts was usually informal within the confines of the community. For the Inupiat and other egalitarian cultures, song duels are a ritual pacific form. The disputants publicly insulted one another until the audience laughed down the loser. If individual disputes were not resolved, then they were settled by community consensus.

The predominant value in small cultures, and then many larger cultures, was harmony. This minimized conflict that might have resulted from inequality. Confucian concepts of ritual and etiquette helped to regulate social conduct and made people feel good about their station. For example: "Inequality is the nature of things" and "seek no happiness that does not pertain to your lot in life."

8.1.2.4.2. Group Conflicts

In a Tiwi example from Australia, a dispute occurred because the elders of one band reneged on their promise to bestow daughters for marriage to the sons of another band—a violation of the norm of reciprocal marital exchange. Two war parties of fifteen warriors each met in adjoining territory. They wore white paint symbolizing anger. Both sides exchanged insults the first day. Then agreed to meet the next day for socializing and renewing acquaintances. The third day the duel resumed; words escalated into wild spear throwing that wounded a few spectators and warriors. That ended the dispute.

Are conflicts different between groups? Why is warfare limited in band societies? Could it be due to a lack of interpersonal competition for status? Are people taught to be restrained? Are conflicts are resolved by ridicule before getting out of hand? Is size a factor in explaining larger conflicts? Population density was controlled by the traditional approaches to resources. In archaic societies, cooperation and consensus, as opposed to competition and individual exaltation, permitted planning to remain informal. Population growth triggered competition and conflict, which lead to positive feedback of the thing that caused the stress.

8.1.3. War

At what point does a conflict become a war? Does every society have war? The anthropologist Carol Ember surveyed band societies and found that sixty-four percent waged some kind of war. The reasons for war have gone from insults and personal conflicts having to do with bride exchange, broken rituals and personal honor to group and external reasons, such as territory, resources and patriotism.

The nation states are closely related to large-scale violence, usually having to do with trying to consolidate their power. They then have a monopoly on power. They specialize on political and economic issues—the Spanish, for instance, ransacked every land they could for gold. Nations enlarged and consolidated their territoriality, which gave them increased capacity for marshaling resources.

Wars, like the recent one in Iraq, are based on weak assumptions: That the war can be waged by blasting away any threats, and that it can be contained by using conventional weapons. But, like most actions, it has affects that can get magnified in the larger system. Furthermore, the war breaks out of the barriers that the participants try to create, destroying properties and civilians—there are no longer any safe buildings or noncombatants, and ruining the social and ecological fabrics.

All wars now are ecological wars that destroy the basis of civilization. There are no nonmonotonic effects with war. There are no side-effects. There are only effects and they can all be measured. Perhaps the next war will be to directly destroy the ecological basis of a community, culture or nation, an extreme scorched earth policy by the aggressors. These wars would be social wars that destroy entire generations and traditional social structures. Perhaps, future wars will have less to do with honor and territory and more to do with crises, such as famine, population, and environmental collapse. Perhaps, wars will have to do with symbols and religion, again, as they have in the past.

8.1.3.1. Advantages & Disadvantages of War

War has advantages and disadvantages. One of the advantages of war is its long tradition, being simpler to understand than rights, the attendant macroeconomics or the workings of enantiodromia. Another advantage is that war is cheaper than ever before, especially for those who attack. Of course, war is also stimulating and fun, at least for the victorious survivors, who are bonded by the danger.

It used to be that it cost more to attack than to defend, with preparations and supply lines. Now, it costs more to defend against a bullet than to attack with one. Cheap bullets destabilized the western U.S., but laws and enforcement, also with bullets, restabilized it. Possibly the same thing will happen with cheap cruise missiles. Maybe not, as the cost of attacking will now continue to be less than defending. This means that the world may become more violent and less stable, especially at select local levels. Can information warfare be cheaper or less destabilizing? No, because information attacks may be even easier to perform.

Unless everyone has the same weapons.

The disadvantages are greater than the sum of all advantages. Ecosystems in disputed and ravaged territories are always disrupted, damaged or destroyed. Human suffering is always made worse than it was before the war. And, war never solves the original problem.

8.1.3.2. Style and Scale of War

Politics is the management of people through equitable distribution of resources, and the management of relations with neighbors or trading partners, using negotiation or force—war if necessary. Wars have changed in style and scale. From disagreements, wars have become a centralized state function. From religious reasons, where Gods lived with people, it has gone to the secularization of state concerns. The style of warfare has changed dramatically over 500 years (See Table 8132-1).

The essence of war is to defeat or destroy an enemy. Governments are efficient killing machines. On the average in the nineteenth century, states killed 3.7% of their subjects. In the twentieth century states killed 7.3% of the world population.

War is related to growth. Often the same factors that allow unsustainable growth allow unsustainable war. Growth is promoted for its advantages, usually without recognition of its disadvantages—or the difference between growth and development.

Table 8132-1. Styles of War

Kind	Reason	Examples
Personal Conflicts	Insults, broken promises	Palouse, Aborigines
Leaders Conflicts Group conflicts	Social insults, food, territory	Uruk
General's wars	Food, Luxuries	Babylon
Psychological wars	Prestige, Glory Idealism Patriotism	Greece, Macedonia
Professional Soldiers wars (Army wars)	Food, territory, unification	China, Rome
Religious wars Royal wars National wars	Territory / conversion	France, Britain
Economic trade wars	Control of resources in colonies	Britain, France, Portugal, Spain
Chemists war, with poison gases and explosives	Territory expansion	World War I
Physicists war, with radar and nuclear weapons	Territory expansion	WW II
Mathematician's wars, with computer-guided missiles.	Resources Potential threats	Iraq Gulf War I
Electronic wars, to destroy computers and databases.	Destruction of information or economic structure	Iraq II
Ecological wars	Destruction of land base Against Nature	Vietnam, Iraq Pesticides, Medicines

Notice the sequence. Competition with other species and groups leads to conflicts, which grow in size and sophistication, gradually including noncombatants, crops, land, other species, and ecosystems, until the war is against large groups of human communities and then against nature in its various aspects from microorganisms to invasive plants. The sequence also parallels the growth of groups from tribes to Nations and Unions.

The historical rhythm between war and peace inevitably leads back to peace. When does peace occur, when it is won? Unlike many forms of war, peace is a process with a less rigid beginning. It can never be won or kept permanently. It does not have the prestige or honor of war, and perhaps this is why so much less time and effort is devoted to peace.

Is there a way to limit war, within the context of peace? Is there another way to humiliate or embarrass an enemy, or let some other kind of balance be found? Is there a way to limit violence to heroes or to leaders, as if either would agree to have their individual expertise be responsible for a whole population.

Perhaps one basic problem is the easiness of war. War has become too easy. It seems easier in the immediate or emotional short-term to use institutional violence to solve any dispute. It is exciting, and it allows the combatants to form close bonds, perhaps much closer than those in business or play.

The largest problem with war, however, is its scale. War is waged between nations and groups of nations against other nations or groups of nations. Even on a local scale, it often involves larger complexes of political links.

8.1.4. *Summary: Inseparable Behaviors*

Conflict cannot be separated from other interactions, such as environmental destruction or inequities. Conflicts have been escalating, regardless of who won a cold war, or whether nuclear disarmament was achieved. The constant conflict of cultures and the loss of life, human and other, is tragic. But, this kind of tragedy results from a failure of cosmology; humans are responsible ultimately, not fate or chance (See Section 8.2.4.). Humans all are equals, not subhuman others. Humans are tragic because they are responsible for their actions. They can choose a tragedy of the commons or of dictatorial control—or they can expand or alter their cosmology.

Militarism, conflict, intolerance, crimes, and health problems are symptoms of the instability of cultures. Confusion and misinformation contribute further to the destruction of cultures. If cultures are lost and new forms cannot fit themselves to the patterns and uncertainty of natural systems, then people may not be able to adapt to the continued development of the planet and its ecosystems.

Each culture develops rules for living together. A common culture provides an ideal framework for public and private decision making. The Sami in northern Scandinavia have institutionalized ways of avoiding conflict, for instance, by shaming those who would impose their will. The Fipa of Tanzania use cooperative exchange rather than competition to keep the peace. The Akawaio of Guyana believe that community disharmony upsets the spirit world, resulting in illness and misfortune, so harmony is worked for.

The individual in a culture is responsible for the style and simplicity of their life and for its effects on nature and society. The individual is responsible for being tolerant of others in the culture, and is free to make many kinds of choices. Individuals could assimilate new knowledge from their experiences without changing if the experiences fit in the framework of their cosmology, that is, if the experiences aligned with their internal representations of the world. This may also allow them to ignore or judge unimportant events that did not fit. Jean Piaget noted that when individual experiences contradicted their internal representations, they had to change their perceptions of the experiences to fit their internal representations. He called this accommodation, the process of reframing personal mental representation of the external world to fit the new experiences.

Accommodation can lead to learning. When the world violates our expectations, we can fail, but by accommodating this new experience and reframing our model of the way the world works, we can learn from the experience of failure. In a scientific view, ethics and cosmology are separate. But in a good mythology, there is a coherence between areas. The individual feeds back into the cosmology in altered form what has been received. It is almost like a closed loop between cosmology, culture and the individual, even if human cosmologies are limited and contradictory.

Conflicts, territorial or symbolic, are symptoms of insecurity. Many of our wasteful conflicts could be more easily resolved through a neutral international power. Conflicts would still occur. Conflicts over prestige or power, as much as for various crusades as for a true state, still lead to human and environmental destruction. The United Nations Security Council would be charged with the responsibility to avoid massively destructive forms of conflict, such as biological war or nuclear war.

8.2. *Design Factors: Commons*

Traditionally, a commons was a territory managed by a community so that individuals could share the land for grazing or farming. This form of use characterized many tribal groups for 40,000 years. The commons were only open to the members of the community, and perhaps to a few who asked permission. Community beliefs or a kind of ecological ethics discouraged overuse, in that overusers were shamed with informal sanctions or forbidden to continue using the land. In a sense, any territory not explicitly claimed by one community is a commons. That would include the atmosphere and oceans especially, or rather air and water in any form, since these are global things and can move between systems, and their character is determined by the globe itself. Some global economic or ethical method of management, not just for human use but for nonhuman use, must be devised.

In a traditional English village, inherited bundled rights provided commoners with rights of grazing and gathering fuel wood nondestructively "by hook or by crook," which indicated the way wood was gathered by shepherds. The form "commons" is plural, and refers to the whole group of commons, subject to these effects.

8.2.1. *Tragedy of the Commons*
Garrett Hardin extends the idea of the commons to rest on the idea of a carrying capacity of the territory, which is a very dynamic concept, susceptible to change by the weather, season, food preference, or other things. When the commons is not tempered by rules, then an individual user gains all the advantage and the disadvantage is shared by all users. Ecological degradation is assured and the system collapses, a tragedy due to the "remorseless working of things" in Hardin's words. In a limited system, the self-interest of one person cannot decide the public good, or the ecological good. So, the "invisible hand" does not provide a long-term solution. The hand needs to be visible. Necessity needs to be recognized by freedom, as Hardin restates Hegel.

The misuses of commons, some of which are alluded to by Hardin, include: Uncontrolled human population growth, depletion biodiversity, transformation of fossil fuels, pollution of waterways and the atmosphere, logging of forests, overfishing of the oceans, private vehicles jamming public roadways, tossing of trash out of automobile windows (littering), graffiti, poaching, noise pollution, and email spamming. Each of these misuses places a burden on most of the other users.

Many of our less wholesome behaviors, such as rapid technological change or wars, can disrupt tradition and lead to misuse of the commons. If the commons were presented as a prisoner's dilemma for a community, then individuals would have two options, cooperation or defection. Defection, or private mismanagement of common areas, can lead to economic madness, as Paul Shepard indicated. If there are unavoidable costs to defection, then it would be the less likely choice.

The tragedy of the commons is not that the traditional commons system did not work—it did, because it was controlled ethically in a community—it is that other kinds of commons have no such rules governing their use. The tragedy occurs especially when there is a transition from an ethical community use to a free market system that encourages overuse.

Most problems are local not global. So, we have to take responsibility for our local environments, and respect the cultural carrying capacity, as a term for the whole integrated concept of capacity that includes human luxury as well as minimal eating and nutrition.

8.2.2. *Actions Profits & Losses*

Modern industrial culture places an emphasis on individualism and competition. Cooperation, with an understanding of rights and responsibilities, is based on cultural understanding. This kind of understanding, once prevalent in many cultures, is the reason why the tragedy of the commons did not always occur with common resources. Cooperation is crucial. B.F. Skinner suggested that a genetically-based "short-term egoism" leads to a global Tragedy of the Commons. We are tragic because we have to accept responsibility for our actions.

Every action entails a gain and a cost (or profit and loss). Profits and losses are distributed privately, socially, or environmentally. Unfortunately, the modern system privatizes the gain and externalizes the loss to the "commons," considered as a pool of "unowned resources," whereas in traditional societies, it was surrounded by rules for use. As long as this is possible, it is profitable to charge as many costs as possible to the environment. Externalizing costs works fine in an uncrowded world, where the costs are negligible and can be absorbed by natural processes. Resources were traditionally seen as free grabbing; air and water were seen as free sinks. Modern economies, embracing the notion that "nature is capital," draw on the accumulated "capital" of ecosystems for production. By ignoring the real cost of the capital, as well as the costs of natural services, such as nutrient recycling, soil building, and atmospheric renewal, these economies create a temporary wealth. Decisions made on short-term economic grounds could lead to material shortages and environmental degradation.

Economists try to bronze the economy in its current structure; but it is a changing system. Since it is changing, strategies that are appropriate at one stage are totally inappropriate at another—this is the remorseless working of tragedy, where successful strategies are applied in inappropriate circumstances. Tragedies imply conflicts larger than the individual or even society. The external order of things or the cosmic plan is challenged, even if the cosmos is just the size of a city. The source of tragedy for economics is a fatal flaw of the world-view.

8.3. *Design Factors: Behavior Constraints Ethics & Religion*

Human behavior was traditionally linked to ecosystem limits, certainly the patterns of plants and animals, by codes of behavior enforced by shamans and the elders, who were responsible for monitoring animal numbers and health. The set of rules governing the conduct of a group of people, based on traditional behavior of that group, came to be called ethics. Over time the rules became separated from ecological constraints. An ethics abstracted from its ecological and historical context, detached from other societies, and alienated from nature becomes academic, insular, and strange. Such ethics depends entirely on utilitarian values.

Although philosophical systems of ethics attempted to be universal, their basis was mostly theoretical. Kantian ethics is unsuitable for treating other beings in nature; deontological ethics has a weak subjective foundation in human conscience. The areas of concern of ethics and bioethics are not broad enough; their foundations are not deep enough. An ethics refounded on ecological knowledge, by comparison, places human behavior in vital social and biological communities in nature—birth, death, illness, and sex all take place within nature. The frame of reference of ethics is enlarged, leading to appropriate behaviors in a larger context. Human health is related to the health of ecosystems. This ethics addresses animal rights and wilderness preservation, as well as human concerns.

8.3.1. *Plain Ethics, Living Alone*
The word ethics is derived from the Greek word meaning 'custom.' Ethics means 'doing together,' which of course one does living together. And, in an anthropometric universe, it is entirely appropriate. The subject of ethics is human conduct, by definition. Ethics are almost exclusively centered on human action. Ethical theory attempts to give fundamental reasons why actions are right or wrong. Most ethical systems are based on utility and control.

8.3.1.1. Utility and Detachment
People in many modern societies are learning to be uncaring, unattached, and uninvolved, without really thinking about the effects on society and on the environment. Perhaps, it is a result of confusion or fear. We still fear nature because it is uncontrolled and unfathomable. We distance ourselves from what is uncontrolled or unknown. This detachment is the greatest threat to the welfare of nature. William James called for the moral equivalent of war against nature. However, nature and civilization are not in an adversary relationship. Nature is not a threat, although the ignorance of nature is a threat, and the false cosmology of industrial progress is a threat. The ambiguity of the word "nature" reflects the doubts and uncertainties that humans feel about nature. Nature is alien: other. As we have isolated ourselves in human artifacts, we seem to have lost touch and grown cold-hearted.

Nature is separated by a wall of incomprehension. We see wilderness through a glass wall, as we see most all things. Sometimes we look through the thick, convex glass of utilitarianism or the tinted glass of romanticism. Allowing wilderness a place to develop might remove the glass, as would recognition of our continuity with nature. The attitude of distance toward beings excludes consideration of necessary information. Science accepts the sentience of plant life, but does not adjust its methods. Human knowledge grants emotions to animals, but uses them badly anyway. When we can control nature, for whatever purpose, the result is not always good. Our style of efficiency and proclamation of mass values destroys places. The environment of significant places becomes a flatscape. It is turned into uses. John Fowles warns that "We shall never fully understand nature (or ourselves) and certainly never respect it, until we dissociate the wild from the notion of usability—however innocent and harmless the use." This attitude of uselessness lies at the root of our hatred and indifference.

The problem with utilitarian ethics is that it permits the use and exploitation of any natural object, including human beings. Based on a limited science, the ethic fails to value those beings and communities for which no use is known. But, the majority of the beings in nature have no human uses. Even ecologists cannot think of uses for many large birds and mammals. This makes a coldness in the heart of our coexistence with other species. The vivisection of the world depletes our ability to feel compassion for it.

8.3.1.2. Human Center or Human Measure?

The recent formal cosmologies, images of human worlds, were anthropomorphic. Human beings saw human forms in every form of nature. With Plato and Aristotle, nature became anthropocentric, it turned around a center, humanity. Humans were the most important beings, at least through the Middle Ages. By 15th century Arabic and European standards, the universe was a rational order. The human place was prominent, and human life had meaning and purpose. Then, the Copernican revolution transformed the universe from geocentric to centerless. The biological universe, however, was still considered to be a great chain of being. Humans resided between the beasts and angels—until Darwin linked them too closely with the beasts, and cosmology became less meaningful. Then the industrial, scientific revolution restored human importance in a cosmology. Nature is seen exclusively in an anthropocentric manner, as a human resource (especially by John Dewey). Even the Biosphere Reserves are justified according to anthropocentric use. To assume that evolution necessarily progressed to humans ascribes an anthropocentric purpose to nature.

But for environmental effects, dinosaurs, birds, or whales could be the dominant species. So humans are not the unique end or goal. In fact, like new dinosaurs, humans are good competitors, suppressing other species and creating their own pseudo-species. All fields of study are trying to confirm that man makes himself, according to Paul Shepard, no matter how the world is made. Geography endorses economic determinism; history studies the rise of Promethean civilization; the arts separate abstract qualities from content; sociology encourages the theme that everything is possible; the sciences present value-free facts. Ideas are no longer connected. All aspects of life have become interchangeable, including soil, water, and land. All concepts of natural seem to turn on the definition of human. The anthropocentric mistake has been the choice of an oppositional logic, instead of a synthetic one. The either/or complex misses the relationship between the terms.

Earlier, some of the pre-Socratic philosophers developed a nonanthropocentric worldview. Zeno the Stoic preached "life in agreement with nature" as the goal of ethics. Chrysippus added that as individual natures were parts of the nature of the whole, therefore, life was to be in accord with human nature as well as nature. The universe is not anthropomorphic, in the image of man. Nor is it anthropocentric, centered around man. But it is measured and valued by man, as indeed it is measured and valued by all beings. This approach, with its consciousness of limits, is productive and useful. When humans evaluate ecological situations, preference is usually given to human values. This is unavoidable. But there are other life-images that are measuring parts of habitats. There are other centers. The concept of reverence allows the center to be everywhere. This idea has reference in earlier philosophical ideas. The atomists advanced the idea of an infinite universe without center or edge. In the 15th century, Nicholas of Cusa argued that the universe was without edge or center—because its creator was infinite and without location. The universe is "a sphere of which the center is everywhere and the circumference is nowhere." The earth was not the center.

8.3.1.3. Reverence for Life

Albert Schweitzer believed that ethical thought had been developing since prehuman history and that it culminated in the principle of reverence for life. Schweitzer challenges us to plunge into the adventure of ethics: "Let it dare, then, to accept the thought that self-devotion must stretch out not simply to mankind but to all creation, and especially to all life in the world within reach of humanity. Let it rise to the conception that the relation of man to man is only an expression of the relation in which he stands to all being and to the world in general."

Our attitudes are grounded in a belief system that constitutes a particular worldview. The system constitutes a coherent whole. With Schweitzer, the system began to shift toward a biocentric outlook. The concept of reverence (*ehrfurcht*, meaning 'honor-fear' in German) offered some respectability for nature through a proper attitude. Schweitzer proposed an ethics derived from Christian ethics—but arguably larger—that affirmed the world. Reverence for life sometimes conflicts with the Christian paradigm, however, which is just one particular manifestation of a reverential ethic, and Schweitzer's version could lead to an instrumental ethic. Furthermore, Schweitzer's reverence entailed a constant effort to make excruciating decisions. His attitude was a noblesse oblige toward lesser species, based on a Christian idealism, on the fear of death, and on a mistaken image of nature as brutish and dumb.

Through human evolution, the circle of responsibilities has widened, from the family, to tribe, nation, humanity, and now toward all life (see Figure 8313-1). Although Schweitzer noted that the circle of knowledge was widening also, he felt the streams were divergent, that ethics had nothing to gain from understanding nature, and furthermore, that there was no hope of finding meaning in natural phenomena. His ethic was not based on ecological knowledge. His reverence for life principle acquires a new aspect when it is restored to ontologically and ecologically firm ground (expanded in *Redesigning the Planet: Regions*).

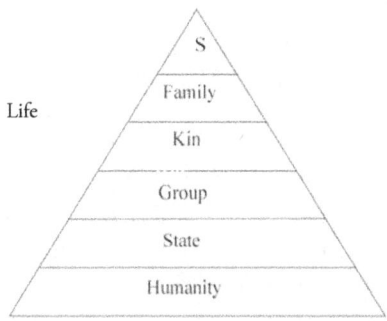

Figure 8313-1. Ethics expanding from Self (S) to all life (context of pyramid)

8.3.2. *Pan Ethics, Living Together*

A new ethics can recover its basis from nature itself, not from an extension of the old anthropocentric ethics, with all its limited assumptions. It is based on ecological knowledge, grounded "in the breadth of being," in Hans Jonas' words, and founded on principles discovered in existence. Many ultrahuman cultures have standards (or codes) of behavior to regulate interactions. In birds and simple mammals, these rules may be very rigid and predictable. With increasing brain complexity, learning takes a larger role.

Animals are ethical already; they live together by rules; they respond to situations to other animals. Ultrahuman ethics are rules for living together. The human community is one of many communities that make up ecosystems. Human relationships embrace other beings as well as other humans. Human ethics describes a small part of the rules, perhaps the only self-conscious part.

An ecocentric view emphasizes human moral responsibility for vulnerable ecosystems and habitats. An ecocentric view recognizes the ends and means of all beings. An ecocentric ethics can place human ethics in proper perspective, in proper relationship with other living, interacting beings. Understanding ecological relationships permits the toleration of fluctuation, irregularity, uncertainty, diversity, spontaneity, flexibility, looseness, and limits. The basic premise of nature is interrelatedness. The interpenetration of boundaries makes humans less discrete, less alone. A shared biology establishes the fellowship of beings.

8.3.2.1. Ecological Ethics

Humanity has started to remember that it is an integral part of food chain and part of an organic cycle of birth and death (Paolo Soleri and others have been celebrating this for decades). Humans need to recognize that they automatically participate in everything, and that they cannot unparticipate by choice. Participation starts at the quantum level and continues through the ecological and cultural. Human nature does not find meaning in an absurd world, but discovers its structure through interaction with the ultrahuman order. Human identity exists partly in relation to nature; the destruction of one involves the other.

A new ethics starts from nature itself, not from an extension of the old anthropocentric ethics, with all its limited assumptions. It is based on ecological knowledge, grounded "in the breadth of being," in Hans Jonas' words, and founded on principles discovered in existence. Many ultrahuman cultures have standards (or codes) of behavior to regulate interactions. In birds and simple mammals, these rules may be very rigid and predictable. With increasing brain complexity, however, learning takes a larger role.

Ethics are assembled inductively, from experience in living in places. Now, because of the uncertainty of human actions, ethics has to encompass the far past and distant future. No one knew that when DDT killed mosquitoes, it would concentrate in the food chain to kill birds. Values are time dependent, and ecological time can be very long indeed. The futures we invent are viable only if compatible with constraints imposed by evolutionary past. An ethics that requires a long-range responsibility also requires a new humility, since technological power exceeds the ability to foresee its consequences. An ecological ethic recognizes the moral obligation to leave the world habitable for future generations.

To address problems with extermination and pollution, Aldo Leopold proposed a conservation ethic, dealing with human relationships to land, plants and animals. The land ethic that Leopold had in mind was a sense of ecological community between humanity and other species. "When we see land as community to which we belong, we will use it with love and respect." Such an ethic would change the human role from master of earth to plain member of it. Predators are members of the community; and no special interest group has the right to exterminate them for the sake of benefit for itself. This attitude is important for habitat protection. Leopold describes the extension of ethics (emphasis added) as "actually a process in ecological evolution. Its sequences may be described in ecological as well as in philosophical terms. An ethic, ecologically, is a limitation on freedom of action in the struggle for existence. An ethic, philosophically, is a differentiation of social from anti-social conduct. These are two different definitions of one thing. The thing has its origin in the tendency of interdependent individuals or groups to evolve modes of cooperation."

The extension of ethics to animals and land is an ecological necessity. Extended ethics defines a social conduct that is a mode of cooperation and, ultimately, symbiosis. Leopold argued that ethics are voluntary limitations of freedom, necessary in a complex world of which we remain incredibly ignorant. Ethics are developed in response to problems that arise from increasing knowledge. Science has phenomenally increased our knowledge of physical and biological processes. It has now become the basis of our moral code, but it cannot very long be a science divorced from feeling and art if that code is to help us survive. To do this science requires aesthetic perception as well as disciplined thinking and feeling. As there is a rational component to ethical judgments, so there is an intuitive and emotional one, also. An evolutionary ethic suggests that humans avoid tampering with complex evolved systems, not because they are good, but because they are the basis of life at this stage of development. Ecological ethics is situational because ecology is the study of changing systems. The morality of the act is determined by the current state of the system. Adaptive modes should conform to ecological patterns. An ecological ethics is based on attributes of ecosystems and human compliance with ecological laws. The aim of an ethic must be harmonious to the idea of the world's population of living beings.

8.3.2.2. Natural Value

Humans have stripped the world of qualities and significance and claimed them for themselves. By valuing humans alone, value is made subjective, and ends are without value. However, the human perspective realizes only a small part of the spectrum of rich possibility of experience. Humans use an incomplete source of value for ultrahuman beings; that source is human need. Abraham Maslow established a hierarchy of human needs, beginning with food and continuing through social acceptance to self-actualization. But human needs are based on the health of the earth. Nature, which is self-supporting and self-managing, is the human life-support system. Human systems depend on natural ones, for recycling of wastes, water, and air. But, as human growth is logarithmic, so is human need, and need shapes facts, like kind and quality of resources.

As Goethe recognized, all fact is theory, a blend of perception, imagination, and needs. Granting this, the need for wilderness is as much a fact as the need for food. Wilderness is ecologically important, for it accounts for ninety percent of the energy trapped by photosynthesis from the sun; it is crucial in the global energy system. Trees and plants are good sources of energy. Energy and materials can be produced through photosynthesis. In fact, wilderness is the greatest producer of renewable sources of energy and materials. As habitat for incredible number of species, it is insurance against the dangers of simplified agricultural systems; it is a depository for genetic types. Biological species are essential to the maintenance of ecosystems. Wilderness is the source of evolutionary process.

Rich sensory and emotional experience can be derived from contact with the wilds. These values are not often mentioned by economists and planners. But values usually encode information having survival or prestige importance. Perhaps the most valuable thing is living time. This may be why humans value walking in the woods or observing the production of art. The value of wild nature is its independence and wildness.

We find that kind of knowing more in literature than in science. Whitehead turned to literature because it corrected the excess of objectivity that occurred in science. Poetry carries an expression of the organic character and value of nature. Value refers to the in-itselfness and for-itselfness of the process of realization. Besides having value for itself, it shares value with the rest of the universe. Everything has signification in the universe, says Whitehead: "Remembering the poetic rendering of our concrete experience, we see at once that the element of value, of being valuable, of having value, of being an end in itself, of being something which is for its own sake, must not be omitted in any account . . ."

Perhaps there is a hierarchy of value from simple forms to complex, corresponding to richness of experience, as John Cobb suggests. Although there is no hierarchy of being, there may be one of richness or value, depending on the frame of reference. There is value for each, and for the whole. And the whole may invert the value. Ecology can expand the narrow anthropocentric evaluation and see things from the perspective of the whole. Lovejoy's 'Great Chain of Being' traces the deductive order in classical nature from Greeks to German idealism. But Lamarck inverted the chain in his theory of transformism; mind is immanent and can determine transformations. Although the hypothesis of inherited characteristics was rejected by Darwin, who shared that hypothesis but denied mind as an explanatory principle, both Lamarck and Darwin inverted the value of life. By inverting the great chain of being, Lamarck escaped the directive that the perfect must precede the imperfect. The result of Elton's food chain was the realization that the bottom link—plants—is the most important.

Ecological value starts at the broad base of the pyramid. Our ethics and legal system is species specific (Singer 1981), but it occurs in a larger, normative order. Ecological science has a very normative component. Even a set of genetic codes is normative; genes show what is valuable for a life-image. So value is morphic on all levels of being. A sort of natural relativity of frames of reference encompasses the smaller human system.

Perhaps we should not argue that things have value in the human system. Let us just respect the existence of the ultrahuman system. Every being has an intrinsic value, before any utilitarian value to humanity. The value of wild nature is its independence and wildness. Perhaps our astonishment contributes to its value. Bees have bee value; wolves have wolf value. Wolves are not efficient at binding nitrogen; neither are humans. Lichen are poor predators, but they break apart rock better than bighorns. The world would not be a better place without sharks, silverfish, cockroaches, rats, hyenas, or whales. Their existence has value; they have functions. From a functional point of view, all beings are equal. In an ecocentric perspective, all beings have intrinsic value and are equally important.

8.3.2.3. Reverence for Being

The recovery of implicit natural values can be expressed as a reverence for natural systems. The systems scientist Ervin Laszlo proposed a social ethic for the age of humanity, calling for reverence for the level-structure of the microhierarchy, including all systems on all levels, from atoms to an emerging planetary culture and ecology. "We can express the recovery of our implicit natural values in requesting a reverence for natural systems," he said. This reverence expresses the insight that humanity is in nature, a part of an embracing network of dynamic self-regulating and self-creating processes. Our thoughts and ideas are nature, as much as clouds and waves. Marveling at humanity is marveling at nature, the matrix from which we arose. Although there are others, humans are remarkable examples of dynamic order brought forth by the universe in its history.

Reverence must include all natural and artificial beings and things and fields, from atoms to weeds, computers to galaxies. Atoms, molecules, and organic cycles are parts of humanity. We must revere all the arrangements of earth stuff. The greatest human dignity follows from the respectfulness of everything as meaningful as ourselves. Such a reverence treats all substances of the earth as precious, to be used carefully, if at all—and certainly not for a flood of mass-produced consumer items. It includes all human artifacts, manufactures, and societies. It promotes a society where individuals would live in close contact with natural support systems. It provides each with the aesthetic necessities of life to develop all capacities.

The basis of all value is being (the verb form). It is reality undistorted by human needs.

Of course, humanity can still enhance nature with its presence, in a nonexploitable manner. The reverence for beings as they are results in the rule of noninterference. In nature, the rule of noninterference means 'letting be' (Heidegger), 'letting alone' (Wilson), and 'not killing for pleasure' (Fox). A formal presentation of the rule states that human beings ought to avoid behavior that disrupts essential ecological processes or destroys biotic communities. As Paul Taylor states his rule of noninterference, it requires a "hands-off policy" for whole ecosystems and biotic communities; the rule stated here is concerned with limited and sustainable exploitation of ecosystems already shaped to some extent by human activities. Noninterference is not indifference, which is diffuse. It is caring. Noninterference will not lead to chaos, poverty, and stagnation. The technocratic vision strives for "life under control," but the earth is self-managing, productive, efficient, and orderly. Reverence can only be felt at the alienness of nature, not at its comfortable conquest. With so little wilderness left, however, formal rules may be necessary.

8.3.3. *Relating Being to Rights*
An ecological ethical model is not distorted by human needs and wants when it argues for the preservation of animals and habitats themselves, because they are, as they are. Paul Shepard says the argument is not new, and that its application is ambiguous because "unlimited rights" will conflict with human interest. But, there are two bad assumptions: that human interests are not ambiguous—they are—and that animals will be granted unlimited rights—they will not. Rights seem to follow the expansion of the sphere of ethics, as formal statements of intuitive knowledge. But codifying rights is more difficult, especially for philosophers, who tend to limit rights with a series of restrictions. For example, a contractual theory assumes a perfect detachment and a rational debate of rules. Animals and imbeciles are left out. Scott Lehman argues that natural objects are not the subjects of experience (certain animals excepted) and so cannot possess rights. He limits experience to mental states (perhaps he meant nervous system states). Some philosophers maintain that a right is a claim to something (Feinberg); others, that it is an entitlement. Richard Watson takes reciprocity as central to the general concepts of rights and duties; few animals and no natural objects have rights intrinsically. He mentions that some primates and mammals are moral entities because they are self-conscious, have free will, understand principles, and intend to act accordingly. But the assumption of self-consciousness would rule out children and feeble-minded adults, as well as most beings. So, a larger definition of claim or reciprocity is needed, without regard for contracts and mutual duties.

8.3.3.1. Natural Rights
When humanity was divided into citizens and slaves, there was less freedom. When it was divided into governors and governed, freedom was advanced by providing the governed with protection against the tyranny of governors. When people became self-governing, protection was needed against majority opinions, by distinguishing between the individual and society. New contrarities—public and private—provide clarification of rights. Rights protect the interests of those holding rights. Natural rights are the rights of an underclass that has not been granted legal rights. These "natural rights" are used by minorities to legitimate their claims against controlling powers. Rights and obligations were first thought of in a political context consisting of customs and practices within and between states. In the 17th century they were thought of in a constitutional context, where forms of government were established to protect natural rights. Now they are thought of in a human context. Freedoms of—speech, worship—depend on institutional protection and are political rights. Freedoms from—want,

fear—are extensions of economic rights. Freedoms for—pleasure, reproduction—are considered biological rights.

Humanity has taken its own opportunities. These opportunities have been codified for centuries as rights. Now, we must allow other beings equal opportunities. The interrelatedness of life dictates the interrelatedness of rights. And these rights are necessary to the integrity of the whole planet. Humanity developed in a community of animals and plants, as part of a clade on the same tree of life. The quality of human life has always depended on the quality of animal life. Animals have sensations and feelings, as important to them as ours are to us. Furthermore, the extension of rights to animals and plants does not deny any traditional human rights. Animals should be accorded higher moral regard and legal standing to reflect the intrinsic worth afforded by their existence and sentience. Welfare laws to conserve species and to guarantee humane treatment in research, transportation, and slaughter indicate a growing concern among people. A new ethic can keep animals free from human intervention, prejudice, or overuse. Animals should be preserved because they are as they are; their existence is moral justification. Their intrinsic worth is independent of the instrumental values imposed on them by humanity.

The strongest argument for rights is interrelatedness in communities. It is a basis for assigning rights to ultrahuman nature. Existence implies intrinsic worth. Garret Hardin considers interrelatedness, but interprets it narrowly. He considers rights as rules of competition; every right is a ploy in the struggle for existence, and every right implies an obligation to furnish it. This is good as far as it goes. Life is more than competition, however; it involves cooperation and play. Rights are formal rules for living together. Society should be organized on the basis of functions, not rights. Ecological rights (customs) could be based on functions. It is foolish not to assign rights to animals, plants, and the earth because of contractual formalities. The reverence for all beings is concerned with the right functioning and right numbers in the right places, according to standards of health and quality of life.

One problem with the current legal system is that all nonhuman beings are given the status of inferior human beings, legal incompetents, thus keeping humans in a guardian role. A new legal category is needed that would respect the existence, competence, and excellence of natural beings. Christopher Stone recognizes that the judicial system has granted rights to a variety of inanimate holders, trusts, corporations, and nations, for instance. The legal system already operates with fictions, so the extension to natural entities should not present an insurmountable problem. To be sure, formal rules should to be altered to account for unconscious, interdependent beings. Current legislation on animal experimentation and protection implicitly recognizes the right to life and to a healthy habitat. Laws are needed to protect entire habitats of animals and plants from human interference.

8.3.3.2. Rights in Justice & Law

Socrates argued in *The Republic* against the conception of justice as giving every man his due, and proposed a definition of justice as every man performing his proper function. This proportion of reward to function in the community was named distributive justice by Aristotle. It describes the right to participate in the benefits of science and culture. If justice is a proportion set up in the community between men and goods, justice is also the restoration of the relationships of men and goods, when disturbed. Aristotle called this justice rectificatory. This has constituted the business of recent laws and courts. During this stage of universal rights, world order transferred the criterion from the nature of man to the community of men; both law and justice, obligations and rights were reduced to equity. Thus, new nations demanded to participate in a common justice, as opposed to the extension of natural rights.

A principle of justice based on need can be extended to the ultrahuman community. To be sure, it needs to be altered to account for unconscious, interdependent beings. The right to use nature is a right to share. Current legislation on animal experimentation and protection implicitly recognizes the right to life and a healthy habitat.

One problem with the current legal system is that all ultrahuman beings are given the status of inferior human beings, legal incompetents, thus keeping humans in a guardian role. A new legal category is needed that would respect the existence, competence, and excellence of natural beings. Christopher Stone recognizes that the judicial system has granted rights to a variety of inanimate holders, trusts, corporations, and nations, for instance. The legal system already operates with fictions, so the extension to natural entities should not present an insurmountable problem.

Formal law tends to seek guidance on normative issues from the general population, rather than from legal experts. People care for animals and wilderness. Natural rights are defined by positive laws and by negative restraints on behavior. Laws are needed to protect ecosystems and wilderness, now. Legal or religious action almost always precedes the general shift in conscience. The obligation to treat others equally includes an obligation to change human social patterns in the direction of equality.

We act by intuition and feelings. Like the inductive creation of cultural ethics, we are building a framework of intuitions, feelings, theories, and principles. The whole is recognized as a valuable end by hunters and actors as well as by scientists and politicians. The framework is supported by principles and theories developed by ecologists and philosophers, by the working rules of conservationists and activists, and by specific instances from cultural traditions as well as from the industrial paradigm (determined to become its own worst enemy as well). Stone considers that these things are only part of the framework. In The Laws, Plato has the Athenian say to a youth that all things are ordered with a view to the preservation of the whole, each portion contributes to the whole, and every other creature is for the sake of the whole. Ethics has expanded in wholes, from the family, to the human community, and to nature, on which everything depends.

To protect the whole, we need laws to encourage practicing the 'Rule of Noninterference.' A machine metaphor approach, with its assumptions of interchangeability and quantity, has difficulty distinguishing between exploitation (the normal use of resources) and interference (destruction for temporary or no gain). An ecological metaphor, that is more receptive and reverential, may be more appropriate to understanding organisms and nature in general. Such an approach would stress noninterfering observation rather than controlling manipulation.

8.3.4. *Binding Humanity to Places*

The problems of ignorance and inappropriate images are multicultural, ecological, and cosmological, and must be solved on the level of culture—the entire activity of culture is guided by metaphors. Metaphors emphasize likenesses between living things and languages (or human constructs). A metaphor furnishes a label and emphasizes similarities. It not only defines and extends new meanings, but redescribes domains seen already through one metaphoric frame.

An understanding of ecological metaphors, with an emphasis on limits, can lengthen the life of a culture, but ecology is not enough. Cultural metaphors are necessary, as are good rules of behavior. Culture can provide the rules. Ethics and economics, for instance, are rules of behavior; and politics is the practice of changing the rules as society changes. Religion is the practice of binding us to constant rules and values regardless of temporal changes.

G.W. Leibniz, concerned for the purpose of restoring the unity of Christianity, worried that without a common belief in God, secular religions would emerge that would be tyrannical and suppress the freedom of the individual. Later, Rousseau thought that society could and should control the individual. Then, Marxism offered its advanced secular religion. But, it collapsed, although it and the secular religion of capitalism exalted the freedom of the individual without linking it to a belief in God, compounding rampant individualism.

Now, what will replace the failed communism and the failing capitalism? Could religion, for instance Christianity, replace the secular attempts? Is that what religion should do? Form a rebirth of belief and shape the person in a 'knowledge' society? Will there be new religions, like the pastoral ones in the US, which will try to do this? Peter Drucker thinks that the end of the salvation through society will make possible a renewed emphasis on the individual. But, an inspired religion could correct that continued damaging distortion, and allow a resurgence of culture that would encourage individual responsibility in the cultural context. James Lovelock considers Gaia as a religious and scientific concept. As sacredness and wonder are essential to any belief system, including science, an inspired religion could incorporate a global model of the planet into its universal beliefs.

Religion asks us to realize that we are obligated to respect larger harmonies in nature, not just be restricted by rules to not disrupt them. Religion can balance some of the other tendencies of a culture. Consuming is much easier than restraint. Maybe that is what religion helps with, reasons for restraint. It provides a supportive social environment, where people practice restraint. E. O. Wilson states that religion possesses strength to the extent it codifies in enduring poetic form the highest values of humanity, which are consistent with empirical knowledge. Only a cultural consistency can provide compelling moral leadership.

Religions have developed in the past 40,000 years. There were long period of animal and force gods. Then, there was a cosmologization of thought systems from 500 BC to 300 AD. Religion has gradually widened the sphere of inclusion, from kin and tribe to nation, animals, humanity, and ecosystems, to maybe the planet and solar system. Religion has its own gigatrends, from Gods of natural forces to war gods, regional gods, global gods, and universal gods. Maybe it will be concerned with machines and hybrids, now. Religion has changed forms, from allowing anyone to contact spirits or ancestors, to limiting authority to shaman and priests. It has been used to justify institutional theft as it became larger and more formal. The original goal of religion was truth, followed by wealth and attendance or conformation. But, meaning went missing. So, religion went to popularization next. People responded by seeking personal religions. But that got replaced by trends such as Scientology. What is replacing those trends? Loose associations?

Religious organizations that have persisted for centuries have done well at preserving individual trees or stands; both Chinese yellow pine and dawn redwood, long thought to be extinct, were found in temple gardens. By contrast, the Queensland Australia management system for moist tropical forests was closed down by a political decision after a state-federal struggle for control. Long-term stability, with secure ownership by state, culture, or tribe is a necessity for permanent forests. This means intergenerational management.

Perhaps ownership will reflect a change from individual and corporate to land trusts and community. We need to find the best combination of ownership, trusts, easements, and plans to ensure that land will be cared for in the long-term.

As a cultural form, religion has potential to help humans adapt to the constraints of a wild planet. Religion could help control disruptive forces, especially those that result from distribution and power. Religion coerces people into a social contract. Religion and storytelling reduce the variability of individuals wants in a group. The introduction of new stories and myths, concerned with partnership and equality and with the value of other places, species and people, could change our attitudes about them into a healthier context.

A start has been made. Ethics now considers almost every human being and human interaction. The restriction of ethics to exclusively human modes of existence, however, leads to a troublesome isolation. Human beings are not separate from their social and biological communities and these communities are embedded in ecological contexts with biogeochemical processes. Ethics must be extended to the framework and to the nonhuman communities in the framework, without which there would be no human health or wealth. Through science as well as through mysticism, we understand that communities of other beings have their own values and rules for living together. It remains for us to integrate and codify human rules that recognize the values and rights of other beings.

8.4. *Design Factors: Politics or Ecocybernics*

For Aristotle, politics was the science of the possible. The city (or *polis*) was a human artifact whose structure could be modified by reason; it was potentially a work of art in which only the capability of the artist limited the expression. The *polis* was made for the amateur; and it produced more complete men (women were erroneously considered lesser beings at the time). The city was necessary for politics then, as face to face communication. Now, communities are different, larger and dispersed, so politics has to change. As a science, politics, or ecocybernics (a neologism meaning 'governing the house'), was concerned with two things basically, a way of distributing power and luxuries internally within society, and a way of surviving contact with other cultures.

Archaic nations governed their areas independently. Their political principles were similar: All land was communally owned by the tribe, although household goods may be personally held; all decisions were made by consensus in which everyone participated; chiefs were not coercive rulers, but teachers and leaders with specific duties limited to their realm—medicine, war, or ceremonies for example.

When the Europeans or Chinese settled many areas, they brought their centralized governments. The original goal of the U.S. republic, according to Jefferson, was to make each person a participant in the everyday affairs of government. But, the government (state or federal) has become gigantic, managing the area from remote locations of power, and participation has dwindled. Despite a recent emphasis on personal responsibility and international cooperation, our political institutions have not responded.

Ultimately, politics defines reality itself. Social reality is created. Different societies have had different realities. The Athenians despised the merely wealthy, for instance, and respected community service. North Americans, on the other hand, revere wealth; community service is romanticized on television but hideously underpaid. Politics is also the art of creating new possibilities for human progress. The current system is defective, however. Although it is admirable to work within the system to prevent further environmental degradation, it might be necessary to produce a change in consciousness that would lead to a new political paradigm.

Politics is the science of government concerned with the regulation of a nation or state, the preservation of its safety, peace, and prosperity, the defense of its existence and rights against foreign control or conquest, the augmentation of its strength and resources, and the protection of its citizens and their rights, with the preservation and improvement of their morals. Politics depends on common rights, trust, information, and consciousness.

The function of politics is to ensure that decisions are made at the right level. A state protects individual freedoms, guards national culture (values and identity), and holds groups accountable for their use of power. The logic of individualism creates conditions that require constraints. Politics has to make them palatable. Although we realize that nothing in nature is without some limit or cost, we may dislike giving up what we now consider (wrongly) as rights. Humans may have to expect much less than they want, even though many expectations are rising.

Central politics usually overwhelms local politics. It dominates the process of decision-making. Politics deals with words, which are arbitrary symbols for events or things. The wrong relationship of things and symbols can result in misguided politics and violence. Decisions are made on narrow political grounds. Citizenship in industrial cultures is the abandonment of responsibility on the assumption that others know how to manage things; government is the assumption of responsibility, without knowledge, that leads to immense and interrelated problems. Government promises to judge disputes in values and protect its citizens from external attack. Modern government, however, is the abandonment of responsibility on the assumption that others know how to manage.

Besides size and power, there are other things that make governing difficult: Division of labor (are there professional citizens?), centralization, and technology. The interrelation of these things necessitates discussing them at the same time. Political institutions are not givens or timeless. Taking from the strengths of earlier forms, it might be possible to modify government to be more effective.

Table 841-1. Political Systems (after Aristotle)

Number of Rulers	Kind of Ruler	Ruling in Interest of	Chosen by
One	Leader (headman)	Band	Consensus
	Big man	Tribe	Consensus
	Chief	Chief Nobles	Ancestry Gods
	King Emperor Dictator	Monarchy City-State Tyranny of Few Self/Power structure	Self Parent Self Military
Few Collective	Triumvirate Co-ruler	Aristocracy Oligarchy	Citizens of State Few (Rich)
Many	Prime Minister President	Polity Voters Democracy Ochlocracy	Voters People or College Poor People
All	People Anarchy	Direct democracy Community anarchy	All People

8.4.1. Forms of Governing & Leading

There are many ways of leading people. Some offer models to follow; others represent people in group decision making; and, a few make all decisions themselves. There are formal and informal ways. Almost every way, however, involves more prestige or status for the leader. Sometimes that translates into more meat, more goods or more wealth.

Leadership is the process of leading, that is, providing direction to others. Political leadership, in the sense used here, is often formal and refers to a person in a position of authority, as a result of possessing skills at directing. The number and kinds of leaders are compared in Table 841-1. Some political systems have only one ruler. For instance, in some band societies, Leaders or 'headmen' were chosen because of their skills. In some tribes, 'Big men' predominated by their generosity. In some Polynesian groups, 'Chiefs' became hereditary. Kings were able to rule by force and then tradition. Emperors sometimes ruled conglomerates of cultures. Dictatorships needed great leaders, or at least powerful ones, to be successful. A dictator may be self-appointed, popularly backed or appointed by a military force.

Other political systems have several leaders. Rome once had a ruling triumvirate. Bosnia and Herzegovina have a three-member Presidency, each of which are elected by a different constituent nation, where the position of President rotates between the three members. Modern Switzerland invests leadership collectively in a seven-member Swiss Federal Council, where the President is a member of the Federal Council elected by the Swiss Federal Assembly for a year and is merely *primus inter pares* (first among equals).

Some states have a Parliamentary system of government, in which the President is the head of state, sometimes with only ceremonial duties, depending on the constitution, and the Prime Minister is the head of government. Countries with this systems include Germany, India, Ireland, Italy, and Singapore.

In nations with a Presidential system of government, the President is the head of government and the head of state. The United States and Venezuela have this system, where the President has more power than in other political systems. Democracies need effective and educated citizens, where governing is the work of the people (remember the Negro spiritual: "We are the ones we've been waiting for").

Leaders themselves may be influenced by choosers or controllers. Controlling groups such as the military, political parties, ruling elites, or religious elites sometimes place higher expectations on the leader, such as transformational change or special favors. The leaders may encourage their followers and believers to worship leadership. The followers may become uncritically obedient.

Leaders, with choosers and controllers, are part of the governing process. The forms of government range from tribal leaders and chiefs to anarchy and direct democracy. These forms work best with small populations where everyone is known, even if they do not share the same culture. Other forms of government, such as authoritarian dictatorships, monarchies, and republics, have worked with larger populations. At the largest sizes, the bureaucracies and infrastructures have more in common, regardless if the government is socialist, communist, parliamentary, or representative democratic.

Larger nations require larger governments, to make laws to accommodate differences in cultures. In this case, specialists are required to resolve conflicts. These specialists often have a monopoly on weapons and on the spectrum of information. Therefore, a bureaucracy is needed to run the government more effectively. The types of governments and leaders are compared in Table 841-2.

Table 841-2. Governments

Leadership / Government	Egalitarian Leader	"Big man"	Chief Hereditary	King Hereditary	Representative
Bureaucracy	None	None	None Crony level	Many levels	More levels
Monopoly on Force/Information	No	No	Yes	Yes	Immense
Resolution Of conflict	Informal	Informal	Central	Laws, judges	Laws, judges, specialists

At the largest sizes, the bureaucracies and infrastructures have more in common, regardless if the government is socialist, communist, parliamentary, or representative democratic.

8.4.1.1. Traditional Small-scale Leaders (Elders, Chiefs, Kings)

As an example, the Coeur d'Alene people had strong leaders where each village had a council with male and female members. Each large village had a headman or woman who regulated economic and religious affairs. Their only real power was their persuasive abilities and the public esteem that they built up. Leaders, who regulated basic resources, were elected or deposed by this council. Next in authority were a war leader, hunting leader, and shaman, chosen for their respective skills.

Marvin Harris shows how leaders were associated with hunting societies, while 'big men' were associated with larger horticultural societies, and finally chiefs with still larger societies, which had started to accumulate goods. Headmen were found in more informal bands of 50 or villages of 150, where everyone knew and understood others and reciprocity was the typical interaction. Headmen were leaders without power, while 'big men' gave away extra wealth, and Chiefs accumulated it. Chiefs had proven ways to keep the underlings happy or at least resigned. They armed the elite and disarmed the public. They used their monopoly of force to maintain order and reduce violence. They redistributed some of the wealth in public displays and games, a kind of informal tax refunds. Chiefdoms were successful in Kwakiutl, Rome and Hawaii.

Meanwhile Kings enlisted ideology or religion to justify the transfer of wealth to the king and rich retainers. Their shared religion also encouraged strangers act more peacefully without kinship and it gave people a reason to sacrifice their lives for an institution. The king was the central human representative of power. For instance, the Mayan king offered his blood with other nobles and prisoners. Since blood was the home of the soul, it was the most important gift a king could give.

Mesopotamian cities had councils of elders to make decisions and appoint a leader to lead in war or trade. The word for this leader, lugal, eventually came to mean king. Agriculture gave the people a surplus, which with increased luxuries, led to conflict and war with those less fortunate. In China, in the Warring States period, the nature of war changed: From a monopoly of volunteer soldiers to a standing army, with professional leaders and peasant soldiers. In China, new towns were planned and built by ranked lords at the edges of the state. The Emperor gave them land, titles, ritual things, and a clan name. Because he could contact his ancestors, the Emperor served as his own priest and held more power.

8.4.1.2. Modern Large-scale Government

As leaders attracted more followers or constituents, they needed help addressing problems of income and payout. They needed a large bureaucracy to maintain records and make local decisions. For example, large-scale irrigation appears in Egypt first in 7100 YBP. In Mesopotamia, large-scale water systems were first managed by religious leaders, then later by secular leaders. As problems developed, temporary military leaders became permanent hereditary kings. Palaces were built and staffs numbered thousands. Power became centralized in the king. As the populations became larger, the king had to have representatives. Centralization and representation became necessary properties of large-scale government.

8.4.1.3. The Importance of Leadership

Leadership can refer both to the process of leading, and to those entities that do the leading. Leadership can have a formal aspect, as in most political or business leadership, or an informal one, as in most friendships. A leader is assumed to have special skills or competencies. Leaders of bands or tribes were often specialized in specific activities such as hunting or social integration. Although leadership can be exhibited by an individual, either a group of people or a heroic character can also show leadership.

Leadership is often confused with transformational change, although usually leaders are the last to change. Leadership may be associated with respect, obedience, or worship. National leaders may be presidents or queens who make the decisions for their countries. Global leaders, however, will have to address a wider spectrum of needs and balance them.

Even in representative systems, leaders may have ceremonial or reserved powers. Leaders need to have people with ideas (like F.D. Roosevelt had), but they also need to have jesters and clowns to mock their mistakes and pride. In most every nation, leaders have too much power, as well as too little information, humor, humility, or wisdom.

8.4.1.4. Forms of Governments

People have chosen, or allowed, or suffered, many kinds of governments and institutions, including rule by a king or co-consuls, presidents or oligarchs, communism or democracy, and by aristocrats or anarchy. Societies are constituted by people and their ideas about government and justice, among other things. Ideas and behavior continue to evolve in a natural and then cultural contexts. The individual human being is a result of a family and local community, but law evolves in a larger culture. Law is now an important part of social structure.

Humans consider themselves to be self-creating and self-ordering, because they are part of a context of self-ordering processes. Government is part of that self-ordering process, wherein specialists make decisions based on their expertise. Specialization is built on trust. We trust that others will produce enough grain for themselves and us if we make a plow for them. Government has to balance giving and receiving of things and services, since it preserves the balance of society.

But, governments do not always balance giving and receiving very well. Since they are run by people, and people tend to put their own interests first, governments can be tyrannical about refusing to fairly distribute power and goods. Are some tyrannies more acceptable than others? The abuse of power by kings or the abuse of power by corporations? Unjust executions or cancer from toxic wastes? Government mismanagement or the abandonment of communities by capital interests?

Governments are, and should be, accountable to the laws that they make. Enforcing the law may keep governments more balanced.

8.4.1.4.1. Excellence & Law

What is government? Tribal government was based on customs and who was best at certain things, from hunting to resolving conflict. In tribes and empires, laws were formed to overcome differences in customs. Government now is based on laws, but it also needs to be based on information as well as on general common human values (and shared human rights).

8.4.1.4.2. Separation of Powers & Functions

Aristotle, the Greek philosopher of 2350 years ago, recognized the importance of a "rule of law" for states, as well as a central government with a separation of powers. The framers of the U.S. Constitution wanted to form a government that did not allow one person to have too much authority or control. While under the rule of the British king they learned that this could be a bad system. Yet government under the Articles of Confederation taught them that there was a need for a strong centralized government.

With this in mind, the framers wrote the U.S. Constitution to provide for a separation of powers, or three separate branches of government: A legislature, executive, and judicial. Each has its own responsibilities and at the same time they work together to assure that the rights of citizens are not ignored or disallowed during the running of the nation. This is done through checks and balances. A branch may use its powers to check the powers of the other two in order to maintain a balance of power among the three branches of government. The checks and balances, combined with a separation of powers, gives responsibility to each branch to oversee the others, but the branches must be equal. Partly this is done by having overlapping responsibilities.

8.4.1.4.2.1. Legislative Branch
The legislative branch of government is made up of the Congress and government agencies, such as the Government Printing Office and Library of Congress, that provide assistance to and support services for the Congress. Article I of the Constitution established this branch and gave Congress the power to make laws. Congress has two parts, the House of Representatives and the Senate. Three functions of the legislature are: To make laws, to check the executive branch by confirming appointments, and to investigate activities of the executive branch.

8.4.1.4.2.2. Executive Branch
The executive branch of Government makes sure that the laws of the United States are obeyed. The President of the United States is the head of the executive branch of government. This branch is very large so the President gets help from the Vice President, department heads (Cabinet members), and heads of independent agencies. Typically, the President is the Leader of the country and commands the military. The Vice President is President of the Senate and becomes President, if the President can no longer do the job. The department heads advise the President on issues and help carry out policies. Independent agencies help carry out policy or provide special services.

The executives could include a monarch, prime minister or cabinet. Six functions are: To execute laws, to execute policies, to control policies, to appoint officials, to command the military, and to veto legislation.

The executive branch could be headed by an Executive Council of seven, with these specialties: Internal Coordination, External coordination, Ecological affairs, Cultural Affairs, Religious affairs, Economic Affairs, and Communication (Press). This Council would elect a President for Internal affairs and a Prime Minister for external affairs.

8.4.1.4.2.3. Judicial Branch
The judicial branch of government is made up of the court system. The Supreme Court is the highest court in many nations. In the US, Article III of the Constitution established this Court and all other Federal courts were created by Congress. Courts decide arguments about the meaning of laws, how they are applied, and whether they break the rules of the Constitution. Three functions of the judicial branch are: To maintain the integrity of the Constitution, to interpret laws, and to check the executive branch by questioning enforcement of those laws.

8.4.1.4.2.4. Other Branches
Are three enough? Or too many? All three deal with the rule of law: To make, execute, and interpret laws. There is no reason there cannot be a fourth or fifth branch. Some have sug-

gested that the "people" or the press are a fourth branch. The Philippines have three branches plus three regulatory commissions, one for civil service, one on elections, and one for auditing all accounts. The Netherlands have a Water Authority Board. Iran has two equals, a religious leader and a Board of Guardians, although they may have more power than the traditional three branches. Some countries in Africa recognize the role of tribal leaders in their government.

Is the rule of law enough for nations? Should there be rule of information or rule of religious beliefs? The formal power of beliefs might be separated into knowing, expressing, and wise application. Are those enough to be separate? Informational power supports law, but it might also be more important than law under some circumstances, and law should be allowed to support information. An informational power might be separated as well, into research, applications for planning, and wise assessments.

Is one form of government better than another? How could we tell? Some dictatorships have been more benign than some democracies. Should we forbid some forms? Should we search for a better kind of representation?

8.4.2. *Functions of Politics*
The functions of politics are internal and external. Internal functions are characterized by activities like the fair distribution of food and rewards. External functions are: To coordinate interaction with other nations, and to decide matters of exchange of people through emigration and immigration.

8.4.2.1. Internal Functions of Politics
Internal functions include: To promote the survival of the nation and communities; To preserve the balance of society; To maintain the affairs of the nation and communities; To guarantee fair distribution of food, goods, and necessities; To manage the distribution of power; To preserve the safety of the nation with laws and defense; To represent the citizens, to educate them, protect their rights, and remind them of their morals; To resolve conflicts between citizens; To regulate the activities of citizens regarding safety of the environment; To encourage conversation and communication; and To moderate the forces of change (internal and external).

8.4.2.1.1. To Promote Survival of Communities & the Nation
Politics has to be successful enough for a new generation to take over ruling the culture and nation. If rulers promote the wrong images, that do not fit the environment, eventually a nation will collapse.

8.4.2.1.1.1. *To Preserve the Operation of the Environment.* Nations are embedded in places. If the place itself, the environment, fails, then no political assurances will be able to feed hungry people. So, political decisions should not interfere with the operation of global cycles or local renewal.

8.4.2.1.1.2. *To Preserve the Balance of Society.* Place sustains government. Land sustains government. Land produces food and resources, but the land is not equal or consistent. Poor land does not produce as much as rich land. Agriculture in general does not seem to produce as much wealth as business. Business does not seem to produce as much wealth as artists (symbols and abstractions have fewer limits on productivity). So, government needs to preserve some balance and equality by giving and taking. Individuals are produced by a community. So, the community in place seems to be the source of most of the wealth. Restoring balance in a small community can dislocate people. Leon MacLaren states that balance can be restored by depression, war, or revolution, and it can be painful. Of course, planning and conscious adjustment might be less painful and less risky.

It is important to encourage conversation and communication. We can best understand the social aspect of culture by realizing that the central function of human symbolization is communication, and it requires adherence to understood conventions. Constant communication allows a culture to be coherent. Lack of communication can lead to misunderstanding, unhappiness or violence.

Government has to manage individual civic relations and conflicts. Many cultures have built-in limits to local kinds of conflict, but conflict between cultures requires some laws or understandings. Behavior has to be understood in a context as meaningful. Sometimes conflicts are the result of language, behavior patterns , or simple geography. There has to a path to resolve conflicts, either through consensus, mediation, or neutral judgment.

Government has to decide limits to growth, development or movement. Keynesian economic theory, the predominant theory in industrial countries, holds that the full utilization of resources is necessary to ensure full employment and the maximum social good. This economics depends on economic growth to avoid crisis. The major premises assume that: Population will grow, that social good is related to the equitable distribution of material products, and that if resources are limited, technology can erase the limits. The economist Kenneth Boulding referred to this as a cowboy economy, an economy that has yet to bump against the limits of wealth.

One solution is to work within limits of sustainability. The population could be stabilized. Consumption could be stabilized, especially with a shift to recycling and solar energy generation. The technology already exists, but educational and political problems are more difficult.

8.4.2.1.2. To Manage the Affairs of the Communities & Nation

Hutterite communities were able to manage their community as long as it was less than about 150 people, Garrett Hardin noted. With more than that, distribution of goods failed, due to some doing less than their share or getting more than their share. So, the Hutterites split communities into two, when they got too large. The scale of community can make some things work better. The force that keeps individuals from laziness or greed seems to be shame, which seems to work as a force only in smaller communities. In a small community, a person can be shamed into working harder or into being less greedy, but in larger communities shame does not work as well. Maladaptive behavior may be less visible or the malcontents may form a subgroup that justifies their behavior. Hardin suggests that most utopias do not consider such a change in scale when describing sizable communities.

Many archaic cultures mismanaged their natural resources, but got away with it because their impact was relatively small. Some cultures were not as lucky, the people of Ur, for instance. Luck, as well as size, has a lot to do with the success of some human cultures. Many other cultures mismanaged the affairs of communities, but they were able to survive because of brute control.

The industrial cultures are mismanaging their resources and affairs, but may not be able to control or organize—or luck—their way out of their problems. Modern society has benefited from modern means of management. In fact, management has made industrial operations, from science to agriculture, possible. But, to continue without disaster, it needs to re-empower local voices, especially indigenous peoples, poor or displaced people, and women. It needs to encourage unique local management solutions, such as the water management on Bali. It needs to adjust to optimum scales that are more efficient and just.

8.4.2.1.2.1. *To Guarantee Fair Distribution of Food & Necessities.* With the complexities of civilization, with various levels of responsibility and duty, politics has to make sure that necessities are distributed in a timely manner. The infrastructure, that is the bureaucracy, the coordination, storage areas, trucks, and other things, has to be in place and functioning.

Government has to manage the Commons, resources and private properties. Traditionally, a commons was a territory managed by a community so that individuals could share the land for grazing or farming. This form of use characterized many tribal groups for 40,000 years. The commons was only open to the members of the community, and perhaps to a few who asked permission. Community beliefs or a kind of ecoethics discouraged overuse, in that overusers were shamed with informal sanctions or forbidden to continue using the land. In a sense, any territory not explicitly claimed by one community is a commons. That would include the atmosphere and oceans especially, or rather air and water in any form, since these are global things and can move between systems, and their character is determined by the globe itself. Some global economic or ethical method of management, not just for human use but for nonhuman use, must be devised.

Hardin extends the idea of the commons to be the carrying capacity of the territory, which is a very dynamic concept, susceptible to change by the weather, season, food preferences, or other things. When the commons is not tempered by rules, then an individual user gains all the advantage and the disadvantage is shared by all users. Ecological degradation is assured and the system collapses, a tragedy due to the "remorseless working of things" in Hardin's words. In a limited system, the self-interest of one person cannot decide the public good, or the ecological good. So, the "invisible hand" does not provide a long-term solution. The hand needs to be visible. Necessity needs to be recognized by freedom, as Hardin restates Hegel.

The misuses of commons, some of which are alluded to by Hardin, include: Uncontrolled human population growth, depletion of biodiversity, transformation of fossil fuels, pollution of waterways and the atmosphere, logging of forests, overfishing of the ocean, private vehicles jamming public roadways, tossing of trash out of automobile windows (littering), graffiti, poaching, noise pollution, and email spamming. Each of these misuses places a burden on most of the other users.

Privatization is one possible solution. However, unless individual ownership includes some regional cycles or global resources, then there might not be incentives for sustainability. One problem is that global things cannot be enclosed like land. Regulation is possible, as is payment by the polluter, a form of taxation. Hardin considers these as forms of enclosure also. One solution would be the return of many commons to the cultures that adapted to them and have had cultural rules governing their use. Hardin's suggestion was "mutual coercion, mutually agreed upon."

Many of our less wholesome behaviors, such as rapid technological change or wars, can disrupt tradition and lead to misuse of the commons. If the commons were presented as a prisoner's dilemma for a community, then individuals would have two options, cooperation or defection. Defection, or private mismanagement of common areas, can lead to economic madness, as Paul Shepard indicated. If there are unavoidable costs to defection, then it would be the less likely choice.

The tragedy of the commons is not that the traditional commons system did not work—it did, because it was controlled ethically in a community—it is that other kinds of commons have no such rules governing their use. The tragedy occurs especially when there is a transition from an ethical community use to a free market system that encourages overuse.

In a larger sense the whole planet is the common, and its limits are obvious, between the sun and the greater vacuum of interstellar space. A photograph of the whole earth made a logical connection between the common earth and the necessity of conservation. So, if the earth is a commons, and humans are threatening the health or existence of the commons, then humans need a common ecological ethic, or some set of rules that they can respect.

In another sense, most problems are local not global. So, at the same time, we have to take responsibility for our local environments, and respect the cultural carrying capacity, as a term for the whole integrated concept of capacity that includes human luxury as well as minimal eating and nutrition.

Government also has to control or limit goods. People made their own goods in foraging societies. For the Desana, goods from the gardens support people, although small numbers of extra goods are traded for clothing, machetes, soap, salt, aluminum pots, fish hooks, and rarely a gun. With farming came specialization, and more goods that could be made and accumulated. Goods became a form of wealth, especially in groups like the Kwakiutl. In Crete, written records show how goods were directed to the palaces and redistributed from there. The demand for goods between Europe and China resulted in the introduction of the camel in the Trans-Saharan trade and boosted the amount of goods that could be transported. Donkey caravans from the Mediterranean brought obsidian and other goods to Mesopotamia, and ivory combs from the Indus river. The value of goods was determined by supply and demand.

Industrial farming depended on a new kind of market, a large market for cheap goods. The English encouraged trade to increase the number of goods. The industrial revolution decreased contact with the natural world and objectified what was left. As a result of drastic changes in the production of economic goods, other political, social and even psychological changes occurred. Other kinds of order were de-emphasized. Human relationships became based on economic allegiances instead of kinship — and were formed in societies, not communities. Money became a symbolic representation for the value of labor and land. Land and labor became commodities.

As the automated production of goods became more efficient, goods-producing jobs declined. The goal of production moved from the production of goods to efficiency. Needs are almost completely defined in terms of commodities. Once a need is defined and partly satisfied, such as air travel, it becomes treated as a right.

The material goods of human societies has been increasing. How many of us can carry everything we own on a bicycle or horse? Goods are privately owned. Items that used to be shared in neighborhoods or communities, such as lawnmowers or radios, have many separate individual owners. While this is good for the modern economic system and for the idea of convenience, it weakens socials bonds, dependencies, and trust. Private goods, like gardens, are also public goods, especially in a city where gardens renew the air. Private good, like a loaf of bread, can be owned and not shared. Public goods are those that can be enjoyed in common, since one consumption does not subtract the possibility of another consumption. Air is a good example, although there might be a time when air is rare.

Stung by the suffering of people with new unsatisfiable needs, modern governments have offered to produce more, safer goods, instead of ecologically analyzing the relationships of needs to satisfactions. People have to imagine and construct new frameworks so that they can develop satisfactions that are not dependent on commodities.

Surprisingly, once basic needs are met, for food shelter, respect, and confidence, according to Abraham Maslow, then happiness is not increased much by additional material goods or money. The things that make people unhappy are when their higher needs, such as self-esteem, are not met. Lack of love or security, and lack of communication or appreciation, can lead to unhappiness. People try to balance their lacks by acquiring more money or things. The culture has to become meaningful, as well as secure and equitable, with lower extremes between wealth or status.

8.4.2.1.2.2. *To Guarantee Fair Distribution of Power & Rewards.* Power in physics is the capacity to move. Socially it is the capacity to act. Power is the capability of making things move or happen. Power is not only making things happen, it is a way of controlling which things happen. Having power determines who gets to decide. Power allows dominance. If power is concentrated in one person or a few people, then there are fewer opportunities to challenge or limit it. Concentrated power can lock or gridlock patterns of movement; this is not always healthy, since many problems need many different solutions. Absolute power is no longer accountable to lesser power.

In archaic societies everyone has some power. Power is given to those who have better abilities, at hunting, healing, or coercing. The real power is the ability to persuade others. First persuasion yielded power, then strength, then knowledge. Power can come from strength or a connection to another form of power, ancestral or holy. Power can be derived from the permission or weakness of others.

The creation of large dense communities required new forms of power, due to size and organizational problems. No matter how big the bureaucracy, for a while it only controlled human muscle power. That limited their reach, regarding armies or builders (of pyramids). Traditional states had trouble controlling regional potentates or their armies completely. New forms of energy expanded human power and control.

Corporations increase their power temporarily because of our failure to grasp their nature. Misunderstanding of power or nature can allow a temporary expression of power. The competitive way of life distorts the meaning of power and demand, and it causes imbalances of power and demands.

Scale of governments or corporations gives them more power. Management techniques, or technology, can augment power. Power allows more waste, but power diminishes the ability to see or feel, or suffer, the consequences. Power often reduces the perceptions of those who have it, such that they use power to simplify ecosystems rather than imagine working inside the systems. Having rewards gives people more power. Such power can be expressed in buying patterns as well as bribes.

Governments with concentrated power can be tyrannical about redistributing power or rewards. Inequality in rewards is maintained by the concentration of power. Rewards and power are treated as primary needs, which results in imbalances. Power can lead to detachment from primary needs and natural wealth. Rewards give more power, so that the two form a positive feedback loop that sometimes cannot be controlled, by the user or the less powerful. The cycle becomes self-reinforcing.

If power is spread more evenly, through many leaders and many kinds of leaders, then they can check each other and power is balanced. Under many constitutions power is dispersed by dividing it into separate branches. Humans are momentarily powerless to replace or transcend the circuitry of natural or cultural systems. People often have the power to deny using the power, especially for activities that can destroy place or human values.

8.4.2.1.2.3. *To Promote Safety & Security.* Without safety and security of the primary needs of food, clothing, and shelter, there will not be a cultivation of secondary needs that promote higher culture. Ultimately security is the availability of food, materials and energy. There are levels of security, starting with a healthy ecosystem, and the ability to use it, protect it, and restore it. Security also requires that resources be available to be used, conserved and substituted. Socioeconomic security requires a healthy culture that can distribute basic needs equitably and efficiently. For people to be secure in their own skins, all of their needs, such as self-actualization, have to be encouraged.

Safety is a basic human need, physiological and psychological. Safety is increased when people accept limits on their social behavior, when natural disasters are anticipated, and when technological extremes, such as nuclear accidents and wastes, are safeguarded. Safety does not mean wiping out large predators, such as alligators or sharks, to ensure than no one ever gets bitten or eaten; it means reducing exposure to wild animals, by reducing the overlap or reduction of territories. Animals have to balance their own safety with migration and food-getting. Safety does not mean killing every form of bacteria or virus; it means limiting exposure, being healthy, and preventing the spread of pandemics through accelerated travel and sharing. Ecologically, safety means leaving functioning ecosystems outside the circle of human domination.

Safety is increased with higher standards for sanitation as well as for technology. It means minimizing the potential for harm in a work or play situation. This occurs when governments and corporations take higher risks than the communities and cultures, accept a burden of proof for new developments, and accept higher margins of error and certainty. Safety is increased with laws, such as those concerned with cheating, thieving or killing. Laws that limit dangerous things, such as alcohol or guns, expand safety.

8.4.2.1.2.4. *To Adapt to Internal Forces of Change.* The principle of change indicates that nature is in flux, culture is in flux. Politics is concerned not only with how power and authority are exercised but with how these relationships get transformed. We are interested in the forces that sustain consensus as well as in the forces that bring about change. Intensive change is the development of consciousness or social sophistication; it is characterized by consciousness, connection, and communication. This should apply to cultures also. Extensive change results in the development of cities and technology; it is characterized by conquest, colonialization and consumption.

As humans stayed in place, they tried to extract more resources from the same area. This required new ideas and technologies, which resulted in denser settlements and expanded social organization. This is extensification. Innovation is influenced by the growth and intensity of population, by the expanding and intensified activities of states or cities, and increasing trade and commercialization. The ease of communication also increases rates of innovation. Perhaps that is why it went east-west in Europe and Asia and north-south in the Americas. Accidental innovation became a culture of innovation, that is, a part of the culture that was encouraged and used.

Civilization is shaped by extensification and intensification, and by complication and complexification, although on a local level there were booms and busts of individual civilizations. Why do countries lag behind in industrialization? Why do they need to join the race? Economists argue that their policies or attitudes are at fault, or that their environment might be poor, that they are unorganized or that they are fearful of change or exploration. This does not seem likely though, since often the problem can be resources or domination, but sometimes it is the social predisposition to change, as a result of historical or cultural factors.

8.4.2.1.3. To Represent Citizens
In archaic cultures, when people could not present themselves at every meeting, it was useful for someone to represent them. Some degree of household autonomy was sacrificed to some

larger order group in return for greater security against attacks by enemies or from starvation. Their government promised to judge disputes in values and to protect its citizens from external attack.

The people of a nation-state were first given full sovereignty officially in 1648, with the Peace of Westphalia. Representative governments still use that sovereignty today to claim responsibility for their actions, without recognition of any international body. They also use it to represent citizens in matters of economic opportunities and trade, to attempt to guarantee access and fairness.

8.4.2.1.3.1. *To Educate Citizens Scientifically & Morally.* Thomas Huxley thought that people in nature were Hobbesian, unfit for civilization unless culture educated them. The same essential belief was held by Sigmund Freud, but the contexts were, and are, different. The thought of Huxley was dominated by ideas of competition and fitness, and that of Freud by individuals in society, from hysterical women and conflicted children to selfish businessmen and power-blinded leaders. The context is expanded to drones in an industrial flatscape.

The modern state has an educated bureaucracy to manage information. It requires compulsory education, resulting in mass education and mass literacy. Mass literacy disenchants the cosmos by undermining traditional and magical ways of thinking. Testing traditional knowledge became a habit. The difficulties of dealing with a half-hearted, half-educated public, of dealing with a hard-hearted, hard-headed professional elite, only add to the problems of a culture.

Education may be a necessity, but it has to be a freely offered broad education, based on rich philosophies and sciences. It has to be adaptive. Tribes in the Brazilian Amazon, for instance, have stopped using chainsaws and tobacco, to live more traditionally. They also set up educational centers to show others how it is done.

8.4.2.1.3.2. *To Protect the Rights of Citizens.* In a traditional English village, inherited bundled rights provided commoners with rights of grazing and gathering fuel wood non-destructively "by hook or by crook," which indicated the way wood was gathered by shepherds. The form "commons" is plural, and refers to the whole group of commons, subject to these effects.

As rights have expanded, they have also been made more explicit in laws and codes. There is a form of an International Bill of Rights, covering human and ambihuman rights. All citizens expect to have equal rights and opportunities.

8.4.2.1.3.3. *To Outline the Duties of Citizens.* In addition to having rights, citizens also have responsibilities to participate in government and to live as wisely as possible, to make good places. Often, duties have to be made known through education and communication from the government.

8.4.2.2. *External Functions of Politics*
External functions are those outside the boundaries of the nation, but those functions may have dramatic influences on the shape and course of a nation.

8.4.2.2.1. To Coordinate Interactions with Other Nations and Cultures
There are only a limited number of behaviors that people can take to interact with others in their culture or nation. They can: Ignore each other; exist separately with trade or contact; compete for resources and people; cooperate with each other; fight for dominance or territory; or destroy the other.

For interactions leading to violence, a number of trends that can be seen. These are feedback loops that loop around to the beginning, also. For instance, the failure of neighboring economies can lead to the failure of trade, or the failure of negotiation. The failure of defense, the failure of contextual system, and the failure of a structure for individual participation can lead to the collapse in the meaning of participating. The fragmentation of social responsibility can led to an isolated self-image (unrelated to the external), to lack of self-confidence, and to a decline in participation. Conspiracies for societal control can lead to international conflict, government intervention, guerrilla warfare, and massacres.

On the other hand, alienation can lead to selfishness, gangs, and anarchy, then neighborhood control by criminals, psychological stress of urban environment, substance abuse, family breakdown, lack of control, and violence. Prejudice can lead to segregation, discrimination against indigenous populations, destruction of cultural heritage, ethnic disintegration, an inadequate sense of identity, psychological alienation, and violence.

Conflict starts with people, but extends to animals, natural events, and nations, leading to the unfortunate metaphor of war. What is a definition of war? Does it have to do with the number of battles or dead? Or with having a professional army? Does it have to have a beginning and an end? Does it have to be a certain scale? Does it have to be agreed on by both sides or all parties? The whole process and its effects are complex. As violence and conflict occurred on larger scales, they were called wars.

8.4.2.2.2. To Decide Matters of Emigration & Immigration

National governments have the right to determine their own immigration policies, even though the policies are influenced by many other factors, from disasters to invasions, outside the control of governments. Governments need to balance emigration and immigration in a context of a satisfactory population suited for its environment.

8.4.2.2.3. To Adapt to External Forces of Change

Extensive evolution is the horizontal spread of species. Extensive change is the spread of cultures through many ecosystems. Human migration was a form of extensification. That is, when the size of foraging communities made hunting and living problematical, some humans moved away. Only when they could not or would not move, did it become intensification, which required different strategies to live, such as intensive food-gathering strategies. Exploitation of new areas shows the ecological power of the species. Intensive development of a place shows the creativity of the species.

Politics has to adapt a culture to deal with external forces, from climate change to invasions of exotic plants and animals. A culture cannot escape the rhythms of nature. Some events, such as earthquakes or floods, can be anticipated with plans and architectural designs.

Ultimately these functions have to be based on an ecological approach. Ecological, economic, social, and religious phenomena are part of the broad responsibility of politics. The basic goal of such a politics is the "survival of the community" as William Ophuls identified it. Politics is the interactive means of providing the basic food and necessities of a community. As survival is survival in nature, politics rests on an ecological foundation. The organization of a community must be in accord with natural laws and limits. Political participation depends on information, much of which can be provided by observers and scientists.

Politics occurs, it is now recognized, in an ecological context. There can be no separation of politics and ecology. Every political act has ecological consequences and every ecological decision is a political demand for control over use of the environment. Ecological consciousness can complement political consciousness. Politicians need to think about the hunger and squalor of billions of human beings and the destruction of habitats with billions of ambihuman lives, before concentrating on missiles and private fortunes.

The ecological, social, and political problems of today do not have simple disciplinary solutions. The problems are cosmological and must be solved on that level. But a single cosmology cannot solve all problems in all places. Where human understanding is still underdeveloped, humanity cannot afford to suppress the diversity of thought necessary for adaptation to the diversity of environments, or to eliminate ecosystems and the societies adapted to them, which explains why archaic cultures are valuable. Practicing a holistic or metapolitical approach is the recognition that humans are part of a larger community, a larger whole that includes humanity and all beings on earth.

8.4.3. *Designing a Good Society*

As the basic goal of politics is the survival of the community, politics has to strive for a good society—that is, a society that is based on the properties of a good culture in a good place. These basic properties develop into the properties of a good society (Table 843-1). We need to design for limits for conflict; maybe that means a limited arena for conflict or maybe a form of competition that would resolve conflict.

Table 843-1. Contrasted Properties of Different Levels of Patterns

— Nature —		— Culture —		— Design —	
Field	*Ecosystems*	*Place*	*Culture*	*Good Places*	*Good Society*
Process	Course	Dynamicism	Conduct	Action	*Method*
Autopoesis	Self-making	Identity	Wholeness	Individuality	*Extension*
Differentiation	Diversity	Uniqueness	Flexibility	Richness	*Variety*
Integration	Construction	Investment	Adaptation	Conviviality	*Cooperation*
Constancy	Stability	Regularity	Endurance	Consistency	*Loyalty*
Development	Productivity	Renewal	Vitality	Health	*Harmony*

8.4.3.1. Method

Method is a way of considering process, course, dynamic change, conduct, and action. More than just courses or actions, ecological designs are methods to create good societies in good places. A science like ecology may be limited by its method, but in ecological design, method is a way of addressing limits to create the conditions for a good society. Play is the method of learning for most juvenile animals, but in ecological design, play is a way of creating imaginative experiences that can describe and test experiences scientifically and aesthetically.

8.4.3.2. Extension

Design combines the self-making of a place with the wholeness of a culture, in context of the identity of an ecosystem, to create good places and extend those properties into a good society. Humans identify with places. Identity becomes an extension of the self to a place. This identity is a form of rootedness in place. The extension of identity creates an ecological democracy that fits the self to the larger Self that extends through local animals and plants and supporting ecosystems. The creation of good societies can be expanded to include all residents of the territory of supporting ecosystems, including people, domestic plants and animals, wild associations, and even plants and animals that can not live well with people, such as large carnivores. The uniqueness of place gives belonging and identity. The whole community gives meaning and richness to life. The human component means all of the people, not just the majority, or the friends and families of the representatives. All people have to represented according to minimum standards. People do have shared common interests, including a healthy environment, meaningful employment, education, security, and health standards, but they also have personal interests, such as roads or factories, that may not be shared by a minority or majority.

8.4.3.3. Variety

Variety is based on differentiation, diversity, uniqueness, flexibility, and richness. Diversity at the ecosystem level and uniqueness of a place promote flexibility at the cultural level and richness in good places. Animation and ecological value change the differentiation of the field to the diversity of the ecosystem. The addition of communication and cultural values to that characteristic of the ecosystem results in the richness of place. And, the addition of social values and awareness of the uniqueness of a place leads to variety in a good society. The design of a good society requires the property of variety.

How do we incorporate this variety into design? Should we ban interference activities entirely? What do we design for? Wild animals? Good domestic or industrial systems? Because the operation of the universe tends to change systems, the design of a good place and good society should be open to the types of processes that could, if ignored or denied, destroy the ecological design. Maintaining community diversity means maintaining or restoring, in previously degraded areas, the variety of ecosystem types that result from natural disturbances at a variety of scales through short and long time frames in a landscape. Because energy and material cycles need to fit the ecosystems, designers need to increase and scale the variety of energy sources.

8.4.3.4. Cooperation

The construction of an ecosystem contributes to an investment, which encourages the adaptation of a culture to place. A culture uses conviviality to make a good place and cooperation to design a good society. The culture and the ecosystem (environment) co-adapt. The process of adaptation involves a self-presentation that offers the possibility of new symbiotic relations. If our old human brain is not adaptive and does not perceive or respond to gradual environmental deterioration—as Robert Ornstein worried—then that may explain why we are surprised by slow, invisible or long-term catastrophes. Cooperation, which can be enhanced through design, can increase conviviality so that we start to shape a good society that diversifies in response to catastrophes that are regular and disruptive.

8.4.3.5. Loyalty

The stability of an ecosystem contributes to regularity, which sets the opportunity for the endurance of a culture to place. A culture uses consistency to make a good place and loyalty to design a good society. Most environments, however, are characterized by fluctuations, irregularities, and uncertainties. As an adaptive system, a culture can change as its ecosystem changes. If a place is regular and displays structural constancy, design can increase stability by creating intermediate structures in a series of levels in an order of complexity—what Arthur Koestler calls holons. Human intervention can make a place more regular and thus predictable. Designs for a good society have to be based on the regularity of a place and the endurance of a culture in the context of ecosystem stability. Design takes the consistency of a good place and lets loyalty emerge from the process of dwelling. Dissent occurs within the context of loyalty, and freedom is expressed within a context of law that limits it and protects it. This avoids runaway feedback.

8.4.3.6. Harmony

In Chinese medical tradition, the highest good is harmony, especially social harmony or good relations. A good person is one who creates and maintains harmony. Harmony is related to wholeness—indeed, health can be defined as harmony in a whole context. The flowing movement of the implicate order (David Bohm's word) is harmony. But, since the flow and order of the movement are imprecise and uncertain, perfect health or perfect harmony is not possible. We recognize breaks in harmony as disease or disorder. C. S. Holling considers that local pockets of chaos keep ecosystems stable by forcing the evolution of new forms to create new niches. Ecosystems reach their fitness near the edge of chaos, therefore crashes and explosions occur.

In medicine, that chaos seems to be a feature or a sign of ill health. That is to say, health is a form of harmony, not a characteristic of things or bodies. A culture uses the health of the system as a basis to make a good place and the harmony of the system to design a good society. Health is the overall ability of a system to maintain itself under a normal range of environmental conditions. Ecosystem health is one of the goals of design, not an end point that can be reached once, but a continual striving and balancing. This means we need to consider long-term balance in order to have any kind of sustainable pattern. Natural processes are concerned with long-term harmony building, as a fundamental property of that design at that level of value. In ecological design, harmony is good fitness with the environment that results in a meaningful and flexible order on the time frame and scale of living processes. The wisdom of harmony, harmosophy, requires respecting the limits of control and certainty, and striving for the most satisfactory balance possible.

8.4.3.7. Imperfect Design in Incomplete Harmony

Human survival is not guaranteed. The perfection of a place or society is not possible. The properties of a good society—method, extension, variety, cooperation, loyalty, and harmony—are indefinite and incomplete. Although the properties can inform ecological design, which can improve our situation on a developing planet, they cannot be used to create ideal, permanent utopias.

Cultural system health is a measure of the harmony of the overall physical and ecological degenerative and regenerative processes, which may recycle thousands of human dwelling places, especially those badly sited in earthquake or tsunami zones. The ecological designer should take care to create good places, but be aware of short or long-term destructive processes at work.

Design has to participate in the political actions of a society. Every political act has ecological consequences, and every ecological decision is a political demand for control over use of the environment. The restriction of freedom, either through tyranny, cultural uniformity, crowding, or mutual constraints, results in a decrease of variety, which is created by spontaneous play, and which is necessary for flexible and enduring social systems. If we wish to advance through design, to harmonize on higher levels of development, we must preserve and promote variety. Ethical, aesthetic, and utilitarian reasons all support efforts to conserve the variety in nature.

Local ecological design is the creative modification of ecosystems to repair or enhance their ability at self-organization and maintenance of their complexity and diversity. The pattern has to be small cells under a unit larger than the largest cell. That allows harmony and management by ensuring a physical and numerical balance. Then, central authority, for design or politics, can be relatively weak.

8.5. *Survival Requisites & Design*

Historically, we humans have used our skills and images to improve our lives and places. We have used our inventiveness to increase our ability to survive and to reproduce, as well as for our comfort and wealth. We have persevered despite many challenges and problems that have killed or impoverished many people or whole cultures at times. We have survived ecosystem collapses and cultural collapses. We have survived droughts and freezes. We have survived diseases and wars of conquest.

We have used tools to changes places and some of the processes of nature. We have used technologies to protect our plants from insects and our towns from invasion. We have created the possibilities of unprecedented luxuries and wealth. We have made designs for stimulation or profit, more than for safety or elegance.

But, we have failed to understand much of the detail and scale of nature. We have failed to use our imaginations for things beyond luxury or excitement. We have failed to share new wealth with most others or to share much of the planet with ultrahuman beings. We have failed to create designs that would incorporate us within the limits and cycles of ecological processes. We have failed to have the nerve or courage to try, since it would inconvenience some or disrupt the wealth of a few.

We have the conceptual tools. We have religions that bind us to cultures and places—that teach us simplicity and charity. We have sciences that allow us to learn ever more about ourselves and nature—that teach us. We have words and metaphors that can produce strong flexible images to guide our learning and decisions. We have the ability to create thought experiments and fundamental analyses of problems and catastrophic situations. We have the creativity to make ecological designs on a regional and global scale, to protect the ecological services that result from a living planet. We have the ability to expand our consciousness to include an ecological perspective of our situation. We have the possibility to share and open ourselves in a eutopian effort to balance the use of a common planet.

But, we have been seduced by our brilliance. We have become trapped by the changes we made to eating and living—by our close adaptations to agriculture and cities. We have become too comfortable in our habits and pleasures. We have allowed growth and inequity, dominance and slavery, to continue because it has become part of our perceived patterns of economic success. We have allowed intolerance, conflict and war to continue because they are perceived as unavoidable products of our success.

8.5.1. *Humility & Wildness*

Now, we are beginning to recognize that the scale of problems has become regional and global. We intuit that our economic styles, expanded to a global level, may not be healthy or sustainable. We allow corporations to enjoy unprecedented benefits from a minimum of regulations, because we think that they are necessary to continue the growth we think we need to be prosperous and happy. We start to question whether our modern cosmology—in the image of a machine—is appropriate to understand or relate to living beings, systems, and cycles. We are starting to realize that traditional and archaic cultures were more attuned to their ecological places, and to suggested limits that allowed them to fit into their surrounding ecosystems. We are beginning to realize that the scale of our changes is affecting the atmosphere and oceans, that we are making things change out of the range of desirable conditions. We worry that the entire planet is wobbling and fragile and could change into a configuration that is not pleasant for us or for the large charismatic mammals we love—from polar bears and whales to tigers and wolves. Maybe, Gaia will turn out to be the kind of wicked mother who may not favor human children above other children, or mammals above insects or reptiles.

We have started to save animals and plants, species and habitats. We have started to design ecovillages and more efficient automobiles. We have started to create standards for more intelligent buildings and roads. We have started to talk about and sketch whole communities or cities that may fit into specific places.

But, we are not thinking about all cities or all roads, not the planetary patterns that overwhelm places and movements of species. We are not paying attention to changes in scale or meaning. We are not considering some kind of economic equalization or the limits to our economic growth. We are not saving entire systems.

We need to understand the spheres and cycles of a living planet. Then we need to start to redesign our activities so that they are in line with these cycles. We need to identify areas to preserve or restore as wilderness, by making sure that they are the right shape and size to continue as self-making entities. We especially need to consider forests and many kinds of animals. We need to understand the factors that may act as triggers of change: Our conversion of ecosystems into domestic lands, the use of commons, or the scale of industrial manufacturing. We need to analyze our adaptive patterns, from agriculture and technology to urbanization and industrialization, and then understand how they developed and became dominant forms of exploitation of places and systems. Then we need to redesign them for the appropriate local or global scales. We need to design ecological cultures based on archaic one, without giving up the benefits of some urban civilizations.

We need to rethink design, to make it ecological and global. We need to apply ecological design to local ecosystems, whether forests and streams or cities and fields. We need to consider the design of regional areas, from common areas that are used for many cultures, and where natural processes, migrations, and exploitations are allowed to continue. We need to design areas that are independent of national boundaries, where animals and plants can migrate or move freely.

We need to rethink culture, to analyze the weaknesses and strengths of traditional cultures. We need to create a framework to fit cultures with the limits of human activity as well as within the limits of ecosystems and regions. We need to redesign economics and politics through an understanding of what their functions and goals really are. We need to create adequate legal forms for nations and nonhuman nations and the planet. We need to approach the planet through one framework of decisions, a framework based on understanding and shared goals.

We need a wild design to solve these large global wicked problems. We need to be surveying and monitoring so we know enough to design and act. We need to be preserving and restoring as much as we think the planet needs. We need a wild design that is based on traditional wisdom and known operating limits to address these problems all at once. We need a wild design to actually try to create good places using all of our geniuses and strategies over the whole of the planet. We need to do it together and we need to do it now!

8.5.2. *Advertising & Excitement*

Most of all, we need to make design and ecology exciting. The other day, tired of writing, I went to visit friends, who were watching a football game. They were drinking, shouting and throwing things to each other. It was nice; I got to play for a while, too. It occurred to me that only with entertainment industries is there so much technical fireworks and coordinated enthusiastic teamwork. Imagine all that energy and enthusiasm directed to appropriate technology for reforestation or the proper use of fossil fuels. Imagine television coverage of wetland restoration work with the same amount of attention and detail. Why not have a competition for the most beautiful productive forest or teams working to restore devastated city areas—broadcast by a major network as an important event?

Some people have noticed that this remorseless entertainment is an anesthetic against fear, emptiness, self-searching, or death. Continuous entertainment is a kind of guarantee of health, riches, and long-life. Everything that is pleasurable, thought George Orwell, seems to be an attempt to destroy consciousness. Television seems intent on proving him correct. Ecology or design cannot ever compete with entertainment if it raises troubling questions or difficult expectations. As long as the industry can guarantee many happiness and wealth through the arithmetic of fantasy, others of us will always seem to be complainers and false prophets—until it's too late, and then it is likely to be we who will be blamed for not avoiding the catastrophes.

Good design, good ecology, ecological design, and global designs require new images and new language. Getting new images and new language to engineers, politicians, financiers, and landowners, as well as to scientists and industrialists, is problematic. Perhaps poetry, art and design can help, although they are undervalued as forms of communication, not to mention as ways of shaping and making. Business has transformed much of art and poetry into advertising, to match the style and attention span of the people in industrial cultures. Advertising, quite literally from the Wall Street Journal to college textbooks, refers to its activities as "shaping the American dream" or the World Dream. Like art, advertising creates an image of a way of experiencing. Unlike art, it limits its focus for a specific goal—profit. Like art, it mirrors us. Unlike art, it intensifies and glorifies only the positive aspects of culture, ignoring the dark, negative aspects or the complex nonhuman framework.

The simplicity of advertising is irresistible. Our environment deteriorates according to ecologists, but always improves according to economists (who require it to improve to make a profit). And their pictures are much prettier. People want to hear that it is getting better. Advertising tells them it is. People want to act stupid, greedy, and selfish, and spend the inheritance of their children on themselves. Advertising tells them these actions are rewarded. The real issues of life and death, destruction and hope, make people feel helpless and anxious, so advertising draws their consciousness to comfortable trivia.

Despite the limits of the dreams of progress and growth, of waste and stylistic frenzy, advertising, using sophisticated techniques and narrowing the focus out of context, makes the dreams desirable and irresistible. People in agricultural and hunting cultures interiorize the abstract industrial vision. African farmers are convinced to buy inorganic fertilizers, even though it degrades the soil; women to buy powdered milk for their children, even if it kills them; tractors replace draft animals in the paddies in the Philippines, even though they are costly and less energy-efficient; French winter fashions are found desirable in tropical Brazil, even if they can only be worn in air-conditioned villas. People in industrial societies are convinced that their children will be ruined without personal computers, even if they become isolated game-players, unconcerned with wildness in remote locations.

Advertising has been serving the dream of progress, but progress is leading to catastrophe, a long, slow, global catastrophe. When people experience local, sudden catastrophe, they usually respond immediately, with heroism and sacrifice, aiding the victims of earthquakes or floods, or sometimes famines. Advertising could bring to consciousness the slow catastrophes of erosion and the destruction of entire forests, and, perhaps, invoke the same altruistic responses to them.

Advertising may be the most effective means to reshape desires and reform buying habits, as well as where and how we live. Advertising presents the symbols of modern experience effectively, even if they are just the trivial ones. It could present healthy symbols equally well. Advertising does incorporate traditional values, like family, friendship, and love, although to sell beer and cereal and, sometimes, churches and hospitals. And, like art, advertising lies, although Jules Henry thought it might be a new kind of truth—"pecuniary pseudo-truth"—not intended to be believed, or certainly, proved).

Advertising is beginning to support more informational functions, such as the dangers of drug abuse and smoking. Advertising also can create values—fur coats, fast cars, dark beer, slim cigarettes are certainly recent and artificial values—but it could be used to create positive ecological values and new identities that show that our needs for prestige, esteem, and belonging can be met without stylistic waste at mindless speeds. Advertising could promote new attitudes about appropriate technology, the rights of other cultures, and the place of people in nature. Good advertising could be as subversive and conservative as design or ecology (Pablo Picasso said art was subversive; Paul Shepard called ecology subversive). Good ecological advertising could avoid confrontation with people's values and emphasize positive aspects without negative ones. A good ad could capture and carry the most self-indulgent viewer—after all, for the most part, ads don't require effort, literacy, or consciousness, just attention.

Maybe we can present images that rival the industry images. Maybe we too can speak the languages of euphemism that large corporations use to conduct their businesses of larceny and fraud. Positive images and pleasing language skills are everything these days; no one really looks for substance. The devotion to money, beauty and youth is the old focus. Maybe that could be subverted with the beauty and wealth of nature.

To work towards this goal, ecological design groups, with conservation groups, preservationists, sportspersons, politicians, and actors, could define and promote an integrative mythology as the basis for the framework of diverse efforts to protect life and the environment in the whole planet. Ecological organizations could provide a meaningful philosophical foundation, as well as coordination for other humane, social, and conservation programs. But, the approach must be egalitarian: Respect for life cannot neglect human life and suffering. The approach must be eutopian: A new cosmology cannot ignore adaptive cultural traditions that arose in place over centuries. Furthermore, in addition to formal education, groups could fund and provide re-education through the most effective means, such as advertising. Wildlife groups could spend money advertising "humane consciousness," moderation, and the joy of living—instead of just consuming or winning. Ecological design ads would be unique and compelling, simple and effective. They could present our goals and successes, the importance of our efforts to redesign out human structures to protect and restore the planet. They would advertise not a product, but a way, not for a profit, but for a dream. And, we have to start now!

8.5.3. *Learning to Love a Place Region or Planet?*

Do we start by loving the planet? Obviously, we can love ourselves and families and friends, even other animals and plants, but is it meaningful to say that we love the planet? Is it some sense of topophilia, or biophilia, or ecophilia? Are those labels too abstract? Perhaps we can understand them philosophically, historically or ecologically.

The philosopher Empedocles, trying to observe the tenets of the Eleatics, explained all natural processes in terms of the combination or separation of the four elements (as in Anaximander), due to the opposing influences of love and strife (similar to Heraklitus). Love and strife were not only dynamic principles, but were physical masses, thereby making the mind and body homogenous. Nothing came into being or perished; there was only mixture and separation.

Later the philosopher Alfred North Whitehead combines Aristotle's unmoved mover, directing the process through love, with Plato's perfect forms that existed in a perfect reality, to presuppose the cosmic process.

Abraham Maslow thought that humans had a hierarchy of needs, beginning with physiological needs and moving up to safety, love (belonging), esteem, and self-actualization. The higher do not take over until the lower are satisfied.

Early foragers became one with the animal hunted or seeds collected, to realize that they became part of you, living through you. On the other hand, when the relevant social unit was the tribe or nation, it was possible for the local mythology to represent others outside the bounds as inferior and its local inflection of the universal heritage of mythical imagery as the one, true, supreme image. The young of the group would be trained to love home and hate outsiders. Xenophobia was once adaptive; now it is anachronistic. But there are no outsiders, we are all passengers, inhabitants. All dividing horizons on the planet have been shattered. We cannot hold our loves at home and project hatred outward.

The concept of tribe and state is growing toward an ecumene, an inhabited earth. Today, there is no outward on earth. Our mythology has to grow also, to include the whole planet. There is no practical elsewhere anymore. A global mythology cannot afford to teach of elsewheres. It must teach of a multiplicity of cosmologies. The difference between cosmologies is not due to the number of phenomena taken into account; it is due to the difference in basic postulates of thought. The difference is not a matter of truth or falsity. Truth and falsity are meaningless for cosmologies. Primordial water is no less true than six-dimensional phase space or primordial ylem. No one group's image is more true or accurate than any other's. Each group views and reconstructs the world through their experiences and values.

During the Renaissance, Christianity became affected by the spirit of affirmation, especially with the discovery and popularization of the teachings of the Stoic and Epicurean schools: Love of man was the virtue above all virtues.

Albert Schweitzer believed that even if we despaired of comprehending the phenomenal world or the plans of God, we would not need to confront the problem of life with utter perplexity, because the ethics of Jesus, reinforced by reason, lead to the reverence for life, whose edict is the rule of universal love. This same reason would find the bridge between love for God, love for man and love for all creatures, and express reverence for all being, however dissimilar to our own, reverence and compassion for all that is called life. Such a foundation for morality forces the realization that when we establish gradations of values between lives, we only judge them in relation to ourselves and that is wholly subjective. How are we to know the importance of each? The principle of reverence for life rejects relativism— "it recognizes as good only the preserving and benefiting of life." Individuals must transform themselves from blind men into seeing ones by following the new commandment: Revere life. The quality of personal existence depends more on it than on laws and prophets; it comprises the whole ethic of love in the deepest sense; it is the source of constant renewal for humanity.

Schweitzer's love was abstract, yet he did not love nature. Part of Schweitzer's dilemma came from the prevailing myth of the opposition of life and death, strife and love. Opposites are necessary conditions for each other.

Leopold has proposed a conservation ethic, dealing with the human relationships to land, plants and animals. The land ethic Leopold had in mind was a sense of ecological community between human and other species. When we see land as community to which we belong, we will use it with love and respect. Such an ethic would change the human role from master of earth to plain member of it. Predators are members of the community; no special interest has the right to exterminate them for the sake of benefit for itself.

The aim of an ethic must be harmonious with the world's population of living beings. Michael W. Fox proposes a biospiritual ethic as a unifying set of principles, ethics and values that will bring about a nonconflicting state of one earth, one mind. The ethic is based on the biological fact that all humans and living beings are kin and that life is spiritual—love is stronger than violence. It arises from seeing humanity in an ecological perspective (from knowledge of evolution).

Other modes of knowledge must be accommodated along side science. Roszak proposed to define true knowledge as "gnosis,"—Gnosis is used in a generic way here, not the way the Gnostics used it—of which scientific rationality is only a small part. It is gnosis that is needed to perfect the universe and soul, mutually, with the spirit of love. Gnosis is the whole spectrum: the hard, bright lines of science, the hues of art, and the dark voids of religion. Gnosis is augmentative knowledge (intuition and revelation), in contrast to the reductivity of science. Tillich calls gnosis "knowledge by participation"—and that is what humans need to do—recognize that they automatically participate in everything, and that we cannot unparticipate by choice.

The principle of reciprocity in ecology is that no entity can exist by and for itself; everything is connected to everything else. In religious terms, this is the golden rule (according to Aldous Huxley's *Perennial Philosophy*). Reciprocity is the recognition of mutual obligation. All things are bound in bonds of mutual dependency. Ecology has become a philosophic viewpoint as well as a systematic discipline. The counsel of ecology is caution. Caution is an expression of love. Love is treated as the basis of religious feeling.

There are pictures called anamorphoses, which are fragmentary deformities to the naked eye, but reveal perfect forms in a conic mirror. Like the mirror, culture organizes the fragments of life. The life of an individual, often won at great cost, receives its complement from culture. But the conic mirror is lost to modern cosmologies; the world is fragmented. The pieces of experience can be reconstituted by a gestalt, a perception of the total pattern. Nature, myth or love can bestow the mirror again. Nature has an inherent order. Myth creates an order. Love combines orders.

Every cosmology includes ideas of the past, present and future. An ontology of temporal process would provide a global perspective on world order, on the coming to be of higher levels of meaning in the universe as a whole. As mysticism, with its synchronal sight, regards all beings as equal, science, with its diachronal sight, regards all equal through time. A complete cosmology needs science and mysticism. The proper attitude of such a cosmology is care, by which Heidegger meant the abandonment of the neutral attitude of consciousness. Care is an attitude that moves beyond its own preoccupations towards the beings of the world. It means letting be. It means reacting spontaneously to nature; loving, fearing or fighting when appropriate.

Gregory Bateson defines wisdom as knowledge of the larger interactive system. Wisdom is a perception of relationships and relativity, an awareness of the wholeness of things without losing sight of the unique particularities. It joins the left and right brains in a union of logic, poetry and feeling. It reintegrates knowledge with values. It implies making judgments in advance, infusing elements of older wisdom into a new expression. The wisdom of cultures forms a perennial philosophy of the human race. Gary Snyder discerned an undercurrent in civilization since the late Paleolithic. He considered Buddhist Tantrism to be its finest and most modern statement: "that Mankind's mother is Nature and Nature should be tenderly respected; that man's life and destiny is growth and enlightenment in self-disciplined freedom; that the divine has been made flesh and that flesh is divine; that we not only should but do love one another . . . these values seem almost biologically essential to the survival of humanity" (in *Earth House Hold*).

Myths (with transformations) and metaphors (with structure of integrated differences) are modes for conveying ecological wisdom; they are less concerned with survival than the survival value of a good fit between dualisms of life.

But we fear to try to create a just world order, a new design for living in the planet. But, fear casts out love, and with love, goodness, beauty, truth, and intelligence, until all that remains is fear of other beings and the unknown; fear of the smiling science and technology that take away more than is given; fear of fellow human beings, who are trying to regain what was taken. But what can cast our fear? Love? Love is the problem of an animal who must find the world compelling and symbolic. We confuse our symbols with the natural world, the present with the represented. Sometimes the image of love is loved more than the actual.

Certain elements—care, responsibility, respect, and knowledge according to Eric Fromm—define a loving relationship. The inexhaustibility of a being or of our relationships constitutes much of the nature of love. Human beings are compelled to seek other beings and love is the only approach.

Maslow presents "love-knowledge" as unlimited. Love creates an openness to experience, without judgment. Beings unfold. Contradictions abound in love's completeness. Love expands the awareness of the self and other beings. And its intensification of feeling pulls the frame through the focus, yet preserves the original distance. Love binds space and time in miniature. Its intimacy permits distance. Its duration reaches future generations of beings. Love personalizes the universe, but keeps it free.

The world is a living symphonic poem. The score does not fully determine the music, the passion and intensity. The present movement gives no clue as to how it will finish. Each voice fits in with others, but is not limited. We can be aware of the symphony as a whole and the individual voices. When we are attuned to it, control is not a problem. If we can trust ourselves to improvise as the need arises, we can trust nature, and not follow some rigid plan. Every cosmology has limits, of incompleteness, indeterminacy, and of locality. Our new ecological cosmology, expressed through our intentions and designs, recognizes its limits with love, and continues to act.

8.5.4. *Brief Summary of Ecological Design*

We are weak and arrogant, clever and clumsy. We ignore gigatrends and hide behind bad metaphors and the momentum of recent industrial history—and the financial size of that current human system is $55 trillion in output in 2004. We do not see the long-term, slow, invisible or other catastrophes such as extinctions and ecosystem collapses, although climate change is being noticed. We neglect to question our adaptations to past conditions, such as agriculture or cities, even as they require prodigious amounts of energy that push basic ecological systems towards collapse. We chant our desired needs for luxury, money and growth. Then, we complain that the current local and global systems have no historical or cultural analogs. Global environmental change is considered beyond our management or control, so business can continue as usual, greenwashed or not. The massive destruction of ranges by livestock are considered an unavoidable by-product of economic growth. People sliding into poverty, privation and starvation are considered acceptable losses. The failure of ecosystems and landscapes is dismissed as being less important than discovering further deposits of fossil fuels. The decline of two-thirds of planetary ecosystem services does not get mentioned in the excitement at the behavior of actors and politicians. Nothing makes us happy.

Virtually all of our actions are unsustainable in the short-term and the long-term. Human civilization is living on borrowed time and stolen assets and services. The novelty and complexity of global interactions lead to thresholds that when exceeded will trigger unwelcome surprises.

Central to this effort to outline and initiate a participatory framework is the recognition that our catastrophic situation requires immediate emergency actions on a global scale, from expanding personal consciousness to reforming the very character of human civilization and its coevolution with wild nature. First, an international organization such as the United Nations has to assume control of the planetary commons for limited utilization as well as protection. It would manage the commons in the interests of all living communities. By charging usage, extraction or loss fees for water, fossil carbon and other resources, it would become self-supporting. By international agreement, it would have the largest and only standing army, with some large-scale nonnuclear weapons (nations and communities maintaining police forces for public safety and control). It would allow all nations—even minor landless ones or business entities like corporations—to join with equal votes. It would protect traditional and modern cultures in a loose framework, encouraging cultural health related to environmental health, stressing economic equity through a variety of measures, and trying to direct the emerging global culture. Through this organization, we could consider incorporating the planet as an interested party.

This international organization, as well as nations, communities and individuals, would start to question things; to learn to apply an ecological perspective to knowledge and processes; to understand and direct trends and gigatrends; to understand fields, systems and limits; to understand flows and connections; to understand the history of the planet, especially spheres and cycles; to learn how elements and materials interact in patterns and how patterns restrain them (and how this produces uniqueness and diversity); and, to understand how communities and ecosystems work. Every person and group needs to understand how challenging events can become problems, especially with water flows on land and air and with atmospheric change; to recognize others, such as cultural collapse, earthquakes, and diseases; to face the global problems of a chaotic changing planet; and, to comprehend the limits and weaknesses of cultures, especially bad images, incomplete cosmologies, maladaptations, drift, and traps. Creating an ecological cosmology, with new metaphors and logic, is an important step.

We need to emphasize science, especially addressing systems and synthetic thought. We need to end the thoughtless, large-scale experiments on ourselves, living communities and the planet, and imagine real thought experiments. We need to reduce our massive interference with ecosystems and landscapes, especially through conversion. We need to reduce and recollect many kinds of pollution, from plastic nurdles to carbon dioxide gas. We need to rethink and rework design at every level, from personal and local to regional and global, through principles and actions, then use design to recover our tremendous losses, to save and restore wild and common areas, especially extents of forests. We need to create surveying programs to inventory places and the planet, then monitoring programs to keep informed as well as to reconnect many plant and animal patterns. Human populations should be related to ecosystem productivity, as well as to rarity and diversity, even if this means a large reduction in numbers. Failed cities should be removed or refitted. New cities should be built with psychological and ecological prospects (perhaps as arcologies). Technology could be reduced and integrated to fit needs and limits, instead of being automatic. Many predominantly technological problems, such as carbon production, fossil fuel extraction, nitrogen pumping, phosphorus loss, and pollution, can be solved through a combination of conservation and balancing. New smaller scale forms of technology can be designed using precautionary principles. Industry could be reformed as limited artificial ecosystems. Appropriate technology could be directed and certified.

We need to learn to be healthy individuals in living communities, to reduce antibiotic use and unnecessary luxuries. In fact, we need to place health in the context of ecosystems and global cycles. We need to learn to adjust to our own psychological and social limits. This means addressing problems with patterns of wealth, especially growth, inequality, poverty, dominance, and slavery. We need to link economics to ecologics, especially at the global level. We need to redefine the nature and limits of corporations, especially with new goals and community responsibilities. We have to consider the problems of intolerance, conflict and war within a frame of ecological ethics, and to reconsider the goals and responsibilities of individual nations and a global community. Failed nations and communities will have to be rebuilt, after restructuring international politics.

We have to create new surveys, inventories, monitoring and planning, not only of local resources, but of global properties, not just of the atmosphere and ocean, but of bioregions and deep continents. We have to monitor all global threats, from earthquakes to meteors.

The design of new forms of human settlements and activities can be accomplished through ecological design. We can design local systems and regional patterns, we can design the entire planet as a whole. We can integrate traditional adaptive patterns, from agriculture to technology and cities, to natural processes. We can consider our civilization and planet as part of the solar system and local space. We can strengthen religions with common understanding to bind people to their places and planet. We can make politics conscious and fair, with traditional goals and limits. We can make global designs, mostly to restrain humanity and restore balances of systems, but also to coevolve with the development of natural systems. We can create a planetary framework with clearly delineated functions at each level from personal to planet, based on an integrated approach of koinomics. And, we can advertise this to everyone so they can participate. Then we, they, all of us, can say what else should be done.

Perhaps our efforts will be as naïve as continuous war, or as utopian as ecosystem destruction, but we have to make the attempt, and to continue without pause or stop. Perhaps the ideas of global commons and small nations are as unrealistic as an infinitely growing economy or as warring for peace, but we have to try new patterns. We have to promote the appreciation of places, to awaken the delicacies and qualities of designs, to plan frameworks for development, and to allow everyone to participate in creating good, personal and social designs. Each single step is easy. Each challenge can lead to an adventure. Each change will have some effect. Each vision could inspire more efforts.

8.6. *Foundation Summary*

The purpose of this book has been to provide simple descriptions of the kinds of challenges and problems that are proving to be unmanageable or insurmountable in a globalizing industrial, urban civilization. The book has also presented the basics of cultural categories, from technology and economics to politics and religion, that are being used to adapt human societies to a planet that itself is responding to various disruptive and deadly human pressures on wild species and ecosystems, as well as on the largest global geological, oceanic and atmospheric cycles.

The weaknesses (and strengths) of traditional cultural responses and designs are outlined. Their results include failures, losses, and flatscapes. Their unwanted effects include dominance, inequity, slavery, and violence. Their paths may lead to traps and collapses.

The following series of three volumes, *Redesigning Local Systems in the Planet*, *Redesigning Regions in the Planet*, and *Redesigning the Planet*, are an attempt to answer the challenges and problems through various levels of ecological design. Natural and traditional design are expressly compared to ecological and global ecological design. The basic tools of design, including metaphor, questioning, education, thought experiments, and forms of analysis and synthesis, are related to the important concepts of limits, scales and fields.

General questions about humanity and the spectrum of cultures are raised, especially human patterns of living and whether a global culture is possible or desirable. Larger adaptive patterns, including agriculture, technology and urbanization, are examined for their negative effects and long-term sustainability. Serious problems, from abstract machine images to collapse, are also noted.

For consideration by ecological design, the operations of nature are broken down into elements, processes, and flows. Communities, ecosystems, landscapes, cycles, and planetary spheres, including the hydrosphere and atmosphere, are also design factors. Many current problems, from water flow and atmospheric change to wilderness conversion and human slavery, are also discussed. Kinds of ownership, from private property to commons, are considered as design factors. Forms and functions of politics are related to ecological design.

Design is based on properties and principles that are developed into standards and actions that fit the requirements of wild living systems as well as human cultural systems. Ecological design, placed in a eutopian framework, offers clues for good management and monitoring of these systems as well. It offers ways to avoid insurmountable problems involving drought, climate, crime, and globalization. Specific approaches, from ecosystem medicine to a global framework, are introduced, to be expanded in the remaining volumes.

9.0. Appendix

This section contains: Notes from the various chapters are listed by chapter. The glossary contains common words from all chapters. The bibliography combines references from the chapters. A short biography and contact for author is listed.

9.1. *Notes*
(Being edited)

9.2. *Special Glossary*

Accommodate. To settle, 'conform with right measure.'

Acculturation. Cultural change that results from contact between different cultures; usually referring to loss of traditional culture after it adopts elements from more aggressive or larger culture.

Adaptability. The ability of an organism to change itself to the limits of the system. The ability to adjust to different circumstances, which can increase survival potential.

Agriculture. A form of increasing food supply by simplifying and controlling an ecosystem.

Agrotopias. Wild places (not to be confused with the Latin-derived agriculture).

Allele. Different form of a gene, resulting in genetic variability.

Allozyme. (Gr. for different ferment) Measure of genetic distance.

Ambiguous. Wandering about. Uncertain, being open to multiple interpretations.

Ambihuman. Everything that is not human; nonhuman; ultrahuman; 'surrounding the human.'

Analogy. Correspondence based on similarity; in computers, based on similar electric charges.

Anamorphosis. A distorted image that can only be viewed from a certain perspective or with a special instrument; 'to form anew'

Anarchy. Government by the voluntary association of cooperative individuals and groups. Iceland was an anarchy before accepting the King of Sweden to represent them.

Animism. The belief that all things have a soul; in Pythagoras, an immaterial force.

Anthropocentrism. The assumption that humanity is the center of the universe or final aim.

Anthropomorphism. The attribution of human characteristics to natural phenomena; inclusive of feelings.

Archigraph. Basic design, writing.

Art. Art is the creation of public object or events, by manipulating a medium, to create qualitative experience. Art is also a form of play, or a process of putting experience in symbols that others, with overlapping experiences, can unfold and enjoy in quick time.

Assimilate. Absorb, incorporate, 'to make similar'

Attributes. Characteristics or qualities of a thing.

Autarchy. Similar to anarchy, but meaning 'self-building,' rather than 'without a ruler.'

Autologic. Self-logic.

Autopoesis. Self-making, as used by F. Varela.

Biodiversity. The variation within and among different species.

Biophilia. Generally, the love for life. As coined by E. Wilson, the "connection" between humans and other living forms.

Biotopology. A neologism meaning the study of living in place (what sociology does).

Black box. Mario Bunge used black boxes as metaphors. The idea of the black box was originally conceptualized by electrical engineers to describe certain unknown systems devoid of structure. The black box approach is useful for all theories whose variables are external and global, that is, are simple and have a high degree of generality.

Bottleneck, genetic: An observable and dramatic collapse in numbers. Although only 2 individuals can retain 75% of heterozygosity, the loss of alleles is significant; the total genetic variation lost depends on the speed of recovery.

Boundedness. The boundary is produced by the system. An organism makes its own boundary, such as skin. A partial process requires cuts in the whole, boundaries, identifiable components organized internally with structural boundaries.

Bradytrope. Slow turning. Also, bradytrophe.

Care. This meaning is based on Martin Heidegger's term for 'letting be,' similar to Goethe's nonintervention.

Carrying capacity. The number of people who can be supported indefinitely in a territory, with a given culture, with a given style, and using a given technology (may refer to any organism, as well as to population or use).

Catastrophe. A down-turning or disruptive event. A slow down-turning is a bradycatastrophe.

Cause. A relationship between events, such that the occurrence of an earlier event must be sufficient or necessary for a subsequent event to happen. For example, food scarcity is a necessary but not sufficient event to raise food prices, which also depend on social conditions like demand. As another example, drought can be a sufficient event to cause crop failure.

Characteristics. Qualities that distinguish unique individuals, systems, or patterns. Gregory Bateson calls them differences that make a difference. (See 'Properties')

City. A high-order, multiple-loop, feedback system, characterized by large buildings, created by large groups of humans living in place.

Closure, e.g., well-defined boundaries, with steady input output low rate. Also locality, isolation, self-boundedness.

Coenotopia. Common place

Coevolution is mutual adaptation of species in a system.

Cognitive dissonance. Failure to reconcile conflicting positions, ending in a form of self-deception; Festinger's term.

Collapse is the rapid significant loss of an established level of sociopolitical complexity. For a culture, collapse is a rapid, significant reduction in an established level of sociopolitical complexity, such that there is less specialization and stratification.

Commotion. Moving with, moving together, turbulence, confusion.

Communism. An economic system based on ownership of property by the community. A revolutionary form of government and stage of socialism that emphasizes the requirements of the state before the individual, and which is characterized by the equal distribution of economic goods in a classless society.

Complementarity. Bohr's term for the idea that two descriptions are needed to understand the whole reality; mutual completion.

Complexity. The quality of being complex. It means having interrelated parts or ideas. Complexity is a relationship between system and observer, not a measurable property of a system. Complexity emerges from the interpenetration of processes of differentiation and integration, processes running simultaneously from top and bottom and shaping the hierarchy from both sides.

Compliance. The acceptance of partial freedom and partial conflict, rather than all of either. It is cooperation enlarged to involve an aspect of play, within the framework of freedom and determination.

Components. Elements or ingredients of a whole; constituent parts.

Conciliance. A form of ecological stability. The ability to absorb change and still maintain the system identity

Connectivity. In an ecosystem, functional connectivity is measured by the potential for movement, dispersal and interchange between populations of species, especially those subject to fragmentation. The connectivity of a component of a system is a measure of the number of direct connections between it and the rest of the system.

Conscension. Climbing with.

Consilience. The unity of knowledge, or the test of truth for a theory. Updated by E.O. Wilson to link the arts and humanities to science.

Constraint. A constraint is a form of confinement or restriction.

Contemplative Nonintervention is the Goethean ideal of respectful participation in knowing, in contrast to an observerless and valueless science, which can only tell us how nature can be made to behave.

Contension. Stretching together, struggle.

Continuity, The system is stable and persistent in time. The process is the same as acquiring information. The ecosystem 'learns' the changes, e.g., seasons, of the environment.

Conversation. The communication between two or more beings; 'turning with'.

Conversion. Turning with, transforming, exchanging value.

Conviviality. Social, jovial (from the Latin *convivium,* living together).

Corridor. A narrow strip of land that differs from the matrix on either side. Biological movement corridors can stem inbreeding depression, lessen demographic stochasticity, and fulfill the need for movement

Cosmology: An ideological system that explains the order and meaning of the universe and the place of people in it (a collective image of the universe).

Critical Mass. The minimum amount or number for something to happen. Can refer to chain reaction, control sequence or simple quantity.

Culture. Culture is an adaptive form of human expression, responsive to the environment, developed through materials, ideas and symbols.

Data. An abstraction of knowledge.

Deep Ecology. A movement, as formulated by Arne Naess, that goes beyond a concern with pollution and resource use, with its man-in-the-environment image, to consider humanity in a relational, total-field image.

Democracy. A form of government in which all the people hold ruling power directly or through elected representatives.

Development. An event related to structure (not necessarily to growth) or a stage in growth.

Design. Victor Papanek defines Design as goal-directed play. Design is a future-oriented, trial and error process for making meaningful order. Design is a conscious and intuitive effort to impose significant order (and perhaps this is a basic human drive to order a chaotic environment)

Dialectic. The process of change whereby two opposites are resolved at a higher unity, Hegel's term; 'gathering across' (warp and woof).

Dictatorship. Rule by one person with absolute power.

Difference. The condition of being unlike or dissimilar; 'carrying apart'.

Disturbance. An external disruption of the balance of a species or ecosystem. It is an event that can be caused by climate, biological entities, or other actors. Disturbances in nature are regular but unpredictable events, caused by geological events, climatic events, physical processes, and biological agents.

Diversity (as in biological diversity). Alpha diversity—number of species in a habitat (local diversity). Beta diversity—change in diversity along habitat gradients. Gamma diversity—the composite diversity across a whole region. Also, diversity is the variety and richness of the planet's genetic heritage.

Domination. In politics, the achievement of power within a group of states (economic, political, religious, cultural or military). Ecologically, of one species by another.

Domiture. The system of culture embedded in nature (has to be a larger term to enclose the previous nature/culture dualism. It has to include human reason and human emotion, rather than simply putting them on opposite columns, as with order and chaos, higher and lower, linear and cyclic), as well as agriculture and wilderness.

Drymoperipheral. The nonecocentricity of the forest, which has no center, and must be approached sideways, from the edge.

Earth. The entire planet, including all life forms.

Ecocybernics. A neologism meaning 'controlling the house' or governing the house (politics is a subcategory, a way of distributing power and luxuries in the cityscape).

Ecodeontics. A neologism meaning 'binding to the house.' (This is similar to what the word religion means).

Ecohygiology. The science of the 'health of the house.' Ecosystem medicine.

Ecology. The science that deals with the relations of organisms to their environment (which includes other organisms). From the Greek meaning "study of the house."

Economics. Literally, "the management of the house" or law of the house. The formal study of how people use and share resources to make a living.

Ecophilia. Generally, the love of the house (where house is metaphor for ecosystem or earth) expressing the need for meaning and significance in a place.

Ecopoetics. A neologism meaning 'making the house, or expressing the house. This is how the invisible or power is represented in art, which is a subcategory.

Ecosystem. An adaptively organized system of living beings and elements that recycle important elements within the system (after Lynn Margulis); the smallest unit that recycles biologically important elements. Margulis states that elements recycle faster within ecosystems than between them. Also, an ecosystem is a natural system where the exchange of materials between living and nonliving parts follows circular paths; the ecosystem is the level of integration and the unit of organization undergoing a directional development (after Eugene Odum). Or a set of communities of species receiving an influx of energy and matter in the same time and place (after Daniel Botkin). Or, the functioning of a community of living organisms with their environment so that a flow of energy leads to clearly defined structures, cycles, and diversity within the system.

Edge. The physical boundary between two plant communities or between a patch and a matrix. Human edges tend to be sharper and straighter.

Ektropy. The fundamental ordering process of the universe; Hirth's term 'turning in.' Similar to negentropy or syntropy.

Emergence. The unpredictable appearance of new characteristics in a process. The word 'emergent' is borrowed from Lloyd Morgan (1912), who first used it in this manner in *Instinct and Experience.* Emergence and ideas of organism came in rough form from Herbert Spencer (circa 1880). Morgan later set forth a view of the world as evolutionary process in *Emergent Evolution.* The word "emergent" was used to show that higher orders of being are not mere resultants of what went before and were not contained in them as an effect is in its efficient cause. Jan Smuts (1926) stated a principle of emergence: Nature is permeated by an impulse towards the creation of wholes; each stage of evolution is marked by emergence of a new type of individuality embracing and transcending the previous parts of itself.

Enmotion. Turning out (cf. Emotion, feeling, course of feeling).

Energy. Energy is inherent power or capacity for action or movement. Or the capacity for doing work.

Entropy. The fundamental disordering process of the universe; Clausius' term 'turning out'.

Environment. Surroundings. Or, the complete range of external conditions for an organism, including soil, climate, social, and cultural patterns.

Escension. Climbing out, emerging.

Ethics. A description of human conduct, or set of rules, determined by custom and culture. Ecologically, a limit on one's freedom of action.

Eutopia. Good place (as used by Thomas More).

Eutopias. Good places that can be made culturally and ecologically.

Event. A change of state, described by two points, an initial state and a subsequent or final state.

Eversion. Exversion or ektropy, the spontaneous generation of order.

Evolve. Roll out, unfold.

Evolution is the process by which species reorganize their structures to adapt to the environment. Evolution is an integrated, partly-open process that selects whole individuals in whole environments. Evolution flows upwards and outwards as well as inwards and downwards, from the simple to the complex, but also back again.

Exploitation: The normal use of a resource by a species. It may also create diversity in ecological systems.

Extension. Stretching towards.

Facts. Facts are data transformed by emotion, theory, and imagination, as Goethe noted. As Kenneth Boulding notes, "For any organism or organization, there are no such things as 'facts.' There are only messages filtered through a changeable value system."

Feedback. Feedback is the signal that is looped back to control a system within itself—a feed-back loop. In cybernetics and control theory, feedback is a process whereby some pro-portion of the output signal of a system is passed (fed back) to the input. This is often used to control the dynamic behavior of the system. There are two kinds of feedback: Negative feedback (or Deviation-reducing, or stabilizing, or centripetal), and Positive feedback (or deviation-amplifying, or error-amplifying, or cumulative causation, or destabilizing, or centrifugal).

Flexibility. The potential for change within a system. It is not being over connected. Not being too rigid or efficient. Able to slough off species or living forms and incorporate new. A measure of unused potential (once the potential is used, the system becomes less flexible) or the capacity to adapt to larger systems.

Footprint. An ecological footprint is a two-dimensional measure of resource use by human individuals and groups. The idea was derived from Georg Borgstrom's "ghost acreage (1961) and Eugene Odum's required acreage (1969).

Form. The contour or structure of something; mode; essence; 'shape'

Frame. Something that encloses or encircles; equal to vehicle or ground.

Gaia. The hypothesis of the living earth; Homer's term; James Lovelock's theory.

Gestalt. A unified configuration having properties that cannot be derived from the parts.

Gigatrend. A long-lasting pattern in a human culture, such as accumulation of goods, or the height of early agriculturalists.

Global Powers. The term used by George Modelski to describe states that had significant but limited power in the world. Examples: France, Spain, Germany, Japan.

Gnosis. Augmentative knowledge; intuitive apprehension; 'to know'.

Government. A system of ruling a country, concerned with the distribution of work, prestige, and materials.

Harmony. The simultaneous occurrence of tones; agreement in feeling.

Harmosophy. The wisdom of harmony. Similar to pansophy.

Health. Vitality. The ability of a system to continue to develop.

Heterarchy. A hierarchy of values; ambiguous order.

Heterozygosity. (Gr. for other yoke) Different alleles; measure of genetic diversity.

Hierarchy. A sacred order; a vertical arborizing process.

History. A sequence of states. In complex systems, the earlier state, also called the 'past,' contributes to or constrains a subsequent state, such as the magnetic hysteresis of fer-romagnets or the behavior of organisms having a form of memory.

Holarchy. A whole order.

Holocosmology. The whole framework of human cosmologies.

Hologram. A pattern produced on photosensitive medium; a three-dimensional image of an object; equal to holograph; 'whole writing.'

Holomovement. The motion of the universe as a whole; D. Bohm's term.

Holon. A holon is any structural/functional subsystem in a biological, social hierarchy that manifests rule-governed behavior and/or structural Gestalt constancy, according to Koestler. Each holon has considerable autonomy. In considering the relations between wholes and parts, Arthur Koestler presents the Janus principle: Every holon is two-faced, a coherent whole related to its parts, but itself part of a larger assembly.

Holonomy. Naming the whole process. George Leonard suggests the word holonomy for the study of wholeness. Holonomy is a frame for individual wholes.

Holopoetic. Referring to making wholes.

Holotopia. Whole place, filled.

Home. A place for one or more beings, invested with emotion and time.

Identity. Identity is that persistent quality that serves nothing; it is. A system's identity can be described apart from its performance in interactions, but not isolated. The state of being complete or whole. The relationship between identity and wholeness is a rhythm.

Ideology. A body of ideas reflecting the needs or aspirations of a culture or individual.

Image. The reproduction of an appearance of something; 'imitation'.

Immotion. Movement.

Industrialization. The use of machines and technology to satisfy needs. A form of social and economic organization characterized by large industry, mechanization, fossil-fuel use, and concentrations of workers.

Information. The act of imparting form, as in 'in-form-ation.' Some consider it the basic substance of the universe.

Infrastructure. The underlying foundation of a structure (itself something that is organized or built).

Interaction. Reciprocal effects by two living beings (on nonliving things) on each other.

Interference. Exploitation or disturbance on a scale that destroys the ability of organisms or their ecosystem to continue. The complete disruption of an ecosystem, as a result of a large event such as a meteor strike, volcanic eruption or complete human conversion. It is an activity that can degrade, destabilize, or destroy entire ecosystems. Interference is not a form of disturbance, exploitation, or competition; it is destruction without gain to any species; sometimes it is caused by planetary events, but in the case of human interference, it is the destruction of the structures and processes of evolution for short-term gain.

Indigenous. Existing naturally in a region; born in a region.

Industrialism. A form of social and economic organization characterized by large industry, mechanization, fossil-fuel use, and concentrations of workers.

Interrelatedness. A state of mutual connection.

Inversion. Turning in, entropy.

Knowledge. The act of impressing an 'object' into the mind to make it in the form of knowledge. The storage of learned information.

Koinomics. The equal apportionment of resources for all beings in the planet. The kind of economics that must be employed before a human ecological economics can be.

Kondratieff cycle. A long-term trade fluctuation named after the Russian economist.

Landscape. A heterogeneous land area composed of a cluster of interacting ecosystems that are repeated in similar form throughout. Landscapes vary in size, from bioregions to a few kilometers in diameter.

Law. A law is a stable pattern inherent in things; it has to be perceived or discovered (like gravity). Example of a cultural law: "Higher culture does not emerge in society until the basic needs of a segment of its members have been satisfied."

Leader. A limited or temporary position in Bands.

Legatism. The doctrine or theory of recognizing the legacies of the evolution of life.

Limit. The word limit comes from the Latin word for boundary, border or frontier. It means boundary, restriction, or utmost extent, either largest or smallest. A limit is the point or line where something must end.

Logic. The science of reasoning. A way of describing relationships among propositions.

Love. An intense feeling; mysterious, emotive force.

Matrix. The most extensive and connected landscape element type, which plays a significant role in landscape functioning. Or, that element surrounding a patch.

Maturity. A biological organism grows to maturity, which is a stopping point for size. Growth serves as a mechanism of evolutionary adaptation, by carrying out genetic instructions in environment; but growth is also conservative and stabilizing rather than innovative and reorganizing. In an ecological system, maturity is the ability to move to a greater efficiency over time. Ramon Margalef suggests replacing word climax with high maturity.

Membrane. A permeable layer or boundary that separates or filters an organism from its environment.

Metalysis. The process of renewal by discrete units; 'loosening change'.

Metaphor. Based on the Greek words meaning "carrying beyond." A metaphor is a figure of speech that combines two frames of reference. It is, as defined in Webster's Seventh New Collegiate Dictionary as a "figure of speech," a "form of expression," an act "of representing in words."

Metapolitics. Discussion of civilization as more than cities.

Military-industrial complex. Large defense-related industries that developed in the U.S. after WWII and became a major economic and political force.

Miniature. A complete assimilation in a whole.

Model. A stylized or hypothetical representation of an object. A generalized description based on analogy or metaphor.

Moral. Acting in accord with standards; individual behavior; 'doing together'.

Morals. Customs or manners. The "way of going together" (from the Latin root).

Motion. Movement.

Nation. A people or tribe. A people living in a territory united by a single government. A stable, historically-developed community of people with a distinct culture occupying a common territory.

Nature. The wild environment of humanity, the self-making process of living. Nature is a feeling system, with mutual experience and history.

Net. An openwork fabric of woven threads forming a mesh in various sizes.

Optimum. What is an optimum? Must it have a less than maximum in every case? An optimum for rats includes challenges and dangers or bad behavior. Simon uses the term satisficed, also used by Varella.

Order. A condition of comprehensible arrangement among elements of a group; a number of successive differentiations.

Organicism goes back to 6th century B.C. China, with the foundation of Taoism. Things are what they are and act upon one another by virtue of their position within a system of patterns. For L. Bertalanffy, organicism was necessary to accomplish three specific jobs in biology: appreciation of wholeness (regulation), organization (hierarchy and level laws), and dynamics (process, behavior of open systems).

Overshoot is that growth beyond the carrying capacity of an ecosystem leading to collapse or die-off.

Ownership rights: The right to claim a territory and the right to grant others permission to use it.

Panarchy. A term referring to a specific form of governance chosen by individuals in a locale. C.S. Holling and L. Gunderson's definition of the structure in which the systems of nature and human nature are interlinked in never-ending adaptive cycles of change.

Panecology. The study of all forms of ecosystems, including human ideational.

Panethic. The human recognition of the importance of interrelations of all beings.

Panocracy. The rule of all beings. In the human legal system, humans represent the interests of all other beings, much as they are starting to do now.

Participation. The state of being part of a system. Joining or sharing with others. The system is embedded in larger systems and global cycles, cyclicity. Relationality to the other systems.

Patch. A nonlinear surface area differing in appearance/composition from its surroundings.

Path. An event is described by two points, an initial and a final state. A process is a sequence of states, also called a history. A process creates a path, described by a trajectory of states. Paths are also structures that enable people move easily, knowledgeably and efficiently from and to special areas.

Pattern. A pattern is an arrangement of form of elements. Process applied to the components of a system yields pattern. Nature is composed of patterns. A pattern can be a covariation or relation of two or more processes. Regularities in systems are patterns. Patterns can be seen in things or even cultures.

Perception. Perception is where patterns are grasped and called stable or complex. The interactions of systems is mutual; the secondary system of a metaphor imposes a reorganization on the concepts of the primary, but the use of the metaphor alters perception of the secondary as well as the primary. Simple behavior seems to work, with local feedback, and more sophisticated behavior "trickles up" to approximate a global perception. Gestalt thinking (Wolfgang Kohler) emphasizes the importance of wholes and tries to identify organizing principles of perception.

Permaculture. The conscious design and maintenance of agricultural systems in the context of natural ecosystems.

Perturbation. Change to a system beyond the normal range of variation.

Phenomenological spiral. A way of questioning and investigating "things as they are" by repeating under different circumstances, with new knowledge or perspectives.

Physiocryptology. The study of nature.

Place. Emotionally-invested space, living space.

Practice. Actual doing, habitual action, custom, repeated performance (from the Greek, *praktikos*, fit for action, from *prassein*, to do); the Greek *poieein*, means to make.

Principles. Fundamental rules that we can use to create images or models.

Process. A sequence of states, called a history, or a trajectory of states, describing a path. A process is described as evolutionary if it involves emergence and the creation of new things, as in general speciation; to be a species, however, the novelty has to reproduce, multiply or diffuse. The concept of state precedes the concept of process, although a process can define a state.

Productivity. The conversion of energy into flesh.

Properties. Qualities (or characteristics) common to all members of a class. A property is an attribute proper to a thing or a characteristic quality. Every concrete thing has properties. Properties can be essential or accidental. If an essential property is lost, then the thing becomes another species; an accidental is one that makes less difference to essential properties. There are also general properties, which all things can share.

Proposition. That which is expressed in a statement, having logical constants, asserting or denying something. Propositions can be tested for truth.

Proxemics. The study of the perception and use of space by people.

Qualitative. Referring to unique (nonstatistical) information or data that is expressed in words or pictures.

Quality. Refers to a basic nature or characteristic element. A characteristic of something; a feature; 'of what kind'

Quantitative. Referring to data that can be expressed as numbers.

Quantity. An amount of something; the character of a proposition.

Quantum. A quantity of something; indivisible unit; 'how great.'

Radical. Referring to root.

Rate. The degree of anything in relation to units of something else.

Reciprocity (general): The distribution of goods by direct sharing, without comparison or accounts, but with the understanding of an eventual balance.

Renewability. The system is autopoetic, self-creating. It renews itself as contents change, as disturbance change the parameters.

Resilience. A measure of how fast variables return to equilibrium after a perturbation in a system.

Resistance. The degree to which a variable is changed after a perturbation.

Retension. Power for remaining.

Reverence. Schweitzer's 'honor-fear' a feeling of respect, awe or love.

Reversion. Turning back, or out, returning.

Revolution. One turning. Sudden change in culture, especially government.

Rhythm. Durational quality; a regular movement

Rights. Claims in accordance with human law, rule or morality.

Ritual. A religious ceremony; the technique of constituting an order.

Rule. A social convention set up by the people in a culture.

Sacred. Worthy of reverence; made holy.

Satisficing. An amount that leads to satisfaction, that may be more then the minimum, less than a maximum and different from an optimum (or satisficing).

Scale. A ratio between dimensions.

Scension. Climbing. The universal process of emergence.

Science. Knowledge. A form of systematic knowledge derived from observation and experiment.

Sign. The suggestion of a presence of a quality; a gesture to convey meaning; 'seal.'

Signal. A sign serving as a means of communication.

Socialism. A system of ownership by the people or community.

Society. An interacting group of people sharing a common culture.

Stability. Stability is the ability to provide a constant internal environment, or to maintain the identity of a system under the flow of external forces and disturbances. Stability can be refined through the specifics of constancy, resistance, resilience, and accommodation.

Standard. A model of quality that can be repeated.

State. A condition in time. The state of an organism, for instance, depends on its history and its environment.

STEM. The universal field of Space-Time-Energy-Matter.

Succession. The slow sequence of changes in the development of an ecosystem.

Sustainable. The ability of a system to maintain its structure and function indefinitely.

Symbol: Anything with a culturally-defined meaning.

Synthesis. The putting together of parts to form a whole.

System. A complex unit (in STEM) whose components keep structure and function stable despite changes and disturbances. A system as a complex object, every part of which is connected with other parts of the object in such a way that the whole possesses emergent properties that the parts lack. A concrete system is composed of concrete things linked together by real physical ties, often moving in circular paths described as feedback loops.

SynGeo ArchiGraph. Basic Designs With the Earth. A company generalizing in ecological design.

Tao. Way; way of the universe (sometimes uncapitalized).

Technology. The sum of techniques and methods for modifying and controlling materials in the environment.

Tension. Stretching.

Theory. The dictionary definition states that a theory is a reasoned expectation, as opposed to practice. A scientific theory is a statement that postulates ordered relationships among natural phenomena and explains some aspect of the world. It allows one to ask certain kinds of questions, some as specific hypotheses.

Tragedy. A dramatic work of struggle ending in ruin; 'goat song.'

Topophilia. Generally, love of place. As coined by Y-F. Tuan: the "effective bond" between people and place.

Topopoetics. The art and science of place-making.

Trap. The use of resources by a people, where the replenishment rate is constant and the rate of use exceeds it, resulting in ecosystem degradation that is less reversible. Agriculture is an energy trap, because it allows the concentration of energy, e.g., higher yields, but then it requires more energy put into the system to maintain it.

Trend. A temporary pattern in a human culture, such as human infertility or globalization of capital, which can be reversed.

Umwelt. A perceived world; von Uexkull's term for animal world.

Unity. A condition of accord; the state of being one.

Universe. All existing things; 'turning into one.'

Urban ecosystems. Specialized human ecosystems where production and living are concentrated, as an adaptation to environmental conditions, such as drought. Because of this concentration, most energy and materials are imported from outside the system.

Use rights. The right to use territory owned by others.

Value. The worth or amount of a thing. A standard by which a society defines a desirable or undesirable.

Variables. Qualities with no fixed value, changeable.

Version. The universal process of turning (see ektropy, entropy, evolution).

Vitality. Activity, health.

War. The condition of open armed conflict between factions or countries. A formal non-monotonic condition of destruction (that is, all living beings are affected).

Wealth. Having rare things. Having more than the minimal requirements to live. Having great quantities of things that are valuable as measured by price. Having extra needs met, from physical to self-actual.

Wild. Pre-reflective, undomesticated.

Wisdom. The art of disciplined use of imagination dealing with facts with respect to knowledge of alternatives, exercised at the right time and in the right measure (after Jonas Salk).

World. Our planet and its inhabitants; or the perspective of the inhabitants (from the German word for "man-image").

World economy. An economically autonomous part of human existence, wherein economic functions are integrated wholes.

World orders. Stable structures that define certain periods of world history, such as the Cold War era.

Zonagraphy. Similar to geography, it emphasizes relative isolation of wild and artificial areas.

9.3. *Partial Bibliography*

Aberle, Doug, ed. 1994. The Future by Design: The Practice of Ecological Planning. Gabriola Island, BC, CA: New Society Publishers.

Alexander, Christopher, et al. 1977. A Pattern Language. New York: Oxford University Press.

Allen, Barry. Artifice and Design.

Anderson, E. N. 1996. Ecologies of the Heart. New York: Oxford University Press.

Andrewartha, H. G. and L. C. Birch. 1984. The Ecological Web. Chicago: University of Chicago Press.

Aplet, Gregory H., N. Johnson, J. Olson, and V. Sample, eds. 1993. Defining Sustainable Forestry. Washington: Island Press.

Appadurai, Arjun. 1996. Modernity at Large. Cultural Dimensions of Globalization, Minneapolis/London, University of Minnesota Press

Arbib, Michael. 1972. The Metaphorical Brain. New York, John Wiley and Sons.

Aristotle. 1952. The Works of Aristotle. trans. I. Bywater. W. Ross, ed. Oxford: Clarendon Press.

Avise, John C. 1994. "Conservation genetics," in Molecular Markers, Natural History and Evolution. New York: Chapman & Hall.

Axtmann, Roland (1995), Kulturelle Globalisierung, kollektive Identität und demokratischer Nationalstaat, Leviathan 23 (1), 87-101.

Bachelard, Gaston. 1971. On Poetic Imagination and Reverie. trans. C. Gaudin. Indianapolis: Bobbs-Merrill Co. Inc.

_____. 1969. Poetics of Space. trans. M. Jolas. Boston: Beacon Press.

Bacon, Francis. 1901. Novum Organum. J. Devey, ed. New York: P. F. Collier.

_____. 1974. .The Advancement of Learning and New Atlantis. Oxford: Clarendon Press.

Badgley C, et al, Organic agriculture and the global food supply. Renewable Agriculture and Food Systems, 22:86-108, 2007

Bain, M. B. and J. T. Finn. 1988. Streamflow regulation and fish community structure. Ecology 69(2):382-392.

Baker, Richard St. Barbe. 1980. "A man of the trees: Ed Goldsmith interviewing Richard St. Barbe Baker." Coevolution Quarterly 25: 66-70.

Barfield, Owen. 1964. Poetic Diction: A Study in Meaning. New York: McGraw-Hill.

Barabasi, A-L. 2002. Linked. Cambridge: Perseus Publishing.

Barnes Michael and Twila Jacobsen. 1996. Ecoforestry Institute Papers.

Barton, Hugh, Editor. 2002. Sustainable Communities: The Potential for Eco-Neighborhoods. London: Earthscan.

Bateson, Gregory. 1987. Steps to an Ecology of Mind. Northvale, NJ: Jason Aronson Inc.

Beatley, Timothy. 1999. Green Urbanism. Washington: Island Press.

_____. 2004. Native to Nowhere. Washington: Island Press.

Bell, J. S. 1964. On the Einstein Podolsky Rosen Paradox. Physics 1:195-200.

Bellah, Robert N. et al. 1991. The Good Society. New York: Alfred A. Knopf.

Bergstraesser, A. 1962. Goethe's Image of Man and Society. Freiburg: Herder.

Berlinski, David. 2008. Infinite Ascent. New York: Modern Library.

Berry T., The Dream of the Earth, San Francisco, CA, Sierra Books, 1988

Berry W., The Unsettling of America: Culture and Agriculture, New York, Avon, 1978.

Bertalanffy, Ludwig von. 1952. Problems of Life: An Evaluation of Modern Biological Thought. New York: John Wiley and Sons.

Bertalanffy, Ludwig von. 1975. Perspectives on General Systems Theory. New York: G. Braziller.

Bhabha, Homi K. (1996), Culture's In-Between, in Stuart Hall/Paul du Gay (ed.) Cultural Identity, London, Sage, 53-60.

Birch, Charles, and Cobb, Jr., John B. 1981. The Liberation of Life. Cambridge: Cambridge University Press.

Birch, Thomas H. 1982. "Man the Beneficiary?: A planetary perspective on the logic of wildland preservation." International Dimensions of the Environmental Crisis. R. Barrett, editor. Boulder: Westview Press.

Birkeland, Janis. 2004. Design for Sustainability: Sourcebook of Integrated Eco-logical Solutions. London: Earthscan.

Black, Max. 1962. Metaphor In: Models and metaphors. M. Black, ed. Ithaca: Cornell University Press.

Bohm, David. 1980. Wholeness and the Implicate Order. London: Routledge & Kegan Paul.

Boltzmann, Ludwig. 1974. The Second Law of Thermodynamics. Populare Schriften.

Borgstrom, George. 1965. The Hungry Planet. New York: Macmillan Co.

_____. 1973. Harvesting the Earth. New York: Abelard-Schuman.

Boulding, Kenneth E. 1956. The Image: Knowledge in Life and Society. Ann Arbor: University of Michigan Press.

_____. 1968. Beyond Economics. Ann Arbor: University of Michigan Press.

Brand, Stewart. 2009. Whole Earth Discipline. New York: Viking.

Brown, George Spencer. 1969. The Laws of Form. London, Allen and Unwin.

Brown, Lester. 1979. "Crossing the threshold?—Pressures on earth's biological systems." Environment 21(8):12-37.

_____. 2003. Plan B. Rescuing a Planet Under Stress and a Civilization in Trouble. New York: W.W. Norton.

Bryan, Frank and John McClaughry. 1991. The Vermont papers. Post Mills: Chelsea Green.

Buchanan, Richard. 1992. Wicked problems in design thinking. Design issues. 8:5-21.

Bunge, Mario. 1973. Method, Model and Matter. Boston: Synthese Library.

Burch, Jr., E.S. 1972. Amer. Antiquity 37:339-368.

Burlingh, P. et al. 1975. Computation of the Absolute Maximum Food Production of the World. Wageningen, Netherlands: Agriculture University.

Butler, Tom, ed. 2002. Wild Earth. Minneapolis: Milkweed.

Carlson, Christine and Dennis Canty. 1986. A Path for the Palouse. Seattle: National Park Service.

Carson, Rachel. 1962. Silent Spring. Boston: Houghton Mifflin.

Catton, William R. 1982. Overshoot: The Ecological Basis of Revolutionary Change. Urbana: University of Illinois Press.

Cheng, T.C. 1970. Symbiosis. New York: Pegasus.

Chew, Geoffrey. 1970. Hadron bootstrap. Physics Today Oct., 1970: 23.

Chew, Sing C. 2006. World Ecological Degradation. New York: Altamira Press.

Cherfas J., and M. & J. Fanton, The Seed Saver's Handbook, Chichester, UK, Grover books, 1996. See also International Pant Genetic Resources institute www.ipgri.cgiar.org

Chernushenko, David. 2008, July 28. A model for real community energy self-sufficiency. Times Colonist, (op ed section). See his project at www.livinglightly.ca/film.

Clark, Colin. 1958. "World Population." Nature 181:1235-1236.

_____. 1967. Population Growth and Land Use. London: Macmillan.

Cobb, John B., Jr. 1965. A Christian Natural Theology. Philadelphia: Westminster.

_____. 1971. Is It Too Late? A Theology of Ecology. New York: Glencoe.

_____. 1992. Sustainability. Eugene: Wipf & Stock.

_____. 1994. For the Common Good. Boston: Beacon Press.

_____. 2010. Spiritual Bankruptcy: A Spiritual Call to Action. New York: Abington Press.

Conservation International and Agrupación Sierra Madre. "Wilderness: Earth's Last Wild Places."

Cooper J., Leifert C, and Niggily U., eds. 'Handbook of Organic Food Safety and Quality', Cambridge, UK 2007.

Costanza, Robert, ed. 1991. Ecological Economics: The Science and management of Sustainability. New York: Columbia University Press.

Costanza, Robert, Bryan G. Norton, and Benjamin D. Haskell, eds. 1992. Ecosystem Health: New Goals for Environmental management. Washington: Island Press.

Costanza, Robert, L.J. Graumlich, and W. Steffen, eds. 2007. Sustainability or Collapse? Cambridge: MIT Press.

Cummins, K. W. 1979. "The natural stream ecosystem," in J. V. Ward and J. A. Stanford, eds. The Ecology of Regulated Streams. New York: Plenum.

Daly, Herman. E. 1977. Stead-State Economics. San Francisco: Freeman.

_____. 1968. Economics as a life science. Journal of Political Economy 76:392-401.

Daly, H.E. and Cobb, J. B., Jr. 1989. For the Common Good: Redirecting the Economy Toward Community, the Environment and a Sustainable Future. Boston: Beacon Press.

Dansereau, Pierre. 1957. Biogeography: An Ecological perspective. New York: Ronald Press.

Darwin, Charles. 1962. The Origin of the Species by Means of Natural Selection or the Preservation of Favoured Races in the Struggle for Life. New York: Collier.

Dasmann, Raymond. 1972. Environmental Conservation. New York: Wiley.

Daubenmire, Rexford. 1942. An ecological study of the vegetation of southeastern Washington and adjacent Idaho. Ecol. Mono. 12:53-79.

DeWit, K. 1967. Incomplete reference.

Diamond, J. M. 1975. The island dilemma: Lessons of modern biogeographic studies for the design of natural preserves. Biol. Conserv. 7:129-146.

_____. 1997. Guns, Germs, and Steel: The Fate of Human Societies. New York: W. W. Norton.

Diamond J. 2005. Collapse: How Societies Chose to Fail or Succeed. New York, Penguin.

Dobben, W.H. van, and R.H. Lowe-McConnell, eds. 1975. Unifying Concepts of Ecology. Report of the Plenary Sessions of the First International Congress of Ecology. The Hague.

Dobson, A. P. et al. 1991. "Conservation biology: The ecology and genetics of endangered species," in Genes in Ecology, ed. R. J. Berry et al. London: Blackwell Scientific.

Domingo J.L. Toxicity Studies of Genetically Modified Plants: A Review of Published Literature, Critical Reviews in Food Science and Nutrition 47:721-733, 2007.

Doolittle, W. F. 1981. Is nature really motherly? Coevolution Quarterly 29:58-62.

Doxiades, C. A. 1975. Building Entopia. New York: Norton.

_____. 1977. Ecology and Ekistics. Gerald Dix, ed. London: Elek Books.

Drengson, Alan. 1980. Shifting Paradigms: From Technocrat to Planetary Person. Environmental Ethics, 3, 221-240. Revised in Drengson and Inoue, 1995.

_____. 1989. Beyond Environmental Crisis: From Technocrat to Planetary Person. New York: Peter Lang.

Drengson, Alan et al. 1994. "The ecoforester's way: An oath of ecological responsibility," International Journal of Ecoforestry 10(1):48.

Drengson, Alan. (1995). The Practice of Technology. Albany NY: SUNY Press.

Drengson, Alan and Yuichi Inoue (Eds). 1995. The Deep Ecology Movement: An Introductory Anthology. Berkeley, CA: North Atlantic Books. (Also in Japanese.)

Drengson, Alan and Duncan Taylor. 1997. Ecoforestry: The Art and Science of Sustainable Forest Use. Gabriola Island, BC: New Society Pub.

Drucker, Peter. 1990. The New Realities. New York: Dutton.

_____. 1993. The Post-Capitalist Society. New York: HarperCollins.

_____. 1995. Managing in a Time of Great Change. New York: Dutton.

Dubos, Rene. 1965. Man Adapting. New Haven: Yale University Press.

Dubos, Rene. 1976. Symbiosis between the earth and humankind. Science 193:459-462.

_____. 1980. The Wooing of Earth. New York: Charles Scribner's Sons.

Dyson, Freeman. 1979. Disturbing the Universe. New York: Harper & Row.

Easterbrook, D. J. and D. A Rahm. 1970. Landforms of Washington: The Geological Environment. Bellingham: Union Printing Co.

Eckholm, Eric P. 1976. Losing Ground: Environmental Stress and World Food Prospects. New York: W. W. Norton.

Edie, James. 1972. New Essays in Phenomenology. New York: Quadrangle.

Ehrlich, P. and A. Ehrlich. 1972. Population, Resources, Environment. San Francisco: Freeman.

Ehrlich, P. 1981. An ecologist standing up among seated social scientists. Coevolution Quarterly 31:24-35.

Eibl-Eibesfelt, I. 1970. Ethology: The Biology of Behavior. New York: Holt, Rinehart and Winston.

Eigen, Manfred, and P. Schuster. 1979. The Hypercycle: A Principle of Natural Self-Organization. Berlin: Springer-Verglag.

_____. 1981. Laws of the Game. New York: Alfred A. Knopf.

Einstein, Albert and Leopold Infeld. 1966. The Evolution of Physics. New York: Simon and Schuster.

Eisenberg, Evan. 1998. The Ecology of Eden. New York: Vintage.

Eliot, T.S. 1948. Notes towards a Definition of Culture. London: Faber & Faber.

Ellen, Roy and Katsuyoshi Fukui, eds. 1996. Redefining Nature. Oxford: Berg.

Elton, Charles. 1966. Animal Ecology. New York: October House.

Emmons, H. 2006. The Chemistry of Joy. New York: Simon and Schuster.

Eschenbach, Ted G. and G. A. Geistauts. 1986. Alaska's Future: Commentary on a Delphi Perspective. Alaska Pacific University Press.

Evernden, N. 1981. Out of Place. unpublished manuscript.

Ewald, P.W. 1993. The Evolution of Virulence. Scientific American, April, p 86-93.

Eyre, Samuel. 1978. The Real Wealth of Nations. London: E. Arnold.

Fiedler, Peggy L. and Subodh K, Jain, eds. 1992. Conservation Biology. New York: Chapman & Hall.

Flannery, Tim. 2002. The Eternal Frontier. New York: Grove Press.

_____. 2010. Here on Earth. New York: Atlantic Monthly Press.

Forestry Commission, 1994. Forest Landscape Design. London: HMSO.

Foreman, Dave. 1985. Ecodefense. San Francisco: Earth First! Books.

_____. 2004. Rewilding North America. Washington: Island Press.

Formann, R. T. T. and M. Godron. 1986. Landscape Ecology. New York: John Wiley.

Formann, Richard T. T. 1997. Land Mosaics. Cambridge: Cambridge University Press.

Fox, Michael W. 1974. Concepts in Ethology: Animal and Human Behavior. Mineapolis: U. of Minnesota Press.

_____. 1976. Between Animal and Man. New York: Coward, McCann and Geoghehan Inc.

_____. 1980. One Earth, One Mind. New York: Coward, McCann and Geoghehan Inc.

_____. 1978. Personal Communication.

—————. 1980. One Earth, One Mind. Coward, McCann and Geoghehan Inc., NY, pp. 174-234.

_____. 1980 Returning to Eden: Animal Rights and Human Responsibility. Viking Press, NY, pp. 19-141.

—————. 1984. Farm Animals: Husbandry, Behavior and Veterinary Practice' Baltimore MD.

—————. 1988. Agricide: The Hidden Farm and Food Crisis that Affects us All. New York: Schocken books.

—————. 1997. Eating With Conscience: The Bioethics of Food. Troutdale OR: New Sage Press.

—————. 2001. Bringing Life to Ethics: Global Bioethics for a Humane Society. Albany, NY State University of New York Press.

—————. 2004. Killer Foods: What Scientists Do To Make Better Is Not Always Best. Guilford CT: The Lyons Press.

_____. 2011. Healing Animals and the Vision of One Health. Golden Valley: One Health Vision Press.

Frankel, Otto. 1975. Crop Genetic Resources for Today and Tomorrow. O. Frankel and J. Hawkes, eds. New York: Cambridge University Press.

Frankel, O. H. and M. E. Soule. 1981. Conservation and Evolution. Cambridge: Cambridge University Press.

Friedman, Thomas L. 2005. The World is Flat. New York: Farrar, Straus & Giroux.

Fromm, Erich. 1956. The Art of Loving. New York: Harper.

_____. 1976. To Have or To Be. New York: Bantam Books.

Frome, Michael. 1974. Personal Communication.

Fowles, J. 1979. Seeing Nature Whole. Harper's. 259:49-56.

Fukuoka, Masanobu. 1978. One-Straw Revolution. Emannus, PA: Rodale Press.

_____. 1985. The Natural Way of Farming: The Theory and Practice of Green Philosophy. New York: Japanese Publications Inc.

Fuller, R. Buckminster. 1969. Operating Manual for Spaceship Earth. Lars Muller Pub.

—————. 1970. Utopia or Oblivion. Lars Muller Pub.

—————. 1981. Critical Path. New York: St. Martin's Press.

—————, with E.J. Applewhite. 1982. Synergetics. New York: Macmillan Pub Co.

Gabel, Medard. 1979. HO-PING: Food for Everyone. Garden City, NY: Doubleday.

Georgescue-Rogen, Nicholas. 1971. Entropy and the Economic Process. Cambridge: Harvard University Press.

Gezon, Lisa. 2006. Global Visions, Local Landscapes. New York: Altamira Press.

Goldsmith, Edward et al. 1972. A Blueprint for Survival. Boston: Houghton Mifflin Co.

Goodland, R. 1997. Environmental Sustainability in Agriculture: Diet Matters. Ecological Economics, 23: 189-200.

Golley, Frank B., K. Petrusewicz, and L. Ryszkowski, eds. 1975. Small Mammals: Their Productivity and Population Dynamics. New York: Cambridge University Press.

Goodman, Paul. 1962. Utopian Essays and Practical Proposals. New York: Vintage.

Gould, S.J. 1977. Ever Since Darwin. New York: W.W. Norton.

Gray, Russell D. 1988. Metaphors and methods. In Mae-Wan Ho and S. W. Fox, eds., Evolutionary Processes and Metaphors. New York: Wiley.

Greene, Patricia and Dean Apostol. 1994. Design for biodiversity. Landscape Architecture 85 (4):63-65.

Gunderson, Lance H. and C.S. Holling, eds. 2002. Panarchy: Understanding Transformations in Human and Natural Systems. Washington: Island Press.

Hall, Edward T. 1969. The Hidden Dimension. Garden City, NY: Doubleday & Co.

Hall, Stuart (1992), The Question of Cultural Identity, in Stuart Hall et al., eds., Modernity and its Futures. London, Polity Press, 273-325

Hampden-Turner, C. 1981. Maps of the Mind. New York: Macmillan.

Hardin, Garrett. 1969. Population, Evolution, and Birth Control. San Francisco: Freeman.

_____. 1977. The Limits of Altruism. Bloomington: Indiana University Press.

_____. 1987. Cultural carrying capacity. Fourth World Wilderness Congress.

_____. 1993. Living within Limits. New York: Oxford University Press.

Harper, J. L. 1977. Population Biology of Plants. New York: Academic Press.

Harris, Larry D. 1984. The Fragmented Forest: Island Biogeography Theory and the Preservation of Biotic Diversity. Chicago: UC Press.

Hart, Richard. 1994. "Monitoring for ecosystem management," International Journal of Ecoforestry 10(2):74-75.

Hartmann, Thom. 2004. The Last Hours of Ancient Sunlight. New York: Three Rivers.

Hatch, L.U. and M.E. Swisher, eds. 1999. Managed Ecosystems. New York: Oxford.

Hebb, D.O. 1958. Alice in Wonderland or psychology among the biological sciences. In The Biological and Biochemical Bases of Behavior. H. Harlow and C. Woolsley, eds. Madison: University of Wisconsin Press.

Henberg, M. 1984. "Wilderness as playground." Environmental Ethics 6:253-263.

Henderson, Hazel. 1980. Creating Alternative Futures: The End of Economics. New York: Putnam.

Higgs, Eric. 2003. Nature by Design: People, Natural Process and Ecological restoration. Cambridge: MIT Press.

Ho, Mae-Wan and S. W. Fox. 1988. Processes and metaphors in evolution. In Mae-Wan Ho and S. W. Fox, eds., Evolutionary Processes and Metaphors. New York: Wiley.

Hoffman, David. 1976. Inuit land use on the barren grounds. Inuit Land Use and Development Project, Vol. 2. Ottawa: Dept. of Indian and Northern Affairs.

Holling, C.S. 1973. "Resilience and stability of ecological systems." Annual Review of Ecology and Systematics. R.F. Johnston et al., editors. 4:1-24.

Homer-Dixon, Thomas. 2006. The Upside of Down. Washington: Island Press.

Hornborg, Alf and Carole Crumley, eds. 2007. The World System and the Earth System. Walnut Creek: Left Coast Press.

Hornborg, Alf et al. 2007. Rethinking Environmental History. New York: Altamira Press.

Houghton R. A. and D. L. Skole. 1990. "Changes in the global carbon cycle between 1700 and 1895, In The Earth Transformed by Human Action, B. Turner, ed. Cambridge: Cambridge University Press.

Hu Frank B. and Walter C. Willett. 1998. The Relationship Between Consumption of Animal Products ... and Risk of Chronic Disease: a Critical Review. Report for the World Bank. Cambridge, MA: Harvard School of Public Health.

Hughes, J. Donald. 1983. American Indian Ecology. El Paso: Texas Western Press. Indian use of forests, Pp. 98-100, Sacredness of the forest (and fires), Pp. 52-57, Use of Indian forests, Pp. 124-127.

Hulet, H.R. 1970. "Optimum world population." Bioscience 20(3):160-161.

Huston, Michael A. 1994. Biological Diversity. Cambridge: Cambridge University Press.

Huxley, Aldous. 1945. The Perennial Philosophy. New York: Harper.

_____. 1956. "Knowledge and Understanding." In Adonis and the Alphabet. London: Chatto and Windus.

_____. 1977. The Human Situation. P. Ferrucci, ed. New York: Harper & Row.

Illich, Ivan. 1973. Tools for Conviviality. New York: Harper and Row.

Imhoff D., and Baumgartner J.A., eds. 2006. 'Farming and the Fate of Wildlife Healdsburg, CA: Watershed Media.

Jackson D.L., and Jackson L.L., eds. 2002. The Farm as Natural Habitat: Reconnecting Food Systems With Ecosystems. Washington DC: Island Press.

Jackson, Wes. 1980. New Roots for Agriculture. San Francisco: Friends of the Earth.

_____. 1994. Becoming Native to this Place. Lexington: University of Kentucky.

_____. 2010. Consulting the Genius of Place. Berkeley: Counterpoint.

Jackson, Wes, W. Berry, and B. Colman. 1984. Meeting the Expectations of the Land: Essays in Sustainable Agriculture. San Francisco: North Point.

Jacobs, Jane. 1969. The Economy of Cities. New York: Random House.

James, A., K. Gaston and A. Balmford. 1999. "Balancing the earth's Accounts." Nature 401:323-324.

Jantsch, E. 1975. Design for Evolution. New York: Braziller.

_____. 1980. The Self-Organizing Universe. New York: Pergamon Press.

Johnson, Steven. 2001. Emergence. New York: Scribner.

Jordan III, W. R., M. E. Gilpin, and J. D. Aber, eds. 1989. Restoration Ecology: A Synthetic Approach to Ecological Research. Cambridge: Cambridge University Press.

Jouvenel, Bertrand de. 1968. The stewardship of the earth. In: The Fitness of Man's Environment. Smithsonian Annual II. New York: Harper Colophon Books.

Julier, Guy. 2008. The Culture of Design. New York: Sage Publications.

Kaplan, Rachel. 1983. "The role of nature in the urban context," in I. Altman and J. F. Wohlwill, eds. Behavior and the Natural Environment. New York: Plenum Press.

Kaplan, Robert D. 2000. The Coming Anarchy. New York: Vintage Books.

Kaplan, S. 1973. "Cognitive maps; human needs and the designed environment." In W. F. E. Preiser, ed., Environmental Design Research. Stroudsberg: Dowden, Hutchinson & Ross.

Kauffman, Stuart A. 1993. Origins of Order: Self-organization and Selection in Evolution. Oxford: Oxford U. Press.

Keeling, C. D. and T. P. Whorg. 1992. Muana Loa: Atmospheric CO_2—modern record. In T. A. Boden et al., eds., Trends 91: A Compendium of Data on Global Change. Oak Ridge: Oak Ridge National Laboratory.

Kellert, S.R. 1996. The Value of Life: Biological Diversity and Human Society. Washington: Island Press.

Kellert, Stephen R. and E.O. Wilson. 1993. The Biophilia Hypothesis. Washington: Island.

Kelly, Kevin. 1998. New Rules for the New Economy. New York: Viking.

Kelsall, J. P. 1968. The Migratory barren-ground caribou of Canada. Can. Wildlife Serv.

Monograph, No. 3.

Kirk, G. S. and J. E. Raven. 1957. The Pre-Socratic Philosophers. Cambridge University Press.

Klein, David. 1972. "Toward an ecophilosophy." Tomte Symposium on Ecology and Land Use, Steinsgard, Norway.

Klein, David. 1983. Personal Communication.

Klein, David R. 1970. IUCN Publ. New Series No. 16:209-242.

Koch, M. 1996. Wildlife people and development." Trop. Ani. Health Prod. 28:68-80.

Koestler, A. and J.R. Smythies, eds. 1969. Beyond Reductionism: New Perspectives in the Life Sciences. London: Hutchinson.

Koestler, A. 1978. Janus: A Summing Up. New York: Random House.

Kohr, Leopold. 1957. The Breakdown of Nations. New York: E. P. Dutton.

Kohr, Leopold. 1977. The Overdeveloped Nations: Diseconomics of Scale. New York: Schocken Books.

Koop, C. Everett. 1988. In the 1988 Report on Nutrition and Health.

Korten, David. 1995. 'The Tyranny of the Global Economy.' West Hartford, CT.

_____. 2006. The Great Turning. San Francisco: Berrett-Koehler.

Kozlovsky, Daniel G. 1974. An Ecological and Evolutionary Ethic New York: Prentice-Hall.

Krebs, Charles J. 1985. Ecology. 3rd ed. New York: Harper & Row.

Kropotkin, P.A. 1972. Mutual Aid: A Factor in Evolution. New York: New York University Press.

Krutch, J. 1970. The Best Nature Writing of Joseph Wood Krutch. New York: Pocket Books.

Kuhn, T. (1970) The Structure of Scientific Revolutions. Chicago: University of Chicago Press.

Kunstler, James H. 1994. The Geography of Nowhere. New York: Free Press.

_____. 2006. The Long Emergency. New York: Grove Press.

Lackner, S. 1984. Peaceable Nature. New York: Harper and Row.

Lamarck, Jean. 1963. Zoological Philosophy. trans. H. Elliot. New York: Hafner Publishing.

Lappe, Frances Moore and Joseph Collins. 1979. Food First: Beyond the Myth of Scarcity New York: Ballantine.

Laszlo, Ervin. 1972. Introduction to Systems Philosophy: Toward a New Paradigm of Contemporary Thought. New York: Harper Torch Books.

Laszlo, Ervin et al. 1977. Goals for Mankind. New York: E. P. Dutton.

Laszlo, Ervin. 1989. The Inner Limits of Mankind. London: Oneworld.

Laughlin, Robert B. 2011. Powering the Future. New York: Basic Books.

Lehmann, Scott 1981. "Do Wildernesses Have Rights?" Environmental Ethics 3:129-146.

Leopold, Aldo. 1945. "The Green Lagoons." American Forests. 51:414.

_____. 1949. A Sand County Almanac. And Sketches of Here and There. New York: Oxford University Press.

Levitt, Stephen D. and S.J. Dubner. 2006. Freakonomics. New York: William Morrow.

Lewin, K. 1951. Field Theory in Social Science. D. Cartwright, ed. New York: Harper & Row.

Liebig, J. von. 1840. Chemistry in its Application to Agriculture and Physiology. London: Taylor and Walton. (cited in Odum 1970)

Lieth, Helmut F H. and Robert Whittaker, eds. 1975. Primary Productivity of the Biosphere. New York: Springer-Verlag.

Likens, Gene E., Ed. 1989. Long-Term Studies in Ecology. New York: Springer-Verlag.

Lincicome, D.R. 1969. The Goodness of Parasitism: A New Hypothesis. Thomas C. Cheng, ed. Aspects of the Biology of Symbiosis. Baltimore: University Park Press.

Loehle, C., 1989. Catastrophe theory in ecology: A critical review and an example of the butterfly catastrophe. Ecol. Modelling 49:125-144.

Lorenz, Konrad. 1952. King Solomon's Ring: New Light on Animal Ways. trans. M. K. Wilson. New York: Crowell.

_____. 1974. Civilized Man's Eight Deadly Sins. trans. M. K. Wilson. New York: Harcourt, Brace, Javonovich.

Lovejoy, A.O. 1964. The Great Chain of Being: A Study of the History of an Idea. Cambridge: Harvard University Press.

Lovejoy, Thomas and Richard Bierregaard. In Soule, Michael. 1986. Conservation Biology: The Science of Scarcity and Diversity. Sunderland: Sinauer Associates.

Lovejoy. Thomas E and D. C. Oren. 1981. "The minimum critical size of ecosystems," in W.D. Billings et al., eds., Forest Island Dynamics in Man-dominated Landscapes. New York: Springer-Verlag.

Lovelock, J. E. 1979. Gaia: A New Look at Life on Earth. Oxford University Press, Oxford.

_____. Ages of Gaia. New York: Bantam Books.

_____. 1991. Healing Gaia: Practical Medicine for the Planet. New York: Harmony Books.

_____. 2007. Revenge of Gaia. New York: Basic Books.

_____. 2009. The Vanishing Face of Gaia. London: Allen Lane.

Lovins, Amory. 1977. Soft Energy Paths. New York: Harper & Row.

Lucas, Oliver. 1990. The Design of Forest Landscapes. New York: Oxford University Press.

Lyle, J. T. 1985. Design for Human Ecosystems: Landscape, Land use and Natural Resources. New York: Van Nostrand Reinhold.

MacArthur, Robert H. 1972. Geographical Ecology: Patterns in the Distribution of Species. New York: Harper & Row.

MacArthur, R.H. and E.O. Wilson. 1967. The Theory of Island Biogeography. Princeton: Princeton University Press.

Maillat, Dennis, Bruno Lecoq, Florian Nemeti, Marc Pfister. 2009. "Technology District and Innovation: The Case of the Swiss Jura Arc." Informaworld.com

Mandelbrot, B. B. 1982. The Fractal Geometry of Nature. New York: W. H. Freeman.

Mander, Jerry. 1991. In the Absence of the Sacred. San Francisco: Sierra Club Books.

Mander, Jerry, and Edward Goldsmith. 1996. The Case against the Global Economy: And a Turn toward the Local. San Francisco: Sierra Club Books.

Mangold, Robert et al. 1993. Tree Planting in the United States—1993. Washington: USDA, Forest Service.

Margalef, Ramon. 1968. Perspectives in Ecological Theory. Chicago: University of Chicago Press.

Margulis, Lynn. 1974. Five kingdoms—classification and the origin and evolution of cells. Evol Biol 7:45-48.

Margulis, Lynn and Dorion Sagan. 1986. Microcosmos. New York: Summit Books.

Margulis, Lynn. 1991. Big trouble in biology: Physiological autopoiesis versus mechanistic neo-Darwinism. In John Brockman, ed., Doing Science. New York: Prentice Hall.

Marsh, G. P. 1964. Man and Nature: Or Physical Geography as Modified by Human Action. Cambridge: Harvard University Press.

Maruyama, Magorah. 1978. Cultures of the Future. The Hague: Mouton.

_____. 1980. "Toward Cultural Symbiosis." In Evolution and Consciousness: Human Systems in Transition, pp. 198-213. E. Jantsch and C. H. Waddington, eds. Reading, MA: Addison-Wesley Publishing Co.

Maser, Chris. 1994. Sustainable Forestry. Delray Beach: St. Lucie Press.

Maslow, A. H. 1968. Toward a Psychology of Being. 2nd ed. New York: D. Van Nostrand Co.

_____. H. 1971. The Farther Reaches of Human Nature. New York: Viking Press.

Mazrui, Ali Al Amin. 1976. A World Federation of Cultures: An African Perspective. New York: Free Press.

McArthur, Robert H. 1972. Geographical Ecology: Patterns in the Distribution of Species. New York: Harper & Row.

MacArthur, R.H. and E.O. Wilson. 1967. The Theory of Island Biogeography. Princeton: Princeton University Press.

McDonough, William and Michael Braungart. 2002. Cradle to Cradle: Remaking the Way We Make Things. New York: North Point Press.

McHarg, Ian. 1969. Design with Nature. Garden City: Natural History Press.

McKibben, Bill. 2008. Deep Economy: The Wealth of Communities and the Durable Future. New York: Holt Paperbacks.

McLuhan, Marshall. 1964. Understanding Media. New York: McGraw Hill.

Meadows, Donella. 2007. Thinking in Systems. Chelsea Green.

Meadows, Dennis. 1982. "Fallacies in resource planning," in Charles Hewett, T. Hamilton, and I. Anderson, eds., Forests in Demand. Boston: Auburn House.

Mech, L. D. 1970. The Wolf: The Ecology and Behavior of an Endangered Species. New York: Natural History Press.

Meeker, Joseph. 1974. The Comedy of Survival. New York: Charles Scribner's Sons.

Merchant, Carolyn. 1980. The Death of Nature: Women, Ecology, and the Scientific Revolution. San Francisco: Harper & Row.

Merleau-Ponty, Maurice. 1968. The Visible and the Invisible. Translated by A. Lingis. Evanston, Northwestern University Press.

Midgley, Mary. 1989. Wisdom Information & Wonder. New York: Routledge.

_____. 2002. Science & Poetry. London: Routledge.

Miller, George. 1956. The magic number seven plus or minus two. Psych. Rev. 6.

Miller, G. T. 1982.Living in the Environment. Belmont: Wadsworth.

Mollison, Bill. 1988. Permaculture: A Designers' Manual. Tyalgum, Australia: Tagari Publications.

Moran, Emilio F., ed. The Ecosystem Approach in Anthropology. Ann Arbor: University of Michigan Press.

More, Thomas. 1982. Utopia. London: Penguin Books.

Morgan, Conwy Lloyd. 1925. Emergent Evolution. New York: Henry Holt and Co.

Mumford, Lewis. 1956. The Transformation of Man. New York: Harper and Row.

_____. 1961. The City in History: Its Origins and Transformations and its Prospects. New York: Harcourt, Brace and World.

_____. 1967. Technics and Human Development. San Diego: HBJ Book.

Myers, Nancy and Carolyn Raffensperger, eds. 2006. Precautionary Tools for Reshaping Environmental Policy. Cambridge: MIT Press.

Myers, Norman. 1984. The Primary Source. New York: Norton.

Naess, Arne. 1972. The shallow and the deep, long-range ecology movement. A summary. Inquiry, 16: 95-100.

_____. 1987. Self-Realization: An Ecological Approach to Being in the World. The Trumpeter: Journal of Ecosophy, 4 (3), 35-42. Retrieved November 21, 2008, from http://trumpeter.athabascau.ca /index.php/trumpet/article/view/623/992.

_____. 2008. The Ecology of Wisdom: Writings by Arne Naess. (Alan Drengson and Bill Devall, Eds.). Emeryville, CA: Counterpoint Press.

Naisbitt, John. 1984. Megatrends: Ten New Directions Transforming Our Lives. New York: Warner Books.

Ndubisi, Forster. 2002. Ecological Planning. Baltimore: Johns Hopkins.

Nicolis, G., and Prigogine, I. 1977. Self-organization in Non-equilibrium Structures. New York: Wiley.

_____. 1989. Exploring Complexity: An Introduction. New York: W. H. Freeman.

Nieman, H. et al, Transgenic farm animals, Rev.sci.Off.int.Epiz, 24:285-298.

Norberg-Hodge, Helena. 2009. Ancient Futures. San Francisco: Sierra Club.

Norberg-Schulz, C. 1971. Existence Space and Architecture. New York: Praeger.

Noss, Reed. And Allen Cooperider 1994. Saving Nature's legacy: Protecting and restoring Biodiversity. Washington, DC: Island Press.

Odum, Eugene P. 1970. "Optimum population and environment: A Georgian microcosm." Current History 58:355-366.

_____. 1971. Fundamentals of Ecology. 3rd Edition. Philadelphia: W.B. Saunders.

Odum, Eugene P., Clyde E. Connell, and Leslie B. Davenport. 1965. "Population energy flow of three primary consumer components of old-field ecosystems." Ecology 43(1):88-96.

Odum, Howard T. 1957. "Trophic structure and productivity of Silver Springs, Florida." Ecological Monographs 27(1).

Odum, Howard T. and Elisabeth C. Odum. 1981. Energy Basis for Man and Nature. New York: McGraw Hill.

Odum, William. 1988. Predicting ecosystem development following creation and restoration of wetlands. In J. Zelazny and J. S. Feierabend, eds., Increasing Our Wetland Resources. Washington: National Wildlife Federation Proceedings.

Olson, Steve. 2002. Mapping Human History. Boston: Mariner Book.

O'Neill, R.V. et al., eds. 1986. A Hierarchical Concept of Ecosystems. Princeton: PUP.

Ophuls, William. 1977. Ecology and the Politics of Scarcity. San Francisco: W. H. Freeman.

Organic salvation down on the farm. 2004. New Scientist, 184: p 9.

Ovington, J.D., Dale Heitkamp, and Donald Lawrence. 1963. "Plant biomass and productivity of prairies, savanna, oakwood, and maize field ecosystems in central Minnesota." Ecology 41(1):52-65.

Owens, Owen D. 1993. Living Waters. New Brunswick: Rutgers University Press.

Passmore, J. 1974. Man's Responsibility for Nature: Ecological Problems and Western Tradition. London: Duckworth.

Pepper, S. 1961. World Hypotheses. Berkeley: University of California Press.

Pedersen, D. 1996. Disease ecology at a crossroads: man-made environments, human rights and perpetual development utopias. Soc Sci Med, 43(5):745-58 1996 Sep.

Perry, David A. 1995. Forest Ecosystems. Baltimore: Johns Hopkins University Press.

Pielou, E. C. 1974. Population and Community Ecology: Principles and Methods. New York: Gordon and Breach.

Pimental, D., Houser J, Preiss E, et al. 1997. Water reserves: agriculture, the environment and society; an assessment of the status of water resources. 'Bioscience' 47: 97-106.

Pimm, Stuart L. 1991. The Balance of Nature. Chicago: University of Chicago Press.

Pirie, N. W. 1976. Food Resources. London: Pelican Books.

Polunin, Nicholas, ed. 1980. Growth without Ecodisasters? New York: Wiley.

Portmann, Adolf. 1964. New Paths in Biology. New York: Harper & Row.

Prigogine, Ilya. 1977. Order Out of Chaos: Man's New Dialogue with Nature. New York: Bantam Books.

Prigogine, Ilya. 1980. From Being to Becoming. San Francisco: Freeman.

Public Health Service, Centers for Disease Control, 1975.

Rapoport, Amos. 1969. House Form and Culture. New York: Prentice-Hall.

Rapoport, Amos. 1982. The Meaning of the Built Environment. Beverly Hills; Sage Pubs.

Rappoport, Anatol. 1960. Fights, Games, and Debates. Ann Arbor: University of Michigan.

Rapport, D.J., C Thorpe, and HA Regier, 1979. Ecosystem medicine. Bul Ecol Soc Am 60:180-182.

Rapport, D.J., et al. 1985. "Ecosystem Behavior Under Stress," American Naturalist 125:617-640.

Rapport, D.J. 1995. "Ecosystem health: An emerging integrative science." In Rapport, DJ, CL Gaudet, and P. Calow, eds. Evaluating and Monitoring the health of large scale ecosystems. pp. 5-34. Heidelberg: Springer.

Rees, Bill and Mathis Wackernagel. 1996. Our Ecological Footprint: Reducing Human Impact on the Earth. Gabriola Island: New Society Publishers.

Reich, Robert B. 2010. Afterhock. New York: Alfred A. Knopf.

Reichel-Dolmatoff, Gerardo. 1971. Amazonian Cosmos. Chicago: University of Chicago Press.

Reinheimer, H. 1910. Evolution by Co-operation: A Study in Bio-economics. NC: NP.

Relph, E. 1976. place and placelessness. London: Pion Limited.

Riewe, R. R. Changes in Eskimo utilization of arctic wildlife. In L. C. Bliss et al., eds., Tundra ecosystems: A comparative analysis. Cambridge: University Press, 1981.

Rifkin, J. 1982. Algeny: The Last Magic.

Ritzer, George 1995. The McDonaldization of Society: An Investigation into the Changing Character of Contemporary Social Life. Thousand Oaks: Pine Forge Press.

Robbins J. 2001. 'The Food Revolution: How Your Diet Can Help Save Your Life and Our World', Newburyport, MA.

Roberts, Neil. 1989. The Holocene: An Environmental History. New York: Basil Blackwell.

Robinson J. 2004. 'Pasture Perfect: The Far-Reaching Benefits of Choosing Meat, Eggs, and Dairy Products from Grass-Fed Animals'. Vashon, WA, Vashon Island Press.

Rodin, L.E., N.I. Bazilevich, and N.N. Rozov. 1975. "Productivity of the world's main ecosystems." In: Productivity of World Ecosystems. Washington, D.C.: National Academy of Sciences.

Rodman, J. 1977a. "The Liberation of Nature?" Inquiry 20:83-145.

Rogers, Richard. 1997. Cities for a Small Planet. London: Faber & Faber.

Rolston, III, H. 1983. "Values Gone Wild." Inquiry 26:181-207.

Roszak, Theodore. 1972. Where the Wasteland Ends. Garden City, New York: Doubleday.

Roszak, Theodore, et al., eds. 1995. Ecopsychology. San Francisco: Sierra Club.

Sachs, Jeffrey D. 2008. Common Wealth. New York: Penguin Press.

Sage, B. 1981. Conservation of the tundra. In L. C. Bliss et al., eds., Tundra ecosystems: A comparative analysis. Cambridge: University Press.

Sahlins, Marshall. 1968. Tribesmen. Englewood Cliffs: Prentice Hall.

_____. 1972. Stone Age Economics. Chicago: Aldine Publishing.

Sale, Kirkpatrick. 1980. Human Scale. New York: Coward, McCann and Geoghegan.

Salk, J. 1973. Survival of the Wisest. New York: Harper & Row.

Santos, Boaventura de Sousa. 1998. The Fall of the Angelus Novus: Beyond the Modern Game of Roots and Options, in Current Sociology 46 (2), 1998, 81-118.

Savory, Alan. 1988. Holistic Resource Management. Washington: Island Press.

Schaller, G.B. 1972. The Serengeti Lion. Chicago: University of Chicago Press.

Schellnhuber, H.J. et al., eds. 2004. Earth System Analysis for Sustainability. Cambridge: MIT Press.

Schlosser E. 2001. Fast Food Nation. Boston MA, Houghton Mifflin, 2001.

Schneider, Stephen H. 1997. Laboratory Earth. New York: Basic Books.

Schonewald-Cox, Christine M. 1983. Conclusions: Guidelines to management. In C. M. Schonewald-Cox et al., eds., Genetics and Conservation. Menlo Park: Benjamin/ Cummings.

Schor, J.B. and B. Taylor, eds. 2002. Sustainable Planet. Boston: Beacon Press.

Schrodinger, Erwin. 1946. What is Life? The Physical Aspect of the Living Cell. New York: Macmillan.

Schulze, E.-D. and H. A. Mooney, eds. 1993. Biodiversity and Ecosystem Function. New York: Springer-Verlag.

Schumacher, E. F. 1973. Small is Beautiful. New York: Harper and Row.

Schweitzer, Albert. 1949. Out of My Life and Thought. New York: Henry Holt and Co.

_____. 1957. The Philosophy of Civilization. trans. C. T. Campion. New York: Macmillan Co.

Searles, H. 1962. The role of the nonhuman environment. Landscape (Winter 1961-1962):31-34.

Shepard, Paul and D. McKinley, eds. 1969. The Subversive Science. Boston: Houghton-Mifflin.

Shepard, P. 1973. The Tender Carnivore and the Sacred Game. New York: Scribner's Sons.

_____. 1974. Animal rights and human rites. The North American Review Winter.

_____. 1978. Thinking Animals. New York: Viking Press.

_____. 1982. Nature and Madness. San Francisco: Sierra Club.

Sierra Club. 1987. Survey. San Francisco: Sierra Club.

Simberloff, Daniel and L. G. Abele, 1976. "Island Biogeography Theory and Conservation Practice." Science 191 4224: 285-6.

Singer, Peter. 1981. The Expanding Circle: Ethics and Sociobiology. New York: Farrar, Strauss & Giroux.

Singer, S. Fred, ed. 1971. Is There an Optimum Level of Population? New York: McGraw-Hill.

Skolimowski, Henryk. 1978. "Eco-philosophy versus the scientific world view." Ecologist Quarterly 3 (Autumn): 227-248.

_____. 1981. Ecophilosophy. Boston: Marion Boyars.

Slater, P. 1974. Earthwalk. New York: Bantam Books.

Smil, Vaclav. 2003. The Earth's Biosphere. Cambridge: MIT Press.

_____. 2008. Global Catastrophes and Trends. Cambridge: MIT Press.

Smith, Anthony D. 1991. The Ethnic Origins of Nations. Cambridge: Blackwell.

Smith J. M. 2006. 'Genetic Roulette: The Documented Health Risks of Genetically Engineered Foods', on website www.seedsofdeception.com.

Smith J.R. 1950. Tree Crops: A Permanent Agriculture. Washington: Island Press.

Smith, Maynard J. 1968. Mathematical Ideas in Biology. Cambridge: Cambridge University.

Smith, P.M., and Watson, R.A. 1979. "New Wilderness Boundaries." Environmental Ethics 1:61-64.

Smuts, Jan C. 1926. Holism and Evolution. Ann Arbor, MI: University Microfilms.

Snyder, Gary. 1969. Earth House Hold. New York: New Directions.

_____. 1983. The Coevolution Quarterly, Fall.

Sokoloff, J. 1985. The Politics of Food. San Francisco, CA: Sierra Books.

Soleri, P. 1969. Arcology: The City in the Image of Man. Cambridge: The MIT Press.

_____. 1978. A response to "Fields of Danger." The North American Review, Spring: pp. 71-72.

_____. 1983. The Food Chain (pamphlet). Scottsdale: Cosanti.

Soule, Michael and Wilcox, B. A., eds. 1980. Conservation Biology: An Evolutionary-Ecological Perspective. Sunderland, MA: Sinauer Associates.

Soule, Michael. 1986. Conservation Biology: The Science of Scarcity and Diversity. Sunderland: Sinauer Associates.

Spellerberg, I.F. 1991. Monitoring Ecological Change. Cambridge: University Press.

Speth, James G. 2008. The Bridge at the Edge of the World. New Haven: Yale.

Stanley, Steven. 1981. The New Evolutionary Timetable. New York: Basic Books.

Steiner, Frederick. The Productive and Erosive Palouse Environment. Pullman: Extension.

_____. 1991. The Living Landscape. New York: McGraw-Hill.

Stevens, Peter. 1974. Patterns in Nature. Boston: Little Brown.

Stone, Christopher D. 1975. Should Trees Have Standing? New York: Avon Books.

Stroganov, S.U. 1969. Carnivorous Mammals of Siberia. Springfield: US Dept. of Commerce.

Sturm, Andreas et al. The Winners and Losers in Global Competition: Why Eco-efficiency Reinforces Global Competitiveness: A Study of 44 Nations.

Tainter, Joseph A. 1990. The Collapse of Complex Societies. Cambridge: Cambridge University Press.

Tansley, A.G. 1935. The use and abuse of vegetational concepts and terms. Ecology 16:284-307.

Tapscott, Don and A. D. Williams. 2010. Wikinomics. New York: Portfolio.

Thackera, John. 2005. In the Bubble: Designing in a complex world. Cambridge: MIT Press.

Thom, Rene. 1975. Structural Stability and Morphogenesis: An Outline of a General Theory of Models. trans. D. C. Fowler. Reading, MA: W. A. Benjamin.

Thomas, Keith. 1983. Man and the Natural World. New York: Pantheon.

Thomas, Lewis. 1975. Lives of a Cell. Toronto: Bantam Books.

Thompson, William I. 1974. Passages About Earth. New York: Harper & Row.

_____. 1976. Evil and World Order. New York: Harper & Row.

_____. 1981. The Time Falling Bodies Take to Light. New York: Harper & Row.

_____. 1987. Gaia: A New Way of Knowing. Political Implications of the New Biology. Great Barrington, MA: Lindisfarne Press.

Todd, N. J. and J. Todd. 1984. Bioshelters, Ocean Arks, City Farming: Ecology as the Basis of Design. San Francisco: Sierra Club Books.

_____. 1994. From Eco-Cities to Living Machines: Principles of Ecological Design. Berkeley: North Atlantic Books.

Todd, Nancy Jack. 2005. A Safe and Sustainable World. Washington: Island Press.

Toynbee, Arnold J. 1976. Mankind and Mother Earth: A Narrative History of the World. New York: Oxford University Press.

The Trumpeter. Various articles on ecosophy and ecoforestry. BC, Canada: Trumpeter.athabascau.ca.

Tuan, Yi-Fu. 1974. Topophilia. Englewood Cliffs: Prentice-Hall.

Uexkull, J. von. 1957. A Stroll Through the World of Animals and Men. Instinctive Behavior. C. Schiller, ed. New York: International Universities Press Inc.

Ulanowicz, Robert E. 1986. Growth and Development: Ecosystems Phenomenology. New York: Springer-Verlag.

Ulanowicz, R. E. and W. M. Hemp. 1979. Toward canonical trophic aggregation. American Naturalist 114:871-883.

United Nations Development Programme. 1996. 'Urban Agriculture: Food, Jobs and Sustainable Cities', New York, UNDP.

USDA Soil Conservation Service. 1980. Soil Survey of Whitman County, Washington. Pullman: USDA.

Van der Ryn, Sim and Stuart Cowan. 1996. Ecological Design. Washington: Island Press.

Varela, Francisco. 1976. "Not one, not two." Coevolution Quarterly. Fall.

_____. 1979. Principles of Biological Autonomy. New York: North Holland.

Waddington, C.H. 1960. The Nature of Life. New York: Atheneum.

_____. 1971. The Evolution of an Evolutionist. Ithaca: Cornell.

Waldrop, M. Mitchell. 1992. Complexity. New York: Touchstone.

Wallace, Anthony F. C. 1966. Religion: An Anthropological View. New York: Random House

Waltner-Toews, D. and E. Wall. 1997. Emergent perplexity: in search of post-normal questions for community and agroecosystem health. Soc Sci Med, 45(11):1741-9 Dec.

Webb, Warren L., William K. Lauenroth, Stan R. Szarek, and Russell S. Kinerson. 1983. "Primary production and abiotic controls in forests, grasslands, and desert ecosystems in the United States." Ecology 64(1):134-151.

Weizsacker, Carl F. von. 1951. The History of Nature. London: Routledge & Kegan Paul.

Wells, Malcom. 1981. Gentle Architecture. New York: McGraw-Hill.

Westing, Arthur H. 1981. "A world in balance." *Environmental Conservation* 8(3):177-183.

Wheelwright, P. 1962. *Metaphor and Reality*. Bloomington: Indiana University Press.

Whittaker, R.H., F.H. Bormann, G.E. Likens, and T.G. Siccama. 1974. "The Hubbard Brook ecosystem study: Forest biomass and production." Ecological Monographs 44:233-252.

Whitehead, A.N. 1929. Process and Reality. New York: Macmillan.

_____. 1967. Science and the Modern World. New York: Free Press. P. 136.

_____. 1958. The Function of Reason. Boston: Beacon Press.

Willard, B. E. et al. 1977. Ethics of Biospheral Survival: A dialogue. In Growth Without Ecodisasters? pp. 505-535. N. Polunin, ed. New York: John Wiley & Sons.

Williams C. 2000. The Environmental Threat to Human Intelligence a study funded by Britain's Economic and Social Research Council in its Global Environmental Change Programme, April 24.

Wilson, A.K, J.R. Latham, and R.A. Steinbrecher. 2006. 'Transformation-induced mutations in transgenic plants: Analysis and biosafety implications. Biotechnology and Genetic Engineering Reviews, 23, p 209-226.

Wilson, E.O. 1975. Sociobiology: The New Synthesis. Cambridge: Belknap Press.

Wittbecker, A. E. 1970. *Eutopias: A Poetic Commonwealth of the Earth*. Newark: Shamrock Press.

_____. 1976. *The Poetic Archaeology of the Flesh*. Wilmington: Mozart & Reason Wolfe, Ltd.

_____. 1976. "The psychology of catastrophe: Environmental deterioration and rapid social change." *Proc. Marsh Inst.* 1:1-17.

_____. 1983. An optimum human population based on NCP. Ecological Society of America annual meeting, Fargo.

_____. 1986. "The place of human society in wilderness." *The Trumpeter*. 3(3):34-38.

_____. 1986. "Deep anthropology: Ecology and human order." *Environmental Ethics* 8(3):261-270.

Wittbecker, A. E. 1990. "Metaphysical implications from physics and ecology." *Environmental Ethics* 12(3):276-281.

_____. 1991. "An empowered United Nations: Proposals for cooperation and survival." *Common Voice*: 1(1):1-8.

_____. 1995. "Saving common places: The Palouse," *Wild Earth* 5(1):54-58.

_____. 1995. "Gigatrends in Forestry," *International Journal of Ecoforestry* 11(2/3):69-78.

_____. 1997. "Waldgedankenexperiment—Forest thought experiment," *International Journal of Ecoforestry* 12 (3/4):1-4.

_____. 1999. "Varieties of Interaction in nature," *The Trumpeter*, Spring (Web Edition: www.athabascau.ca/trumpeter).

_____. 2001. "Ecological Thought Experiments" Sofia Echo Vol. 5, Issue 31, Aug 3-9, p. 12 (and 8 others in a series, 2001-2002).

_____. 2002. REviewing REthinking REturning: Essays. Baltimore: Cambridge

Books and www.ebooksonthe.net.

_____. 2003. *Good Forestry From Good Theories & Good Practices*: Essays. Baltimore: Cambridge Books and www.ebooksonthe.net

_____. 2007. *Global Emergency Actions* (For a Small Urban Industrial Planet). Sarasota: Urania Science Press.

Wolfe, L. M. 1945. *Son of the Wilderness*: The Life of John Muir. New York: Knopf.

Woodwell, George M. and Robert Whittaker. 1968. "Primary production in terrestrial ecosystems." *American Zoologist* 8:19-30.

Worldwatch Institute. Annual. *State of the World*. Washington, DC: Worldwatch Institute.

Zadeh, L. A. 1965. "Fuzzy Sets" *Information and Control* Vol. 8:338-353.

9.4. *Brief Biography*

Alan Wittbecker is an ecologist with SynGeo ArchiGraph, LLC in Florida, where he works on a variety of wildlife projects. He is also Professor of Environmental Science at Ringling College in Sarasota.

9.5. *Contacts*
Alan Wittbecker: aw@syngeo.org *or* aw@redesignpla.net

9.6. *Index*
(Being prepared)

Colophon

Typeface: Adobe Garamond Pro, 11/14
Hardware: Imac, HP3310
Software: Pages, Indesign, Acrobat
Design: Rian Garcia Calusa
Editing/Graphics: Alan Wittbecker (unless noted)
Place: Overlooking a square in Arlington, VA
Music: Arvo Part & Neil Young

www.ingramcontent.com/pod-product-compliance
Lightning Source LLC
Chambersburg PA
CBHW080227180526
45167CB00006B/2236